Algebraic homotopy

T0312107

Algebraic homotopy

HANS JOACHIM BAUES

Max-Planck-Institut für Mathematik, Bonn

The right of the University of Cambridge to print and sell all manner of books was granted by Henry VIII in 1534. The University has printed and published continuously since 1584.

CAMBRIDGE UNIVERSITY PRESS

Cambridge

New York New Rochelle

Melbourne Sydney

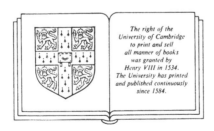

CAMBRIDGE UNIVERSITY PRESS
Cambridge, New York, Melbourne, Madrid, Cape Town, Singapore, São Paulo

Cambridge University Press
The Edinburgh Building, Cambridge CB2 8RU, UK

Published in the United States of America by Cambridge University Press, New York

www.cambridge.org
Information on this title: www.cambridge.org/9780521333764

First published 1989
This digitally printed version 2008

A catalogue record for this publication is available from the British Library

Library of Congress Cataloguing in Publication data

Baues, Hans J., 1943–
Algebraic homotopy.

Bibliography: P.
Includes index.
1. Homotopy theory. 2. Categories (Mathematics)
I. Title.
QA612.7.B385 1986 514′.24 86–32738

ISBN 978-0-521-33376-4 hardback
ISBN 978-0-521-05531-4 paperback

Für Barbara
und für unsere Kinder Charis und Sarah

CONTENTS

PREFACE

This book gives a new general outlook on homotopy theory: fundamental ideas of homotopy theory are developed in the presence of a few axioms so that they are available in a broad variety of contexts. Many examples and applications in topology and algebra are discussed; we consider the homotopy theory of topological spaces, the algebraic homotopy theory of chain algebras, and rational homotopy theory.

The axiomatic approach saves a lot of work in the various fields of application and offers a new way of organizing a course of modern homotopy theory. A fruitful interplay takes place among the various applications.

This book is also a research monograph on homotopy classification problems. The main new result and our principal objective is the '**tower of categories**' which approximates the homotopy category of complexes. Such towers turn out to be a useful new tool for homotopy classification problems; they complement the well-known spectral sequences. The theory on complexes is a continuation of J.H.C. Whitehead's combinatorial homotopy. In fact, some of Whitehead's results can be derived readily from the properties of the towers.

In a later chapter (Chapter IX) we describe the simplest examples of towers of categories from which nevertheless fundamental results of homotopy theory can be immediately deduced.

Most of the material in the book does not appear in any textbook on algebraic topology and homotopy theory.

As prerequisites the reader should be familiar with elementary topology and the language of categories. The book can also be used by readers who have only a little knowledge of topology and homotopy theory, for example when they want to apply the methods of homotopical algebra in an algebraic context. The book covers the elementary homotopy theory in an abstract way.

The nine chapters which comprise the book are subdivided into several sections, §0, §1, §1a, §1b, §2, etc. Definitions, propositions, remarks, etc., are consecutively numberd in each section, each number being preceded by the section number, for example (1.5) or (1a.5). A reference such as (II. 5.6) points to (5.6) in Chapter II, while (5.6) points to (5.6) in the chapter at hand. References to the bibliography are given by the author's name, e.g. J.H.C. Whitehead (1950).

I lectured on the material presented in this book in Bonn (1982), Lille (1982), Berlin (1985), Zürich (1986) and on several conferences. There are further applications of the results which cannot be described in a book of this size. In particular, we obtained an algebraic classification of $(n-1)$-connected $(n+3)$-dimensional polyhedra for $n \geqq 1$. The invariants are computable; for example, there exist exactly 4732 simply connected homotopy types which have the homology groups $(n \geqq 4)$

$$H_i(X) = \begin{cases} \mathbb{Z}/4 \oplus \mathbb{Z}/4 \oplus \mathbb{Z} & i = n \\ \mathbb{Z}/8 \oplus \mathbb{Z} & i = n+1 \\ \mathbb{Z}/2 \oplus \mathbb{Z}/4 \oplus \mathbb{Z} & i = n+2 \\ \mathbb{Z} & i = n+3 \\ 0 & \text{otherwise.} \end{cases}$$

Further details will appear elsewhere. However, the basic machinery for these results is developed in this book.

I would like to acknowledge the support of the Sonderforschungsbereich 40 Theoretische Mathematik, of the Max-Planck-Institut für Mathematik in Bonn, and of the Forschungsinstitut für Mathematik ETH Zürich.

Moreover, I am very grateful to A. Grothendieck for a series of letters concerning Chapters I and II. I especially thank my colleagues and friends S. Halperin, J.M. Lemaire, and H. Scheerer for their interest and for valuable suggestions during the years that I worked on this book; in fact, their work influenced and inspired the development of the ideas; I remember with pleasure the discussions in Bonn, Toronto, Nice, and Berlin and also in Lille and Louvain la Neuve where J. Ch. Thomas and Y. Felix organized wonderful meetings on rational homotopy. I am also very grateful to students in Bonn, Berlin, and Zürich; in particular, to W. Dreckmann, M. Hartl, E.U. Papendorf, M. Hennes, M. Majewski, H.M. Unsöld, and M. Pfenniger who read parts of the manuscript and who made valuable comments.

I am equally grateful to the staff of Cambridge University Press and to the typesetter for their helpful cooperation during the production of this book.

H.J. Baues
Zürich, im Mai 1986

INTRODUCTION

In his lecture at the international congress of mathematicians (1950) J.H.C. Whitehead outlined the idea of algebraic homotopy as follows:

> In homotopy theory, spaces are classified in terms of homotopy classes of maps, rather than individual maps of one space in another. Thus, using the word category in the sense of S. Eilenberg and Saunders Mac Lane, a homotopy category of spaces is one in which the objects are topological spaces and the 'mappings' are not individual maps but homotopy classes of ordinary maps. The equivalences are the classes with two-sided inverses, and two spaces are of the same homotopy type if and only if they are related by such an equivalence. The ultimate object of **algebraic homotopy** is to construct a purely algebraic theory, which is equivalent to homotopy theory in the same sort of way that 'analytic' is equivalent to 'pure' projective geometry.

This goal of algebraic homotopy in particular includes the following basic *homotopy classification problems:*

> Classify the homotopy types of polyhedra $X, Y \ldots$, by algebraic data. Compute the set of homotopy classes of maps, $[X, Y]$, in terms of the classifying data for X and Y. Moreover, compute the group of homotopy equivalences, $\mathrm{Aut}(X)$.

There is no restriction on the algebraic theory which might solve these problems, except the restriction of 'effective calculability'. Indeed, algebraic homotopy is asking for a theory which, a priori, is not known and which is not uniquely determined by the problem. Moreover, it is not clear whether there is a suitable purely algebraic theory for the problem better than the simplicial approach of Kan. For example, in spite of enormous efforts in the last four decades, there is still no successful computation of the *homotopy groups of spheres*

$$\pi_m S^n = [S^m, S^n],$$

which turned out to have a very rich structure. This example shows that the difficulties for a solution of the homotopy classification problems increase rapidly when, for the spaces involved, the

$$\textit{range} = (\text{dimension}) - (\text{degree of connectedness})$$

grows. On the other hand by a classical result of Serre, the *rational homotopy groups of spheres*

$$\pi_m(S^n) \otimes \mathbb{Q} = \begin{cases} \mathbb{Q}, & m = n > 0 \\ \mathbb{Q}, & m = 2n - 1,\ n \text{ even} \\ 0, & \text{otherwise,} \end{cases}$$

are indeed simple objects. These remarks indicate two suitable restrictions for the homotopy classification problem: consider the problem in a small range, or consider the problem for rational spaces.

Quillen (1969) showed that a *differential Lie algebra* is an algebraic equivalent of the homotopy type of a simply connected rational space. Sullivan (1977) obtained the 'dual' result using *commutative cochain algebras* and the de Rham functor.

On the other hand, it is surprising how little is known on homotopy types of finite polyhedra. J.H.C. Whitehead (1949) showed that the cellular *chain complex* of the universal covering is an algebraic equivalent for a 3-dimensional polyhedron. Moreover, he classified simply connected 4-dimensional polyhedra by his '*certain exact sequence*'.

Using towers of categories we obtain in this book new proofs and new insights for these results of Quillen, Sullivan, and Whitehead respectively.

Algebraic models of homotopy types are often obtained by functors which carry polyhedra to algebraic objects like chain complexes, chain algebras, commutative cochain algebras, and chain Lie algebras. The categories defined by these objects yield homotopy categories in which the 'mappings' are not individual maps but homotopy classes of maps. There are actually many more algebraic homotopy categories, some of which not related to spaces at all. In each of them one has homotopy classification problems as above. It turns out that there is a striking similarity of properties of such homotopy categories (compare, for example, Chapter IX). This fact and the large number of homotopy categories make it necessary to develop a theory based on axioms which are in force in most of the homotopy categories.

The idea of axiomatizing homotopy is used implicitly by Eckmann–Hilton in studying the phenomena of duality in homotopy theory. Hilton (1965, p. 168) actually draws up a program by mentioning:

> Finally we remark that one would try to define the notions of cone, suspension, loop space, etc. for the category C and thus *place the duality on a strict logical basis*. It would seem therefore that we should consider an abstract system formalizing the category of spaces, its homotopy category and the homotopy functors connecting them.

To carry out this program is a further objective of this book. We develop homotopy theory abstractly in the presence of only four axioms on cofibrations and weak equivalences. These axioms are substantially weaker than those of Quillen. Many applications of the abstract theory and numerous examples in topology and algebra are described.

There is a wide variety of contexts where the techniques of homotopy theory are useful. Therefore, the unification due to the abstract development of the theory possesses major advantages: *one proof replaces many*; in addition, an interplay takes place among the various applications. This is fruitful for many topological and algebraic contexts. We derive from the axioms a theory which in topology can be compared with combinatorial homotopy theory in the sense of J.H.C. Whitehead.

Hence a few axioms on cofibrations and weak equivalences in a category imply a rich homotopy theory in this category. Moreover, such theories can be compared by use of functors which carry weak equivalences to weak equivalences. This leads to a wider understanding of homotopy theory and offers methods for the solution of the homotopy classification problems.

LIST OF SYMBOLS

Numbers in brackets are page numbers.
□ indicates end of proof.
|| indicates end of definition or end of remark.

(i) Notation for categories

Boldface letters like **C**, **F**, **K**, **A**, **B**..., denote categories

Ob **K**, class of objects	(229)
$\mathbf{K}(A, B) = \mathrm{Hom}(A, B)$, set of morphisms	(229)
$\cong, A \cong B$, isomorphism = equivalence	(6)
$1 = 1_A = \mathrm{id}$, identity	(6)
$f: A \to B$, morphism = map	(3, 229)
$f \mid Y$, restriction	(92)
$\tilde{f}$, extension	(87)
$\{f\}$, equivalence class	(7, 230)
$\mathbf{K}/\sim$, $\mathbf{C}/D$, quotient category	(229, 231)
$E_{\mathbf{A}}(A) = \mathrm{Aut}_{\mathbf{A}}(A)$, group of automorphisms,	(231)
$\varnothing$, initial object	(5, 7)
e, final object	(11)
$\lim = \varinjlim$, colimit	(34, 182)
$\mathrm{Lim} = \varprojlim$, limit	(182)
$\mathbf{F}^{\mathrm{op}}$, opposite category	(10)
$\mathbf{C}^Y$, $\mathbf{Top}^Y$, under Y	(31, 86)
$\mathbf{C}_D$, $\mathbf{Top}_D$, over D	(31)
$\mathbf{C}(C \to D)$, under C and over D	(30)
$S^{-1}\mathbf{C}$, $Ho(\mathbf{C})$, localization	(99)
$\lambda: \mathbf{A} \to \mathbf{B}$, functor	(242)
$\lambda\mathbf{A}$, image category	(230)
$\mathrm{Real}_\lambda(B)$, class of realizations	(249)
$D \xrightarrow{+} \mathbf{A} \xrightarrow{\lambda} \mathbf{B} \xrightarrow{O} H$, exact sequence	(242)
$D: F(\mathbf{C}) \to \mathbf{Ab}$, natural systems	(233)

(ii) Topological and algebraic notation

(iii) List of 'dual' notations

Cofibration category C		***Fibration category F***	
$Y \xrightarrow{\sim} X \in we$	(5)	$X \xrightarrow{\sim} Y \in we$	(10)
$Y \rightarrowtail X \in \mathrm{cof}$	(6)	$X \twoheadrightarrow Y \in \mathrm{fib}$	(11)
$(f: B \to A) = (A, B) \in \mathbf{Pair}\ (\mathbf{C})$	(85)	$(f: A \to B) = (A \mid B) \in \mathbf{Pair}(\mathbf{F})$	(153)
$\simeq$ push	(8)	$\simeq$ pull	
$A \bigcup_B Y \xrightarrow{(\alpha,\beta)} U$	(6, 83)	$U \xrightarrow{(\alpha,\beta)} A \times_B Y$	(11)
$\alpha \cup \beta$	(84)	$\alpha \times \beta$	
$\varnothing = *$	(7)	$e = *$	(11)
$A + Y = A \vee Y = A \bigcup_{\varnothing} Y$	(8)	$A \times Y = A \times_e Y$	
$IX = I \times X$	(2, 18)	$PX = X^I$	(3, 28, 152)
$I_Y X$	(8, 20)	$P_Y X$	
$X \cup I_B A \cup X$	(97)	$X \times P_B A \times X$	
$\Sigma_Y^0 X, \Sigma_Y X, \Sigma_Y^n X$	(107, 115, 137)	$\Omega_Y^0 X, \Omega_Y X, \Omega_Y^n X$	(110)
$\Sigma \cdot f, \Sigma f$	(107, 177, 133)	$\Omega \cdot f, \Omega f$	
$O_X = 0: X \to *$	(115)	$O_X = 0: * \to X$	(152)
CX	(119)	WX	(152)
$\Sigma X, \Sigma^n X, \Sigma_B^n$	(20, 116, 135)	$\Omega X, \Omega^n X, \Omega_B^n$	(152)
Z_f	(8, 23)	W_f	(28)
$C_f = CA \bigcup_f B$	(125)	$P_f = B \times_f WA$	(154)
$f_0 \simeq f_1$	(3, 20)	$f_0 \simeq f_1$	(3, 28)
$\alpha \simeq \beta$ rel X, under X	(8, 92)	$\alpha \simeq \beta$ over X	(12)
$\mathrm{Hom}\,(X, U)^u$	(92)	$\mathrm{Hom}\,(U, X)_u$	
$[X, U]^Y = [X, U]^u$	(92)	$[X, U]_Y = [X, U]_u$	(110)
$[X, U]^* = [X, U]$	(3, 92, 116)	$[X, U]_* = [X, U]$	
g_*, f^*	(94, 118, 137)	g^*, f_*	
$\pi_n^A(U) = [\Sigma^n A, U]$	(117, 137)	$\pi_B^n(U) = [U, \Omega^n B]$	(153)
$\pi_{n+1}^A(U, V)$	(120)	$\pi_B^{n+1}(U \mid V)$	(153)
$\pi_n^X(A \vee B)_2$	(142)	$\pi_X^n(A \times B)_2$	(153)
$E_g, E, (\pi_g, 1)$	(143)	$L_g, L, (\pi_g, 1)$	(154)
$\nabla f, \nabla, \nabla_F$	(145, 146, 206, 314)	$\nabla f, \nabla, \nabla_F$	
$\nabla(v, f), \nabla^{n+1}(u, f)$	(151)	$\nabla(u, f), \nabla^{n+1}(u, f)$	
w^+	(126)	w^+	
$\mathrm{d}(u_0, H, u_1), \mathrm{d}(u_0, u_1)$	(127)	$\mathrm{d}(u_0, H, u_1), \mathrm{d}(u_0, u_1)$	
$\Sigma(w, f)$	(127)	$\Omega(w, f)$	
B/A cofiber	(128)	fiber	
$\Sigma: [A, X]_0 \to [\Sigma A, \Sigma X]_0$	(133)	$\Omega: [X, A]_0 \to [\Omega X, \Omega A]_0$	
α_L	(111, 118, 141)	α_L	
$H_Y(x, y) = [I_Y X, U]^{(x,y)}$	(105)	$H_Y(x, y) = [U, P_Y X]_{(x,y)}$	
$O, -G, H + G$	(93, 105)	$O, -G, H + G$	
$G^\#$	(106, 138)	$G^\#$	
$[\Sigma_Y X, U]^u = \pi_1(U^{X \mid Y}, u)$	(107)	$[U, \Omega_Y X]_u = \pi_1((X \mid Y)^U, u)$	
$U^{X \mid Y}$	(108, 137)	$(X \mid Y)^U$	
$u^+(H) = u + H = H^\#(u)$	(109)	$u^+(H) = u + H = H^\#(u)$	
$\pi_n(U^{X \mid Y}, u)$	(136)	$\pi_n((X \mid Y)^U, u)$	
$\mathrm{Ob}_f, \mathbf{C}_c, \mathbf{C}_{cf}$	(7, 99)	$\mathrm{Ob}_c, \mathbf{F}_f, \mathbf{F}_{fc}$	
$\mathrm{Ho}\,(\mathbf{C}) = \mathbf{C}_{cf}/\simeq$	(99)	$\mathrm{Ho}\,(\mathbf{F}) = \mathbf{F}_{fc}/\simeq$	
$R: \mathbf{C}_c \to \mathbf{C}_{cf}/\simeq$	(100)	$M: \mathbf{F}_f \to \mathbf{F}_{fc}/\simeq$	

(iv) Further symbols in cofibration categories

I

Axioms for homotopy theory and examples of cofibration categories

Axiomatic homotopy theory is the development of the basic constructions of homotopy theory in an abstract setting, so that they may be applied to other categories. And there is, indeed, a strikingly wide variety of categories where these techniques are useful (e.g. topological spaces, differential algebras, differential Lie algebras, chain complexes, modules, sheaves, local algebras....).

The subject is not new and goes back, for example to Kan (1955), Quillen (1967), Heller (1968), and K.S. Brown (1973) each of whom proposes a different set of axioms. In fact, it is not evident what is the most appropriate choice. The best-known approach is that of Quillen who introduces the notion of a (closed) model category, as the starting point for his development of the quite sophisticated 'homotopical algebra'.

The set of axioms which define a model category is, however, quite restrictive. For instance, they do not apply to topological spaces with the usual definitions of fibrations and cofibrations. There are other examples, as pointed out by K.S. Brown, where they give rise to a 'somewhat unsatisfactory' homotopy theory.

We here introduce the notion of a *cofibration category*. Its defining axioms have been chosen according to two criteria:

(1) The axioms should be sufficiently strong to permit the basic constructions of homotopy theory.
(2) The axioms should be as weak (and as simple) as possible, so that the constructions of homotopy theory are available in as many contexts as possible.

They are substantially weaker than the axioms of Quillen, but add one essential axiom to those of Brown. In this chapter we compare the various systems of axioms in the literature. It turns out that if one applies the criteria above to these axioms one is almost forced into the definition of a cofibration category. In the chapters to follow we present some of the homotopy theory which can be derived from the axioms of a cofibration category.

In this chapter we also describe many examples of cofibration categories and of fibration categories. In an introductory section §0 we recall the classical definitions of fibrations and cofibrations in topology. Using the notion of an I-category (I = cylinder functor) we prove that the cofibrations in topology satisfy the axioms of a cofibration category. A strictly dual proof shows that fibrations in topology satisfy the axioms of a fibration category. Moreover, we give a complete proof that the algebraic categories of chain complexes, chain algebras, commutative cochain algebras, and chain Lie algebras respectively satisfy the axioms of a cofibration category.

§ 0 Cofibrations and fibrations in topology

Cofibrations and fibrations are of fundamental importance in homotopy theory. Here we recall their mutual dual definitions which imply many properties which correspond to each other. We will deduce such properties from the axioms of a cofibration category, see § 1. Hence the theory of topological cofibrations and fibrations has two aspects:

(1) The study of all properties which can be derived from the axioms (this is part of homotopical algebra).
(2) The study of properties which are highly connected with the topology, for example local properties as studied in tom Dieck–Kamps–Puppe (1970) or James (1984).

In textbooks on algebraic topology and homotopy theory these two aspects are often mixed. This creates unnecessary complexity. Using the axioms we will see that many results on fibrations and cofibrations, respectively, deserve only one proof.

Let **Top** be the **category of topological spaces** and of continuous maps and let

$$I = [0, 1] \subset \mathbb{R}$$

be the unit interval of real numbers. These data are the basis of usual homotopy theory. The notion of homotopy can be introduced in two different ways, by use of a cylinder, or by use of a path space:

The **cylinder** I is the functor

$$(0.1) \qquad \begin{cases} I: \mathbf{Top} \to \mathbf{Top}, \\ IX = I \times X \text{ with product topology,} \end{cases}$$

for which we have the 'structure maps'

$$X \xrightarrow{i_0, i_1} IX \xrightarrow{p} X$$

with $i_0(x) = (0, x)$, $i_1(x) = (1, x)$, $p(t, x) = x$ ($t \in I, x \in X$). These maps are natural with respect to maps $X \to Y$ in **Top**. The **path space** P is the functor

(0.2) $$\begin{cases} P: \mathbf{Top} \to \mathbf{Top}, \\ PX = X^I, \end{cases}$$

where X^I is the set of all maps $\sigma: I \to X$ with the compact open topology. Now we have the 'structure maps'

$$X \xrightarrow{i} X^I \xrightarrow{q_0, q_1} X$$

with $i(x)(t) = x$ and $q_0(\sigma) = \sigma(0)$, $q_1(\sigma) = \sigma(1)$.

The product topology for IX and the compact open topology for $PY = Y^I$ have the well-known property that a map $G: IX \to Y$ is continuous if and only if the **adjoint map** $\bar{G}: X \to Y^I$ with $\bar{G}(x)(t) = G(t, x)$ is continuous. Therefore we have the bijection of sets

(0.3) $$\mathbf{Top}(IX, Y) = \mathbf{Top}(X, Y^I),$$

which carries G to $\bar{G}$. Here **Top** (A, B) denotes the set of all maps $A \to B$ in **Top**. The bijection shows that the following two definitions of homotopy are equivalent;

(0.4) ***Definition.*** Maps $f_0, f_1: X \to Y$ are **homotopic** ($f_0 \simeq f_1$) if there is a map $G: IX \to Y$ with $Gi_0 = f_0, Gi_1 = f_1$. ||

(0.5) ***Definition.*** Maps $f_0, f_1: X \to Y$ are **homotopic** ($f_0 \simeq f_1$) if there is a map $H: X \to PY$ with $q_0 H = f_0, q_1 H = f_1$. ||

There is a standard proof that the relation of homotopy $\simeq$ is an equivalence relation on **Top**(X, Y). Moreover, this relation is compatible with the law of composition in **Top**, that is, for maps $f_i: X \to Y$, $g_i: Y \to Z$ ($i = 0, 1$) with $f_0 \simeq f_1$, $g_0 \simeq g_1$ we get $g_0 f_0 \simeq g_1 f_1$. Therefore the **homotopy category** ***Top***/ $\simeq$ is defined. The morphisms are the homotopy classes of maps in **Top**. The set of morphism in **Top**/$\simeq$ from X to Y is the set

(0.6) $$[X, Y] = \mathbf{Top}(X, Y)/\simeq$$

of homotopy classes. For a map $f: X \to Y$ let $\{f\} \in [X, Y]$ be the homotopy class represented by f, we also write $\{f\}: X \to Y$.

(0.7) ***Definition.*** A map $f: X \to Y$ in **Top** is a **homotopy equivalence** if (a) or equivalently (b) is satisfied:

(a) $\{f\}$ is an isomorphism in **Top**/$\simeq$.
(b) There is a map $g: Y \to X$ such that $gf \simeq 1_X$ and $fg \simeq 1_Y$. ||

Next we introduce cofibrations and fibrations in **Top** by use of the cylinder and the path space respectively.

(0.8) ***Definition.*** A map $i: A \to X$ has the **homotopy extension property** (HEP) with respect to Y if for each commutative diagram of unbroken arrows in **Top**

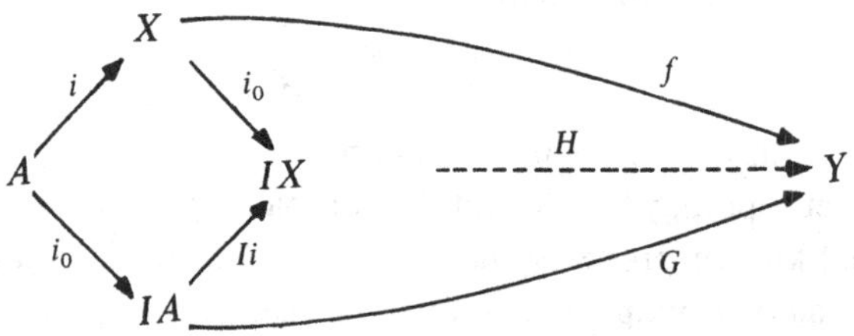

there exists H extending the diagram commutatively. The map i is a **cofibration** in **Top** if it has the homotopy extension property with respect to any space in **Top**. The cofibration i is **closed** if iA is a closed subspace. ||

The following definition of a fibration is **dual** to the definition of a cofibration in the sense that the cylinder is replaced by the path space and all arrows are replaced by arrows in the opposite direction.

(0.9) ***Definition.*** A map $p: X \to B$ has the **homotopy lifting property** (HLP) with respect to Y if for each diagram of unbroken arrows in **Top**

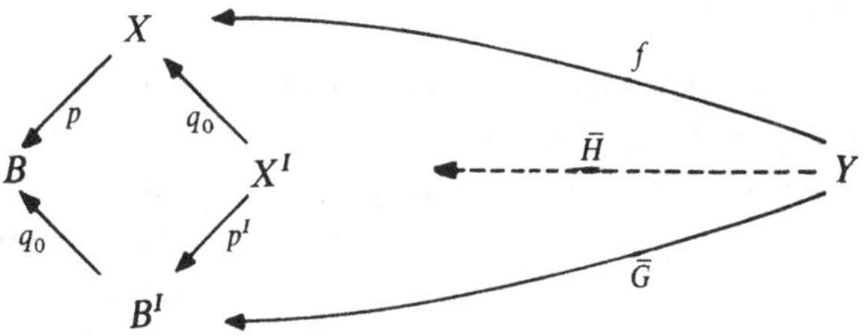

there exists $\bar{H}$ extending the diagram commutatively. The map p is a **fibration** in **Top** if it has the homotopy lifting property with respect to any space in **Top**. ||

Using the adjunction (0.3) we can reformulate this definition as follows: The map p has the HLP with respect to Y iff for each commutative diagram of unbroken arrows

(0.10)
$$\begin{array}{ccc} Y & \xrightarrow{f} & X \\ {\scriptstyle i_0}\downarrow & \overset{H}{\nearrow} & \downarrow \\ IY & \xrightarrow[G]{} & B \end{array}$$

there is H extending the diagram commutatively. The map H is called a **lifting** of G.

Let $D^n = \{x \in \mathbb{R}^n;\ \|x\| \leqq 1\}$ be the **disk** in $\mathbb{R}^n$ with boundary $\partial D^n = S^{n-1}$, $n \geqq 0$. For $n = 0$ we have $S^{-1} = \varnothing =$ the empty set.

(0.11) ***Definition.*** A map $p: X \to B$ is a **Serre-fibration** if p has the HLP with respect to D^n, $n \geqq 0$. ||

On the other hand, we obtain by use of the disks the following cofibrations in **Top**.

(0.12) ***Definition.*** We say that $A \subset X$ is given by **attaching a cell** to A if there exists a push out diagram in **Top** ($n \geqq 0$)

$$\begin{array}{ccc} D^n & \longrightarrow & X \\ \cup & & \cup \\ S^{n-1} & \longrightarrow & A \end{array}$$

The inclusions $S^{n-1} \subset D^n$ and $A \subset X$ are cofibrations in **Top**. ||

In the next section we define a cofibration category. A basic example of a cofibration category is the category **Top** with cofibrations as in (0.8) and with weak equivalences given by homotopy equivalences in **Top**, compare (5.1) below. Moreover, we will see that fibrations and homotopy equivalences in **Top** satisfy the axioms of a fibration category which are obtained by dualizing the axioms of a cofibration category, see (1a.1) and (5.2) below.

§1 Cofibration categories

Here we introduce the notion of a cofibration category. This is a category together with two classes of morphisms, called cofibrations and weak equivalences, such that four axioms $C1, \ldots, C4$ are satisfied.

(1.1) ***Definition.*** A **cofibration category** is a category **C** with an additional structure

$$(\mathbf{C}, cof, we),$$

subject to axioms C1, C2, C3 and C4. Here *cof* and *we* are classes of morphisms in **C**, called **cofibrations** and **weak equivalences** respectively. ‖

Morphisms in **C** are also called *maps* in **C**. We write $i: B \subset A$ or $B \rightarrowtail A$ for a cofibration and we call $u|B = ui: B \to U$ the **restriction** of $u: A \to U$. We write $X \xrightarrow{\sim} Y$ for a weak equivalence in **C**. An isomorphism in **C** is denoted by $\cong$. The identity of the object X is $1 = 1_X = id$. A map in **C** is a **trivial cofibration** if it is both a weak equivalence and a cofibration. An object R in a cofibration category **C** will be called a **fibrant model** (or simply **fibrant**) if each trivial cofibration $i: R \overset{\sim}{\rightarrowtail} Q$ in **C** admits a retraction $r: Q \to R$, $ri = 1_R$.

The axioms in question are:

(C1) *Composition axiom*: The isomorphisms in **C** are weak equivalences and are also cofibrations. For two maps

$$A \xrightarrow{f} B \xrightarrow{g} C$$

if any two of f, g, and gf are weak equivalences, then so is the third. The composite of cofibrations is a cofibration.

(C2) *Push out axiom*: For a cofibration $i: B \rightarrowtail A$ and map $f: B \to Y$ there exists the push out in **C**

$$\begin{array}{ccc} A & \xrightarrow{\bar{f}} & A \bigcup_B Y = A \bigcup_f Y \\ \uparrow i & & \uparrow \bar{i} \\ B & \xrightarrow{f} & Y \end{array}$$

and $\bar{i}$ is a cofibration. Moreover:

(a) if f is a weak equivalence, so is $\bar{f}$,
(b) if i is a weak equivalence, so is $\bar{i}$.

(C3) *Factorization axiom*: For a map $f: B \to Y$ in **C** there exists a commutative diagram

$$\begin{array}{ccc} B & \xrightarrow{f} & Y \\ & {\scriptstyle i}\searrow \quad \nearrow{\scriptstyle g}\,\sim & \\ & A & \end{array}$$

where i is a cofibration and g is a weak equivalence.

(C4) *Axiom on fibrant models*: For each object X in **C** there is a trivial cofibration $X \overset{\sim}{\rightarrowtail} RX$ where RX is fibrant in **C**. We call $X \overset{\sim}{\rightarrowtail} RX$ a **fibrant model of** X.

(1.2) ***Remark.*** We denote by Ob_f a class of fibrant models in $\mathbf{C}$ which is **sufficiently large**, this means that each object in $\mathbf{C}$ has a fibrant model in Ob_f. Let $\mathbf{C}_f$ be the full subcategory of $\mathbf{C}$ with objects in Ob_f. By the structure of $\mathbf{C}$ we have cofibrations and weak equivalences in $\mathbf{C}_f$. One can check that $\mathbf{C}_f$ satisfies the axioms (C1), (C3) and (C4) but not necessarily axiom (C2) since the push out of objects in Ob_f needs not to be an object in Ob_f. If, however, push outs as in (C2) exist in $\mathbf{C}_f$ then $\mathbf{C}_f$ is a cofibration category in which all objects are fibrant.

(1.3) ***Remark.*** Let ϕ be an **initial object** of the cofibration category $\mathbf{C}$. We call an object X in $\mathbf{C}$ **cofibrant** if $\phi \to X$ is a cofibration. Let $\mathbf{C}_c$ be the full subcategory of C consisting of cofibrant objects. By the structure of $\mathbf{C}$ we have cofibrations and weak equivalences in the category $\mathbf{C}_c$. One easily checks the axioms (C1),…,(C4). Thus $\mathbf{C}_c$ is a cofibration category in which all objects are cofibrant. We point out that the notion 'cofibrant' is not dual to the notion 'fibrant' in (C4). Therefore we call a cofibrant object in $\mathbf{C}$ as well a **ϕ-cofibrant** object since its definition depends on the existence of the initial object ϕ.

The development of the homotopy theory in a cofibration category is most convenient if all objects in $\mathbf{C}$ are fibrant and cofibrant.

(1.4) **Lemma.** *Let* $\mathbf{C}$ *be a cofibration category. Then* (C2) (a), (C1) *and* (C3) *imply* (C2) (b). *If all objects in* $\mathbf{C}$ *are cofibrant then* (C2) (b), (C1) *and* (C3) *imply* (C2) (a).

Thus axiom (C2) (b) is redundant. We call (C2)(a) the '*axiom of properness*' (compare (2.1) below); many results in a cofibration category actually do not rely on this axiom. If all objects are cofibrant then the axiom of properness is redundant by (1.4).

Proof. We consider the push out diagrams

$$\begin{array}{ccccc} A & \overset{\bar{j}}{\rightarrowtail} & \bar{X} & \overset{\bar{g}}{\longrightarrow} & \bar{Y} \\ {\scriptstyle i}\uparrow & \text{push} & {\scriptstyle i_1}\uparrow & \text{push} & \uparrow{\scriptstyle \bar{i},} \\ B & \underset{j}{\rightarrowtail} & X & \underset{g}{\overset{\sim}{\longrightarrow}} & Y \end{array}$$

were $gj = f$ by (C3). If i is a weak equivalence, so is i_1 *by* (C2) (a). Moreover, since g is a weak equivalence, also $\bar{g}$ is one by (C2) (a). Thus by (C1) also $\bar{i}$ is a weak equivalence.

For the proof of the second part of (1.4) we need the mapping cylinders in (1.8) below. We continue the proof of (1.4) in the appendix §1b of this section.   □

We define cylinders and the notion of homotopy in a cofibration category as follows: Let $B \rightarrowtail A$ be a cofibration. Then we have by (C2) the push out diagram

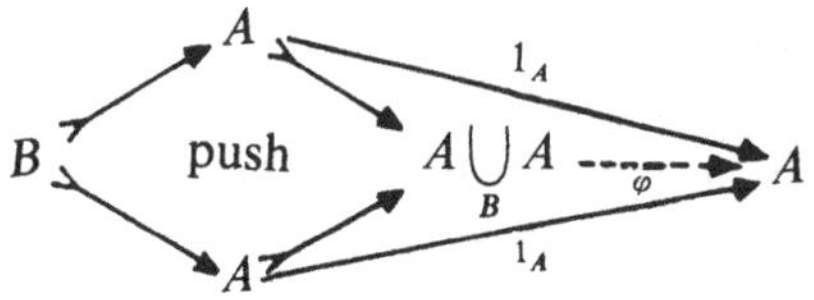

where $\varphi = (1_A, 1_A)$ is called the **folding map**. By (C3) there is a factorization

(1.5) $$A \underset{B}{\cup} A \overset{i}{\rightarrowtail} Z \overset{p}{\underset{\sim}{\to}} A$$

of the folding map φ. We call $Z = I_B A$ together with i and p in (1.5) a **relative cylinder** on $B \rightarrowtail A$. For $i = (i_0, i_1)$ the maps $i_\varepsilon : A \overset{\sim}{\rightarrowtail} Z$ are trivial cofibrations since $p i_\varepsilon = 1_A$; use (C1).

Let X be a fibrant object.

Two maps $\alpha, \beta : A \to X$ are **homotopic relative B** (or **under B**) and we write $\alpha \simeq \beta$ rel B if there is a commutative diagram

(1.6)

$$
\begin{array}{ccc}
A \underset{B}{\cup} A & \overset{i}{\rightarrowtail} & Z \\
& \underset{(\alpha,\beta)}{\searrow} \quad \swarrow H & \\
& X &
\end{array}
$$

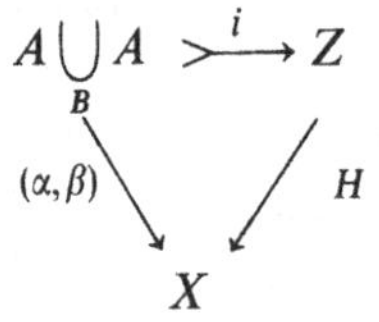

where Z is a relative cylinder on $B \rightarrowtail A$. We call H a **homotopy from α to β rel B**. We will prove that homotopy rel B is an equivalence relation, see Chapter II.

For a ϕ-cofibrant object A there exists the sum $A + Y$ (also denoted by $A \vee Y$). The sum is given via (C2) by the push out

(1.7) $$A + Y = A \underset{\phi}{\cup} Y = A \vee Y$$

where $Y \rightarrowtail A + Y$ is a cofibration by (C2). Also $A \rightarrowtail A + Y$ is a cofibration provided Y is ϕ-cofibrant. We define the **mapping cylinder** Z_f of $f : A \to Y$ by a factorization of the map $(1_Y, f) : Y + A \to Y$ via (C3):

(1.8) $$(1_Y, f) : Y + A \rightarrowtail Z_f \overset{\sim}{\underset{q}{\to}} Y.$$

If Y is cofibrant this yields the factorization $f = qi_1$,

$$f : A \underset{i_1}{\rightarrowtail} Z_f \underset{i_0}{\overset{\longrightarrow}{\longleftarrow}} Y, \qquad (a)$$

where q is a retraction of $i_0 : Y \rightarrowtail Y + A \rightarrowtail Z_f$ and where $i_1 : A \rightarrowtail Y + A \rightarrowtail Z_f$. Moreover, we can use the cylinder $Z = I_\phi A$ in (1.5) for the construction of the mapping cylinder via a push out diagram:

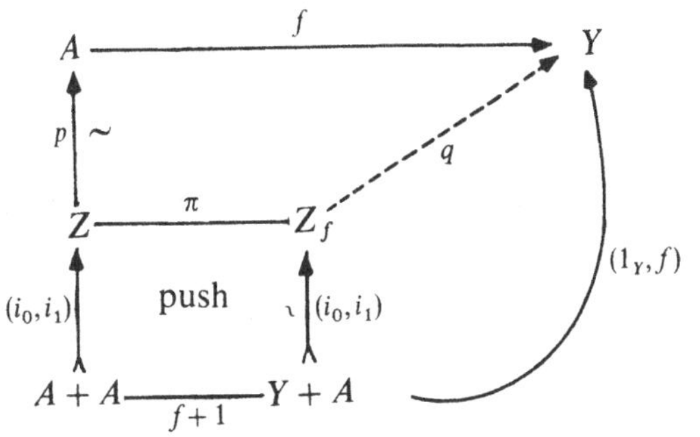

$$(b)$$

Here i_0 is a weak equivalence since $i_0 : A \rightarrowtail Z$ is a weak equivalence. Therefore the retraction q of i_0 is a weak equivalence by (C1).

(1.9) Definition. A commutative square

$$
\begin{array}{ccc}
A & \longrightarrow & C \\
f \uparrow & \simeq \text{ push} & \uparrow \\
B & \underset{g}{\longrightarrow} & D
\end{array}
$$

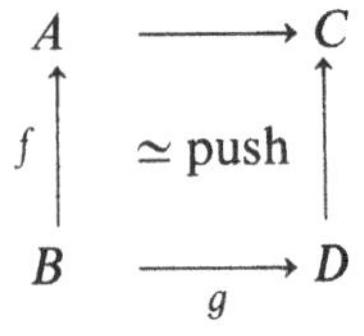

in a cofibration category **C** is a **homotopy push out** (or **homotopy cocartesian**) if for some factorization $B \rightarrowtail W \overset{\sim}{\to} A$ of f the induced map

$$W \bigcup_B D \to C$$

is a weak equivalence. This easily implies that for any factorization $B \rightarrowtail V \overset{\sim}{\to} A$ of f, the map $V \bigcup_B D \to C$ is a weak equivalence. Thus in the definition we could have replaced 'some' by 'any' or used g in place of f. We leave the proof of these remarks as an exercise, compare (II,§1) ‖

Next we consider functors between cofibration categories.

(1.10) Definition. Let **C** and **K** be cofibration categories and let $\alpha : \mathbf{C} \to \mathbf{K}$ be a functor.

(1) The functor α is **based** if **C** and **K** have an initial object (denoted by $*$) with $\alpha(*) = *$.

(2) The functor α **preserves weak equivalences** if α carries a weak equivalence in **C** to a weak equivalence in **K**.

(3) Let

$$\begin{array}{ccc} A & \xrightarrow{\bar{f}} & Y = A \bigcup_B X \\ {\scriptstyle i}\uparrow & & \uparrow{\scriptstyle \bar{i}} \\ B & \longrightarrow & X \end{array}$$

be a push out diagram in **C**. We say that α is **compatible with the push out** $A \bigcup_B X$ if the induced diagram

$$\begin{array}{ccc} \alpha A & \longrightarrow & \alpha(A \bigcup_B X) \\ \uparrow & & \uparrow \\ \alpha B & \longrightarrow & \alpha X \end{array}$$

is a homotopy push out in **K**, see (1.9).

(4) We call α a **model functor** if α preserves weak equivalences and if α is compatible with all push outs as in (3). Hence a model functor α carries homotopy cocartesian diagrams in **C** to homotopy cocartesian diagrams in **K**. ||

We will see that a based model functor is compatible with most of the constructions in a cofibration category described in this book. In general, we do not assume that a model functor carries a cofibration in **C** to a cofibration in **K**.

§ 1a Appendix: fibration categories

By dualizing (1.1) we obtain

(1a.1) ***Definition.*** A **fibration category** is a category **F** with the structure

$$(\mathbf{F}, \mathit{fib}, \mathit{we}),$$

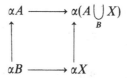

subject to axioms (F1), (F2), (F3) and (F4). Here *fib* and *we* are classes of morphisms in **F**, called **fibrations** and **weak equivalences** respectively. These morphisms satisfy the condition that the **opposite category** $\mathbf{C} = \mathbf{F}^{op}$ is a cofibration category where the structure of **C** is given by

$$\text{(1a.2)}\quad \begin{array}{l} f^{op} \text{ is a cofibration in } \mathbf{C} \Leftrightarrow f \text{ is a fibration in } \mathbf{F}, \\ f^{op} \text{ is a weak equivalence in } \mathbf{C} \Leftrightarrow f \text{ is a weak equivalence in } \mathbf{F}. \end{array}$$

By dualizing the axioms (C1),...,(C4) we obtain the axioms (F1),...,(F4) which characterize the fibration category. ||

We write $A \twoheadrightarrow B$ for a fibration and $A \xrightarrow{\sim}\!\!\!\!\twoheadrightarrow B$ for a trivial fibration. An object X is **cofibrant** (or a **cofibrant model**) in **F** if each trivial fibration $Y \xrightarrow{\sim}\!\!\!\!\twoheadrightarrow X$ admits a section. An object X is **e-fibrant** in **F** if $X \to e$ is a fibration. Here e is a **final object** in **F**. Hence 'e-fibrant' is the notion dual to 'ϕ-cofibrant' in (1.3) and 'cofibrant model in **F**' is the notion dual to 'fibrant model in **C**' in (1.1).

We leave it to the reader to formulate the axioms (F1), (F2) and (F3). Axiom (F4) is given as follows.

(F4) Axiom on cofibrant models: For each object X in **F** there is a trivial fibration $MX \xrightarrow{\sim}\!\!\!\!\twoheadrightarrow X$ where MX is cofibrant in **F**.

Of course, a cofibration category has properties which are strictly dual to the properties of a fibration category and vice versa. It turns out that this is a good axiomatic background for many results which satisfy the Eckmann–Hilton duality. Any result in a cofibration category which follows from the axioms (C1),...,(C4) corresponds to a dual result in a fibration category which follows precisely by dual arguments from the dual axioms (F1),...,(F4). Therefore, we describe our results only for a cofibration category. We leave the formulation of the dual results to the reader.

We obtain path objects and the notion of homotopy in a fibration category as follows: By (F2) there exist pull backs

$$\begin{array}{ccc} A \times_B Y & \longrightarrow & A \\ \downarrow\!\!\downarrow & & \downarrow\!\!\downarrow \\ Y & \xrightarrow[f]{} & B \end{array}$$

in **F**. We denote by

$$\text{(1a.3)}\qquad (1_A, 1_A): A \to A \times_B A$$

the **diagonal map** which is dual to the folding map. A factorization of the diagonal map by (F3), namely

$$\text{(1a.4)}\qquad A \underset{j}{\xrightarrow{\sim}} P \underset{q}{\twoheadrightarrow} A \times_B A$$

with $qj = (1_A, 1_A)$, is called a **path object** for $A \twoheadrightarrow B$. Two maps $\alpha, \beta: X \to A$ are **homotopic over** B ($\alpha \simeq \beta$ over B) if there is a commutative diagram

(1a.5)
$$\begin{array}{ccc} P & \xrightarrow{q} & A \times_B A \\ {\scriptstyle H}\nwarrow & & \nearrow{\scriptstyle (\alpha,\beta)} \\ & X & \end{array} .$$

We call H a homotopy from α to β over B. Here we assume that X is a cofibrant object in **F**. Compare (0.5).

As an example, the category **Top** with fibrations, defined in (0.9), and with homotopy equivalences as weak equivalences is a fibration category. This is proved in §4 below.

Our definition of a cofibration category (resp. of a fibration category) corresponds to the following concepts in the literature.

(1a.6) ***Remark.*** A *category of cofibrant objects* in the sense of K.S. Brown (1973) has the structure ($\mathbf{C}$, *cof*, *we*, ϕ) which satisfies (C1), (C2) (b), (C3) and for which

'all objects are ϕ-cofibrant'. (A)

D.W. Anderson (1978) essentially adopts these axioms but he omits (A). His *left homotopy structure* ($\mathbf{C}$, *cof*, *we*) satisfies the axioms (C1), (C2)(b) and (C3) and the axiom '**C** has finite colimits'. In addition he uses the axiom

'all objects are fibrant models', see (1.1), (B)

which he calls the homotopy extension axiom. Moreover A. Heller (1968) defines the structure of an *h–c–category* ($\mathbf{C}$, *cof*, $e = \phi$, $\simeq$) where $\simeq$ is a natural equivalence relation, compare also Shitanda. If weak equivalences are the maps in **C** which are isomorphisms in $\mathbf{C}/\simeq$, then this structure satisfies the axioms of K.S. Brown above. Recently Waldhausen (1984) used axioms on cofibrations, in particular, in his set up he assumes $e = \phi$ and (A).

§1b Appendix: proof of (1.4)

We now continue the proof of (1.4). Assume all objects in **C** are cofibrant and assume (C2) (b), (C1) and (C3) are satisfied. For the construction of the mapping cylinder in (1.8) we only made use of these axioms. Now consider the push out in (C2) and assume that f is a weak equivalence. We hope to show that also $\bar{f}$ is a weak equivalence. First we get the commutative diagram

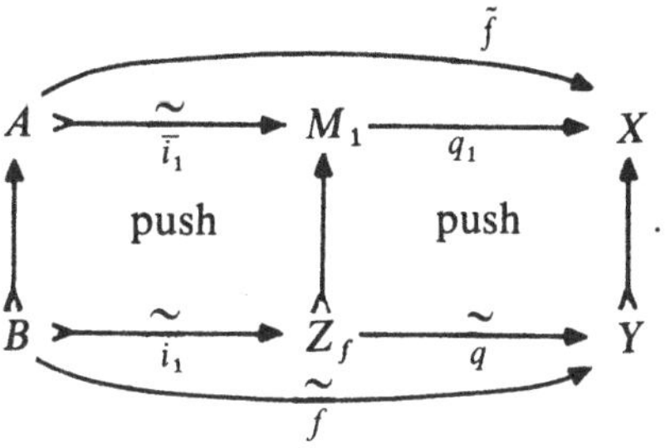

(1)

By (C1) we know that i_1 is a weak equivalence, thus by (C2) (b) also $\bar{i}_1$ is a weak equivalence.

Next consider

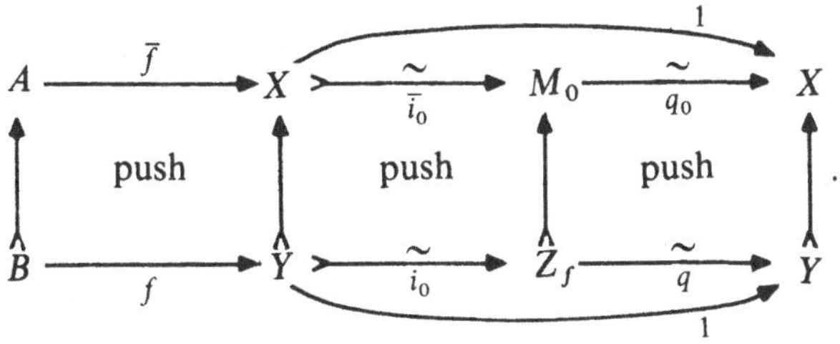

(2)

Since $qi_0 = 1$ we see that $q_0\bar{i}_0 = 1_X$ and therefore q_0 and $\bar{i}_0$ are weak equivalences. Next we obtain for Z in (1.8) the following diagram with $\varepsilon = 0$ or $\varepsilon = 1$:

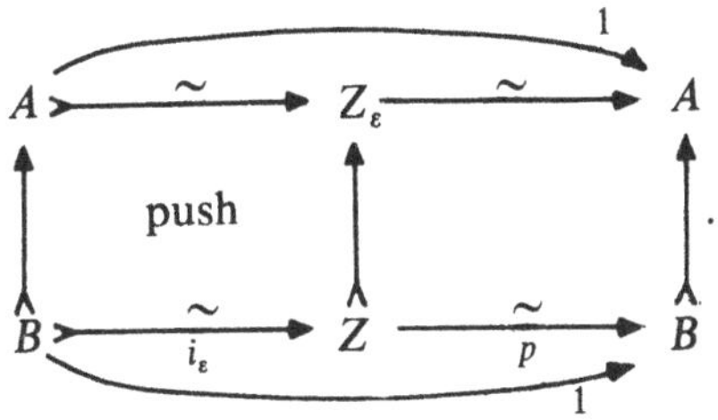

(3)

This yields the commutative diagram:

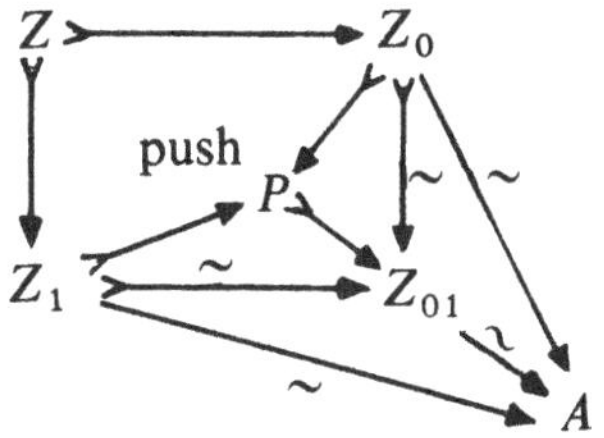

(4)

Here Z_{01} is given by the push out map $P \to A$ and by (C3). Now we have the following commutative diagram:

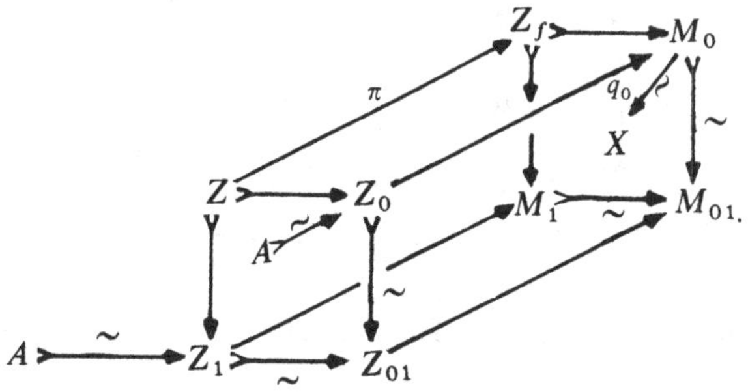

(5)

The top, the bottom, the right-hand side and the left-hand side of this cubical diagram are push outs. The map π is the one in (1.8). Now

$$A \xrightarrow{\sim} Z_1 \rightarrow M_1$$

is the map $\bar{i}_1$ in (1). Therefore $Z_1 \rightarrow M_1$ is a weak equivalence. Thus $Z_{01} \rightarrow M_{01}$ is a weak equivalence. Hence $Z_0 \rightarrow M_0$ is a weak equivalence and therefore by (2) the composition $\bar{f}: A \rightarrowtail\!\!\xrightarrow{\sim} Z_0 \xrightarrow{\sim} M_0 \xrightarrow{\sim} X$ is a weak equivalence. This completes the proof of (1.4). ||

§ 2 The axioms of Quillen

We now compare the notion of a cofibration category with the notion of a model category as introduced by Quillen (1967). The model category is defined by axioms on cofibrations and weak equivalences in a category **M**. This notion is self-dual, that is, the dual category $\mathbf{M}^{op}$ is once more a model category where the roles of fibrations and cofibrations are interchanged. Therefore a model category has two faces which are dual to each other. We show that the notion of a cofibration category extracts from a model category the essential properties of one of these faces.

(2.1) ***Definition.*** A **model category** is a category **M** with the structure

$$(\mathbf{M}, \mathit{cof}, \mathit{fib}, \mathit{we}),$$

which satisfies the axioms (C1), (F1), (C2) (b), (F2) (b) and the following axioms (M0), (M1), (M2):

(M0): **M** is closed under finite limits and colimits.
(M1): Given a commutative diagram of unbroken arrows

$$\begin{array}{ccc} A & \longrightarrow & X \\ {\scriptstyle i}\downarrow & \nearrow & \downarrow{\scriptstyle p}, \\ B & \longrightarrow & Y \end{array}$$

where i is a cofibration, p is a fibration, and where either i or p is a weak equivalence, then the dotted arrow exists such that the triangles commute.

(M2): Any map f may be factored $f = pi$ where i is a cofibration and a weak equivalence, and p is a fibration. Also $f = pi$ where i is a cofibration, and p is a fibration and weak equivalence.

If in addition (C2) (a) and (F2) (a) are satisfied **M** is called a **proper model category**, see Thomason and Bousfield–Friedlander. ||

(2.2) ***Definition.*** A map $f: V \to W$ is called a **retract** of $g: X \to Y$ if there exists a commutative diagram

$$\begin{array}{ccccc} V & \longrightarrow & X & \longrightarrow & V \\ {\scriptstyle f}\downarrow & & {\scriptstyle g}\downarrow & & {\scriptstyle f}\downarrow, \\ W & \longrightarrow & Y & \longrightarrow & W \end{array}$$

in which the horizontal maps compose to the identity. A model category as in (2.1) is a **closed model category** if it satisfies in addition the following axiom (CM).

(2.3) If f is a retract of g, and g is a weak equivalence, fibration, or cofibration, then so is f. ||

(2.4) ***Definition.*** Axiom (M0) implies that a model category **M** has an initial object ϕ and a final object e. Hence ϕ-cofibrant and e-fibrant objects are defined in **M** as in (1.3) and § 1a respectively. Let $\mathbf{M}_c$ (resp. $\mathbf{M}_f$) be the full subcategory of **M** consisting of ϕ-cofibrant (resp. e-fibrant) objects. ||

(2.6) **Proposition.** *Let* **M** *be a model category. Then* $\mathbf{M}_c$ *with cofibrations and weak equivalences as in* **M** *is a cofibration category. Dually* $\mathbf{M}_f$ *with fibrations and weak equivalences as in* **M** *is a fibration category. Moreover, if* **M** *is proper* (**M**, *cof*, *we*) *is a cofibration category and* (**M**, *fib*, *we*) *is a fibration category.*

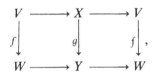

Proof: (C3) follows directly from (M2). We now prove (C4). For any object X we have by (M2) a factorization $X \overset{\sim}{\rightarrowtail} R \twoheadrightarrow e$ of $X \to e$. We claim that R is a fibrant model. In fact, for each trivial cofibration $R \overset{\sim}{\rightarrowtail} Q$ we have by (M1)

the commutative diagram

$$\begin{array}{ccc} R & \xrightarrow{1} & R \\ {\scriptstyle\sim}\downarrow & \nearrow & \downarrow \\ Q & \longrightarrow & e \end{array}$$

This shows that each e-fibrant object in $\mathbf{M}$ is a fibrant model. □

The following lemma shows that our definition of fibrant models in (1.1) is essentially consistent with Quillen's definition of fibrant objects.

(2.6) **Lemma.** *Let X be an object in a model category. Then X is a fibrant model in the sense of* (1.1) *if and only if X is a retract of an e-fibrant object in* $\mathbf{M}$.

By axiom (CM) in (2.3) we derive:

(2.7) **Corollary.** *In a closed model category the fibrant models are exactly the e-fibrant objects.*

Proof of (2.6). Let X be a fibrant model. By (M2) we have the factorization

$$X \overset{\sim}{\rightarrowtail} \bar{X} \twoheadrightarrow e$$

of $X \to e$. Then $\bar{X}$ is e-fibrant. Since X is a fibrant model the trivial cofibration $X \overset{\sim}{\rightarrowtail} \bar{X}$ admits a retraction. Therefore X is a retract of an e-fibrant object.

Now assume X is a retract of an e-fibrant object U with

$$X \xrightarrow{i} U \xrightarrow{r} X, \quad ri = 1_X.$$

For any trivial cofibration $X \overset{\sim}{\rightarrowtail} Y$ we have the push out diagram

$$\begin{array}{ccc} Y & \xrightarrow{\bar{i}} & P \\ j\uparrow{\scriptstyle\sim} & & \bar{j}\uparrow{\scriptstyle\sim} \\ X & \xrightarrow{i} & U \end{array},$$

where $\bar{j}$ is a trivial cofibration by (C2). Since U is e-fibrant we know as in the proof of (2.5) that U is a fibrant model. Therefore there exists a retraction $\bar{r}$ of $\bar{j}$. Now $r\bar{r}\bar{i}$ is a retraction of j since $(r\bar{r}\bar{i})j = r\bar{r}\bar{j}i = ri = 1_X$.

Thus X is a fibrant model in $\mathbf{M}$. □

Since there are many examples of model categories we obtain by (2.5) many examples of cofibration categories and of fibration categories. In the following list of model categories we use the notation: Ob = class of all objects, Ob_C = class of cofibrant objects and Ob_f = class of fibrant objects. We write

(2.8) $$cof = M(fib, we)$$

if *cof* contains precisely all maps i which satisfy (M1) in (2.1) whenever p is a trivial fibration. Dually we write

(2.9) $$fib = M(cof, we)$$

if *fib* contains precisely all maps p which satisfy (M1) whenever i is a trivial cofibration. In a closed model category (see (2.2)) we always have (2.8) and (2.9).

(2.10) *Proper model category* (Strøm (1972)), Compare (4a.5).

C = **Top**,
cof = closed cofibrations in **Top** (see (0.8)),
fib = fibrations in **Top** (see (0.9)),
we = homotopy equivalences,
$Ob_c = Ob_f = Ob$.

(2.11) *Proper closed model category* (Quillen (1967)).

C = **Top**,
cof = M(fib, we),
fib = Serre fibrations (see (0.11)),
we = weak homotopy equivalence (= maps inducing isomorphisms for the functors $[K, \cdot]$ where K is a finite CW-complex), (see (0.6)),
$Ob_f = Ob$, Ob_c contains the class of all CW-complexes.

(2.12) *Proper closed model category* (Quillen (1967)).

C = category of *simplical sets*,
cof = injective maps,
fib = Kan fibrations,
we = maps which become homotopy equivalences is **Top** if the *geometric realization* functor, $X \longmapsto |X|$, is applied,
Ob_f = Kan complexes, $Ob_c = Ob$.

Moreover the *singular set* is a model functor (see (1.10)) from the cofibration categories **Top** in (2.10) or (2.11) to the cofibration category of simplicial sets.

By change of the weak equivalences in (2.12) one obtains the next example.

(2.13) *Closed model category* (Bousfield (1975)). **C** = category of simplicial sets, *cof* = injective maps, *fib* = M (*cof*, *we*) and

> *we* = maps $f:K \to L$ which induce isomorphisms $f_*:k_*(K) \cong k_*(L)$,
> $Ob_f = k_*$-local Kan complexes, $Ob_c = Ob$.

Here k_* is a *generalized homology theory* defined on CW-pairs and satisfying the limit axiom, see (5.10). The functor k_* is transfered to simplicial sets by setting $k_*(K, L) = k_*(|K|, |L|)$.

(2.14) ***Remark.*** Further examples of closed model categories are described in K.S. Brown (1973), Bousfield–Kan (1973), Dwyer–Kan (1983), (1984), Dwyer (1979), Munkholm (1978), Bousfield–Gugenheim (1976), Quillen (1967) (1969), Edwards–Hastings (1976), Bousfield–Friedlander (1978), Jardine (1985). The structure of a model category on the category of small categories is considered by Thomason, Illusie, Fritsch–Latch and Grothendieck (1983), for the category of groupoids see Anderson. Moreover, Varadarajan (1975) studies the category of modules the homotopy theory of which first appears in Eckmann (1956), Hilton (1965). This is one of the first examples in the literature in which homotopy theory is discussed in a non topological context, see also Huber (1961).

§ 3 Categories with a natural cylinder

We here combine the notion of a cofibration category with the concept of abstract homotopy theory as introduced by Kan (1955). This concept relies on a natural cylinder object, see also Kamps. Under fairly weak assumptions on a natural cylinder we can derive the structure of a cofibration category.

(3.1) ***Definition.*** An **I-category** is a category **C** with the structure (**C**, *cof*, I, ϕ). Here *cof* is a class of morphisms in **C**, called cofibrations. I is a functor $\mathbf{C} \to \mathbf{C}$ together with natural transformations i_0, i_1 and p. ϕ is the initial object in **C**. The structure satisfies the following axioms (I1), ..., (I5). ||

(I1) Cylinder axiom: $I:\mathbf{C} \to \mathbf{C}$ is a functor together with natural transformations

$$i_0, i_1 : id_{\mathbf{C}} \to I, \quad p: I \to id_{\mathbf{C}},$$

such that for all objects X $pi_\varepsilon: X \to IX \to X$ is the identity of X for $\varepsilon = 0$ and $\varepsilon = 1$. Compare (0.1).

(I2) Push out axiom: For a cofibration $i:B \to A$ and a map f there exists the push out

$$\begin{array}{ccc} A & \longrightarrow & A \bigcup_B X \\ {\scriptstyle i}\uparrow & & \uparrow{\scriptstyle \bar{i}} \\ B & \xrightarrow[f]{} & X \end{array},$$

where $\bar{i}$ is also a cofibration. Moreover, the functor I carries the push out diagram into a push out diagram, that is $I(A \bigcup_B X) = IA \bigcup_{IB} IX$. Moreover, $I\phi = \phi$.

(I3) Cofibration axiom: Each isomorphism is a cofibration and for each object X the map $\phi \to X$ is a cofibration. We thus have by (I2) the sum $X + Y = X \bigcup_\phi Y = X \vee Y$. The composition of cofibrations is a cofibration. Moreover, a cofibration $i: B \rightarrowtail A$ has the following **homotopy extension property** in **C**. Let $\varepsilon \in \{0, 1\}$. For each commutative diagram in **C**

$$\begin{array}{ccc} B & \xrightarrow{i_\varepsilon} & IB \\ {\scriptstyle i}\downarrow & & \downarrow{\scriptstyle H} \\ A & \xrightarrow[f]{} & X \end{array}$$

there is $E: IA \to X$ with $E(Ii) = H$ and $Ei_\varepsilon = f$. Compare (0.8).

(I4) Relative cylinder axiom: For a cofibration $i: B \to A$ the map j defined by the following push out diagram is a cofibration:

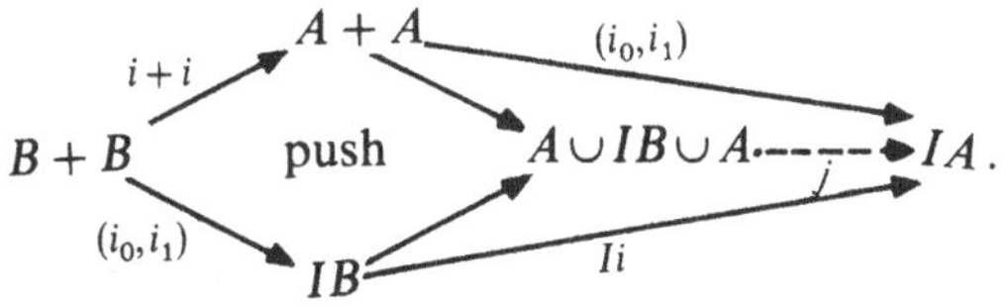

Equivalently $(IB, B \vee B) \to (IA, A \vee A)$ is a cofibration in **Pair(C)**, see (II.1.3) below.

(I5) The interchange axiom: For all objects X there exists a map $T: IIX \to IIX$ with $Ti_\varepsilon = I(i_\varepsilon)$ and $TI(i_\varepsilon) = i_\varepsilon$ for $\varepsilon = 0$ and $\varepsilon = 1$. We call T an interchange map.

We sketch the double cylinder IIX by

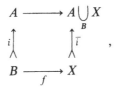

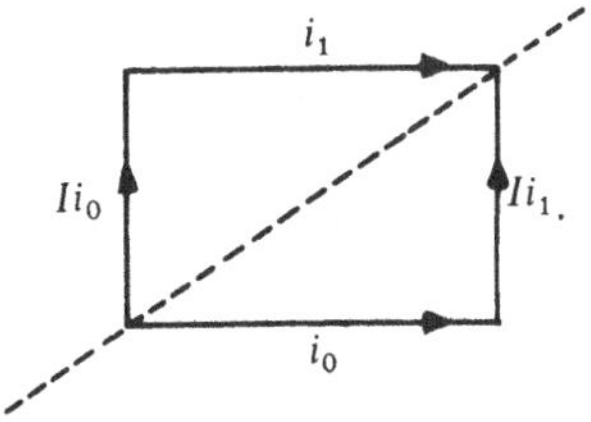

The interchange map T restricted to the boundary $\partial I^2 X$ is the reflection at the diagonal.

Remark. If a functor I is given such that (I1) is satisfied we can define the class of cofibrations by the homotopy extension property in (I3). In this case, (I2), (I4) and (I5) are additional properties of the functor I. An example is the cylinder functor for the category of topological spaces, see §0.

(3.2) ***Definition.*** We say $f_0, f_1: A \to X$ are **homotopic** if there is $G: IA \to X$ with $Gi_0 = f_0$ and $Gi_1 = f_1$. We call G a homotopy and we write $G: f_0 \simeq f_1$. A map $f: A \to X$ is a **homotopy equivalence** if there is $g: X \to A$ with $fg \simeq 1_X$ and $gf \simeq 1_A$. Compare (0.7). ||

(3.3) **Theorem.** *Let* $(\mathbf{C}, cof, I, \phi)$ *be an I-category. Then* $\mathbf{C}$ *is a cofibration category with the following structure. Cofibrations are those of* $\mathbf{C}$, *weak equivalences are the homotopy equivalences and all objects are fibrant and cofibrant in* $\mathbf{C}$.

Remark. Assume the initial object $\phi = *$ is also a final object. Then we obtain the **suspension functor** $\Sigma: \mathbf{C} \to \mathbf{C}$ by the push out diagram

$$\begin{array}{ccc} IA & \longrightarrow & \Sigma A \\ \uparrow & \text{push} & \uparrow \\ A + A & \longrightarrow & * \end{array}$$

in $\mathbf{C}$. This *suspension functor is an example of a based model functor* which carries cofibrations to cofibrations, see (1.10). We leave the proof of this useful fact as an exercise, compare Chapter II.

Before we prove theorem (3.3) we derive some useful facts from axioms (I1), ..., (I5). First we see by (I4) that $j = (i_0, i_1): A + A \to IA$ is a cofibration and by (I3) and (I2) that also i_0, $i_1: A \to A + A$ are cofibrations. Let $i: B \to A$ be a cofibration. We derive from (I4) the push out diagram:

(3.4)

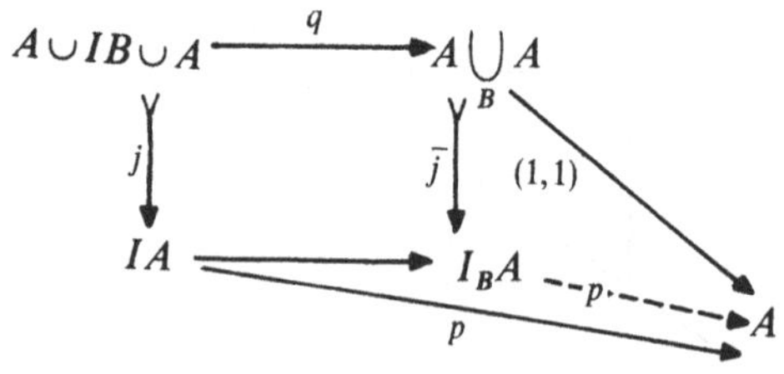

where $A\bigcup_B A$ is the push out of $A \leftarrow B \rightarrow A$ and where $q = 1 \cup p \cup 1$. We call $I_B A$ the **cylinder under** B. We call $H: IA \rightarrow X$ a **homotopy under** B if H factors over $I_B A$.

(3.5) **Lemma.** *Homotopy under B is an equivalence relation.*

Proof. Let $H: f \simeq f_0$ and $G: f \simeq f_1$ be homotopies under B. Then we have by (I2) and (I4) the commutative diagram

$$\begin{array}{ccc} A \cup IB \cup A & \xrightarrow[i_0]{} & I(A \cup IB \cup A) \\ {\scriptstyle j}\downarrow & & \downarrow{\scriptstyle (H, fpp, G)} \\ IA & \xrightarrow[fp]{} & X \end{array} \quad .$$

Since j is a cofibration we obtain the homotopy extension $E: IIA \rightarrow X$ by (I3). Then $-H + G = Ei_1: f_0 \simeq f_1$ under B. Moreover, $f_0 \simeq f$ under B if we set $f_1 = f$ and $G = fp$. □

For a map $u: B \rightarrow U$ let

(3.6) $$[A, U]^u = [A, U]^B$$

be the set of homotopy classes $\{v\}$ under B of all maps $v: A \rightarrow U$ with $v|B = u$. Clearly, a pair map (f, g) for which

$$\begin{array}{ccc} Y & \xrightarrow[g]{} & B \\ \downarrow & & \downarrow \\ X & \xrightarrow[f]{} & A \end{array}$$

commutes, induces

(3.7) $$f^*: [A, U]^u \rightarrow [X, U]^{ug}$$

by $f^*\{v\} = \{vf\}$. f^* is well defined since If gives us the map $If: I_Y X \rightarrow I_B A$, see (3.4). Moreover, a homotopy $G: u \simeq \bar{u}$ determines the function

(3.8) $$G^\#: [A, U]^u \rightarrow [A, U]^{\bar{u}}$$

as follows: For the diagram

$$\begin{array}{ccc} B & \xrightarrow[i_0]{} & IB \\ \downarrow & & \downarrow{\scriptstyle G} \\ A & \xrightarrow[v]{} & U \end{array}$$

we obtain the extension $E:IA \to U$ by (I3). Then we set $G^{\#}\{v\} = \{Ei_1\}$. Here $G^{\#}$ is well defined, since for a homotopy $H:v \simeq v'$ under B we have the commutative diagram (by (I2))

$$\begin{array}{ccc} A \cup IB \cup A & \xrightarrow[i_0]{} & I(A \cup IB \cup A) \\ \downarrow & & \downarrow (E, GIp, E') \\ IA & \xrightarrow[H]{} & U \end{array} .$$

We choose a homotopy extension $\bar{E}$. Then $\bar{E}i_1 : Ei_1 \simeq E'i_1$ under B.

(3.9) **Lemma.** *$G^{\#}$ is a bijection.*

Proof. Since we assume cofibrations to have the extension property for i_0 and i_1, see (I3), we obtain the inverse of $G^{\#}$ if we replace in the construction of $G^{\#}$ the map i_0 by i_1. □

For the cofibration $A + A \to IA$ and for maps $u, v: A \to U$ we have by (3.6) in particular the set of homotopy classes $[IA, U]^{u,v}, (u, v): A + A \to U$, of homotopies from u to v. For $f:X \to A$ the pair map $(If, f + f)$ induces $(If)^*:[IA, U]^{u,v} \to [IX, U]^{uf,vf}$.

(3.10) **Lemma.** *If f is a homotopy equivalence then $(If)^*$ is a bijection.*

Proof. Let $g:A \to X$ be a homotopy inverse of f and let $H:fg \simeq 1, H:IA \to A$, be a homotopy. Then we have the commutative diagram

$$\begin{array}{ccc} [IA, U]^{u,v} & \xrightarrow{(fg)^*} & [IA, U]^{ufg,vfg} \\ & {\scriptstyle 1} \searrow & \downarrow (uH, vH)^{\#} \\ & & [IA, U]^{u,v} \end{array} . \tag{1}$$

Here we have for $G:IA \to U, G:u \simeq v$, the equation

$$(uH, vH)^{\#}(fg)^*\{G\} = (uH, vH)^{\#}\{GI(fg)\} = \{Ei_1\}, \tag{2}$$

where we can choose E to be (see (I5)): $E = G(\mathrm{IH})T:IIA \to IA \to U$. In fact, E is a homotopy extension for

$$\begin{array}{ccc} A + A & \xrightarrow{i_0} & I(A + A) \\ \downarrow & & \downarrow (uH, vH) \\ IA & \xrightarrow[GI(fg)]{} & U \end{array} , \tag{3}$$

since we have $uH = EI(i_0), vH = EI(i_1), GI(fg) = Ei_0$. Moreover, we have $Ei_1 = G$. This proves that the diagram above commutes.

Since $(fg)^* = g^*f^*$ we see by (3.9) that f^* is injective. In a similar way we prove that g^* in (1) is injective. This proves the proposition. □

For a map $f : B \to Y$ we define the **mapping cylinder** Z_f by the push out diagram (Compare (1.8)):

(3.11)

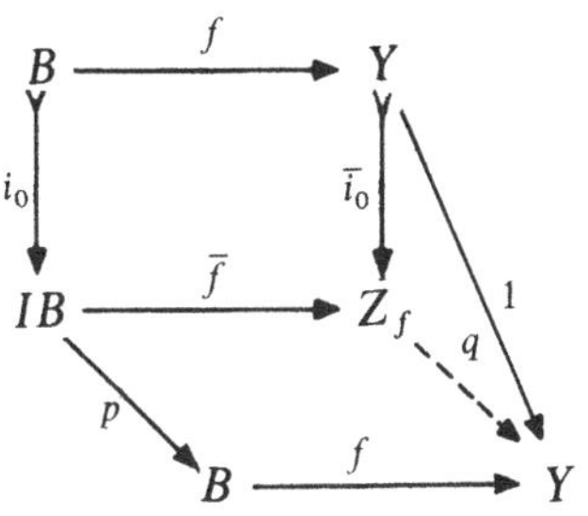

(3.12) **Lemma.** $j = \bar{f} i_1 : B \to Z_f$ is a cofibration and $q : Z_f \to Y$ is a homotopy equivalence with $qj = f$.

In particular we see that $p : IB \to B$ is a homotopy equivalence if we set $f = 1$.

Proof. j is a cofibration since we have the push outs:

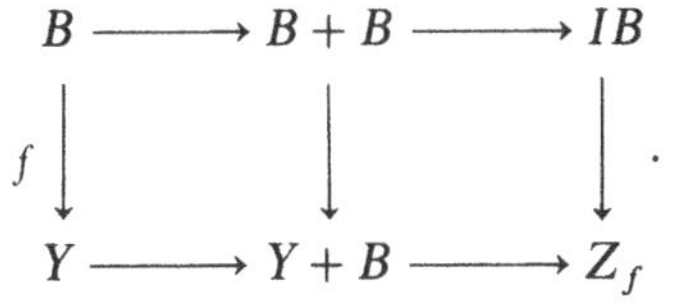

Now $\bar{i}_0$ is the homotopy inverse of q. Clearly, $q\bar{i}_0 = 1$. Moreover, we obtain a homotopy $H : \bar{i}_0 q \simeq 1_{Z_f}$ by the following diagram where we use (I2):

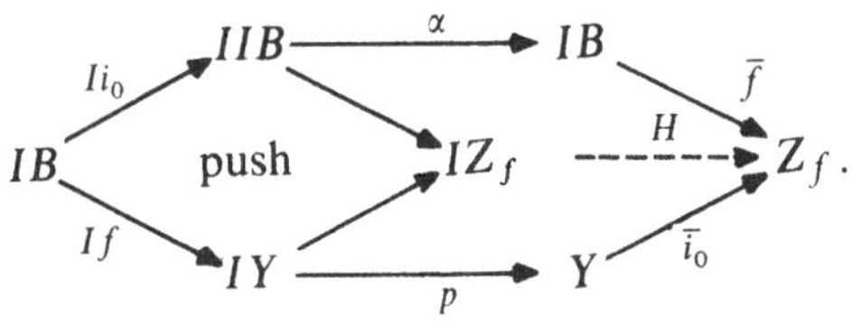

Here we obtain $\alpha = \bar{\alpha}T$ (see (I5)) by the map $\bar{\alpha} : IIB \to IB$ which is the extension for

$$
\begin{array}{ccc}
B + B & \xrightarrow{\;\;i_0\;\;} & I(B + B) \\
{\scriptstyle j}\downarrow & & \downarrow{\scriptstyle (i_0 p, 1)} \\
IB & \xrightarrow{\;\;i_0 p\;\;} & IB
\end{array}
$$

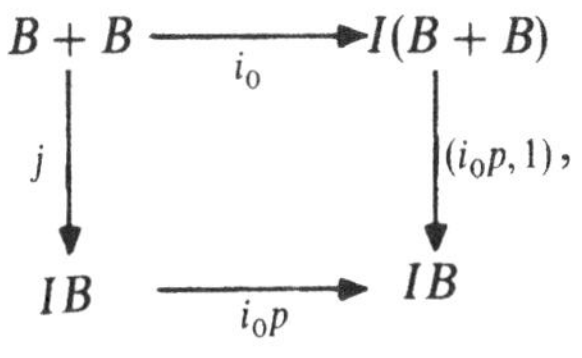

where we use the fact that j is a cofibration.   □

We now are ready for the proof of the theorem:

Proof of (3.3). We have to check the axioms (C1),...,(C4). By (I1) and (3.5) we see easily that (C1) is satisfied. By (3.12) we see that (C3) is satisfied. We now prove (C4) and (C2). We consider

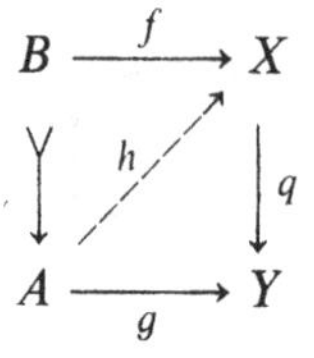

$$\begin{array}{ccc} B & \xrightarrow{\ f\ } & X \\ \downarrow & \nearrow^{h} & \downarrow q \\ A & \xrightarrow[\ g\ ]{} & Y \end{array} \qquad (1)$$

with $qf = g|B$ and where q is a homotopy equivalence. Let $\bar{q}: Y \to X$ be a homotopy inverse of q and let

$$H: \bar{q}q \simeq 1, \quad H: IX \to X \qquad (2)$$

be a homotopy. By (3.10) the map $(Iq)^*: [IY, Y]^{q\bar{q},1} \to [IX, Y]^{q\bar{q}q,q}$ is surjective. We choose $G \in (Iq)^{*-1}\{qH\}$. Then we have a homotopy $\Lambda: IIX \to Y$

$$\Lambda: G(Iq) \simeq qH \quad \text{under } X + X. \qquad (3)$$

We define

$$h_0 = \bar{q}g: A \to X. \qquad (4)$$

Thus we have a homotopy

$$H(If): h_0|B = \bar{q}g|B = \bar{q}qf \simeq f. \qquad (5)$$

Let $\bar{H}: IA \to X$ be the homotopy extension for

$$\begin{array}{ccc} B & \xrightarrow{\ i_0\ } & IB \\ \downarrow & & \downarrow H(If) \\ A & \xrightarrow[\ h_0\ ]{} & X \end{array}$$

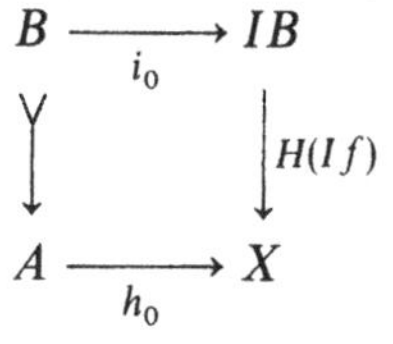

and let $h = \bar{H}i_1$. We see by (5)

$$h|B = f \quad \text{since } \bar{H}|IB = H(If). \qquad (6)$$

Therefore h makes the upper triangle in (1) commutative. Thus we have proven:

$$\left\{ \begin{array}{l} \text{For each diagram of unbroken arrows as in (1) there is a map } h \\ \text{which makes the upper triangle commutative.} \end{array} \right. \qquad (7)$$

We derive (C4) from (7) as follows. We show that all objects are fibrant, that is:

$$\begin{cases} \text{For each trivial cofibration } i: B \overset{\sim}{\rightarrowtail} A \text{ there is a retraction} \\ r: A \to B \text{ with } ri = 1_B. \end{cases} \tag{8}$$

We obtain (8) if we apply (7) to the diagram

$$\begin{array}{ccc} B & \xrightarrow{\ 1\ } & B \\ {\scriptstyle i}\big\downarrow & \overset{r}{\nearrow} & \big\downarrow{\scriptstyle \simeq\ i} \\ A & \xrightarrow[\ 1\ ]{} & A \end{array}. \tag{9}$$

Next we prove (C2). We consider the push out diagram

$$\begin{array}{ccc} A & \xrightarrow[\ t\ ]{} & A \bigcup_X Y \\ {\scriptstyle i}\big\uparrow & & \big\uparrow{\scriptstyle \bar{i}} \\ X & \xrightarrow[\ q\ ]{} & Y \end{array}. \tag{10}$$

Clearly, $\bar{i}$ is a cofibration by (I2). Now let q be a homotopy equivalence. We have to prove that t is a homotopy equivalence. We choose H, G and Λ as in (2) and (3). Since $X \to A$ is a cofibration we have the retraction r which is given by a homotopy extension for the push out diagram

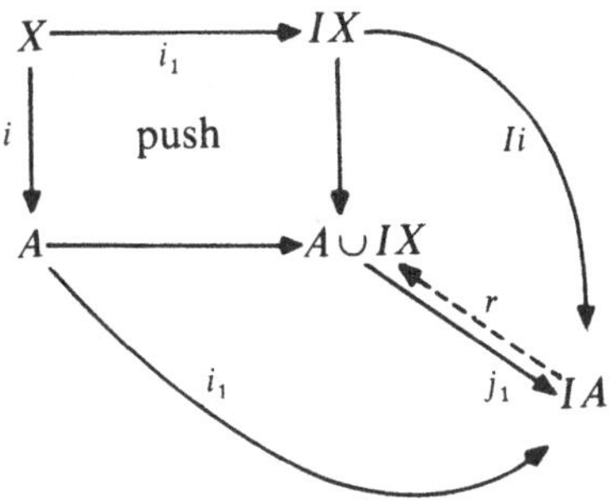

(11)

We have the commutative diagram

$$\begin{array}{ccccc} A & \xleftarrow[\ i\ ]{} & X & \xrightarrow[\ q\ ]{} & Y \\ {\scriptstyle i_0}\big\downarrow & & \big\downarrow{\scriptstyle 1} & & \big\downarrow{\scriptstyle 1} \\ IA & \longleftarrow & X & \xrightarrow[\ q\ ]{} & Y \\ {\scriptstyle r}\big\downarrow & & \big\downarrow{\scriptstyle 1} & & \big\downarrow{\scriptstyle 1} \\ A \cup IX & \xleftarrow{\ i_0\ } & X & \xrightarrow{\ q\ } & Y \end{array}.$$

For the push outs of the rows we obtain the composition $\bar{t}$ by the commutative upper triangle in the diagram

$$\begin{array}{ccccc} A\bigcup_X Y & \xrightarrow{i_0 \cup 1} & IA \bigcup_X Y & \xrightarrow{r\cup 1} & (A\cup IX)\bigcup_X Y \\ \bar{t}\downarrow & & ((1,iH),i\bar{q}) \swarrow & & \downarrow (1,ip)\cup 1 \\ A & & \xrightarrow{t} & & A\bigcup_X Y \end{array} \qquad (12)$$

Here $(1, iH): A\cup IX \to A$ is well defined since for H in (2) we have $Hi_1 = 1$. Moreover, $\bar{t}$ is well defined since $(1, iH)ri_0 i = iHi_0 = i\bar{q}q$. We claim that $\bar{t}$ is a homotopy inverse for t in (10). First we see

$$\bar{t}t = (1, iH)ri_0 \simeq (1, iH)ri_1 = 1_A \qquad (13)$$

since the identity on IA is a homotopy $i_0 \simeq i_1$. We now consider the composition $t\bar{t}$ in (12). First we see

$$\left.\begin{array}{l} ((1, ip)\cup 1)(r\cup 1)(i_0 \cup 1) = R\cup 1, \text{ where} \\ R = (1, ip)ri_0 : A \to A. \end{array}\right\} \qquad (14)$$

Now $(1, ip)ri_1 = 1_A$. Since r is a retraction for j_1 in (11) we see that $(1, ip)r = F$ is in fact a homotopy $F: R \simeq 1_A$ relative X. This gives us the homotopy

$$F\cup p: R\cup 1 \simeq 1. \qquad (15)$$

From (14) and (15) we derive that $t\bar{t} \simeq 1$ since the lower triangle in (12) homotopy commutes. To see this we construct below a map

$$\bar{\Lambda}: IIX \to Y \qquad (16)$$

with the following properties $\bar{\Lambda}(Ii_0) = G(Iq)$, $\bar{\Lambda}i_0 = qH$, $\bar{\Lambda}(Ii_1) = qp$, $\bar{\Lambda}i_1 = qp$. We sketch this by

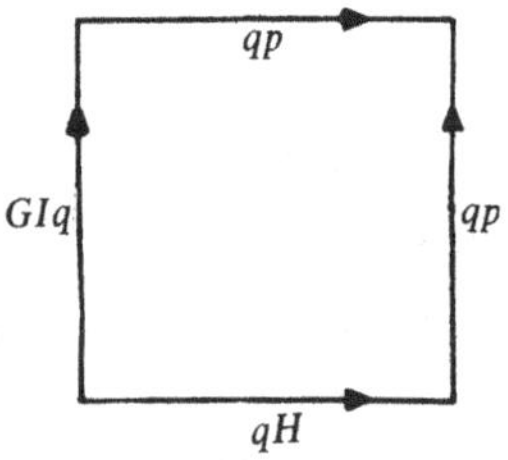

Now $\bar{\Lambda}$ gives us the homotopy $\bar{\bar{\Lambda}}: I((A\cup IX)\bigcup_X Y) \to A\bigcup_X Y$, $\bar{\bar{\Lambda}}|IA = tp$, $\bar{\bar{\Lambda}}|IIX = \bar{i}\bar{\Lambda}$, $\bar{\bar{\Lambda}}|IY = \bar{i}G$. One can check that $\bar{\bar{\Lambda}}$ is well defined by (I2) and that indeed $\bar{\bar{\Lambda}}$ is a homotopy $\bar{\bar{\Lambda}}: t[(1, iH), i\bar{q}] \simeq (1, ip)\cup 1$ so that the lower triangle in (12) homotopy commutes.

We now construct $\bar{\Lambda}$ in (16). For the cofibration $X + X \to IX$ we obtain by

(I4) the cofibration

$$j: \partial I^2 X = IX \cup I(X+X) \cup IX \to I^2 X. \tag{17}$$

We consider the commutative diagram of unbroken arrows:

$$\begin{array}{ccc} \partial I^2 X & \xrightarrow{\ \gamma\ } & IX \\ {\scriptstyle j}\big\downarrow & \Gamma \nearrow & \simeq \big\downarrow p, \\ I^2 X & \xrightarrow{\ pp\ } & X \end{array} \tag{18}$$

where $\gamma(Ii_0) = \gamma i_0 = 1$, $\gamma(Ii_1) = \gamma i_1 = i_1 p: IX \to X \to IX$. Since p is a homotopy equivalence, there is by (7) a Γ such that the upper triangle in (18) commutes. This gives us the commutative diagram

$$\begin{array}{ccc} \partial I^2 X & \overset{}{\underset{i_0}{\rightarrowtail}} & I(\partial I^2 X) \\ {\scriptstyle j}\big\downarrow & & \big\downarrow \bar{\Gamma} \\ I^2 X & \underset{\lambda}{\longrightarrow} & Y \end{array} , \tag{19}$$

where $\lambda = G(Iq)\Gamma$ and where $\bar{\Gamma}$ is defined by $\bar{\Gamma}(Ii_0) = \Lambda$ (see (3)), $\bar{\Gamma}(IIi_0) = \lambda(Ii_0)p = G(Iq)p$, $\bar{\Gamma}(Ii_1) = \lambda i_1 p = qpp$, $\bar{\Gamma}(IIi_1) = \lambda(Ii_1)p = qpp$. We now choose a homotopy extension E for (19), $E: I^3 X \to Y$. Then we see that $Ei_1 = \bar{\Lambda}$ has the properties in (16). □

We check that $I_B A$ in (3.4) is a cylinder on $B \subset A$ as in (1.5).

(3.13) **Lemma.** *$p: I_B A \to A$ is a homotopy equivalence.*

Proof. We have the push out diagram

$$\begin{array}{ccc} & A & \\ \nearrow & & \nwarrow \\ IA & \xrightarrow{\ \bar{p}\ } & I_B A \\ \big\uparrow & \text{push} & \big\uparrow \\ IB & \underset{p}{\xrightarrow{\ \simeq\ }} & B \end{array}$$

Since p is a homotopy equivalence also $\bar{p}$ is one by (C2). □

§ 3a Appendix: categories with a natural path object

The dual of an I-category is a **P-category** ($\mathbf{C}$, *fib*, P, e), where *fib* is the class of fibrations, P is the natural **path object** and e is the final object in $\mathbf{C}$. We write

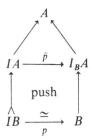

$PX = X^I$ for an object X in $\mathbf{C}$. $P:\mathbf{C} \to \mathbf{C}$ is a functor together with natural transformations

(3a.1) $$X \xrightarrow[i]{} X^I \xrightarrow[q_0, q_1]{} X$$

with $q_\varepsilon i = 1_X (\varepsilon = 0, 1)$, compare (0.2). The P-category satisfies the axioms (P1), ..., (P5) which are dual to the axioms (I1), ..., (I5) in (3.1). The dual of the homotopy extension property of a cofibration in (I3) is the following **homotopy lifting property** of a fibration $A \twoheadrightarrow B$ in *fib*: Let $\varepsilon \in \{0, 1\}$. For each commutative diagram

(3a.2) $$\begin{array}{ccc} X & \xrightarrow{f} & A \\ {\scriptstyle H}\downarrow & & \downarrow\!\!\!\downarrow {\scriptstyle p} \\ B^I & \xrightarrow[q_\varepsilon]{} & B \end{array}$$

there is $E:X \to A^I$ with $(p^I)E = H$ and $q_\varepsilon E = f$. Compare (0.9).

We leave it to the reader to formulate precisely the axioms (P1), ..., (P5) which are dual to (I1), ..., (I5) in (3.1). Two maps $f_0, f_1:X \to A$ are **homotopic** with respect to the natural path object if there is a map $G:X \to A^I$ with $q_0 G = f$ and $q_1 G = g$.

Clearly, the results which are dual to those in § 3 are available for a P-category. The dual of the mapping cylinder in (3.11) is the **mapping path object** W_f given by the pull back diagram

(3a.3)

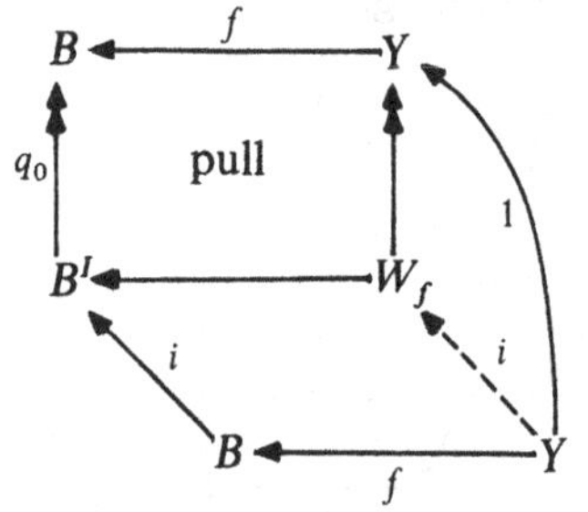

As a corollary of (3.3) we get by strict duality:

(3a.4) **Theorem.** *Let $(\mathbf{C}, fib, P, e)$ be a P-category. Then $\mathbf{C}$ is a fibration category with the following structure: Fibrations are those of $\mathbf{C}$, weak equivalences are the homotopy equivalences and all objects are cofibrant and fibrant.*

§4 The category of topological spaces

We show that **Top** has the structure of an I-category and of a P-category. This result generalizes to the category **Top**$(C \to D)$ of spaces under C and over D. Thus **Top** $(C \to D)$ has the structure of a cofibration category and of a fibration category by the results in §3.

Let I and P be the cylinder and the path space on **Top**, respectively, see (0.1) and (0.2). By (0.3) we see that (I, P) is an adjoint pair in the following sense.

(4.1) ***Definition.*** Let I be a natural cylinder and P be a natural path object on **C**. Then (I, P) **is an adjoint pair** if a natural bijection $\mathbf{C}(IX, Y) = \mathbf{C}(X, PY)$ is given with $i_\varepsilon^* = q_{\varepsilon *}$ for $\varepsilon = 0$, 1 and with $p^* = i_*$. ||

Clearly, for an adjoint pair (I, P) the homotopy relations induced by I and P respectively, coincide.

The empty space $\varnothing$ is the *initial object* in **Top** and the point e is the *final object* in **Top**. Moreover, in **Top** exist push outs and pull backs.

(4.2) **Proposition.** *With the notation in* §0 *the category* **Top** *has the structure of an I-category and of a P-category.*

Proof. Only (I4) is not so obvious. We prove (I4). In a dual way we prove the P-category structure, see page 133 in Baues (1977). Let $\alpha: I \times I \approx I \times I$ be a homeomorphism which is given on the boundary by the sketches:

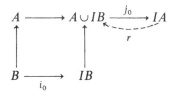

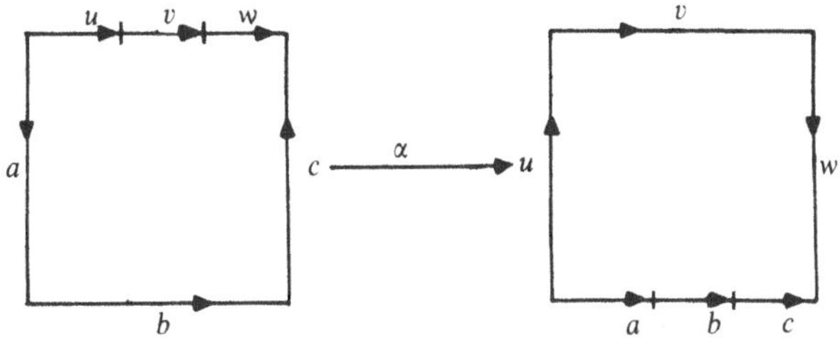

One easily verifies that $i: B \to A$ is a cofibration iff for the push out diagram

$$\begin{array}{ccccc} A & \longrightarrow & A \cup IB & \xrightarrow{j_0} & IA \\ \uparrow & & \uparrow & \underset{r}{\dashleftarrow} & \\ B & \xrightarrow[i_0]{} & IB & & \end{array}$$

with $j_0 = (i_0, Ii)$ there is a retraction r with $rj_0 = 1$. For the map j in (I4) we obtain such a retraction $\bar{r}$ for $j_0: D = (IA) \cup I(A \cup IB \cup A) \to IIA$, by use of α. Let β be the restriction of $\alpha \times 1_A$ such that the diagram

$$\begin{array}{ccc} D & \xrightarrow[j_0]{} & IIA \\ {\scriptstyle\beta}\big\downarrow & & \big\downarrow{\scriptstyle\alpha\times 1} \\ I(A\cup IB) & \xrightarrow[T(Ij_0)]{} & IIA \end{array}$$

commutes. Here $T{:}IIA \to IIA$ is the map in (I5) with $T(t_1, t_2, a) = (t_2, t_1, a)$. We now define $\bar{r} = \beta^{-1}(Ir)T(\alpha \times 1)$. □

(4.3) ***Definition.*** Let **C** be a category and let $u{:}C \to D$ be a fixed map in **C**. We define the category $\mathbf{C}(u) = \mathbf{C}(C \to D)$ **of objects under** C **and over** D. Objects are triples $(X, \check{x}, \hat{x})$ where

$$C \xrightarrow{\check{x}} X \xrightarrow{\hat{x}} D, \quad u = \hat{x}\check{x}.$$

A map $f{:}(X, \check{x}, \hat{x}) \to (Y, \check{y}, \hat{y})$ is a map $f{:}X \to Y$ such that $f\check{x} = \check{y}$, $\hat{y}f = \hat{x}$. ||

$\mathbf{C}(u)$ has the initial object $C \xrightarrow{1} C \to D$ and the final object $C \to D \xrightarrow{1} D$. Pull backs and push outs in $\mathbf{C}(u)$ are given by pull backs and push outs in **C**. We define a natural cylinder I and a natural path space P which are functors

(4.4) $$I, P{:}\mathbf{Top}(u) \to \mathbf{Top}(u).$$

We define $Z = I(X, \check{x}, \hat{x})$ by the push out diagram in **Top**

$$\begin{array}{ccccc} X \times I & \longrightarrow & Z & \xrightarrow{\hat{z}} & D \\ {\scriptstyle\check{x}\times 1}\big\uparrow & \text{push} & \big\uparrow{\scriptstyle\check{z}} & & \\ C \times I & \xrightarrow[p]{} & C & & \end{array},$$

where $\hat{z} = (\hat{x}p, u)$. We define $W = P(X, \check{x}, \hat{x})$ by the pull back diagram in **Top**

$$\begin{array}{ccccc} C & \xrightarrow{\check{w}} & W & \longrightarrow & X^I \\ & & {\scriptstyle\hat{w}}\big\downarrow & \text{pull} & \big\downarrow{\scriptstyle\hat{x}^I} \\ & & D & \xrightarrow[i]{} & D^I \end{array},$$

where $\check{w} = (u, i\check{x})$. The natural transformations i_0, i_1, q_0, q_1, p, i are given similarly as for (0.1) and (0.2). Again (I, P) is an adjoint pair.

(4.5) Some examples are:

$\mathbf{Top}(\varnothing \to *) = \mathbf{Top}$ ($\varnothing$ empty, $* =$ point),

$\mathbf{Top}(* \to *) = \mathbf{Top}^*$; this is the category of basepoint preserving maps,
$\mathbf{Top}(C \to *) = \mathbf{Top}^C$, spaces under C,
$\mathbf{Top}(\varnothing \to D) = \mathbf{Top}_D$, spaces over D,
$\mathbf{Top}(* \to D) = \mathbf{Top}^*_D$, pointed spaces over D,
$\mathbf{Top}(D \xrightarrow{1} D) = \mathbf{Top}(D)$, spaces under and over D = ex-spaces over D.

Let **internal cofibrations** in $\mathbf{Top}(u)$ be the maps in $\mathbf{Top}(u)$ which have the homotopy extension property with respect to I in (4.4). Let **internal fibrations** in $\mathbf{Top}(u)$ be the maps which have the homotopy lifting property with respect to P in (4.4). With these notations we obtain in the same way as in (4.2):

(4.6) **Proposition.** *The category* $\mathbf{Top}(u)$ *with internal cofibrations and internal fibrations has the structure of an I-category and of a P-category respectively.*

The homotopy theory of $\mathbf{Top}(u)$ was studied in many papers in the literature, see for example Mc Clendon (1969), James (1971) and Eggar (1973). In fact, many of the results in these papers rely only on the fact that the axioms of a cofibration category or of a fibration category are satisfied in $\mathbf{Top}(u)$.

§ 4a Appendix: the relative homotopy lifting property

We say $\mathbf{C}$ is an **IP-category** if $\mathbf{C}$ has the structure $(C, cof, fib, I, P, \varnothing, e)$ with the following properties (IP1), (IP2) and (IP3).

(IP1) $(\mathbf{C}, cof, I, \varnothing)$ is an I-category and $(\mathbf{C}, fib, P, e)$ is a P-category.

(IP2) (I, P) is an adjoint pair, see (4.1).

(IP3) The following **relative homotopy lifting property** is satisfied: Let $i: A \rightarrowtail X$ be a cofibration and let $p: Y \twoheadrightarrow B$ be a fibration. Then any solid-arrow diagram

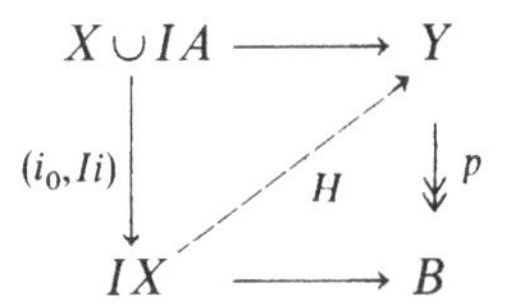

can be extended commutatively by a map H. Here $X \cup IA$ is the push out of $X \underset{i}{\leftarrowtail} A \underset{i_0}{\longrightarrow} IA$.

(4a.1) **Proposition.** *Let* $\mathbf{C}$ *be an IP-category and assume* $\mathbf{C}$ *satisfies* (M0) *and* (M2) *in* (2.1). *Then* $\mathbf{C}$ *is a proper model category in which weak equivalences are the homotopy equivalences.*

Proof. We obtain (M1) by the following argument (see 6.1 in Hastings).

Consider the commutative diagram of solid arrows:

$$\begin{array}{ccc} A & \longrightarrow & X \\ {\scriptstyle i}\downarrow & \overset{f}{\nearrow} & \downarrow{\scriptstyle p} \\ B & \longrightarrow & Y \end{array} \qquad (1)$$

Suppose i is a homotopy equivalence. Then i is a deformation retract by (II.1.12). Let $r:B \to A$ be a retraction, and let $H:IB \to B$ be a homotopy relative A from ir to 1_B. We obtain the commutative diagram of solid arrows:

$$\begin{array}{ccccc} B \cup IA & \xrightarrow{(r,p)} & A & \longrightarrow & X \\ \downarrow & F & \downarrow{\scriptstyle i} & & \downarrow{\scriptstyle p} \\ IB & \underset{H}{\longrightarrow} & B & \longrightarrow & Y \end{array} \qquad (2)$$

By (IP3) the filler F exists. Now $Fi_1 = f$ is a filler for (1). In a dual way we obtain the filler f in (1) if p is a homotopy equivalence since we have:

(4a.2) **Lemma.** *The dual* $\mathbf{C}^{op}$ *of an IP-category* $\mathbf{C}$ *is an IP-category with* $I = P^{op}$ *and* $P = I^{op}$.

In particular, (IP3) is self-dual! □

We now consider the category **Top** with the structure $\overline{\mathbf{Top}} = (\mathbf{Top}, \overline{cof}, fib, I, P, \varnothing, e)$. Here $\overline{cof}$ is the class of closed cofibrations in **Top**. As in (4.4) we see that $\overline{\mathbf{Top}}$ satisfies (IP1) and (IP2). Strøm proved that $\overline{\mathbf{Top}}$ satisfies (IP3). Therefore we get:

(4a.3) **Proposition.** $\overline{\mathbf{Top}}$ *is an IP-category.*

Moreover, Strøm proved that $\overline{\mathbf{Top}}$ satisfies (M2). Therefore we derive from (4.8) the following result of Strøm:

(4a.4) **Corollary.** $\overline{\mathbf{Top}}$ *is a proper model category, compare* (2.10).

This example shows that it is convenient to consider the class *cof* of cofibrations as part of the structure of an I-category. Compare the remark following (I5) in (3.1).

(4a.5) **Proposition.** *Let* $u:C \to D$ *be a map in* $\mathbf{C}$. *Suppose that* $\mathbf{C}$ *is a model category or that* $\mathbf{C}$ *is an IP-category which satisfies* (M2). *Then the category*

$\mathbf{C}(u)$ is a cofibration category with the following **external structure**

$$cof = \textit{maps in } \mathbf{C}(u) \textit{ which are cofibrations in } \mathbf{C},$$
$$we = \textit{maps in } \mathbf{C}(u) \textit{ which are weak equivalences in } \mathbf{C},$$

all objects $C \to X \twoheadrightarrow D$ (for which $\hat{x}$ is a fibration in $\mathbf{C}$) are fibrant in $\mathbf{C}(u)$.

Moreover, $\mathbf{C}(u)$ is a fibration category with the dual external structure. We leave the proof of (4a.5) as an exercise. By (4a.5) for example $\mathbf{Top}(u)$ has external structures; these are different from the internal structures in (4.5).

§ 4b Appendix: the cube theorem

Let $\mathbf{C}$ be a category which has the structure ($\mathbf{C}$, *cof*, *fib*, *we*) such that ($\mathbf{C}$, *cof*, *we*) is a cofibration category and such that ($\mathbf{C}$, *fib*, *we*) is a fibration category. A proper model category or a category which satisfies (IP1) in §4a has such a structure. In particular, the category **Top** in (4.2) (and Top(u) in (4.6)) is an example.

(4b.1) ***Definition.*** We say that the **'cube theorem'** holds in $\mathbf{C}$ if $\mathbf{C}$ has the following property: Consider a commutative cubical diagram in $\mathbf{C}$

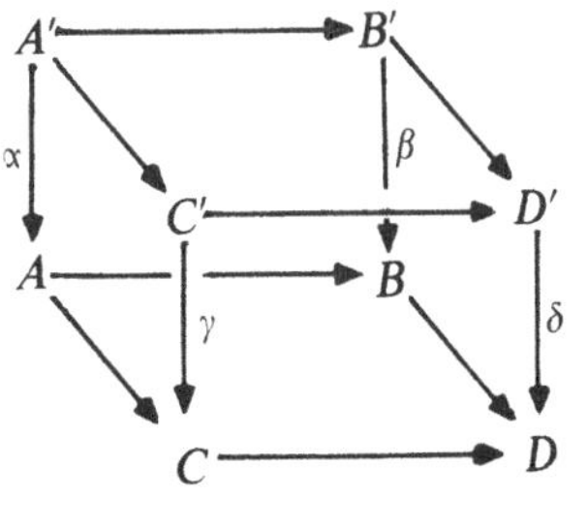

,

in which

(a) the left and rear faces are homotopy pull backs, and
(b) the bottom face is a homotopy push out. Then
(a)′ the front and right faces are homotopy pull backs, if and only if
(b)′ the top face is a homotopy push out. ||

Clearly, homotopy pull backs (= homotopy cartesian diagrams) in a fibration category are dual to homotopy push outs in a cofibration category, see (1.9). There is an obvious dualization of the 'cube theorem' in (4b.1) which we call the '*dual cube theorem*'. We recall the following well known result, compare V. Puppe.

(4b.2) **Proposition.** *Let* (**Top**, *cof*, *fib*, *we*) *be defined by cofibrations, fibrations* *and homotopy equivalences in* **Top** *as in* §0. *Then the cube theorem holds in* **Top**. *The dual cube theorem, however, does not hold in* **Top**.

This is a crucial example which contradicts a global assumption of duality in the structure (**Top**, *cof*, *fib*, *we*). Hence Eckmann–Hilton duality does not hold in general in **Top**.

(4b.3) ***Remark.*** The cube theorem is not a formal consequence of the axioms of a proper closed model category. Indeed in such a category the cube theorem holds iff the dual cube theorem holds. Using (4b.2) the model categories in (2.10) or (2.11) are counter examples.

(4b.4) ***Remark.*** If the top and bottom faces of the diagram in (4b.1) are homotopy push outs and if α, β, and γ are weak equivalences, then also δ is a weak equivalence. This follows from the axioms of a cofibration category as we shall see in (II.1.2).

In addition to the cube theorem we have in **Top** a further compatibility of homotopy colimits and homotopy limits: consider the diagram ($n \geqq 0$)

$$\begin{array}{ccccc} A_n & \rightarrowtail & A_{n+1} & \rightarrowtail & \varinjlim A_n \\ {\scriptstyle\alpha_n}\downarrow & \text{I} & \downarrow{\scriptstyle\alpha_{n+1}} & \text{II} & \downarrow{\scriptstyle\varinjlim \alpha_n} \\ B_n & \rightarrowtail & B_{n+1} & \rightarrowtail & \varinjlim B_n \end{array} .$$

(4b.5) **Proposition.** *Let* (**Top**, cof, fib, we) *be defined as in* (4b.2). *If diagram* I *is a homotopy pull back in* **Top** *for all n, then also* II *is a homotopy pull back for all n.*

Compare V. Puppe. The properties in (4b.2) and (4b.5) can be considered as further basic 'axioms' of homotopy theory concerning both fibrations and cofibrations.

§5 Examples of cofibration categories

We describe some basic topological examples of cofibration categories and of fibration categories respectively. For algebraic examples see the following sections §6, §7, §8 and §9.

(5.1) **Theorem.** *The category* **Top** *of topological spaces with the structure*

$$\begin{cases} \textit{cof} = \textit{cofibrations in } \mathbf{Top}, \textit{ see } (0.8), \textit{ and} \\ \textit{we} = \textit{homotopy equivalences in } \mathbf{Top}, \textit{ see } (0.7), \end{cases}$$

is a cofibration category in which all objects are fibrant and cofibrant.

Proof. **Top** *is an* I-category by (4.2) and hence the result follows from (3.3). □

(5.2) **Theorem.** *The category* **Top** *with the structure*

$$\begin{cases} fib = \textit{fibrations in } \mathbf{Top}, \textit{ see } (0.9), \textit{ and} \\ we = \textit{homotopy equivalences in } \mathbf{Top}, \textit{ see } (0.7), \end{cases}$$

is a fibration category in which all objects are fibrant and cofibrant.

Proof. **Top** is a P-category by (4.2) and thus the proposition follows from (3.3) by strict duality, see (3a.4). This shows that the results (5.1) and (5.2) are strictly dual. □

Next we consider the category **Top*** of base point preserving maps with the following (*exterior*) *structure*:

(5.3)
$$\begin{aligned} cof &= \text{maps in } \mathbf{Top^*} \text{ which are cofibrations in } \mathbf{Top}, \\ fib &= \text{maps in } \mathbf{Top^*} \text{ which are fibrations in } \mathbf{Top}, \\ we &= \text{maps in } \mathbf{Top^*} \text{ which are homotopy equivalences in } \mathbf{Top}. \end{aligned}$$

We say an object X is **well pointed** if $* \to X$ is a cofibration in **Top** and we say X is **very well pointed** if $* \to X$ is a closed cofibration in **Top**, see (4a.3).

(5.4) **Theorem.** **Top*** *with the exterior structure* (cof, we) *in* (5.3) *is a cofibration category in which all objects are fibrant models. The well pointed spaces are the cofibrant objects.*

Proof. This is an easy consequence of (5.1), compare (II.3.2). □

(5.5) **Theorem.** **Top*** *with the exterior structure* (fib, we) *in* (5.3) *is a fibration category in which all objects are fibrant. The very well pointed spaces form a sufficiently large class of cofibrant models, see* (1.2).

Proof. This result is not strictly dual to (5.4). We deduce the proposition from Strøm's theorem in (4a.4). □

Recall that by (4.6) we have a different internal structure of **Top*** as a cofibration category and as a fibration category.

Next we describe some examples for which the weak equivalences are not homotopy equivalences in **Top**. We define the **CW-structure** on **Top** by

cof = inclusions $B \subset A$ for which A is given by a well-ordered succession of attaching cells to B, see (0.12)
we = *weak homotopy equivalences* = maps $f: X \to Y$ which

induce isomorphisms $f_*:\pi_k(X, x_0) \cong \pi_k(Y, f x_0)$
on homotopy groups for $k \geqq 0$, $x_0 \in X$.

(5.6) **Theorem.** *The category* **Top** *with the CW-structure (cof, we) is a cofibration category in which all objects are fibrant models. CW-complexes are cofibrant objects and all cofibrant objects are CW-spaces.*

In the theorem we use the following definitions of CW-complexes and CW-spaces. A relative CW-complex under A is obtained by inductively attaching cells to A as follows:

(5.7) ***Definition.*** The pair (X, A) is a **relative CW-complex** if a filtration

$$A \subset X^0 \subset X^1 \subset \cdots \subset X = \varinjlim X^n$$

of cofibrations in **Top** is given with the following properties: X^0 is the disjoint union of A with a discrete set Z_0 and X^n is obtained by a push out diagram $(n \geqq 1)$ in **Top**

$$\begin{array}{ccc} \dot{\bigcup\limits_{Z_n}} D^n & \xrightarrow[c_n]{} & X^n \\ \cup & & \cup \\ \dot{\bigcup\limits_{Z_n}} S^{n-1} & \xrightarrow[f_n]{} & X^{n-1} \end{array},$$

where $\dot{\cup}$ denotes the disjoint union. The space X^n is the **relative *n*-skeleton** of (X, A) and f_n is the **attaching map** of n-cells, c_n is the **characteristic map** of n-cells. If A is empty we call X a **CW-complex.** A **CW-space** is a topological space which is homotopy equivalent in **Top** to a CW-complex. Clearly, for a relative CW-complex (X, A) the inclusion $A \subset X$ is a cofibration in the CW-structure (5.6). ||

Proof of (5.6). (C1) is clearly satisfied, (C3) and (C4) are typical results of obstruction theory, moreover, (C2) can be proved by the Blakers–Massey theorem. We can deduce the result as well from the model category of Quillen in (2.11) since cofibrations in the CW-structure are also cofibrations in this model category. □

(5.8) ***Remark.*** Cofibrant objects in the CW-structure of **Top** yield the following CW-models of spaces. By (C3) any space X in **Top** admits the factorization $\varnothing \rightarrowtail \bar{X} \xrightarrow{\sim} X$ where $\varnothing$ is the empty space. The cofibrant object $\bar{X}$ is called a *CW-model* or a *resolution of* X. There is a functorial construction of a resolution $\bar{X}$ by using the *singular set* SX of X. The realisation $|SX|$ of the simplicial set SX admits a natural map $|SX| \to X$ which is a weak homotopy equivalence. Since $|SX|$ is a CW-complex we see that $\varnothing \rightarrowtail |SX| \sim X$ is a CW-model. Compare, for example, the appendix of § 16 in Gray.

We will prove in Chapter II that a weak equivalence $X \to Y$ between objects, which are fibrant and cofibrant, is also a homotopy equivalence. This fact, which holds in any cofibration category, yields by (5.6) the following Whitehead theorem:

(5.9) **Proposition.** *Let $f: X \to Y$ be a weak homotopy equivalence. If X and Y are CW-spaces then f is a homotopy equivalence in* **Top**.

Indeed, the concept of 'weak equivalence' originates from this result.

In the next example we define weak equivalences by use of homology groups instead of homotopy groups. Let k_* be a *generalized homology theory* defined on CW-pairs, see for example Gray. A *CW-pair* (X, A) is a cofibration $A \rightarrowtail X$ in **Top** for which A and X are CW-spaces. We assume that k_* satisfies the *limit axiom*, namely, that for all X the canonical map $\varinjlim k_*(X_\alpha) \to k_*(X)$ is an isomorphism where the X_α run over the *finite* subcomplexes of the CW-complex X. Let **CW-spaces** be the full subcategory of **Top** consisting of CW-spaces.

(5.10) **Theorem.** *The category* **CW-spaces** $=$ **C** *with the structure*

$$\begin{cases} cof = \text{maps in } \mathbf{C} \text{ which are cofibrations in } \mathbf{Top}, \\ we = h_*\text{-equivalences} = \text{maps } f: X \to Y \text{ in } \mathbf{C} \\ \qquad \text{which induce an isomorphism } f_*: k_*(X) \cong k_*(Y) \end{cases}$$

is a cofibration category.

In its most general form, as stated, this result is due to Bousfield (1975), compare (2.13). The non-trivial part of the theorem is the existence of the fibrant models (C4).

(5.11) ***Definition.*** A fibrant model of a CW-spaces X in the cofibration category (5.10) is called a **k_*-localization** of X. A fibrant object is called a **k_*-local space**. ||

There are the following special cases of k_*-localizations where $H_*(X; R)$ denotes the *singular homology* of X with coefficients in R. In this case we denote the cofibration category (5.10) by **CW-spaces** (R).

(5.12) ***Remark.*** Let R be a subring of the rationals $\mathbb{Q}$. If X is simply connected (or nilpotent) and $k_* = H_*(\ ; R)$ then the k_*-localization X_R of X is the usual

R-localization of X with $\pi_n(X_R) = \pi_n(X) \otimes R$. As remarked in Bousfield (1975) this case was discovered by Barratt–Moore (1957, unpublished). Subsequently, this case has been discovered and/or studied by various authors, e.g. Bousfield–Kan (1972), Hilton–Mislin–Roitberg (1975), Mimura–Nishida–Toda (1971), Quillen (1969), Sullivan (1970).

(5.13) ***Remark.*** If X is simply connected (or nilpotent) and $k_* = H_*(\ ;\mathbb{Z}/p\mathbb{Z})$ with p prime then the k_*-localization RX of X is the *p-completion* of Bousfield–Kan (1972). If in addition X is of finite type then RX is the *p-profinite completion* with $\pi_n RX$ given by the p-profinite completion of $\pi_n X$, see Sullivan (1970) and Quillen (1969).

The next example relies on results of Quillen (1973) and Loday (1976). This example is significant in **algebraic *K*-theory**. Recall that a group π is called a **perfect** group if $[\pi, \pi] = \pi$. Here $[\pi, \pi]$ denotes the commutator subgroup of π.

(5.14) ***Example.*** **C** is the following category: objects are pairs (X, N_X) where X is a well pointed path connected CW-space and where N_X is a perfect and normal subgroup of $\pi_1(X)$. Maps $f:(X, N_X) \to (Y, N_Y)$ are basepoint preserving maps $f: X \to Y$ in Top* with $f_*(N_X) \subset N_Y$.

cof = maps in **C** which are cofibrations in **Top**,
we = maps f in **C** which induce isomorphisms,

$$\begin{cases} f_*: \pi_1(X)/N_X \cong \pi_1(Y)/N_Y, \text{ and} \\ f_*: \hat{H}_*(X, f_*^* q^* \mathfrak{L}) \cong \hat{H}_*(Y, q^* \mathfrak{L}) \end{cases}$$

for any $(\pi_1(Y)/N_Y)$-module $\mathfrak{L}$. ||

Here $\hat{H}_*$ denotes *homology with local coefficients.* The modules $q^*\mathfrak{L}$ and $f_*^* q^* \mathfrak{L}$ are lifted by the projection $q: \pi_1(Y) \to \pi_1(Y)/N_Y$ and by $f_*: \pi_1(X) \to \pi_1(Y)$ respectively. By definition all objects in (5.14) are cofibrant. There is a sufficiently large class of fibrant objects since we have:

(5.15) **Theorem.** *The structure* (5.14) *is a cofibration category. A fibrant model of* (X, N_X) *is given by Quillen's* (+)*-construction:* $R(X, N_X) = X^+$.

Proof. (C1) and (C3) are clearly satisfied. We use the van Kampen theorem for the proof of (C2). For $X = A \bigcup_B Y$ let N_X be the normal subgroup of the push out $\pi_1 X = \pi_1 A \bigcup_{\pi_1 B} \pi_1 Y$, generated by $N_A \subset \pi_1 A$ and $N_Y \subset \pi_1 Y$. Clearly, the group N_X is normal and perfect since N_A and N_Y are normal and perfect. It is easy to see that (X, N_X) is the push out of a diagram (C2) in the category **C**. Now let f be a weak equivalence. Since

$$\begin{array}{ccc} \pi_1 A/N_A & \xrightarrow{\bar{f}_*} & \pi_1 X/N_X \\ \uparrow & & \uparrow \\ \pi_1 B/N_B & \xrightarrow[\cong]{f_*} & \pi_1 Y/N_Y \end{array}$$

is a push our diagram of groups we see that the top row $\bar{f}_*$ is an isomorphism. Now let $\mathfrak{L}$ be any $(\pi_1 X/N_X)$-module. By the five lemma and the long exact homology sequences we see that $\bar{f}$ is a weak equivalence since (A, B) and (X, Y) have the same homology by excision, compare Spanier (1966) for the properties of $\hat{H}_*$. This proves (C2).

Next we construct fibrant models by theorem (1.1.1) in Loday (1976) which shows that for (X, N_X) there exists a cofibration and a weak equivalence.

$$(X, N_X) \overset{\sim}{\rightarrowtail} (X^+, N_{X^+} = 0).$$

Each object $(Y, N_Y = 0)$ in **C** is fibrant since a trivial cofibration $(Y, N_Y) \overset{\sim}{\rightarrowtail} (Z, N_Z)$ yields by the Whitehead theorem (5.9) the homotopy equivalence f,

$$f: Y \overset{\sim}{\rightarrowtail} Z \overset{\sim}{\rightarrowtail} Z^+ = R(Z, N_Z),$$

which by (5.1) admits a retraction. □

§6 The category of chain complexes

We consider the structure of a cofibration category and of an I-category on the category of chain complexes.

First we have to introduce some standard notation. Let R be a ring of coefficients with unit 1. A **graded module** V is a sequence of left (or right) R-modules $V = \{V_n, n \in \mathbb{Z}\}$. An element $x \in V_n$ has degree $|x| = n$ and we write $x \in V$. A **map of degree** r between graded modules, $f: V \to W$, is a sequence of R-linear maps $f_n: V_n \to W_{n+r}$, $n \in \mathbb{Z}$.

A graded module V is *free*, *projective*, *flat* or of *finite type* if each $V_n (n \in \mathbb{Z})$ is a free, projective, flat or finitely generated R-module respectively.

A graded module V is **positive** (or equivalently non negative) if $V_n = 0$ for $n < 0$. Moreover V is **bounded below** if $V_n = 0$ for $n \ll 0$. On the other hand, V is **negative** if $V_n = 0$ for $n > 0$. We also write $V^n = V_{-n}$, especially if V is negative.

The **suspension** sV of a graded module V is defined by $(sV)_n = V_{n-1}$. Let $s: V \to sV$ be the map of degree $+1$ which is given by the identity, $|sv| = |v| + 1$.

(6.1) ***Definition.*** A **chain complex** V is a graded module V with a map $d: V \to V$

of degree -1 satisfying $dd=0$. A **chain map** $f:V\to V'$ between chain complexes is a map of degree 0 with $df=fd$. Let $\mathbf{Chain}_R$ be the category of chain complexes and of chain maps. The **homology** HV is the graded module

$$HV = \text{kernel } d/\text{image } d.$$

The homology is a functor from $\mathbf{Chain}_R$ to the category of graded modules. We may consider a graded module as being a chain complex with trivial differential $d=0$. ||

Remark. A **cochain complex** V is a graded module $V=\{V^n, n\in\mathbb{Z}\}$ with degrees indicated by upper indices with a map $d:V\to V$ of (upper) degree $+1$ $(d=d^n:V^n\to V^{n+1})$ satisfying $dd=0$. Equivalently, $d:V_n=V^{-n}\to V_{n-1}=V^{-n+1}$, $n\in\mathbb{Z}$, is a chain complex. Clearly, for a cochain complex V we have the **cohomology** $HV=\text{kernel } d/\text{image } d$ with $H^nV=\text{kernel }(d^n)/\text{image }(d^{n-1})$.

(6.2) ***Definition.*** A **weak equivalence** in $\mathbf{Chain}_R$ is a chain map $f:V\to V'$ which induces an isomorphism on homology $f_*:HV\cong HV'$. *A* **cofibration** $i:V\to V'$ in $\mathbf{Chain}_R$ is an injective chain map for which the cokernel V'/iV is a free chain complex. Equivalently, i is a cofibration if there is a free submodule W of V' such that $V\oplus W\cong V'$ is an isomorphism of graded modules. Clearly $W\cong V'/iV$ in this case. ||

(6.3) ***Remark.*** All results in this section remain valid if we define cofibration $i:V\to V'$ by the condition that the cokernel V'/iV is a *projective* module. The proofs below can be generalized without difficulty. Compare the result of Quillen (1967) that $\mathbf{Chain}_R$ is a model category.

Let $\mathbf{Chain}_R^+$ be the full subcategory of $\mathbf{Chain}_R$ consisting of chain complexes which are bounded below.

(6.4) **Proposition.** *By the structure in* (6.2) *the category* $\mathbf{Chain}_R^+$ *is a cofibration category for which all objects are fibrant.*

We prove this result in (6.12) below.

The **direct sum** of graded modules $V\oplus V'$ is given by

$$(V\oplus V')_n = V_n\oplus V'_n. \tag{6.5}$$

The direct sum of chain complexes, $V\oplus V'$, satisfies in addition $d(x+y)=dx+dy$ for $x\in V$, $y\in V'$.

The **tensor product** $V\otimes W$ is given by

$$(V\otimes W)_n = \bigoplus_{i+j=n} V_i \bigotimes_R W_j. \tag{6.6}$$

Here we assume that V_i is a right R-module and that W_j is a left R-module for i,

$j \in \mathbb{Z}$. In case V and W are chain complexes also $V \otimes W$ is a chain complex with the differential

(6.7) $$d(x \otimes y) = dx \otimes y + (-1)^{|x|} x \otimes dy.$$

(6.8) ***Definition.*** Let V be a chain complex. The **cylinder** IV is the chain complex with

$$\left.\begin{aligned} (IV)_n &= V'_n \oplus V''_n \oplus sV_{n-1}, \\ dx' &= (dx)', \\ dx'' &= (dx)'', \\ dsx &= -x' + x'' - sdx. \end{aligned}\right\} \qquad (1)$$

Here V', V'' are two copies of V and $x \longmapsto x'(x \longmapsto x'')$ denotes the isomorphism $V = V'(V = V'')$. Let I be the free chain complex generated by $\{0\}$, $\{1\}$ in degree 0 and by $\{s\}$ in degree 1 with differential $d\{s\} = -\{0\} + \{1\}$. Then we identify

$$IV = I \otimes V, \qquad (2)$$

by $V' = \{0\} \otimes V$, $V'' = \{1\} \otimes V$ and $sV = \{s\} \otimes V$. Now (6.6) shows that (2) is an isomorphism of chain complexes. The cylinder IV has the structure maps

$$V \oplus V \xrightarrow{i_0, i_1} IV \xrightarrow{p} V \qquad (3)$$

with $i_0 x = x'$, $i_1 x = x''$, $p(x') = p(x'') = x$, $p(sx) = 0$. ||

(6.9) ***Remark.*** Let $f_0, f_1: V \to V'$ be chain maps. A homotopy $G: f_0 \simeq f_1$ of chain maps in the sense of (3.2) can be identified with a **chain homotopy** $\alpha: f_0 \simeq f_1$. This is a map $\alpha: V \to V'$ of degree $+1$ with

$$d\alpha + \alpha d = -f_0 + f_1. \qquad (1)$$

Now α yields a chain map $G: IV \to V'$ by $G(x') = f_0 x$, $G(x'') = f_1 x$ and $G(sx) = \alpha(x)$. One easily checks that (1) is equivalent to

$$Gd = dG. \qquad (2)$$

In case G is given we get α by $\alpha(x) = G(sx)$. In particular,

$$S: V \to IV, \quad Sx = sx \qquad (3)$$

is a chain homotopy, $S: i_0 \simeq i_1$, with $\alpha = GS$.

Let $(\mathbf{Chain}_R^+)_c$ be the category of free chain complexes, which are bounded below, and of chain maps. This is the subcategory of cofibrant objects in (6.2). Clearly, the trivial chain complex $V = 0$ is the initial (and the final) object of $\mathbf{Chain}_R$.

(6.10) **Proposition.** *The cylinder functor I in* (6.8) *and the cofibrations in* (6.2) *yield the structure of an I-category on* $(\mathbf{Chain}_R^+)_c$.

(6.11) **Corollary.** *Let $f: V \to V'$ be a weak equivalence between free (or projective, see* (6.3)) *chain complexes which are bounded below. Then f is a homotopy equivalence in the I-category* $(\mathbf{Chain}_R^+)_c$.

The corollary follows from (6.4) by the result in (II.2.12) since V and V' are cofibrant and fibrant objects, see also (5.9).

Proof of (6.10). The cylinder axiom (I1) is satisfied by definition in (6.8). The push out axiom (I2) is satisfied since for graded modules V, W we have $(V \oplus W) \otimes I = V \otimes I \oplus W \otimes I$. Next consider the cofibration axiom (I3). The sum is the direct sum $X + Y = X \oplus Y$ of chain complexes. Let

$$B \rightarrowtail A = (B \oplus W, d)$$

be a cofibration and let $W^n = \{x \in W \mid |x| \leqq n\}$. Then $A^n = (B \oplus W^n, d)$ is a subchain complex of A with $A^n \rightarrowtail A^{n+1}$. We now define inductively the **homotopy extension** $E(\varepsilon = 0)$ by the maps

$$E^n : IA^n \bigcup_{A^n} A \to X.$$

Let E^{n+1} be given by

$$\begin{cases} E^{n+1}(sw) = 0, \\ E^{n+1}(w'') = f(w) + E^n S(dw), \end{cases}$$

for $w \in W_{n+1}$. We check that E^{n+1} is a well-defined chain map: From $dS + Sd = i_1 - i_0$ we derive

$$\begin{aligned} E^{n+1} dw'' &= E^n i_1 dw \\ &= E^n(dSdw + i_0 dw) \\ &= dE^n Sdw + dfw \\ &= d(E^{n+1} w''). \end{aligned}$$

On the other hand, we get

$$\begin{aligned} E^{n+1} dsw &= E^{n+1}(w'' - w' - Sdw) \\ &= (fw + E^n Sdw) - E^n(w' + Sdw) \\ &= 0. \end{aligned}$$

Since we assume that A is *bounded below* we can start the induction and therefore the homotopy extension E exists. The relative cylinder axiom (I4) is an easy consequence of the definitions. Moreover, the interchange axiom (I5) is satisfied since we have the chain map

$$T \otimes 1_X : I \otimes I \otimes X \to I \otimes I \otimes X,$$

with $T(a \otimes b) = (-1)^{|a||b|} b \otimes a$. ||

(6.12) *Proof of* (6.4). The composition axiom (C1) is clearly satisfied. We now prove the push out axiom (C2). Consider the short exact sequences of chain complexes in the rows of the diagram

$$\begin{array}{ccccc} B & \rightarrowtail & A & \longrightarrow & A/B \\ \sim\downarrow f & \text{push} & \downarrow \bar{f} & & \wr\| \\ Y & \rightarrowtail & X & \longrightarrow & X/Y \end{array} \quad . \tag{1}$$

The long exact homology sequence of the rows and the five lemma shows that $\bar{f}$ is a weak equivalence. Compare IV.2. in Hilton–Stammbach. Next we prove the factorization axiom (C3). We construct the *factorization*

$$f : B \rightarrowtail A = (B \oplus W, d) \underset{g}{\overset{\sim}{\to}} Y \tag{2}$$

inductively, where we use the notation in the proof of (6.10). Assume we constructed $g_n : A^n \to Y$ such that g_n is **n-connected**, that is $(g_n)_* : H_i A^n \to H_i Y$ is an isomorphism for $i < n$ and is surjective for $i = n$. Then we choose a free module $V' = V'_{n+1}$ and we choose d such that

$$V' \xrightarrow{d} (ZA^n)_n \twoheadrightarrow H_n A^n \tag{3}$$

maps surjectively onto kernel $(g_n)_*$. Here Z denotes the cycles. Therefore we can choose g' such that

$$\begin{array}{ccc} A^n_n & \xrightarrow{g_n} & Y_n \\ \uparrow d & & \uparrow \\ V' & \xrightarrow{g'} & Y_{n+1} \end{array} \tag{4}$$

commutes, (V' is free). We define the chain map

$$g' : A' = (A^n_n \oplus V', d) \to Y \tag{5}$$

by (4). By definition of d in (3) the map g' induces an isomorphism in homology for all degrees $\leq n$. The map g', however, needs not yet to be $(n+1)$-connected. Therefore we choose a free module V and we choose g'' such that

$$V \xrightarrow{g''} (ZY)_{n+1} \twoheadrightarrow H_{n+1} Y \tag{6}$$

is surjective. Let $dV = 0$ and let $W_{n+1} = V' \oplus V$. Then g' and g'' yield the chain map g_{n+1} which is $(n+1)$-connected. Since Y is *bounded below* we can start the induction. Therefore the factorization in (2) exists. Finally we show that all objects are fibrant. Let

$$i:B \rightarrowtail\!\!\xrightarrow{\sim} A = (B \oplus W, d) \tag{7}$$

be a cofibration and weak equivalence. We choose inductively a retraction

$$f_n: A^n \to B, \quad f_n i = 1_B, \tag{8}$$

and a homotopy $\alpha_n: if_n \simeq g_n$ where $g_n: A^n \subset A$ is the inclusion. Assume f_n and α_n are defined. By

$$\alpha_n d + d\alpha_n = g_n - if_n \tag{9}$$

we have

$$\begin{aligned} if_n d &= g_n d - d\alpha_n d \\ &= d(g_n - \alpha_n d). \end{aligned} \tag{10}$$

Since i is a weak equivalence there is by (10) a map $x: W_{n+1} \to B_{n+1}$ with $dx = f_n d$. Moreover $(g_{n+1} - \alpha_n d - ix)$ carries W_{n+1} to the cycles of A by (9). Again since i is a weak equivalence we can choose maps $z: W_{n+1} \to B_{n+1}$, $y: W_{n+1} \to A_{n+2}$ such that

$$iz + dy = g_{n+1} - \alpha_n d - ix. \tag{11}$$

We now define the extension f_{n+1} of f_n by $f_{n+1} = x + z$ on W_{n+1} and we define the extension α_{n+1} of α_n by $\alpha_{n+1} = y$ on W_{n+1}. This shows

$$\begin{aligned} df_{n+1} &= dx = f_n d = f_{n+1} d, \\ g_{n+1} - if_{n+1} &= g_{n+1} - ix - iz \\ &= dy + \alpha_n d \quad \text{by (11)} \\ &= d\alpha_{n+1} + \alpha_{n+1} d. \end{aligned} \tag{12}$$

We can start the induction since A is bounded below. By (12) the choice of maps f_n yields a retraction of i in (7). □

Let

$$SC_*: \mathbf{Top} \to (\mathbf{Chain}_Z^+)_c \tag{6.13}$$

be the functor which carries a space X to the **singular chain complex** SC_*X of X. The following result is an easy exercise, compare the notation in (1.10).

(6.14) Proposition. *The singular chain functor SC_* is a model functor which carries the cofibration category of topological spaces in* (5.1) *to the cofibration category of chain complexes in* (6.3).

§ 7 The category of chain algebras

We show that the category of chain algebras is a cofibration category and that the category of free chain algebras is an I-category. Moreover, we consider the chain algebra of a loop space.

In this section let R be a *commutative ring of coefficients* with unit 1. Thus a left R-module M is equivalently a right R-module by $r\cdot x = x\cdot r$, $r \in R$, $x \in M$. The tensor product $M \bigotimes_R N = M \otimes N$ of R-modules is an R-module by $r\cdot(x \otimes y) = (rx) \otimes y$.

(7.1) ***Definition.*** A (graded) **algebra** A is a positive (or negative) module A together with a map $\mu: A \otimes A \to A$ of degree 0 and an element $1 \in A_0$ such that the multiplication $x\cdot y = \mu(x \otimes y)$ is associative and 1 is the neutral element, $(1\cdot x = x\cdot 1 = x)$. A (non-graded) algebra is a graded algebra A which is concentrated in degree 0, that is $A_0 = A$. A map $f: A \to B$ between algebras is a map of degree 0 with $f(1) = 1$ and $f(x\cdot y) = f(x)\cdot f(y)$. ‖

The ring of coefficients R is a graded algebra which is concentrated in degree 0. This is the initial object $*$ in the category of graded algebras since we always have $i: * \to A$, $i(1) = 1$.

(7.2) ***Definition.*** An **augmentation** of an algebra A is a map $\varepsilon: A \to *$ between graded algebras. Let $\tilde{A} = \text{kernel}(\varepsilon)$ be the **augmentation ideal**. The quotient module $QA = \tilde{A}/\tilde{A}\cdot\tilde{A}$ is the module of **indecomposables**. Here $\tilde{A}\cdot\tilde{A}$ denotes $\mu(\tilde{A} \otimes \tilde{A})$. An augmentation preserving map f between algebras induces $Qf: QA \to QB$. ‖

(7.3) ***Definition.*** For a positive graded module V we have the **tensor algebra**

$$T(V) = \bigoplus_{n \geq 0} V^{\otimes n},$$

where $V^{\otimes n} = V \otimes \cdots \otimes V$ is the n-fold product, $V^{\otimes 0} = R$. We have inclusions and projections of graded modules

$$V^{\otimes n} \xrightarrow[i_n]{} T(V) \xrightarrow[p_n]{} V^{\otimes n}.$$

The tensor algebra is an algebra with multiplication given by $V^{\otimes n} \otimes V^{\otimes m} = V^{\otimes(n+m)}$. The algebra is augmented by $\varepsilon = p_0$. We clearly have $QT(V) = V$. For a map $\alpha: V \to W$ of degree 0 let

$$T(\alpha): T(V) \to T(W)$$

be given by $T(\alpha)(x_1 \otimes \cdots \otimes x_n) = \alpha x_1 \otimes \cdots \otimes \alpha x_n$. Then $QT(\alpha) = \alpha$. ‖

(7.4) ***Definition.*** An algebra A is **free** if there is a submodule $V \subset A$ with the properties: V is a free module and the homomorphism $T(V) \to A$ of algebras given by $V \subset A$ is an isomorphism. ||

In this case V generates the free algebra A. If E is a basis of V the *free monoid* Mon(E), generated by E, is a basis of the free R-module A. Moreover, the composition $V \subset \tilde{A} \to QA$ is an isomorphism of free modules.

For algebras A and B we have the **free product** $A \coprod B$ which is the push out of $A \leftarrow * \to B$ in the category of algebras. Such free products exist. In particular, the free product $B \coprod T(V)$ of B with the free algebra $T(V)$ is given by

$$B \coprod T(V) = \bigoplus_{n \geqq 0} B \otimes (V \otimes B)^{\otimes n}. \tag{7.5}$$

Here the multiplication is defined by

$$\begin{aligned}&(a_0 \otimes v_1 \otimes \cdots \otimes v_n \otimes a_n)(b_0 \otimes w_1 \otimes \cdots \otimes w_m \otimes b_m)\\ &\quad = a_0 \otimes v_1 \otimes \cdots \otimes v_n \otimes (a_n \cdot b_0) \otimes w_1 \otimes \cdots \otimes w_m \otimes b_m,\end{aligned}$$

with $a_i, b_j \in B$ and $v_i, w_j \in V$. The product $a_n \cdot b_0$ is taken in the algebra B. We have $T(V) \coprod T(W) = T(V \oplus W)$.

If B is augmented by $\varepsilon_B : B \to *$ we obtain the augmentation $\varepsilon : B \coprod T(V) \to *$ by $\varepsilon(b) = \varepsilon_B(b)$ for $b \in B$ and $\varepsilon(v) = 0$ for $v \in V$.

(7.6) ***Definition.*** A **differential algebra** A is a positive (or negative) graded algebra A together with a differential $d : A \to A$ of degree -1 such that (A, d) is a chain complex and such that

$$\mu : A \otimes A \to A$$

is a chain map, see (6.7), that is

$$d(x \cdot y) = (dx) \cdot y + (-1)^{|x|} x \cdot dy.$$

A *map* $f : A \to B$ between differential algebras is an algebra map and a chain map: $f(1) = 1$, $f(x \cdot y) = fx \cdot fy$ and $df = fd$.

If A is positive we call $A = (A_*, d)$ a **chain algebra**. If A is negative we call $A = (A^*, d)$ a **cochain algebra**, in this case (A^*, d) is a cochain complex as in (6.1) and the differential has upper degree $+1$, $d : A^n \to A^{n+1}$.

The algebra $* = R$ (concentrated in degree 0) is also a differential algebra which is the **initial object** in the category of differential algebras. An **augmentation** ε of A is a map $\varepsilon : A \to *$ between differential algebras. ||

In this section and in Chapter IX we study the **category of chain algebras** which we denote by **DA**. Let $\mathbf{DA}_*$ be the category of objects over $*$ in **DA**; this is exactly the **category of augmented chain algebras** and of augmentation preserving maps. Therefore $*$ is the initial and the final object in $\mathbf{DA}_*$. We also use the category **DA** (flat) which is the full subcategory of **DA** consisting of chain algebras A for which A is flat as an R-module. When we replace 'flat' by 'free' we get the full subcategory **DA** (free). Recall that all free R-modules are flat.

The **homology** $HA = H(A, d)$ of the underlying chain complex of a differential algebra is an algebra with the multiplication

(7.7) $$HA \otimes HA \xrightarrow{j} H(A \otimes A) \xrightarrow{\mu_*} HA$$

with $j(\{x\} \otimes \{y\}) = \{x \otimes y\}$ where x and y are cycles in A. If A is augmented also HA is augmented by $\varepsilon = \varepsilon_*: HA \to H_* = *$. We point out that for a chain algebra A we have the canonical map

$$\lambda: A \to H_0 A \tag{1}$$

with $\lambda x = 0$ for $|x| > 0$ and $\lambda x = \{x\}$ for $|x| = 0$ since for a chain algebra each element x in degree zero is a cycle. An augmentation ε of A yields an augmentation ε_* of the algebra $H_0 A$ such that

$$\varepsilon = \varepsilon_* \lambda: A \twoheadrightarrow H_0 A \twoheadrightarrow * = R. \tag{2}$$

This shows that a chain algebra is augmented iff the algebra $H_0 A$ is augmented.

In the category **DA** of chain algebras we introduce the following structure:

(7.8) ***Definition***

(1) A map $f: B \to A$ in **DA** is a **weak equivalence** if f induces an isomorphism $f_*: HB \to HA$ in homology.
(2) A map $B \to A$ in **DA** is a **cofibration** if there is a submodule V of A with the properties: V is a free module and the homomorphism $B \amalg T(V) \to A$ of algebras, given by $B \to A$ and $V \subset A$, is an isomorphism of algebras. We call V a **module of generators** for $B \rightarrowtail A$. The cofibration $B \rightarrowtail A$ is **elementary** if $d(V) \subset B$.
(3) A map $f: B \to A$ in $\mathbf{DA}_*$ is a weak equivalence (resp. a cofibration) if f is a weak equivalence (resp. a cofibration) in **DA**. For a cofibration f in $\mathbf{DA}_*$ we can choose the module of generators V such that $\varepsilon(V) = 0$. ||

A chain algebra A is **free** if $* \to A$ is a cofibration. Thus the full subcategory of free chain algebras is the category $\mathbf{DA}_c$ of cofibrant objects in **DA**. Clearly $\mathbf{DA}_c \subset \mathbf{DA}$ (flat) since a free chain algebra is also free as an R-module.

(7.9) ***Push outs in DA.*** For the cofibration $B \rightarrowtail A = (B \amalg T(W), d)$ and for

$f:B\to Y$ the induced cofibration $Y \rightarrowtail A\bigcup_B Y$, given by

$$\begin{array}{ccc} A & \xrightarrow{\bar{f}} & A\bigcup\limits_B Y=(Y\coprod T(W),d) \\ {\scriptstyle i}\uparrow & \text{push} & \uparrow{\scriptstyle \bar{i}} \\ B & \xrightarrow[f]{} & Y \end{array},$$

is generated by W and $\bar{f}=f\coprod 1$ is the identity on W. The differential d on $A\bigcup_B Y$ is the unique differential for which $\bar{f}$ and $\bar{i}$ are chain maps. In case i is an elementary cofibration we obtain d on $A\bigcup_B Y$ by $d:W\xrightarrow{d}B\xrightarrow{f}Y$.

(7.10) **Theorem.** *Let R be a principal ideal domain. Then the categories* **DA** *(flat) and* $\mathbf{DA}_*$ *(flat) with the structure* (7.8) *are cofibration categories for which all objects are fibrant.*

We prove this result in (7.21) below.

Next we describe an explicit cylinder for a cofibration in **DA**.

(7.11) ***Definition.*** Let $B\rightarrowtail A$ be a cofibration in **DA**. We define a **cylinder**

$$A\bigcup_B A \underset{(i_0,i_1)}{\rightarrowtail} I_BA \underset{p}{\longrightarrow} A$$

as follows: We choose a generating module W of the cofibration $B\subset A$ so that $A=B\coprod T(W)$. The underlying algebra of I_BA is

$$I_BA=B\coprod T(W'\oplus W''\oplus sW). \tag{1}$$

Here W' and W'' are two copies of the graded module W and sW is the graded module with $(sW)_n=W_{n-1}$. We define i_0 and i_1 by the identity on B and by

$$i_0x=x',\, i_1x=x'' \quad \text{for } x\in W. \tag{2}$$

Here $x'\in W'$ and $x''\in W''$ are the elements which correspond to $x\in W=W'=W''$. Moreover, we define p by the identity on B and by

$$\left.\begin{array}{ll} px'=px''=x & \text{for } x'\in W',x''\in W'', \\ p(sx)=0 & \text{for } sx\in sW. \end{array}\right\} \tag{3}$$

The differential d on I_BA is given by

$$\left.\begin{array}{l} dx'=i_0dx,\, dx''=i_1dx \\ dsx=x''-x'-Sdx \end{array}\right\} \tag{4}$$

where $x\in W$. Here $S:A\to I_BA$ is the unique map of degree $+1$ between graded modules which satisfies

$$\left.\begin{array}{ll} Sb = 0 & \text{for } b \in B, \\ Sx = sx & \text{for } x \in W, \\ S(xy) = (Sx)(i_1 y) + (-1)^{|x|}(i_0 x)(Sy) & \text{for } x, y \in A. \end{array}\right\} \tag{5}$$

||

Lemma

(*a*) *The map S is well defined by* (5).
(*b*) *The differential is well defined by* (4) *and* (5) *and satisfies* $dd = 0$.
(*c*) *The inclusions* i_0, i_1 *satisfy* $i_1 - i_0 = Sd + dS$.
(*d*) i_0, i_1 *and p are chain maps with* $pi_0 = pi_1 = \text{identity}$.
(*e*) (i_0, i_1) *is a cofibration.*

Proof. For (*a*) it is enough to check, by (7.5), that S is compatible with the associativity law of the multiplication, that is

$$S((xy)z) = S(x(yz)) = (Sx)y''z'' + (-1)^{|x|}x'(Sy)z'' + (-1)^{|x|+|y|}x'y'(Sz).$$

Moreover, (*b*) follows from (*c*). Now (*c*) is clear on B and on $w \in W$ we obtain (*c*) by $Sdw + dSw = Sdw + dsw = (w'' - w')$, see (4) in (7.11). Assume that (*c*) holds on $x, y \in A$. Then (*c*) holds on the product $x \cdot y$ since

$$\begin{aligned} Sd(xy) + dS(xy) &= (Sdx + dSx) \cdot y'' + x'(Sdy + dSy) \\ &= (x'' - x')y'' + x'(y'' - y') = x''y'' - x'y'. \end{aligned}$$

Moreover, (*d*) is clear since $pS = 0$ by (5) in (7.11). By definition in (1) of (7.11) we directly see that (i_0, i_1) is a cofibration. □

(7.12) ***Definition.*** Let $B \subset A$ be a cofibration and let $f, g: A \to X$ be maps between chain algebras. We say f and g are **DA-homotopic relative** B if $f|B = g|B$ and if there is a map $H: I_B A \to X$ of chain algebras with $f = Hi_0$, $g = Hi_1$. We call H a **DA**-homotopy from f to g rel B. Equivalently, we call the map $\alpha = HS: A \to X$ of degree $+1$ a **DA-homotopy** from f to g (rel B), see (7.11) (5). A map $\alpha: A \to X$ of degree $+1$ is given by a homotopy H iff the following holds:

(1) $\alpha(b) = 0$ for $b \in B$,
(2) $\alpha d + d\alpha = g - f$,
(3) $\alpha(xy) = (\alpha x)(gy) + (-1)^{|x|}(fx)(\alpha y)$ for $x, y \in A$. ||

By (2) we see that a **DA**-homotopy is a chain homotopy. Theorem (7.18) below implies that '**DA**-homotopic rel B' is actually an equivalence relation.

Lemma. *Assume the cofibration* $B \rightarrowtail A = B \coprod T(W)$ *is generated by W.*

Then a map $\alpha: A \to X$ of degree $+1$ is a **DA**-*homotopy from f to g if* (1) *and* (3) *in* (7.12) *hold and if* (2) *in* (7.12) *is satisfied on generators $w \in W$.*

Proof. We prove inductively that (2) in (7.12) is satisfied. Assume (2) is satisfied on x and $y \in A$, that is, $\alpha dx + d\alpha x = gx - fx$, $\alpha dy + d\alpha y = gy - fy$. Then (2) is also satisfied on $x \cdot y$ since we have

$$\begin{aligned}\alpha d(xy) + d\alpha(xy) &= \alpha((dx)y + \bar{x}(dy)) + d((\alpha x)gy + (f\bar{x})\alpha y)\\ &= (\alpha dx + d\alpha x)(gy) + fx(\alpha dy + d\alpha y)\\ &= (gx - fx)gy + fx(gy - fy)\\ &= gxgy - fxfy = g(xy) - f(xy).\end{aligned}$$

Here we set $\bar{x} = (-1)^{|x|}x$. □

(7.13) ***Definition of $T(V, dV)$.*** Let $V = \{V_n\}$ be a positive graded module with $V_0 = 0$. We define the graded module dV by

$$(\mathrm{d}V)_n = V_{n+1}, \text{ equivalently } V = s(dV). \tag{1}$$

Let $d: V \to dV$ be the maps of degree -1 given by (1). This yields the object

$$T(V, dV) = (T(V \oplus dV), d) \tag{2}$$

in $\mathbf{DA}_*$ by defining the differential d on generators via $d: V \to dV$, $d(dV) = 0$, and by defining $\varepsilon(V) = \varepsilon(dV) = 0$. Note that

$$* \rightarrowtail T(dV) \rightarrowtail T(V, dV) \tag{3}$$

are cofibrations in $\mathbf{DA}_*$ (where $T(dV)$ has trivial differential) and that $T(V, dV)$ has the following universal property. Let A be an object **DA** and let $\varphi: V \to A$ be a map of degree 0 between graded modules. Then there is a unique map

$$\bar{\varphi}: T(V, dV) \to A \quad \text{in } \mathbf{DA} \tag{4}$$

which extends φ, that is $\bar{\varphi} | V = \varphi$. Moreover, in case A is augmented and if $\varepsilon\varphi = 0$ then also $\bar{\varphi}$ is augmented. ||

By the next lemma we see that $T(V, dV) = CT(dV)$ is a **cone** for the chain algebra $T(dV)$.

(7.14) **Lemma.** *$T(V, dV)$ is acyclic, that is, $R \rightarrowtail T(V, dV)$ is a weak equivalence. In fact, $R \rightarrowtail T(V, dV)$ is a homotopy equivalence in the I-category* $\mathbf{DA}_c$, *see* (7.18).

Proof. In the I-category $\mathbf{DA}_c$ we have the push out diagram

$$\begin{array}{ccc} I_* T(dV) & \longrightarrow & CT(dV) \cong T(V, dV) \\ {\scriptstyle i_0}\uparrow & & \uparrow{\scriptstyle \bar{i}_0} \\ T(dV) & \xrightarrow[\varepsilon]{} & * \end{array}$$

Since i_0 is a homotopy equivalence by (3.13) also $\bar{i}_0$ is one by (C2)(b) and (3.3). □

The cylinder in the category **DA** is natural since a pair map $f:(A, B) \to (X, Y)$ with $B \rightarrowtail A$, $Y \rightarrowtail X$ induces the map

$$If: I_B A \to I_Y X$$

as follows: Let $A = B \coprod T(V)$. Then we define

$$\left.\begin{aligned} (If) i_0 a &= i_0 f a \quad \text{for } a \in A, \\ (If) i_1 a &= i_1 f a \quad \text{for } a \in A, \\ (If) sv &= S_X f v \quad \text{for } v \in V. \end{aligned}\right\} \tag{$*$}$$

Here $S_X = S: X \to I_Y X$ is the map of degree $+1$ in (7.11)(5).

(7.15) **Lemma.** *There is a unique map If in* **DA** *which satisfies* $(*)$ *above and we have*

$$(If) S_A = S_X f. \tag{$**$}$$

Proof. Clearly, there is a unique map between algebras which satisfies $(*)$. We now check $(**)$. On $b \in B$ equation $(**)$ is true since $S_A b = 0$ and since for $fb \in Y$ also $S_X f b = 0$. Moreover, by definition, $(**)$ is true on $v \in V$. Assume now $(**)$ is true on $x, y \in A$. Then we have

$$\begin{aligned} (If) S(xy) &= (If)((Sx)y'' + (-1)^{|x|} x' Sy) \\ &= ((If)(Sx)) \cdot (fy)'' + (-1)^{|x|} (fx)' \cdot ((If)(Sy)) = Sf(xy), \end{aligned}$$

compare (7.11)(5). This shows that $(**)$ is also true on xy and thus $(**)$ is proven. From $(**)$ we derive that If is a chain map:

$$\begin{aligned} (If) d(sv) &= (If)(v'' - v' - Sdv) = (fv)'' - (fv)' - Sfdv \\ &= dS(fv) = d(If)(sv), \end{aligned}$$

compare (7.11)(4). Now lemma (7.15) is proved. □

(7.16) **Corollary.** *For a composition of pair maps $fg:(D, E) \to (A, B) \to (X, Y)$ in* **DA** *we have $I(fg) = (If)(Ig)$.*

Proof. Let $D = E \coprod T(W)$. Then we have for $w \in W$, $(If)(Ig)(sw) = (If)(Sgw) = S(fgw) = (I(fg))(sw)$ by $(**)$ and by $(*)$. □

(7.17) **Corollary.** *The cylinder $I_B A$ of a cofibration $B \rightarrowtail A$ in* **DA** *is well defined up to canonical isomorphism.*

Let B be an object in **DA**. Then we have the category $\mathbf{DA}_c^B$. The objects are the cofibrations $B \rightarrowtail A$ in **DA**, the maps are the **maps under** B in **DA**. We obtain by (7.16) the functor

$$I: \mathbf{DA}_c^B \to \mathbf{DA}_c^B,$$

with $I(A, B) = (I_B A, B)$. We say a map $f:(A, B) \to (A', B)$ in $\mathbf{DA}_c^B$ is a cofibration if $f: A \to A'$ is a cofibration in **DA**. The initial object of $\mathbf{DA}_c^B$ is the pair (B, B) given by the identity of B. For this structure of $\mathbf{DA}_c^B$ we have the

(7.18) **Proposition.** *The category $\mathbf{DA}_c^B$ is an I-category.*

Here we need no assumption on the commutative ring of coefficients R. Similarly as in (7.18) we see that also $(\mathbf{DA}_*)_c^B$ is an I-category. The cylinder in (7.11) is augmented by εp.

We derive from (7.18) and (3.12) that the projection $p: I_B A \to A$ is a homotopy equivalence in $\mathbf{DA}_c^B$ and thus p is a weak equivalence in **DA**. Therefore $I_B A$ is actually a cylinder for the structure (7.8) as defined in (1.6). The following corollary is an analogue of the corresponding result (6.11) for chain complexes.

(7.19) **Corollary.** *Let R be a principal ideal domain and let $f: A \to A'$ be a map in $\mathbf{DA}_c^B$* (flat). *If f is a weak equivalence in* **DA** *then f is a homotopy equivalence in the I-category $\mathbf{DA}_c^B$.*

A similar result is true for augmented chain algebras. The corollary is a direct consequence of (7.18), (7.10) and (II.2.12).

Consider the functor

$$Q: (\mathbf{DA}_*)_c^B \to (\mathbf{Chain}_R^+)_c,$$

which carries (A, B) to $Q(A/B)$ where A/B is the push out of $* \leftarrow B \rightarrowtail A$ in **DA**. Since A is augmented also A/B is augmented and therefore the chain complex $Q(A/B)$ is defined by the quotient in (7.2). Clearly, if V generates $B \rightarrowtail A$ then $Q(A/B) = V$.

(7.20) **Proposition.** *The functor Q above preserves all the structures of the I-categories, that is: $QI = IQ$, Q carries cofibrations to cofibrations, and Q carries push outs to push outs,*

$$Q(A \bigcup_B X) = QA \bigcup_{QB} QX;$$

also $Q(*) = *$ for the initial objects.

This is an easy consequence of the definition of cylinders in (7.11) and (6.8) respectively. The proposition shows that Q is a based model functor in the sense of (1.10).

Proof of (7.18). The proof is actually very similar to the proof of (6.10). For convenience we prove the result only for $B = *$. We leave the case $B \neq *$ as an exercise. Clearly the cylinder exiom (I1) is satisfied by (7.16) and (7.11). Next we prove the push out axiom (I2). For the cofibration

$$B \rightarrowtail A = (B \coprod T(W), d), \tag{1}$$

and for $f: B \to Y$ the push out $A \bigcup_B Y = (Y \coprod T(W), d)$ is the unique chain algebra for which the canonical map

$$\bar{f} = f \coprod 1 : B \coprod T(W) \to Y \coprod T(W) \tag{2}$$

is a chain map, see (7.9). Now the functor I carries $\bar{f}$ to $I\bar{f}$ with

$$(I\bar{f})sw = S\bar{f}w = Sw = sw \quad \text{for } w \in W. \tag{3}$$

Therefore we get $I\bar{f} = \overline{If}$ and thus I carries push outs to push outs.

Next consider the cofibration axiom (I3). The composition of cofibrations is a cofibration since we have

$$(B \coprod T(V)) \coprod T(W) = B \coprod (T(V) \coprod T(W)) = B \coprod T(V \oplus W).$$

For the cofibration (1) let $W^n = \{x \in W \mid |x| \leqslant n\}$ and let

$$A^n = (B \coprod T(W^n), d) \tag{4}$$

be the subchain algebra of A with $A^n \rightarrowtail A^{n+1}$. We now define inductively the homotopy extension E ($\varepsilon = 0$) by the maps

$$E^n : IA^n \bigcup_{A^n} A \to X. \tag{5}$$

As in the proof of (6.10) we define E^{n+1} on generators $w \in W_{n+1}$ by

$$\left.\begin{aligned} E^{n+1}(sw) &= 0, \\ E^{n+1}(w'') &= f(w) + E^n S dw. \end{aligned}\right\} \tag{6}$$

Here S is the homotopy is (7.11)(5). In the same way as in the proof of (6.10) we see that E^{n+1} is a well-defined map in **DA** which extends E^n. Inductively we get the homotopy extension E. This proves (I3).

The relative cylinder axiom (I3) is clear by the definition of the cylinder in (7.11). Finally, we obtain an interchange map for $X = (T(V), d)$ as follows. We observe that for the graded module I in (6.8) we can write

$$IX = (T(I \otimes V), d), \tag{7}$$

where we use the identification in (6.8)(2). Now we define the interchange map on $IIX = (T(I \otimes I \otimes V), d)$ to be the algebra map $T(t \otimes 1)$, see (7.4), where

$t\otimes 1: I\otimes I\otimes V\to I\otimes I\otimes V$ is given by $t(x\otimes y)=(-1)^{|x||y|}y\otimes x$ for $x, y\in I$. One can check that $T(t\otimes 1)$ is a well-defined chain map. Now also (I5) is proved and therefore the proof of (7.18) is complete. □

Proof of (7.10). The composition axiom is clearly satisfied and push outs as in (C2) exist by (7.9). Moreover, we prove (C2), (C3) and (C4) in (7.26), (7.21) and (7.22) respectively. We prove the results only for $\mathbf{DA}_*$, slight modifications yield the proof for $\mathbf{DA}$ as well. □

(7.21) **Lemma.** *Each map $f: B\to Y$ in $\mathbf{DA}_*$ admits a factorization $B \rightarrowtail A \xrightarrow{\sim} Y$.*

Proof. We construct $A=(B\coprod T(W), d)$ inductively where we use the notation (4) in the proof of (7.18). Assume we constructed an extension $g_n: A^n\to X$ of f such that $g_{n*}: H_i A^n\to H_i X$ is an isomorphism for $i<n$ and is surjective for $i=n$. This is true for $n=-1$. Then we choose $V'=V'_{n+1}$ and d such that (Z denotes the cycles)

$$V' \xrightarrow[d]{} (Z\tilde{A}^n)_n \underset{p}{\twoheadrightarrow} H_n\tilde{A}^n, \quad \tilde{A}=\text{kernel } \varepsilon,$$

maps surjectively onto kernel $(g_n)_*$. Therefore we can choose g' such that

$$\begin{array}{ccc} A^n_n & \xrightarrow{\quad g_n \quad} & X_n \\ \uparrow{\scriptstyle d} & & \uparrow{\scriptstyle d} \\ V' & \xrightarrow[\quad g' \quad]{} & \tilde{X}_{n+1} \end{array}$$

commutes. Now g' yields the map $g': A'=A^n\coprod T(V')\to X$ of chain algebras which extends g_n.

Clearly, $H_iA'=H_iA^n$ for $i<n$. For $i=n$ we have $H_nA'=(ZA^n)_n/d'A'_{n+1}$ where $d': A'_{n+1}=A^n_{n+1}\oplus A^n_0\otimes V'\otimes A^n_0\to(ZA^n)_n$, as follows by (7.5). By construction of V' the map pd' maps surjectively onto kernel g_{n*}. Therefore, g' induces the isomorphism $H_nA'\cong H_nX$. Now we choose $V=V_{n+1}$ and g'' such that the composition

$$V \xrightarrow{g''} (Z\tilde{X})_{n+1} \to H_{n+1}\tilde{X}$$

is surjective. We set $dV=0$. Then the extension $g_{n+1}: A^{n+1}=A'\coprod T(V)\to X$ of g' given by g'' induces an isomorphism H_ig_{n+1} for $i\leq n$ and a surjection $H_{n+1}g_{n+1}$. □

(7.22) **Lemma.** *All objects in $\mathbf{DA}_*$ are fibrant.*

We point out that in (7.21) and (7.22) the ring of coefficients is allowed to be any commutative ring.

Proof of (7.22). Let $i: B \rightarrowtail\!\!\xrightarrow{\sim} A$ be a cofibration and a weak equivalence, $A = (B \coprod T(V), d)$. We construct the commutative diagram in $\mathbf{DA}_*$

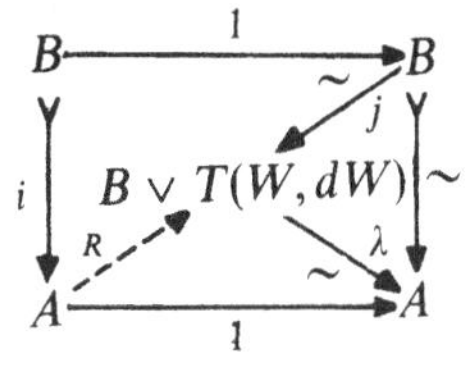

(1)

where $W = \{A_n : n \geqq 1\}$ is given by the underlying module of A and where $T(W, dW)$ is the acyclic object in (7.14). We define j by the inclusion, see (1.7), and we define λ by $W \subset A$, see (7.13)(4). Since j is a weak equivalence also λ is one by $i = \lambda j$. However, one readily checks that λ is a surjective map of modules since $i_* : H_0 B \cong H_0 A$. We have the retraction $r = (1, \varepsilon)$ of j where ε is the augmentation of $T(W, dW)$. Therefore rR is a retraction of i once we have the commutative diagram (1). This shows that B is fibrant.

We construct R inductively. For this we fix a basis J_V of the free module V. Assume $R(\beta)$ is defined for $\beta \in J_V$, $|\beta| < n$, and let $\alpha \in J_V$ with $|\alpha| = n$. Then $R(d\alpha)$ is defined and

$$\lambda R d\alpha = d\alpha$$

is a boundary. Choose $y \in B \vee T(W, dW)$ so that $\lambda y = \alpha$; by the surjectivity of λ this is possible. Then

$$\lambda(Rd\alpha - dy) = d\alpha - \lambda dy = 0.$$

Furthermore, $d(Rd\alpha - dy) = Rdd\alpha = 0$. Now $H(\ker \lambda) = 0$ since λ is a weak equivalence (here we use the long exact homology sequence of the short exact sequence $\ker \lambda \rightarrowtail B \vee T(W_1 dW) \twoheadrightarrow A$). Therefore we can find $w \in \ker(\lambda)$ with

$$Rd\alpha - dy = dw.$$

Now we define R on α by $R\alpha = y + w$. Then $dR\alpha = Rd\alpha$ and $\lambda R\alpha = \lambda y = \alpha$. This completes the inductive construction of R in (1). □

For the proof of the push out axiom (C2) in (7.10) we use the following **spectral sequence of a cofibration:** Let $B \rightarrowtail A = B \coprod T(V)$ be a cofibration generated by V. We introduce the double degree $|x| = (p, q)$ of a typical element

$$x = b_0 v_1 b_1 \ldots v_n b_n \in A,$$

by $p = \sum_{i=1}^{n} |v_i|$ and $q = \sum_{i=0}^{n} |b_i|$, compare (7.5).

Let $A_{p,q}$ be the module of elements in A of bidegree (p,q) and let

$$F_pA = \bigoplus_{\substack{i \leqq p \\ j \geqq 0}} A_{i,j}. \tag{7.23}$$

The modules F_pA form a filtration of subchain complexes in A which is bounded above since $A_n \subset F_nA$. This yields the spectral sequence $\{E^n_{p,q}A, d^n\}$ which converges to HA. We obtain the E^1-term as follows: We have

$$E^0_{p,q} = (F_pA/F_{p-1}A)_{p+q} = A_{p,q}.$$

Using the interchange isomorphism we get

$$A_{p,q} \cong \bigoplus_{n \geqq 0} (V^{\otimes n})_p \otimes (B^{\otimes (n+1)})_q,$$

compare (7.5). Moreover, the differential d^0 is given by the commutative diagram

$$\begin{array}{ccc} E^0_{p,q} & \cong & \bigoplus_{n\geqq 0}(V^{\otimes n})_p \otimes (B^{\otimes(n+1)})_q \\ \downarrow d^0 & & \downarrow \bigoplus_{n \geqq 0} 1 \otimes d_B \\ E^0_{p,q-1} & \cong & \bigoplus_{n\geqq 0}(V^{\otimes n})_p \otimes (B^{\otimes(n+1)})_{q-1} \end{array}$$

Here d_B is the differential on $B^{\otimes(n+1)}$ determined by the differential on B. Since V is a free R-module we derive

$$\begin{aligned} E^1_{p,q} &= H(E^0_{p,q}, d^0) \\ &\cong \bigoplus_{n \geqq 0} (V^{\otimes n})_p \otimes H_q(B^{\otimes(n+1)}). \end{aligned} \tag{7.24}$$

We now consider the push out diagram

$$\begin{array}{ccc} (B \coprod T(V), d) & \xrightarrow{f \coprod 1} & (Y \coprod T(V), d) \\ \| & & \| \\ A & \xrightarrow{\bar{f}} & X \\ \uparrow & \text{push} & \uparrow \\ B & \xrightarrow[f]{} & Y \end{array}$$

in the category **DA**, see (7.20) (2). The map $\bar{f}$ induces a map between spectral sequences such that for the E^1-term the following diagram commutes:

$$(7.25)\qquad \begin{array}{ccc} E^1_{p,q}A \cong \bigoplus_{n\geq 0}(V^{\otimes n})_p \otimes H_q(B^{\otimes(n+1)}) \\ \Big\downarrow \bar{f}_* \qquad\qquad \Big\downarrow \bigoplus_{n\geq 0} 1\otimes H_q(f^{\otimes(n+1)}). \\ E^1_{p,q}X = \bigoplus_{n\geq 0}(V^{\otimes n})_p \otimes H_q(Y^{\otimes(n+1)}) \end{array}$$

(7.26) **Lemma**. *Let R be a principal ideal domain and assume $f:B\to Y$ is a weak equivalence in* **DA**(flat). *Then also $\bar{f}$ is a weak equivalence.*

Proof. Since B is flat as an R-module and since R is a principal ideal domain we can apply the Künneth formula for $H_q(B^{\otimes(n+1)})$ in (7.25). This shows that $\bar{f}_*$ in (7.25) is an isomorphism since f is a weak equivalence. Now the comparison theorem for spectral sequences shows that also $\bar{f}_*:HA\cong HX$ is an isomorphism. Compare Hilton–Stammbach. □

(7.27) ***Example***. A **topological monoid** is a space M with a basepoint $*$ and with an associative multiplication $\mu:M\times M\to M$, $\mu(x,y)=x\cdot y$, in **Top** such that $*\cdot x = x\cdot * = x$. For example a topological group is a topological monoid. The singular chain complex SC_*M in (6.13) is a chain algebra in $\mathbf{DA}_*$(flat) by the multiplication

$$\mu: SC_*(M)\otimes SC_*(M)\xrightarrow{\times} SC_*(M\times M)\xrightarrow{\mu_*} SC_*(M).$$

Here $\times$ is the associative cross-product of singular chains. In case we choose the singular cubical chain complex the cross product is easely given by the product $f\times g$ of singular cubes f,g. The augmentation of the chain algebra $SC_*(M)$ is induced by the map $M\to *$. The homology algebra $H_*(M)=H_*(SC_*M)$ in (7.7) is called the **Pontryagin ring** of M.

(7.28) ***Definition***. For a topological space X with basepoint $*$ we define the **Moore loop space** ΩX. An element $(f,r)\in\Omega X$ is given by $r\in\mathbb{R}$ and by a map $f:[0,r]\to X$ with $f(0)=f(r)=*$. The topology of ΩX is taken from $X^{\mathbb{R}}\times\mathbb{R}$ where the function space $X^{\mathbb{R}}$ has the compact open topology. The addition

$$+:\Omega X\times\Omega X\to\Omega X$$

of loops is defined by $(f,r)+(g,s)=(f+g,r+s)$ with $(f+g)(t)=f(t)$ for $0\leq t\leq r$ and $(f+g)(r+t)=g(t)$ for $0\leq t\leq s$. This shows that ΩX is a topological monoid. A basepoint preserving map $F:X\to Y$ induces the map $\Omega F:\Omega X\to\Omega Y$, $(\Omega F)(f,r)=(Ff,r)$. This shows that Ω is a functor which carries **Top*** to the category of topological monoids. ||

By (7.27) and (7.28) we have the functor

$$SC_*\Omega:\mathbf{TOP}^*\to\mathbf{DA}_*(\text{flat}),$$

which carries the pointed space X to the chain algebra of the Moore loop space of X. Let **Top*** be the cofibration category of pointed spaces with the structure (5.4) and assume that $SC_*\Omega$ has coefficients in a principal ideal domain so that $\mathbf{DA}_*$(flat) is a cofibration category by (7.10). Clearly $SC_*\Omega$ carries a homotopy equivalence in **Top*** to a weak equivalence. This shows that the functor $SC_*\Omega$ preserves weak equivalences. Furthermore, we have with the notation in (3.10):

(7.29) **Theorem.** *The functor $SC_*\Omega$ is compatible with all push outs $A\bigcup_B Y$ for which A, B and Y are path connected and for which $B \rightarrowtail A$ induces an isomorphism on fundamental groups.*

This shows that $SC_*\Omega$ is a model functor on an appropriate subcategory of **Top***, see (1.10), for example on the full subcategory of **Top*** consisting of simply connected spaces. We do not prove the theorem in this book, for simply connected spaces A, B, Y the proposition can be derived from results in Adams–Hilton.

(7.30) ***Remark.*** The functor $SC_*\Omega$ is not compatible with all push outs in **Top***. For example consider the push out of $S^1 \leftarrowtail * \rightarrowtail S^1$ or of $I \leftarrowtail \partial I \to *$ where I is the unit interval and where $S^1 = I/\partial I$ is the 1-sphere.

(7.31) ***Definition.*** Let A and B be graded algebras (positive or negative), see (7.1). Then the **tensor product** $A \otimes B$ is a graded algebra. As a graded module $A \otimes B$ is defined as in (6.6). The multiplication is given by the formula

$$(a \otimes b)\cdot(a' \otimes b') = (-1)^{|b||a'|}(a\cdot a')\otimes(b\cdot b').$$

If A and B are augmented then so is $A \otimes B$ by $\varepsilon(a \otimes b) = \varepsilon(a)\cdot\varepsilon(b)$. Moreover, if A and B are differential algebras as in (7.6) then $A \otimes B$ is a differential algebra by the differential in (6.7). There is a natural isomorphism $T: A \otimes B \cong B \otimes A$ of differential algebras by $T(a \otimes b) = (-1)^{|a||b|} b \otimes a$. $\|$

(7.32) **Proposition.** *For the functor $\tau = SC_*\Omega$ above we have a natural isomorphism of chain algebras in $Ho(\mathbf{DA}_*$(flat)), see (II.3.5),*

$$\tau(X) \otimes \tau(Y) \simeq \tau(X \times Y).$$

This is easily seen by use of the cross-product of singular chains and by use of the natural homotopy equivalence $\Omega(X \times Y) \simeq \Omega X \times \Omega Y$.

§ 8 The category of commutative cochain algebras

We show that the category of connected commutative cochain algebras is a cofibration category and we consider the Sullivan–De Rham functor from spaces to such algebras. In Chapter VIII we shall show that this functor induces an equivalence of rational homotopy theories. This section is based on the excellent notes of Halperin. For further studies we refer the reader to these notes, to Bousfield–Gugenheim, Tanré and clearly to the original work of Sullivan.

Let R be a commutative ring of coefficients with unit 1 and with $\frac{1}{2}\in R$; for most results of this section we actually assume that R is a *field of characteristic zero.*

(8.1) ***Definition.*** A (negative) graded algebra $A=\{A^n, n\geqq 0\}$ is **commutative** if $x\cdot y=(-1)^{|x||y|}y\cdot x$, compare (7.1). ||

(8.2) ***Definition.*** For a graded free module $V=\{V^n, n\geqq 0\}$ we have the **free** commutative algebra over V, denoted by $\Lambda(V)$, with

$$\Lambda(V)=\text{Exterior algebra }(V^{odd})\otimes\text{Symmetric algebra }(V^{even})$$

Here the symmetric algebra generated by V^{even} is also the polynomial algebra on generators in J_V^{even} where J_V^{even} is a basis of V^{even}. For a basis J_V of V we also write $\Lambda(J_V)=\Lambda(V)$. The R-module $\Lambda(V)$ is a free R-module since V is free and since we assume $\frac{1}{2}\in R$. Let $\Lambda(V)^n$ be the submodule of elements of degree n in $\Lambda(V)$ which is generated by the products $x_1\wedge\cdots\wedge x_k$ with $|x_1|+\cdots+|x_k|=n$, $x_c\in V$. Here we denote the multiplication in $\Lambda(V)$ by $x\wedge y$, $x, y\in\Lambda(V)$. The algebra $\Lambda(V)$ is augmented by $\varepsilon:\Lambda(V)\to R$, $\varepsilon(x)=0$, $x\in V$. Thus the composition $V\subset\tilde{\Lambda}(V)=\text{kernel}(\varepsilon)\to Q\Lambda(V)$ is a canonical isomorphism, see (7.2). ||

For commutative graded algebras A, B we have the **tensor product** $A\otimes B$ in (7.31) which is the push out of $A\leftarrow *\to B$ in the category of commutative graded algebras. We have

$$\Lambda(V)\otimes\Lambda(V')=\Lambda(V\oplus V') \tag{8.3}$$

(8.4) ***Definition.*** A **commutative cochain algebra** is a negative differential algebra which is commutative as a graded algebra, see (8.1) and (7.6). Let **CDA** be the *category of commutative cochain algebras*; maps in **CDA** are maps between differential algebras as in (7.6). Moreover let $\mathbf{CDA}_*$ be the *category of augmented commutative cochain algebras*, this is the category of objects over $*=R$ in **CDA**. An object A in **CDA** is **connected** if $*\to A$ induces an isomorphism $R=H^0A$. Let $\mathbf{CDA}_*^0$ be the full subcategory of $\mathbf{CDA}_*$ consisting of *connected augmented commutative cochain algebras.* ||

Clearly, the cohomology $HA=\{H^nA\}$ of an object in **CDA** is a commutative graded algebra which is augmented if A is augmented.

The **tensor product** $A \otimes B$ of objects in **CDA** is an object in **CDA** with the multiplication and the differential in (7.31). If A and B are augmented also $A \otimes B$ is augmented. Moreover, if R is a field, we have the isomorphism of algebras $H(A \otimes B) = HA \otimes HB$.

Remark. Bousfield–Gugenheim introduce the structure of a model category on **CDA** and $\mathbf{CDA}_*$. Cofibrations in this structure, however, are defined only abstractly and we have no explicit cylinder construction for such cofibrations. We therefore consider the following structure of $\mathbf{CDA}_*$ which is also studied by Halperin and Halperin–Watkiss.

(8.5) ***Definition***

(1) A map $f: B \to A$ in $\mathbf{CDA}_*$ is a **weak equivalence** if f induces an isomorphism $f_*: HB \to HA$ in cohomology.
(2) A map $i: B \to A$ in $\mathbf{CDA}_*$ is a **cofibration** (or equivalently a **KS-extension**) if there is a submodule V of A and a well-ordered subset J_V of V with the following properties (a), (b) and (c).

(a) V is a free module with basis J_V and $\varepsilon(V) = 0$.
(b) The homomorphism $B \otimes \Lambda(V) \to A$ of commutative algebras, given by $B \to A$ and $V \subset A$, is an isomorphism of algebras.
(c) For $\alpha \in J_V$ write $V_{<\alpha}$ for the submodule of V generated by all $\beta \in J_V$ with $\beta < \alpha$. Then the differential in A satisfies

$$d(\alpha) \in B \otimes \Lambda(V_{<\alpha}),$$

where we use the isomorphism in (b).

The cofibration i is called **minimal** if in addition $|\beta| < |\alpha| \Rightarrow \beta < \alpha$ for $\beta, \alpha \in J_V$. ||

(8.6) ***Remark.*** A cofibration as in (8.5) is an **elementary** cofibration if $dV \subset B$. For example, if J_V consists of a single element the cofibration $B \rightarrowtail A$ is elementary by (8.5)(2)(c). Also, (8.5)(2)(c) shows that each cofibration is the limit of a well-ordered sequence of elementary cofibrations

$$(B \otimes \Lambda(V_{<\alpha}), d) \rightarrowtail (B \otimes \Lambda(V_{\leq\alpha}), d),$$

with $\alpha \in J_V$. The composition of cofibrations is a cofibration since we have

$$(B \otimes \Lambda(V)) \otimes \Lambda(W) = B \otimes \Lambda(V \oplus W).$$

The basis J_V and the basis J_W yield the basis $J_V \cup J_W$ of $V \oplus W$ which is well ordered by setting $\alpha < \beta$ for $\alpha \in J_V$, $\beta \in J_W$.

(8.7) ***Definition of $\Lambda(V, dV)$.*** Let $V = \{V^n, n \geqq 0\}$ be a graded free module, $V^n = 0$ for $n < 0$. We define the graded module dV by

$$(dV)^n = V^{n-1}, \text{ equivalently } V = sdV. \tag{1}$$

Let $d: V \to dV$ be the map of degree $+1$ given by (1). This map yields the object

$$\Lambda(V, dV) = (\Lambda(V \oplus dV), d) \tag{2}$$

in $\mathbf{CDA}_*$ by defining the differential d on generators via $d: V \to dV$. Clearly $ddV = 0$ and $\varepsilon(V) = \varepsilon(dV) = 0$. If the basis J_V of V consists of a single element t we also write $\Lambda(t, dt)$ for (2). We have the isomorphism in $\mathbf{CDA}_*$

$$\Lambda(V, dV) = \bigotimes_{t \in J_V} \Lambda(t, dt). \tag{3}$$

Here an 'infinite tensor product' has to be interpreted as the direct limit of finite tensor products. Note that

$$* \rightarrowtail \Lambda(dV) \rightarrowtail \Lambda(V, dV) \tag{4}$$

are cofibrations and that $\Lambda(V, dV)$ has the following universal property: let A be an object in $\mathbf{CDA}_*$ and let $\varphi: V \to \ker(\varepsilon) \subset A$ be a map of degree 0 between graded modules. Then there is a unique map

$$\bar{\varphi}: \Lambda(V, dV) \to A \text{ in } \mathbf{CDA}_*, \tag{5}$$

which extends φ, that is, $\bar{\varphi} | V = \varphi$. By the next lemma we see that $\Lambda(V, dV) = C(\Lambda(dV))$ is a **cone** on the algebra $\Lambda(dV)$, which has trivial differential. ||

(8.8) **Poincaré lemma.** *Let R be a field of characteristic zero. Then $\Lambda(V, dV)$ is acyclic, that is, $\eta: R \rightarrowtail \Lambda(V, dV)$ is a weak equivalence.*

Compare 1.3 in Bousfield–Gugenheim.

Proof. By (8.7)(3) and by a limit argument it is enough to prove the proposition for $\Lambda(t, dt)$. If the degree $|t|$ is odd we have $\Lambda(t, dt) = \Lambda(dt) \oplus t \cdot \Lambda(dt)$ and $H\Lambda(t, dt) = \Lambda(dt)/(dt) \cdot \Lambda(dt) = R$. If $|t|$ is even we have $\Lambda(t, dt) = \Lambda(t) \oplus (dt) \cdot \Lambda(t)$. We define a map h of degree -1 on $\Lambda(t, dt)$ by $h(v) = 0$ for $v \in \Lambda(t)$ and

$$h(w) = \sum_{i=0}^{m} \frac{b_i}{i+1} t^{i+1}$$

for

$$w = dt \cdot \sum_{i=0}^{m} b_i t^i \in (dt)\Lambda(t).$$

It is in the *'formal integral'* $h(w)$ that $\mathrm{char}(R) = 0$ is needed. For the augmentation ε and for η in (8.8) one can check $dh + hd = 1 - \eta\varepsilon$. Thus η is a homotopy equivalence of cochain complexes. □

(8.9) ***Remark.*** We recall the following result, (compare Halperin 2.2). Let R be a field of characteristic zero and let $i: B \rightarrowtail A$ be a cofibration in $\mathbf{CDA}^0_*$. Then there exists factorization of i,

$$i: B \rightarrowtail \tilde{\tilde{A}} \overset{j}{\underset{\sim}{\rightarrowtail}} \tilde{\tilde{A}} \otimes \Lambda(V, dV) \overset{\alpha}{\cong} A,$$

where $B \rightarrowtail \tilde{\tilde{A}}$ is a minimal cofibration and where $\tilde{\tilde{A}} \otimes \Lambda(V, dV)$ is the tensor product in the category $\mathbf{CDA}_*$ of $\tilde{\tilde{A}}$ and of an appropriate acyclic object as in (8.7). The canonical inclusion j carries x to $x \otimes 1 = x$ and α is an isomorphism in $\mathbf{CDA}_*$. Clearly, j is a weak equivalence by (8.8).

(8.10) **Lemma.** *Let R be a field of characteristic zero. Then each map $f: B \to Y$ in $\mathbf{CDA}^0_*$ admits a factorization*

$$f: B \underset{i}{\rightarrowtail} A \underset{p}{\overset{\sim}{\to}} Y,$$

where i is a minimal cofibration and where p is a weak equivalence.

Remark. A factorization as in (8.10) is called a **minimal model** of f, or a minimal model of Y in case $B = *$ is the initial object. A minimal model $B \rightarrowtail A$ of f is well defined up to an isomorphism under B, that is: for two minimal models A, A' there is an isomorphism $\alpha: A \to A'$ under B such that $p\alpha \simeq p \operatorname{rel} B$. Compare the notes of Halperin and (II.1.13).

For the convenience of the reader we now recall the proof of lemma (8.10), see 6.4 in Halperin.

Proof of (8.10). We shall construct

$$A = (B \otimes \Lambda X, d_A),$$

where each X^n is decomposed in the form

$$X^n = \bigoplus_{r \geqq 0} X^n_r, \quad n \geqq 0.$$

The differential d_A extends the differential d_B of B and satisfies

$$d_A(X^m_q) \subset B \otimes \Lambda(X^{<m} \oplus X^m_{<q}), \quad m \geqq 0, q \geqq 0. \tag{1}$$

We shall simultaneously construct $p: A \overset{\sim}{\to} Y$ so that

$$pd_A = d_Y p, \quad p|B = f, \tag{2}$$

and

$$pX^0_q \subset \ker(\varepsilon), \quad q \geqq 0. \tag{3}$$

We now construct the modules X_r^0. Set

$$X_0^0 = \ker\{H^1(B) \xrightarrow{f^*} H^1(Y)\}.$$

Define d_A on X_0^0 so that $d_A(x)$ is a cocycle in B representing $x \in X_0^0$. Extend p to X_0^0 so that (2) and (3) holds. Suppose X_q^0 has been defined for $q < r$ and p and d_A have been extended to $B \otimes \Lambda X^0_{<r}$ so that (1), (2) and (3) hold. Let

$$X_r^0 = \ker\{H^1(B \otimes \Lambda X^0_{<r}) \xrightarrow{p^*} H^1(Y)\}$$

and further extend p and d_A just as we did when $r = 0$. Then we get

$$H^0(B \otimes \Lambda X^0) = R \tag{4}$$

$$p^*: H^1(B \otimes \Lambda X^0) \to H^1(Y) \text{ is injective.} \tag{5}$$

For the proof of (4) we need only to show that $H^0(B \otimes \Lambda X^0_{\leqslant r}) = R$ for all r. Assume this holds for $q < r$; then $H^0(B \otimes \Lambda X^0_{<r}) = R$. We write

$$\begin{cases} D = B \otimes \Lambda X^0_{<r}, \\ B \otimes \Lambda X^0_{\leqq r} = D \otimes \Lambda X_r^0. \end{cases}$$

If $\phi \in (D \otimes \Lambda X_r^0)^0$ is a cocycle write $\phi = \phi_0 + \phi_1 + \cdots + \phi_m$ with $\phi_j \in D^0 \otimes \Lambda^j X_r^0$, $\phi_m \neq 0$. Since $d_A \phi = 0$ we see that

$$(d_D \otimes 1)\phi_m = 0. \tag{6}$$

In fact, d_A carries $D^0 \otimes \Lambda^j X_r^0$ to $D^1 \otimes \Lambda^{j-1} X_r^0 \oplus D^1 \otimes \Lambda^j X_r^0$ by the derivation formula where the second coordinate is $d_D \otimes 1$. Thus for $j < m$ the elements $d_A \phi_j$ do not meet $D^1 \otimes \Lambda^m X_r^0$ and thus $d_A \phi = 0$ implies (6). It follows from our induction hypothesis and from (6) that $\phi_m \in \Lambda^m X_r^0$.

Now by construction, d_A injects X_r^0 onto a space of cocycles in D^1 which does not meet $d_D(D^0)$. On the other hand, $d_A \phi = 0$ and (6) show with arguments as in (6) that

$$d_A \phi_m + (d_D \otimes 1)\phi_{m-1} = 0.$$

Since $\phi_m \in \Lambda X_r^0$ this implies

$$d_A \phi_m = 0.$$

Consider the acyclic object $\Lambda(X_r^0, dX_r^0)$ in (8.8). Since $d_A: X_r^0 \to D$ is injective, it follows that $d\phi_m = 0$ in $\Lambda(X_r^0, dX_r^0)$. Since $\Lambda(X_r^0, dX_r^0)$ is acyclic by (8.8), the only cocycles in ΛX_r^0 of $\Lambda(X_r^0, dX_r^0)$ are scalars. Hence $m = 0$ and $\phi = \phi_m \in R$. This completes the proof of (4).

For the proof of (5) assume $\phi \in (B \otimes \Lambda X^0)^1$, $d_A \phi = 0$, $p\phi$ is a coboundary. Then $\phi \in (B \otimes \Lambda X^0_{\leqq r})^1$, some r, and so (by definition of X^0_{r+1}) for some

$x \in X^0_{r+1}$, $\phi - d_A x \in d_A(B \otimes \Lambda X^0_{\leqq r})$. In particular, ϕ is a coboundary in $B \otimes \Lambda X^0$. This proves (5).

Next we construct the modules X^n_r, $n > 0$. Suppose that for some $n > 0$ the spaces X^m_q are defined for $m < n$ and $q \geqq 0$, and that d_A and p are extended to $B \otimes \Lambda X^{<n}$ so that (1) and (2) hold. Assume as well that

$$p^*: H(B \otimes \Lambda X^{<n}) \to H(Y) \text{ is } (n-1) \text{ regular.} \tag{7}$$

Here we say that a map λ (of degree 0) between graded modules is ***n*-regular** if it is an isomorphism in degrees $\leqq n$ and injective in degree $n+1$. Now define spaces $W^n_0 = \operatorname{coker}(p^*)^n$ and $V^n_0 = \ker(p^*)^{n+1}$ where $(p^*)^i = p^*: H^i(B \otimes \Lambda X^{<n}) \to H^i(Y)$, $i = n, n+1$. Let

$$\left.\begin{array}{l} X^n_0 = W^n_0 \oplus V^n_0, \\ d_A(w) = 0 \quad \text{for } w \in W^n_0, \\ d_A(v) = \text{a cocycle in } B \otimes \Lambda X^{<n} \text{ representing } v, v \in V^n_0. \end{array}\right\} \tag{8}$$

Extend p in the obvious way to X^n_0 so that 2 holds. Then p^* is surjective in degree n on $B \otimes \Lambda X^{<n} \otimes \Lambda X^n_0$.

Next, if X^n_q is defined for $0 \leqq q < r$ and if the map p and the differential d_A are extended to $B \otimes \Lambda X^{<n} \otimes \Lambda X^n_{<r}$ so that (1) and (2) hold, set (for $r \geqslant 1$)

$$X^n_r = \ker\{p^*: H^{n+1}(B \otimes \Lambda X^{<n} \otimes \Lambda X^n_{<r}) \to H^{n+1} Y\}.$$

Extend d_A to X^n_r so that

$$d_A(x) \in B \otimes \Lambda X^{<n} \otimes \Lambda X^n_{<r} \text{ is a cocycle representing } x, x \in X^n_r. \tag{9}$$

Extend p to X^n_r so that (2) holds. Then we get

$$p^*: H(B \otimes \Lambda X^{\leqq n}) \to H(Y) \text{ is } n\text{-regular.} \tag{10}$$

In fact, since $(B \otimes \Lambda X^{\leqq n})^m = (B \otimes \Lambda X^{<n})^m$, $m < n$, it follows from (7) that p^* in (10) is an isomorphism in degrees less than n. We show next that it is injective in degree n.

Suppose ϕ is a cocycle in $(B \otimes \Lambda X^{\leqq n})^n$ and $p\phi$ is a coboundary. Note that $\phi \in B \otimes \Lambda X^{<n} \otimes \Lambda X^n_{\leqq r}$, some r. Let $D = B \otimes \Lambda X^{<n} \otimes \Lambda X^n_{<r}$ and let d_D be the restriction of d_A. Write

$$\begin{cases} B \otimes \Lambda X^{<n} \otimes \Lambda X^n_{\leqq r} = D \otimes \Lambda X^n_r, \\ \phi = \psi + \Omega, \quad \psi \in D^0 \otimes X^n_r, \quad \Omega \in D^n. \end{cases}$$

Since $d_A \phi = 0$ we conclude $(d_D \otimes 1)\psi = 0$. But since $H^0(Y) = R$, (7) shows that $H^0(D) = R$. Hence $\psi \in X^n_r$. It follows that $d_A \psi \in d_D(D^n)$. In view of our construction of X^n_r, this implies $\psi = 0$.

Hence $\phi \in D$. Continuing in this way we eventually obtain $\phi \in B \otimes \Lambda X^{<n}$. Now (7) shows that ϕ is a coboundary. Thus p^* in (10) is injective in degree n.

Finally, p^* in (10) is surjective in degree n by the definition of $W^n_0 \subset X^n_0$. It is injective in degree $n+1$ by the same argument as used in the proof of (5). This completes the proof of (10).

By induction we obtain A and the factorization in (8.10). Clearly, (10) shows that $p: A \to Y$ is a weak equivalence and (1) shows that $i: B \rightarrowtail A$ is a minimal cofibration. Moreover, (3) shows that p is augmentation preserving. By definition $pi = f$. Therefore the proof of (8.10) is complete. □

(8.11) **Lemma.** *Let the ring of coefficients be a field of characteristic zero. Then all objects in* **CDA**$_*$ *are fibrant with respect to the structure in* (8.5).

Proof. The arguments are similar as in the proof of (7.22). Let $i: B \overset{\sim}{\rightarrowtail} A$ be a cofibration and a weak equivalence. We construct the following commutative diagram in **CDA**$_*$ where $B \otimes \Lambda(W, dW) = B \vee C(\Lambda(dW))$ is a sum of B and a cone.

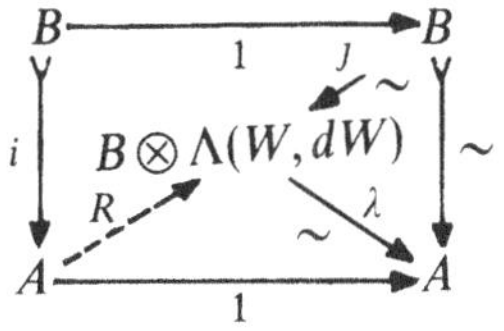

(1)

Here W is the underlying module of A, $A = W$, and $\Lambda(W, dW)$ is the acyclic object in (8.8). We define j by $j(b) = b \otimes 1$ and we define λ by $\lambda(b) = i(b)$ for $b \in B$ and $\lambda(v) = v$ for $v \in W = A$. Since $\Lambda(W, dW)$ is acyclic we see that j is a weak equivalence, therefore also λ is a weak equivalence since $i = \lambda j$ is one. We have a restriction r of j by $r = 1_B \otimes \varepsilon$ where ε is the augmentation of $\Lambda(W, dW)$. Therefore rR is a retraction of i once we have the commutative diagram (1). This shows that B is fibrant.

Now we construct R inductively. For this we fix a bases J_V as in (8.5) (c). Assume $R(\beta)$ is defined for $\beta \in J_V$, $\beta < \alpha$. Then also $Rd_A\alpha$ is defined and

$$\lambda R d_A \alpha = d_A \alpha$$

is a coboundary. Choose $y \in B \otimes \Lambda(W, dW)$ so that $\lambda y = \alpha$. This is possible since λ is a surjective map of modules. Then

$$\lambda(Rd_A\alpha - dy) = d_A\alpha - \lambda dy = 0.$$

Here d is the differential of $B \otimes \Lambda(W, dW)$. Furthermore,

$$d(Rd_A\alpha - dy) = Rd_A d_A \alpha = 0.$$

Since $H(\ker \lambda) = 0$ we can write (for some $w \in \ker(\lambda)$)

$$Rd_A\alpha - dy = dw.$$

Extend R to $B \otimes \Lambda(V_{\leqq\alpha})$ by setting

$$R\alpha = y + w.$$

By definition $dR\alpha = Rd_A\alpha$, while

$$\lambda R\alpha = \lambda y = \alpha.$$

This completes the inductive construction of R in (1). □

(8.12) ***Push outs in CDA$_*$.*** In the category **CDA**$_*$ there exist push outs

$$\begin{array}{ccc}
(B \otimes \Lambda(V), d_A) & \xrightarrow{f \otimes 1} & (Y \otimes \Lambda(V), d) \\
\| & & \| \\
A & \xrightarrow{\bar{f}} & A \underset{B}{\otimes} Y \\
{\scriptstyle i}\uparrow & \text{push} & \uparrow{\scriptstyle \bar{i}} \\
B & \xrightarrow[f]{} & Y
\end{array} \quad .$$

Here $A \otimes_B Y = (Y \otimes \Lambda(V), d)$ is the unique cochain algebra in **CDA** for which $f \otimes 1$ and $\bar{i}$ are chain maps. In case $B \rightarrowtail A$ is an elementary cofibration we obtain the differential on $Y \otimes \Lambda(V)$ by the composition $d = fd_A: V \to B \to Y$.

(8.13) **Lemma.** *The structure* (8.5) *of* **CDA**$_*$ *satisfies the push out axiom* (C2). Here $\mathrm{char}(R) = 0$ is not needed.

Proof. Since the homology is compatible with limits it is enough to check axiom (C2)(a) for elementary cofibrations, see (8.6). In this case we have the filtration of cochain complexes

$$B = F_0 \subset F_1 = B \otimes \Lambda(\alpha) \quad (|\alpha| \text{odd}),$$
$$B = F_0 \subset \cdots \subset F_n = B \otimes \Lambda_n \subset \cdots \subset B \otimes \Lambda(\alpha)$$

($|\alpha|$ even), where Λ_n is the submodule of $\Lambda(\alpha)$ generated by $1, \alpha, \ldots, \alpha^n$. The quotient $F_n/F_{n-1} = B \otimes \alpha^n$ is a cochain complex which up to a change of degree, is isomorphic to B. Thus the exact cohomology sequence of the pair (F_n, F_{n-1}) and the five lemma show inductively that $\bar{f}$ is a weak equivalence provided f is one. □

(8.14) ***Push outs in CDA$_*^0$.*** Let R be a field of characteristic zero. In the category **CDA**$_*^0$ of connected cochain algebras, see (8.4), there exist push out diagrams of the form

$$\begin{array}{ccc} A & \xrightarrow{\bar{f}} & A \bigcup_B Y \\ \uparrow & \text{push} & \uparrow \\ B & \xrightarrow[f]{} & Y \end{array}, \tag{1}$$

where the *connected push out* $A \bigcup_B Y$ is a quotient of $A \bigotimes_B Y$ in (8.12). In case $H^0(A \bigotimes_B Y) = R$ we actually have $A \bigcup_B Y = A \bigotimes_B Y$. In general, however, $A \bigotimes_B Y$ may not be connected even though A, B and Y are; consider for example

$$R \leftarrow \Lambda(dt) \rightarrowtail \Lambda(t, dt) \quad \text{with } |t| = 0.$$

For the construction of the push out $A \bigcup_B Y$ in the category $\mathbf{CDA}_*^0$ we use the method of Halperin–Watkiss: By use of (8.9) we factor the inclusion $i: B \rightarrowtail A$ as a sequence of cofibrations in $\mathbf{CDA}_*^0$

$$i: B \rightarrowtail \bar{B} = (B \otimes \Lambda(V), d) \rightarrowtail \bar{\bar{A}} \rightarrowtail A, \tag{2}$$

where V is concentrated in degree 0 and where $\bar{\bar{A}} = \bar{B} \otimes \Lambda(W)$ with $W^0 = 0$. Choose a basis J_V of the minimal cofibration $B \rightarrowtail \bar{B}$ as in (8.5). We shall define a surjective homomorphism in $\mathbf{CDA}_*$

$$\pi: \bar{B} \bigotimes_B Y = (Y \otimes \Lambda(V), d) \twoheadrightarrow (Y \otimes \Lambda(\bar{V}), \bar{d}) = \bar{Y} \tag{3}$$

where $H^0(\bar{Y}) = R$ and we set

$$A \bigcup_B Y = A \bigotimes_{\bar{B}} \bar{Y}, \tag{4}$$

where we use $\bar{B} \to \bar{B} \bigotimes_B Y \to \bar{Y}$. The map π is given inductively and extends the identity of Y. For each $\alpha \in J_V$ we shall define a surjective map in $\mathbf{CDA}_*$ $\pi_\alpha: Y \otimes \Lambda V_{\leqq \alpha} \to Y \otimes \Lambda \bar{V}_{\leqq \alpha}$ such that $H^0(Y \otimes \Lambda(\bar{V}_{\leqq \alpha})) = R$, and such that if $\beta < \alpha$ the inclusion $Y \otimes \Lambda V_{\leqq \beta} \to Y \otimes \Lambda V_{\leqq \alpha}$ factors to yield a (unique) inclusion $Y \otimes \Lambda \bar{V}_{\leqq \beta} \to Y \otimes \Lambda \bar{V}_{\leqq \alpha}$. Assume π_β constructed for $\beta < \alpha$. Consider

$$\pi_{<\alpha} = \lim_{\beta < \alpha} \pi_\beta : Y \otimes \Lambda V_{<\alpha} \to Y \otimes \Lambda \bar{V}_{<\alpha} = \lim_{\beta < \alpha} Y \otimes \Lambda \bar{V}_{\leqq \beta}$$

In the case $\pi_{<\alpha}(d\alpha) \in \bar{d}(Y \otimes \Lambda \bar{V}_{<\alpha})$ define $Y \otimes \Lambda \bar{V}_{\leqq \alpha} = Y \otimes \Lambda \bar{V}_{<\alpha}$. Since

$$H^0(Y \otimes \Lambda \bar{V}_{<\alpha}) = \varinjlim H^0(Y \otimes \Lambda \bar{V}_{\leqq \beta}) = R,$$

there is a unique w in the augmentation ideal of $Y \otimes \Lambda \bar{V}_{<\alpha}$ with $\bar{d}w = \pi_{<\alpha}(d\alpha)$. Extend $\pi_{<\alpha}$ to π_α by setting $\pi_\alpha(\alpha) = w$.

Otherwise set $Y \otimes \Lambda \bar{V}_{\leqq \alpha} = Y \otimes \Lambda \bar{V}_{<\alpha} \otimes \Lambda(\alpha)$, $\bar{d}\alpha = \pi_{<\alpha}(d\alpha)$ and extend $\pi_{<\alpha}$

to π_α by setting $\pi_\alpha(\alpha)=\alpha$. In this case we need to verify that $H^0(Y\otimes\Lambda\bar{V}_{\leqq\alpha})=R$. We show this similarly as (4) in the proof of (8.10). Suppose $z=\phi_0+\phi_1\alpha+\cdots+\phi_n\alpha^n$ is a $\bar{d}$ cocycle of degree 0, $\phi_k\in Y\otimes\Lambda\bar{V}_{<\alpha}=D$, with $\phi_n\neq 0$. Then $\bar{d}\phi_n=0$ and $n\phi_n\bar{d}(\alpha)+\bar{d}(\phi_{n-1})=0$. The first equation shows that ϕ_n is a scalar $\lambda\neq 0$, since by the hypothesis $\bar{d}\alpha\notin\bar{d}(Y\otimes\Lambda\bar{V}_{<\alpha})$, the second equation shows that $n=0$. Hence $z=\lambda\in R$. Finally, set $\pi=\lim_\alpha\pi_\alpha$ with $Y\otimes\Lambda\bar{V}=\lim_\alpha Y\otimes\Lambda\bar{V}_{\leq\alpha}$ in $\mathbf{CDA}^0_*$. This completes the definition of π. If E is in $\mathbf{CDA}^0_*$ and if $\phi:A\bigotimes_B Y\to E$ is a map in $\mathbf{CDA}_*$ it is evident from the construction that ϕ factors to give a map $A\bigotimes_{\bar{B}}\bar{Y}\to E$. Thus (4) is the desired push out in $\mathbf{CDA}^0_*$. □

(8.15) ***Remark.*** The construction of $A\bigcup_B Y$ in (8.14) has a geometric analogue in the definition of pull backs in the category $\mathbf{Top}^*_0$ of path connected spaces with basepoint. The usual pull back $X x_Y Z$ in $\mathbf{Top}^*$ needs not to be path connected even though X, Y and Z are objects in $\mathbf{Top}^*_0$. Therefore we can use the *path component* $(X x_Y Z)_0$ *of the basepoint* $(*, *)$ *in* $X x_Y Z$ to be the correct pull back in the category $\mathbf{Top}^*_0$.

(8.16) **Lemma.** *Let R be a field of characteristic zero. The structure* (8.5) *of* $\mathbf{CDA}^0_*$ *satisfies the push out axiom* (C2).

Proof. By (8.14) push outs exist. Moreover (8.13) shows that $A\bigotimes_B Y$ is in $\mathbf{CDA}^0_*$ provided f is a weak equivalence. Hence $A\bigotimes_B Y=A\bigcup_B Y$ and thus (8.13) also shows that (C2)(a) is satisfied. □

From (8.16), (8.11), (8.10) and (8.6) we derive the following result:

(8.17) **Theorem.** *Let R be a field of characteristic zero. Then the category* $\mathbf{CDA}^0_*$ *with the structure in* (8.5) *is a cofibration category in which all objects are fibrant. Push outs in* $\mathbf{CDA}^0_*$ *are the connected push outs described in* (8.14).

(8.18) ***Remark.*** Cofibrations and weak equivalences in $\mathbf{CDA}^0_*$ are also cofibrations and weak equivalences in the model category $\mathbf{CDA}_*$ defined by Bousfield–Gugenheim. Therefore $\mathbf{CDA}^0_*$ can be considered as an explicit substructure of this model category. Cofibrations in the model category $\mathbf{CDA}_*$ are more complicated than our cofibrations in (8.5); in particular, they need not to be injective maps between algebras, for example, if V is concentrated in degree 0 then $\varepsilon\otimes 1:\Lambda(V)\otimes A\to A$ (where A is any object in $\mathbf{CDA}_*$) is a cofibration in the sense of Bousfield–Gugenheim. ||

Next we define for each cofibration $i:B\rightarrowtail A=(B\otimes\Lambda(V),d_A)$ in $\mathbf{CDA}^0_*$

an explicit **cylinder object**

(8.19) $$A\bigcup_B A \overset{i_0,i_1}{\rightarrowtail} I_B A \xrightarrow[\sim]{p} A$$

in the cofibration category $\mathbf{CDA}_*^0$ of (8.17), compare (1.5). Here $A\bigcup_B A$ is the connected push out constructed in (8.14), let $V^+ = V/V^0$ and let sV^+ be the graded module defined by

$$(sV^+)^{n-1} = (V^+)^n, \quad \text{compare } \S 6. \tag{1}$$

We have the map of upper degree -1

$$\bar{s} = sp: V^n \to (V^+)^n \to (sV^+)^{n-1}$$

(with $\bar{s}V^0 = 0$) where p is the quotient map and where s is given by the identity (1). For sV^+ we have the acyclic object $\Lambda(sV^+, dsV^+)$ as in (8.8). We now define the object $I_B A$ in (8.19) by the tensor product in the category $\mathbf{CDA}_*$.

$$\begin{aligned} I_B A &= A \otimes \Lambda(sV^+, dsV^+) \\ &= A \vee \Lambda(sV^+, dsV^+), \end{aligned} \tag{2}$$

which is a sum of A and a cone. The weak equivalence p in (8.19) is defined by $p = 1 \otimes \varepsilon$ where ε is the augmentation. Furthermore, we define i_0 in (8.19) by the canonical inclusion

$$i_0(a) = a \otimes 1 \quad \text{for } a \in A. \tag{3}$$

The definition of i_1, however, is more complicated, compare §5 in Halperin. We note that as an algebra

$$I_B A = B \otimes \Lambda(V) \otimes \Lambda(sV^+) \otimes \Lambda(dsV^+). \tag{4}$$

Thus a degree -1 derivation, S, is defined by

$$\left.\begin{aligned} &S(B) = S(sV^+) = S(dsV^+) = 0, \\ &S(v) = \bar{s}(v) \in sV^+, \quad \text{for } v \in V, \text{ see (1)}, \\ &S(x \cdot y) = (Sx)\cdot y + (-1)^{|x|} x \cdot Sy, \ (x, y \in I_B A). \end{aligned}\right\} \tag{5}$$

We define a degree zero derivation, θ, on $I_B A$ in (4) by

$$\theta = dS + Sd, \tag{6}$$

where d is the differential on $I_B A$. Note that

$$d\theta = \theta d, \quad pS = 0, \quad p\theta = 0. \tag{7}$$

Therefore S and θ preserve kernel(ε).

Lemma. *For each $\phi \in I_B A$ there is some N (depending on ϕ) with $\theta^N(\phi) = 0$,* ($\theta^N = \theta \cdots \theta = N$-fold iteration of θ). (8)

Proof of (8). For $\alpha \in J_V$, see (8.5), we have $\theta\alpha - dS\alpha = Sd\alpha \in I_B A_{<\alpha}$ where $A_{<\alpha} = B \otimes \Lambda(V_{<\alpha})$. Hence

$$\theta^2\alpha \in I_B A_{<\alpha}$$

since $\theta dS\alpha = dSdS\alpha + SddS\alpha = 0$ by (5); in fact, $S\alpha = \bar{s}\alpha \in sV^+$ for $\alpha \in J_V$ and thus $SdS\alpha = 0$ since $d\bar{s}\alpha \in dsV^+$, see (2). Inductively we get (8) since $\theta(B) = 0$ and since $\theta\bar{s}v = \theta d\bar{s}v = 0$ for $\bar{s}v \in sV^+$. □

By (8) we can define an automorphism e^θ of $I_B A$ in $\mathbf{CDA}_*$, namely

$$\exp(\theta) = e^\theta = \sum_{N=0}^{\infty} \frac{\theta^N}{N!} \quad \text{with inverse } e^{-\theta}. \tag{9}$$

Finally, we obtain i_1 in (8.19) by

$$i_1 = e^\theta i_0 = (\exp(dS + Sd))i_0. \tag{10}$$

This completes the definition of the cylinder object (8.19) provided we can prove:

Lemma. *The map* (i_0, i_1) *in* (8.19) *defined by* (3) *and* (10) *is a cofibration; in fact* (i_0, i_1) *is a minimal cofibration if* $B \rightarrowtail A$ *is minimal.* (11)

Proof. First assume that $B \rightarrowtail A = (B \otimes \Lambda(V), d_A)$ is minimal. Then we have the factorization

$$B \rightarrowtail B_0 = (B \otimes \Lambda(V^0), d_A) \rightarrowtail A \tag{12}$$

and one can check that (12) induces the canonical isomorphism in $\mathbf{CDA}_*^0$

$$A \bigcup_B A \cong A \bigotimes_{B_0} A = \left(B \otimes \Lambda\left(V' \bigoplus_{V_0} V'' \right), d \right) \tag{13}$$

Here $V' = V'' = V$ are isomorphic copies of V and V_0 is the submodule of V' and V'' respectively with $(V_0)^0 = V^0$ and $(V_0)^n = 0$ for $n \neq 0$. Let $V' \bigoplus_{V_0} V''$ be the push out of $V' \supset V_0 \subset V''$ in the category of graded modules; (clearly $V' \bigoplus_{V_0} V'' \cong V \oplus V^+$). We now obtain an isomorphism g of algebras for which the following diagram commutes and for which $g|sV^+$ is the canonical inclusion $sV^+ \subset I_B A$ given by (4).

$$\begin{array}{ccc} B \otimes \Lambda(V' \bigoplus_{V_0} V'' \oplus sV^+) & \xrightarrow[g]{\cong} & I_B A \\ \uparrow j & & \uparrow (i_0, i_1) \\ B \otimes \Lambda(V' \bigoplus_{V_0} V'') & = & A \bigcup_B A \end{array} \tag{14}$$

Here j is the canonical inclusion. One obtains the inverse g^{-1} of g inductively similarly as in the proof of (9.18) (9) below. Note that (i_0, i_1) is a well-defined map in $\mathbf{CDA}_*^0$ since $i_0|B = i_1|B$, actually $i_0|B_0 = i_1|B_0$ since $B \rightarrowtail A$ is minimal. Let J_V be a well-ordered bases of V as in (8.5). Then we obtain the well ordered basis J_{sV^+} of sV^+ by the elements $\bar{s}(\alpha)$ $(\alpha \in J_V, |\alpha| > 0)$ with $\bar{s}(\alpha) < \bar{s}(\beta)$ if $\alpha < \beta$. Now one can check that

$$g^{-1}d(\bar{s}\alpha) \in B \otimes \Lambda\left(V' \bigoplus_{V_0} V'' \oplus (sV^+)_{<\bar{s}\alpha} \right). \tag{15}$$

This proves that (i_0, i_1) in (14) is a minimal cofibration. Compare 5.28 in Halperin.

Next assume that $B \rightarrowtail A$ is not minimal. In this case we can use the factorization (8.9) and one can check that the induced map q in the following commutative diagram is a cofibration:

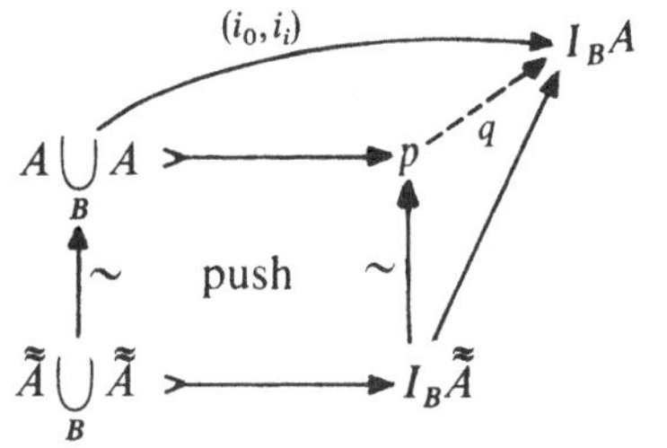

□

(8.20) ***Remark on cofibers.*** Let $B \rightarrowtail A = (B \otimes \Lambda(V), d_A)$ be a cofibration in $\mathbf{CDA}_*^0$. Then we have the push out diagram in $\mathbf{CDA}_*$

$$\begin{array}{ccc} A & \longrightarrow & A/B = A \bigotimes_B * = (\Lambda(V), d) \\ \uparrow & & \uparrow \\ B & \xrightarrow{\varepsilon} & * = R \end{array}, \tag{1}$$

where ε is the augmentation. The sequence $B \rightarrowtail A \rightarrow A/B$ is exactly a *KS extension in the sense of Halperin.* It is easy to check that $* \rightarrowtail A/B$ is a minimal cofibration in case $B \rightarrowtail A$ is minimal. On the other hand, we have the push out diagram in $\mathbf{CDA}_*^0$:

$$\begin{array}{ccc} A & \xrightarrow{g} & (A/B)_0 = A \bigcup_B * \\ \uparrow & & \uparrow \\ B & \xrightarrow{\varepsilon} & * = R \end{array}, \tag{2}$$

where $(A/B)_0$ is the '**connected component**' of A/B, compare (8.15). There is a quotient map $A/B \twoheadrightarrow (A/B)_0$ by the construction in (8.14), this is an isomorphism in case $H^0(A/B) = R$. In general, A/B may not be connected even though A and B are.

Now suppose that $B \rightarrowtail A$ is a minimal cofibration in $\mathbf{CDA}_*^0$. Then Halperin 3.10 shows that the generators V of $A = (B \otimes \Lambda(V), d_A)$ can be chosen such that for $A/B = (\Lambda(V), d)$ and $V = V^0 \oplus V^+$, $V^+ = \{V^n, n \geqq 1\}$, we have

$$d(V^0) = 0, \quad d(\Lambda(V^+)) \subset \Lambda(V^+). \tag{3}$$

In this case we get

$$A/B = \Lambda(V^0) \otimes (A/B)_0 \quad \text{in } \mathbf{CDA}_*, \tag{4}$$

where $(A/B)_0 = (\Lambda(V^+), d)$ is a connected minimal cochain algebra. Therefore we have

$$d\Lambda(V^+) \subset \tilde{\Lambda} \cdot \tilde{\Lambda}, \tag{5}$$

where $\tilde{\Lambda} = \tilde{\Lambda}(V^+)$ is the augmentation ideal, see (8.2) and (7.2). By (5) the differential of $Q(A/B)_0 = V^+$ is trivial.

(8.21) ***Definition.*** For a cofibration $B \rightarrowtail A$ in $\mathbf{CDA}_*^0$ we define the **ψ-homotopy groups** by

$$\pi_\psi^n(A, B) = H^n(Q(A/B)). \tag{1}$$

Here $Q(A/B)$ is the cochain complex with the differential induced by the differential of A/B. In case $B \rightarrowtail A$ is a minimal cofibration with the property (8.20) (3) we get

$$\pi_\psi^0(A, B) = V^0 \tag{2}$$

and

$$\pi_\psi^n(A, B) = V^n = H^n(Q(A/B)_0) \quad \text{for } n \geqq 1. \tag{3}$$ ||

(8.22) ***Example.*** Let M be a $\mathfrak{C}^\infty$-manifold with basepoint $*$. Then the de Rham algebra $A_{DR}^*(M)$ of $\mathfrak{C}^\infty$-forms on M is a commutative cochain algebra (with coefficient ring $R = \mathbb{R}$). The augmentation ε is given by the inclusion $* \to M$ of the basepoint which induces $\varepsilon: A_{DR}^*(M) \to A_{DR}^*(*) = \mathbb{R}$.

For a field R of characteristic zero there is the Sullivan–de Rham functor

(8.23) $$A_R: \mathbf{TOP}^* \to \mathbf{CDA}_*,$$

which is contravariant and which carries a topological space X with basepoint to the commutative cochain algebra $A_R(X)$ of simplicial differential forms on the singular set of X; for a definition of the functor A_R compare Halperin or Bousfield–Gugenheim, since $A_R(X) = A_{\mathbb{Q}}(X) \otimes_{\mathbb{Q}} R$ it is enough to define the functor $A_{\mathbb{Q}}$. We now describe some properties of the functor A_R.

For the coefficients $R = \mathbb{R}$ (given by real numbers), and for a $\mathbb{C}^\infty$-manifold M as in (8.22) one has a natural isomorphism

$$A_{DR}^*(M) \simeq A_{\mathbb{R}}(M) \tag{1}$$

in $Ho(\mathbf{CDA}_*)$, see (II.3.5). The Sullivan–de Rham theorem shows that 'integration of forms' yields a natural isomorphism

$$H^*(X, R) \cong H^*(A_R X) \tag{2}$$

for X in $\mathbf{TOP}^*$. This implies that $A_R X$ is connected ($H^0(A_R X) = R$) in case X is path connected. Let $\mathbf{Top}_0^*$ be the full subcategory of $\mathbf{Top}^*$ consisting of path connected spaces. Then A_R in (8.23) gives us

$$A_R : (\mathbf{Top}_0^*)^{op} \to \mathbf{CDA}_*^0 \tag{3}$$

Here $\mathbf{CDA}_*^0$ is a cofibration category. Moreover, $\mathbf{Top}_0^*$ is a fibration category with the structure in (5.5); pull backs in $\mathbf{Top}_0^*$ are given as in (8.15).

Since $A_R(*) = R = *$ we see that A_R is a based functor. Moreover, (2) shows that A_R preserves weak equivalences. For pull backs in $\mathbf{Top}_0^*$ we have the next result which is proved in 20.6 of Halperin. Let

$$Y \xrightarrow{f} B \twoheadleftarrow{}^{p} A$$

be maps in $\mathbf{Top}_0^*$ and assume that also the fiber F of $A \twoheadrightarrow B$ is path connected. Then the pull back $A \times_B Y$ is path connected too and we get:

(8.24) **Theorem.** *The functor A_R in (8.23) is compatible with the pull back $A \times_B Y$ provided $\pi_1 B$ acts nilpotently on the homology $H_*(F, R)$ and provided either $H_*(F, R)$ or both $H_*(Y, R)$ and $H_*(B, R)$ have finite type.*

Compare the notation in (8.26) below.

By definition in (1.10)(3) compatibility with $A \times_B Y$ means that the induced map q in the commutative diagram

(8.25)

$$
\begin{array}{ccc}
A_R(A) & \longrightarrow & A_R(A \times_B Y) \\
\uparrow^{p^*} \ \ \nwarrow^{\sim} & & \nearrow^{q} \ \uparrow^{\sim} \\
\ & M \to M \underset{A_R B}{\otimes} A_R Y & \\
\ \uparrow & & \searrow \\
A_R(B) & \xrightarrow[f^*]{\text{push}} & A_R(Y)
\end{array}
$$

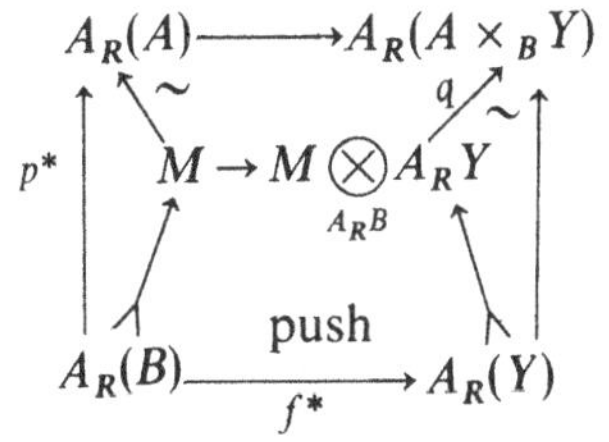

is a weak equivalence.

The result (8.24) shows that $\alpha = A_R$ in (8.22)(3) is a model functor on appropriate subcategories of $\mathbf{Top}_0^*$, see (1.10).

(8.26) *Notation.* For a group G the lower central series $\Gamma_q G \subset G$ is inductively defined by letting $\Gamma_{q+1}G$ be the subgroup of G generated by the commutators $-x-y+x+y$, $x \in G$, $y \in \Gamma_q G$. A group G is **nilpotent** if $\Gamma_q G = 0$ for some $q \geqq 1$. For a group G and a G-module (i.e. $\mathbb{Z}[G]$-module) N the lower central series $\Gamma_q N \subset N$ is inductively defined by letting $\Gamma_{q+1}N$ be the sub-G-module of N generated by $-n+n^g$, $g \in G$, $n \in \Gamma_q N$. A G-module N is **nilpotent** if $\Gamma_q N = 0$ for some $q \geqq 1$. A path connected topological space X is **nilpotent** if $\pi_1 X$ is a nilpotent group and $\pi_n X$ is a nilpotent $\pi_1 X$-module, $n \geqq 2$. In particular simply connected spaces are nilpotent.

(8.27) ***Remark***. Let $fN\mathbb{Q}$ be the class of all nilpotent CW-spaces X for which the homology $H_*(X, \mathbb{Z})$ is a rational vector space of finite type (thus X is a $\mathbb{Q}$-local space in the sense of (5.12)) and for which the inclusion $* \to X$ of the base point is a closed cofibration in **Top**. Let $\mathbf{Top}_0^*(fN\mathbb{Q})$ be the full subcategory of $\mathbf{Top}_0^*$ consisting of objects in $fN\mathbb{Q}$. Then we get the following commutative diagram of functors

$$\begin{array}{ccc} Ho(\mathbf{Top}_0^*) \xrightarrow{A_\mathbb{Q}} Ho(\mathbf{CDA}_*^0) & \underset{\sim}{\xrightarrow{M}} & (\mathbf{CDA}_*^0)_c/\simeq, \\ \cup & & \cup \\ \mathbf{Top}_0^*(fN\mathbb{Q})/\simeq & \underset{MA_\mathbb{Q}}{\xrightarrow{\sim}} & \mathbf{CDA}_*^0(fM\mathbb{Q})/\simeq. \end{array}$$

Here we use the induced functor $A_\mathbb{Q}$ on homotopy categories in (II.3.6) and we define the functor M as in (II.3.9) by choosing minimal models $* \rightarrowtail MA \xrightarrow{\sim} A$. Let $fM\mathbb{Q}$ be the class of all objects A in $\mathbf{CDA}_*^0$ for which $* \rightarrowtail A = (\Lambda(V), d)$ is a minimal cofibration and for which V^n is a finite dimensional rational vector space, $n \geqq 1$, $V^0 = 0$. The category $\mathbf{CDA}_*^0(fM\mathbb{Q})$ is the full subcategory of $\mathbf{CDA}_*^0$ consisting of objects in $fM\mathbb{Q}$. It is a fundamental result of Sullivan that the functor $MA_\mathbb{Q}$ in the bottom row of the diagram is actually on *equivalence of categories*, compare also 9.4 in Bousfield–Gugenheim. We prove a variant of this result, see Chapter VIII, by using towers of categories. Our proof relies only on (8.24) and (8.23)(2) and does not use the realization functor of Sullivan.

§9 The category of chain Lie algebras

The homotopy theory of chain Lie algebras is similar to the homotopy theory of chain algebras. The universal enveloping functor gives the possibility to compare both homotopy theories. We show that the category of chain Lie algebras is a cofibration category (we do not assume that the Lie algebras are 1-reduced as in Quillen (1969)) and we consider the Quillen functor from simply connected spaces to chain Lie algebras. In Chapter IX we will show that this functor induces an equivalence of rational homotopy theories.

Let R be a commutative ring of coefficients, for most results in this section we assume that R contains $\mathbb{Q}$ or that R is a *field of characteristic zero.*

(9.1) ***Definition.*** A (graded) **Lie algebra** L is a positive graded module together with a map

$$[\ ,\ ]: L \otimes L \to L, \quad x \otimes y \longmapsto [x, y],$$

of degree 0 such that (1) and (2) holds:

$$\text{Anticommutativity}: [x, y] = -(1)^{|x||y|}[y, x]. \tag{1}$$

$$\text{Jacobi identity}: [x, [y, z]] = [[x, y], z] + (-1)^{|x||y|}[y, [x, z]]. \tag{2}$$

A **map** $f: L \to L'$ between Lie algebras is a map of degree 0 with $f[x, y] = [fx, fy]$. The Lie algebra L is **1-reduced** if $L_0 = 0$. Let $[L, L]$ be the image of $[\ ,\ ]$ above, then the quotient $QL = L/[L, L]$ of graded modules is the module of **indecomposables** of L. Clearly, f induces $Qf: QL \to QL'$. ||

(9.2) ***Example.*** Let X be a simply connected space. Then the homotopy groups $\pi_*(\Omega X)$ form a 1-reduced Lie algebra with the Whitehead product as Lie bracket, see (II.15.19).

(9.3) ***Example.*** Let A be a positive algebra as in (7.1). Then A is a Lie algebra by the **Lie bracket associated to the multiplication in** A given by

$$[x, y] = x \cdot y - (-1)^{|x||y|} y \cdot x. \tag{1}$$

The corresponding functor, $A \longmapsto (A, [\ ,\ ])$, from algebras to Lie algebras has a 'left adjoint' U which carries a Lie algebra L to its **universal enveloping** algebra $U(L)$. Here $U(L)$ is an algebra together with a map $i: L \to U(L)$ between Lie algebras such that the following universal property holds:

For any algebra A and any Lie algebra map $f: L \to (A, [\ ,\ ])$ there is a unique algebra map $\bar{f}: U(L) \to A$ with $f = \bar{f} i$. (2)

We obtain the augmented algebra $U(L)$ by the quotient

$$U(L) = T(L)/J, \tag{3}$$

where J is the two-sided ideal of the tensor algebra $T(L)$ generated by the elements

$$(x \otimes y - (-1)^{|x||y|} y \otimes x) - [x, y] \tag{4}$$

with $x, y \in L$. The canonical map $i: L \to U(L)$ is given by $L \subset T(L)$ and the augmentation of $T(L)$ yields the augmentation of $U(L)$. One gets by (7.2):

$$QU(L) = Q(L) \tag{5}$$

By use of the Poincaré–Birkhoff–Witt theorem the map of modules

$$i: L \to U(L) \text{ is injective and has a natural retraction } r: U(L) \to L, \tag{6}$$

provided R contains $\mathbb{Q}$, see the remark 3.8 of appendix B in Quillen (1969). For further properties of U we refer the reader to Quillen (1969) and Milnor–Moore.

We say that $L(V)$ is a **free Lie algebra** if V is a free R-module. By the universal properties one readily gets

(9.4) $$U(L(V)) = T(V).$$

If the coefficient ring R contains $\mathbb{Q}$ then the injection (9.3)(6) shows that $L(V)$ is the sub Lie algebra of the tensor algebra $T(V)$ generated by V.

The initial and final object in the category of Lie algebras is the trivial module $*$ which is 0 in each degree. Let $L \amalg L'$ be the **free product** of Lie algebras L and L', this is the push out of $L \leftarrow * \rightarrow L'$ in the category of Lie algebras. The universal properties imply the formulas:

(9.5) $$L(V) \amalg L(V') = L(V \oplus V')$$

and

(9.6) $$U(L \amalg L(V)) = (UL) \amalg T(V).$$

Therefore (7.5) yields a good formula for $L \amalg L(V)$ by the inclusion $L \amalg L(V) \subset (UL) \amalg T(V)$ provided we assume that R contains $\mathbb{Q}$, see (9.3)(6); in fact, in this case $L \amalg L(V)$ is the sub Lie algebra of $(UL) \amalg T(V)$ generated by L and V.

(9.7) ***Definition.*** A **chain Lie algebra** L is a graded Lie algebra together with a differential $d: L \to L$ of degree -1 such that (L, d) is a chain complex and such that

$$[\ ,\ ]: L \otimes L \to L$$

is a chain map, see (6.7), that is

$$d[x, y] = [dx, y] + (-1)^{|x|}[x, dy].$$

A *map* between chain Lie algebras is a map between Lie algebras which is also a chain map. Let **DL** be the category of chain Lie algebras. ||

The homology of a chain Lie algebra L is a Lie algebra with the bracket

(9.8) $$[\ ,\ ]: HL \otimes HL \xrightarrow[j]{} H(L \otimes L) \xrightarrow[[\ ,\ ]_*]{} HL,$$

compare (7.7). Here j is an isomorphism by the Künneth theorem provided R is a field. From (9.3)(3) we derive

(9.9) **Lemma.** *For a chain Lie algebra L there is a unique differential d on $U(L)$ such that $(U(L), d)$ is an augmented chain algebra and such that $L \to U(L)$ is a chain map.*

For a chain Lie algebra L the natural retraction $r: U(L) \to L$ in (9.3)(6) is a chain map between chain complexes, (here we assume that R contains $\mathbb{Q}$), compare 3.6 in appendix B of Quillen (1969).

Using the Poincaré–Birkhoff–Witt theorem and the Künneth formula Quillen (1969), appendix B, shows:

(9.10) **Theorem.** *Let R be a field of characteristic zero then the natural map $U(H(L)) \to H(U(L))$ is an isomorphism.*

Similarly as in (7.8) we define the following structure for **DL**.

(9.11) ***Definition.***

(1) A map $f: B \to A$ in **DL** is a **weak equivalence** if f induces an isomorphism $f_*: HB \cong HA$ in homology.
(2) A map $B \to A$ in **DL** is a **cofibration** if there is a submodule V of A with the following properties:
    (a) V is a free module, and
    (b) the map $B \amalg L(V) \to A$ of Lie algebras, given by $B \to A$ and $V \subset A$, is an isomorphism.

We call V a **module of generators for** $B \rightarrowtail A$. The cofibration $B \rightarrowtail A$ is **elementary** if $d(V) \subset B$. The cofibrant objects in **DL** are the 'free' chain Lie algebras. ||

(9.12) ***Push outs in DL.*** For the cofibration $B \rightarrowtail A = (B \amalg L(W), d)$ and for $f: B \to Y$ the induced cofibration $Y \rightarrowtail A \bigcup_B Y$, given by

$$\begin{array}{ccc} A & \xrightarrow{\bar{f}} & A \bigcup_B Y = (Y \amalg L(W), d) \\ {\scriptstyle i}\uparrow & \text{push} & \uparrow{\scriptstyle \bar{i}} \\ B & \xrightarrow[f]{} & Y \end{array},$$

is generated by W and $\bar{f} = f \amalg 1$ is the identitity on W. The differential d on $A \bigcup_B Y$ is the unique differential for which $\bar{f}$ and $\bar{i}$ are chain maps. In case i is an elementary cofibration we obtain d on $A \bigcup_B Y$ by $d: W \xrightarrow{d} B \xrightarrow{f} Y$.

(9.13) **Theorem.** *Suppose the ring R of coefficients is a field of characteristic zero. Then the category* **DL** *with the structure* (9.11) *is a cofibration category for which all objects are fibrant. Moreover, the universal enveloping functor* $U:\mathbf{DL}\to\mathbf{DA}_*$ *is a based model functor which carries cofibrations to cofibrations.*

We prove this result in (9.16) below. As in (7.13) we define

(9.14) ***Definition of*** $L(V,dV)$. Let V be a positive graded module with $V_0=0$ and let dV with $sdV=V$ and $d:V\to dV$ be given as in (7.13). Then we obtain the object

$$L(V,dV)=(L(V\oplus dV),d) \tag{1}$$

in **DL** by defining the differential d on generators via $d:V\to dV$, $ddV=0$. Note that

$$UL(V,dV)=T(V,dV), \tag{2}$$

compare (7.13), and that

$$*\rightarrowtail L(dV)\rightarrowtail L(V,dV) \tag{3}$$

are cofibrations in **DL** (where $L(dV)$ has trivial differential). Moreover, $L(V,dV)$ has the following universal property. Let A be an object in **DL** and let $\varphi:V\to A$ be a map of degree 0 between graded modules. Then there is a unique map

$$\bar{\varphi}:L(V,dV)\to A \quad \text{in } \mathbf{DL} \tag{4}$$

which extends φ, that is $\bar{\varphi}|V=\varphi$. ||

By the next lemma we see that $L(V,dV)=CL(dV)$ is a **cone** for the chain Lie algebra $L(dV)$ provided the ring R of coefficients is nice:

(9.15) Lemma. *Let R be a field of characteristic zero then $L(V,dV)$ is acyclic, that is, $*\to L(V,dV)$ is a weak equivalence in* **DL**.

Proof. We use (7.14) and (9.10), so that

$$\begin{aligned} UH(L(V,dV)) &= HU(L(V,dV)) \\ &= HT(V,dV)=R, \end{aligned}$$

and hence $H(L(V,dV))=0$. □

(9.16) *Proof of* (9.13). The composition axiom is clearly satisfied, see (9.5), and push outs as in (C2) exist by (9.12). We now prove (C2)(a) by use of (9.10) and by use of (C2)(a) in $\mathbf{DA}_*$. Here we use the fact that

$$UB\rightarrowtail UA=(UB\textstyle\coprod T(W),d) \tag{1}$$

is a cofibration in $\mathbf{DA}_*$, so that a weak equivalence $f:B\xrightarrow{\sim}Y$ in **DL** yields a weak equivalence

$$U(\bar{f}): U(A) \xrightarrow{\sim} U\left(A \bigcup_B Y\right) \quad \text{in } \mathbf{DA}_*. \tag{2}$$

Now (9.10) shows that $A \to A \bigcup_B Y$ is a weak equivalence since we can use (9.3)(6). For the proof of (C3) we use the same inductive construction as in the proof of (7.21). Moreover, we show that all objects in **DL** are fibrant by the same arguments as in (7.22) where we replace $T(W, dW)$ by $L(W, dW)$ and where we use (9.15). This completes the proof of (9.13). □

(9.17) ***Remark.*** Let $\mathbf{DL}_1$ be the full subcategory of **DL** consisting of chain Lie algebras which are 1-reduced. Then Quillen (1969) shows with similar arguments as above that $\mathbf{DL}_1$ is a closed model category for which cofibrations and weak equivalences are defined as in (9.11) and for which fibrations are given by maps which are surjective in degree $\geqq 2$.

Next we define for each cofibration

$$i: B \rightarrowtail A = (B \coprod L(V), d_A) \quad \text{in } \mathbf{DL}$$

an explicit **cylinder object**

(9.18)
$$A \bigcup_B A \overset{i_0, i_1}{\rightarrowtail} I_B A \underset{\sim}{\overset{p}{\longrightarrow}} A$$

in the cofibration category **DL** in (9.13), we assume that R is a field of characteristic zero. The construction is similar to the one in (8.19). For sV, see § 6, let $L(sV, dsV)$ be the acyclic cone in (9.14) with $dsV = V$ and let

$$\begin{aligned} I_B A &= (A \coprod L(sV, dsV), d) \\ &= A \vee L(sV, dsV) \end{aligned} \tag{1}$$

be the sum in **DL**. The weak equivalence p in (9.18) is defined by $p = (1, 0)$ where $0: L(sV, dsV) \to *$. We define i_0 as in (9.18) by the canonical inclusion $A \subset A \vee L(sV, dsV)$. Clearly, $pi_0 = 1$. The definition of i_1 is more complicated. We note that as a Lie algebra

$$I_B A = B \coprod L(V \oplus sV \oplus dsV). \tag{2}$$

Therefore we can define a degree $+1$ derivation S of this Lie algebra by

$$\left.\begin{aligned} &S(B) = S(sV) = S(dsV) = 0, \\ &S(v) = sv \quad \text{for } v \in V, \\ &S([x, y]) = [Sx, y] + (-1)^{|x|}[x, Sy]. \end{aligned}\right\} \tag{3}$$

Now a degree zero derivation θ on $I_B A$ is given by

$$\theta = dS + Sd, \tag{4}$$

where d is the differential in (1). Note that

$$d\theta = \theta d, \quad pS = 0, \quad p\theta = 0. \tag{5}$$

Lemma. *For each $\phi \in I_B A$ there is some N (depending on ϕ) with $\theta^N(\phi) = \theta \cdots \theta(\phi) = 0$.* (6)

One can prove this lemma along the same lines as in (8.19)(8). By (6), (5) we can define an automorphism $e^\theta = \exp(\theta)$ in **DL** of $I_B A$, namely

$$e^\theta = \sum_{N=0}^{\infty} \frac{\theta^N}{N!}; \tag{7}$$

the inverse is $e^{-\theta}$. To this end we can define i_1 in (9.18) by the composition

$$i_1 = e^\theta i_0 = (\exp(dS + Sd))i_0 \tag{8}$$

This completes the definition of the cylinder object (9.18) provided we can show

Lemma. *The map (i_0, i_1) is a cofibration in* **DL**. (9)

Proof. We have

$$A \bigcup_B A = (B \coprod L(V' \oplus V''), d), \tag{10}$$

where $V' = V'' = V$ are isomorphic copies of V. We obtain an isomorphism g of Lie algebras for which the following diagram commutes and for which $g|sV$ is the canonical inclusion $sV \subset I_B A$ given by (1).

$$\begin{array}{ccc} B\coprod L(V' \oplus V'' \oplus sV) & \xrightarrow[g]{\cong} & I_B A \\ \uparrow j & & \uparrow (i_0, i_1) \\ B\coprod L(V' \oplus V'') & = & A \bigcup_B A \end{array} \tag{11}$$

Here j is the canonical inclusion. Note that (i_0, i_1) is a well-defined map since $i_0|B = i_1|B$. We now show inductively that g is an isomorphism. Assume this is true for $I_B A_{<n}$ where $A_{<n} = (B \coprod L(V_{<n}), d)$ is a sub Lie algebra of A given has $V_{<n} = \{V_i, i < n\}$. Then for $v \in V_n$ we know $i_1(v) - v - dsv = w \in I_B A_{<n}$. Hence we can define the inverse g^{-1} by $g^{-1}v = v'$, $g^{-1}sv = sv$ and

$$g^{-1}(dsv) = v'' - v' - g^{-1}(w). \tag{12}$$

We can use g to define a differential d on $B \coprod L(V' \oplus V'' \oplus sV)$ so that g is an isomorphism of chain Lie algebras. □

For the field $R = \mathbb{Q}$ of rational numbers there is the Quillen functor

$$\lambda : \mathbf{Top}_1^* \to \mathbf{DL}_1. \tag{9.19}$$

Here $\mathbf{Top}_1^*$ is the full subcategory of $\mathbf{Top}^*$ of simply connected spaces and $\mathbf{DL}_1$ is the full subcategory of $\mathbf{DL}$ consisting of 1-reduced chain Lie algebras. Let $Ho_{\mathbb{Q}}\mathbf{Top}_1^*$ be the localization of $\mathbf{Top}_1^*$ with respect to $H_*(-,\mathbb{Q})$-equivalences, see (5.10) and (II.3.5). Then Quillen (1969) proves that λ induces an equivalence of localized categories

$$Ho(\lambda) : Ho_{\mathbb{Q}}(\mathbf{Top}_1^*) \xrightarrow{\sim} Ho(\mathbf{DL}_1). \tag{9.20}$$

This corresponds to the result of Sullivan in (8.27) above, compare Neisendorfer. In Chapter IX we give a new proof for the equivalence $Ho(\lambda)$ (restricted to spaces with finite dimensional rational homology). Our proof relies only on the properties of the functor λ described in the following theorem which is due to Quillen.

(9.21) **Theorem.** *The Quillen functor λ in (9.19) is a based model functor between cofibration categories and induces a bijection*

$$\lambda : [S^n_{\mathbb{Q}}, X_{\mathbb{Q}}] \xrightarrow{\approx} [\lambda S^n_{\mathbb{Q}}, \lambda X_{\mathbb{Q}}] \cong H_{n-1}\lambda X_{\mathbb{Q}}$$

of homotopy sets.

Here $S^n_{\mathbb{Q}}$ is the rational n-sphere, $n \geqq 2$, and $X_{\mathbb{Q}}$ is a rational space. We show in (IX.§3) that any functor λ which has the properties in (9.21) induces an isomorphism of towers of categories which approximate $Ho_{\mathbb{Q}}(\mathbf{Top}_1^*)$ and $Ho(\mathbf{DL}_1)$ respectively. We can derive (9.21) and the next result from Quillen (1969), see (II.4.1).

(9.22) **Theorem.** *There is a natural weak equivalence of functors $U\lambda \sim SC_*\Omega(\cdot) \otimes \mathbb{Q}$ in $\mathbf{DA}_*$ where U is the universal enveloping functor and where $SC_*\Omega(\cdot) \otimes \mathbb{Q}$ is the rational chains on the loop space functor (7.29).*

By (9.21) and (9.22) we obtain the commutative diagram of degree 0 maps

$$\begin{array}{ccc} \pi_*(\Omega X) \otimes \mathbb{Q} & \overset{\lambda}{\cong} & H_*(\lambda X_{\mathbb{Q}}) \\ \downarrow{\scriptstyle h} & & \downarrow{\scriptstyle i_*} \\ H_*(\Omega X, \mathbb{Q}) & \overset{\bar{\lambda}}{\cong} & H_*(U\lambda X_{\mathbb{Q}}). \\ \| & & \| \\ U(\pi_*(\Omega X) \otimes \mathbb{Q}) & \underset{U(\lambda)}{\cong} & UH_*(\lambda X_{\mathbb{Q}}) \end{array} \tag{9.23}$$

Here h is the Hurewicz map and i is the inclusion (9.3)(6). The isomorphism λ of Lie algebras is induced by the bijection in (9.21) and the isomorphism $\bar{\lambda}$ of algebras is induced by the natural weak equivalence in (9.22). Now (9.10) gives us the isomorphism in the columns which corresponds to the Milnor–Moore theorem.

II

Homotopy theory in a cofibration category

Much of the standard and classical homotopy theory for topological spaces can be deduced from the axiom of a cofibration category. We derive in this chapter basic facts of homotopy theory from the axioms. We introduce homotopy groups, the action of the fundamental group, homotopy groups of function spaces, and homotopy groups of pairs. Moreover, we describe the fundamental exact sequences for these groups and we prove the naturality of these exact sequences with respect to functors between cofibration categories. This leads further than the results previously obtained in the literature. We deduce from the axioms of a cofibration category various results which are new in topology.

§ 1 Some properties of a cofibration category

Let **C** be a fixed cofibration category. In the commutative diagram of unbroken arrows in **C**

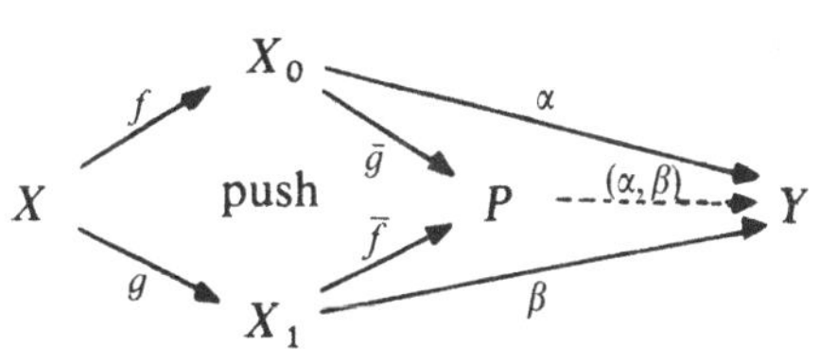

the subdiagram 'push' is called **cocartesian** or a **push out** if for every Y and every pair α, β there exists exactly one map (α, β) in **C** extending the diagram commutatively. We also write $P = X_0 \bigcup_X X_1$ if the choice of f and g in (1.1) is clear from the context. Consider the commutative diagrams

$$
\begin{array}{ccc}
W & \xrightarrow{f} & W' \\
{\scriptstyle\beta}\uparrow & D_2 & \uparrow{\scriptstyle\beta'} \\
V & \longrightarrow & V' \\
{\scriptstyle\alpha}\uparrow & D_1 & \uparrow{\scriptstyle\alpha'} \\
U & \xrightarrow[g]{} & U'
\end{array}
\qquad\qquad
\begin{array}{ccc}
W & \xrightarrow{f} & W' \\
{\scriptstyle\beta\alpha}\uparrow & D_3 & \uparrow{\scriptstyle\beta'\alpha'} \\
U & \xrightarrow[g]{} & U'
\end{array} .
$$

(1.1)(a) Let D_1 be a push out. Then D_2 is a push out if and only if D_3 is a push out.

(1.1)(b) Let D_1 be a homotopy push out. Then D_2 is a homotopy push out if and only if D_3 is a homotopy push out.

We leave the proof of (1.1)(a) and (1.1)(b) as an exercise. In fact, (1.1)(b) is an easy consequence of (*b*) in the next lemma (1.2). Consider the commutative diagram

$$
\begin{array}{ccccc}
Y_0 & \xleftarrow{f'} & Y & \xrightarrow{g'} & Y_1 \\
{\scriptstyle\alpha}\downarrow & & \downarrow{\scriptstyle\gamma} & & {\scriptstyle\beta}\downarrow \\
X_0 & \xleftarrow[f]{} & X & \xrightarrow[g]{} & X_1
\end{array} ,
$$

where in each row one of the maps is a cofibration. Then the push outs of the rows exist in **C** by (C2). We get the map $\alpha\cup\beta\colon Y_0\bigcup_Y Y_1\to X_0\bigcup_X X_1$ with $\alpha\cup\beta=(\bar{g}\alpha,\bar{f}\beta)$ which satisfies the following *gluing lemma.*

(1.2) **Lemma**

(*a*) *Assume* α,β,γ *and the induced map* $(g,\beta)\colon X\bigcup_Y Y_1\to X_1$ *are cofibrations, then also* $\alpha\cup\beta$ *is a cofibration.*

(*b*) *If the columns* α,β,γ *are weak equivalences then also* $\alpha\cup\beta$ *is a weak equivalence.*

Remark. For the cofibration categpry **Top** the result (1.2) (*b*) is proved by Brown–Heath (1970). This is just an example for the numerous results in the literature which are covered by the abstract approach.

Proof. Consider the following diagram in which all squares are push outs and in which $P'\to P$ is the map $\alpha\cup\beta$.

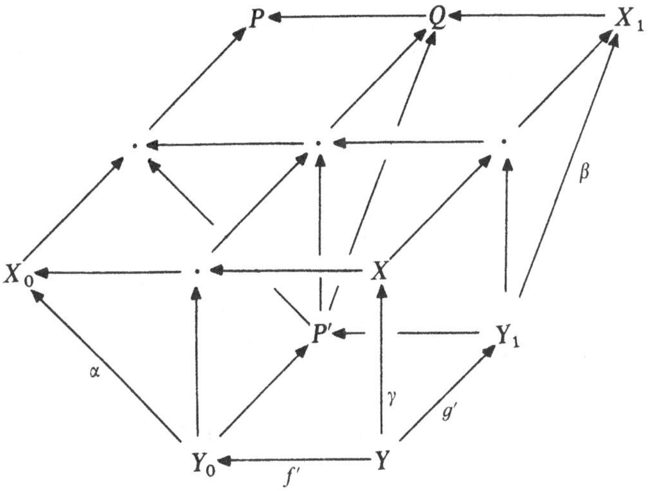

For (*a*) the diagram is well defined, for (*b*) the diagram is well defined if f' and g' are cofibrations. A diagram chase shows that in this case (*a*) and (*b*) hold.

Next we assume for the proof of (*b*) that $Y \to Y_1$ is not a cofibration. Then we have the factorization $Y \rightarrowtail \bar{Y}_1 \xrightarrow{\sim} Y_1$ by (C3). This leads to the commutative diagram

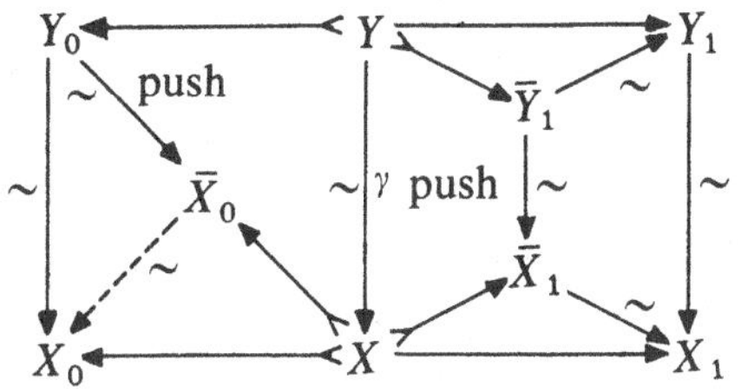

and thus to the commutative diagram

$$\begin{array}{ccc} Y_0 \bigcup\limits_{Y} \bar{Y}_1 & \xrightarrow{\ r\ } & Y_0 \bigcup\limits_{Y} Y_1 \\ {\scriptstyle s}\big\downarrow & & \big\downarrow{\scriptstyle \alpha\cup\beta} \\ \bar{X}_0 \bigcup\limits_{X} \bar{X}_1 & \xrightarrow{\ t\ } & X_0 \bigcup\limits_{X} X_1 \end{array} .$$

Therefore we have to show that r, s and t are weak equivalences. This is proved above. □

(1.3) **Definition.** Let **Pair (C)** be the following category. Objects are morphisms $i_X : Y \to X$ in **C**. We denote i_X also by $i_X = (X, Y)$, in particular, if i_X is a cofibration in **C**. Morphisms $(f, f') = f : i_A \to i_X$ are commutative diagrams

$$\begin{array}{ccc} A & \xrightarrow{f} & X \\ {\scriptstyle i_A}\uparrow & & \uparrow{\scriptstyle i_X} \\ B & \xrightarrow{f'} & Y \end{array}$$

in **C**. The morphism (f, f') is a **weak equivalence** if f and f' are weak equivalences in **C**. Moreover, (f, f') is a **cofibration** if f' and $(f, i_X) : A \bigcup_B Y \to X$ are cofibrations in **C**. We call (f, f') a **push out** if the diagram is a push out diagram with $i_A : B \rightarrowtail A$ a cofibration. ||

We also consider the following subcategory of **Pair (C)**.

(1.4) ***Definition.*** Let Y be an object in **C**. A **map** f **under** Y is a commutative diagram

$$\begin{array}{ccc} & Y & \\ \swarrow & & \searrow \\ X & \xrightarrow[f]{} & A \end{array}$$

in **C**. Let $\mathbf{C}^Y$ be the category of maps under Y. Objects are the maps $Y \to X$. Cofibrations and weak equivalences in $\mathbf{C}^Y$ are the same as in **C**. The identity of Y is the initial object in $\mathbf{C}^Y$. Thus an object $(Y \to X)$ in $\mathbf{C}^Y$ is cofibrant iff $Y \to X$ is a cofibration in **C**. An object $(Y \to X)$ is fibrant iff X is fibrant in **C**. With these notations one easily verifies that $\mathbf{C}^Y$ has the structure of a cofibration category with an initial object. ||

Next we prove for **Pair (C)** the *relativization lemma*:

(1.5) **Lemma.** *The category* **Pair (C)** *with cofibrations and weak equivalences as in* (1.3) *is a cofibration category. An object* $i_A : B \to A$ *is fibrant in* **Pair (C)** *iff* B *and* A *are fibrant in* **C**.

If **C** has the initial object ϕ, then also **Pair (C)** has an initial object given by the identity of ϕ. The object $i_X : Y \to X$ is cofibrant in **Pair (C)** iff $\phi \rightarrowtail Y \rightarrowtail X$ are cofibrations in **C**.

For the proof of (1.5) we use the following *extension property of fibrant models*:

(1.6) **Lemma.** *Let* i *and* f *be given as in the diagram*

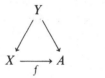

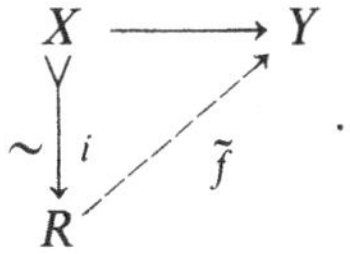

Then, if Y is fibrant, there is $\tilde{f}$ with $\tilde{f}i = f$. Moreover, two extensions $\tilde{f}, \tilde{f}_1$ of f are homotopic rel X.

Proof. We consider the commutative diagram of unbroken arrows

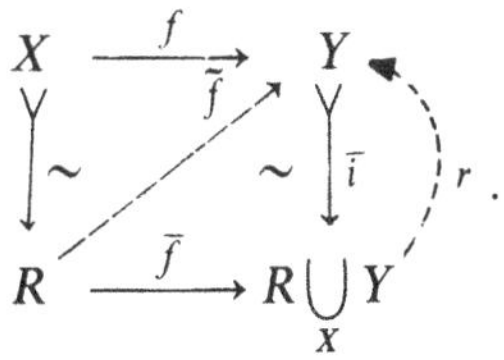

Since Y is fibrant there is a retraction r for $\bar{i}$, let $\tilde{f} = r\bar{f}$.

Now let $\tilde{f}, \tilde{f}_1$ be extensions of f. We have for a cylinder Z on $X \rightarrowtail R$ the diagram

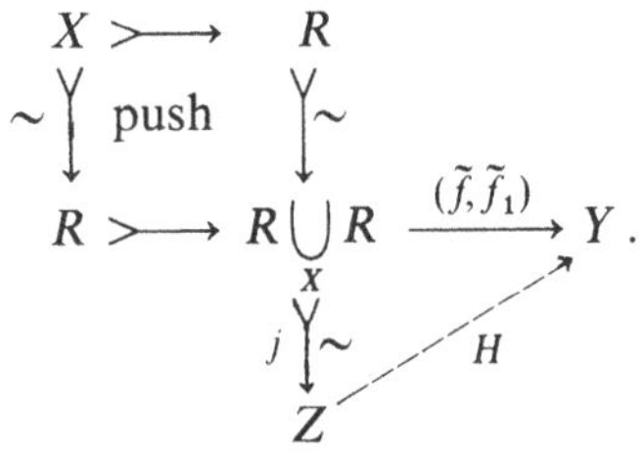

Here j is a weak equivalence since $i_0 : R \to Z$ is one, see (I.1.5). We now obtain the extension H in the same way as above. □

(1.7) *Proof of* (1.5). We have to check the axioms (C1),...,(C4). First we observe that a cofibration (i, i') in **Pair (C)** corresponds to the commutative diagram in C

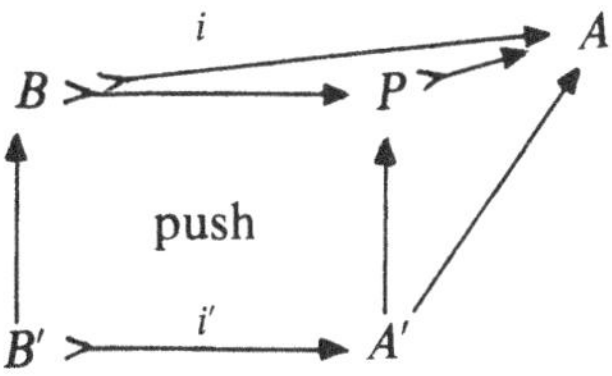

(1)

in which i, i' are cofibrations. It is clear that the composition axiom (C1) is satisfied in **Pair (C)**, see (1.2).

For the proof of the push out axiom (C2) in **Pair (C)** consider maps

$$(X_0, Y_0) \longleftarrow\!\!\!< (X, Y) \longrightarrow (X_1, Y_1) \text{ in } \textbf{Pair (C)}, \qquad (2)$$

as in (1, 2). The diagram in the proof of (1, 2) shows that (P', P) is the push out of (2) in **Pair (C)** with $P = X_0 \bigcup_X X_1$ and $P' = (Y_0 \bigcup_Y Y_1)$. Moreover, the cofibration $Q \rightarrowtail P$ in this diagram shows that $(X_1, Y_1) \rightarrow (P, P')$ is a cofibration in **Pair (C)**. It is easy to check that $(X_0, Y_0) \rightarrow (P, P')$ is a weak equivalence if $(X, Y) \rightarrow (X_1, Y_1)$ is one. Hence (C2) is proved.

Next we prove (C3) in **Pair (C)**. For (f, f') we consider the diagram

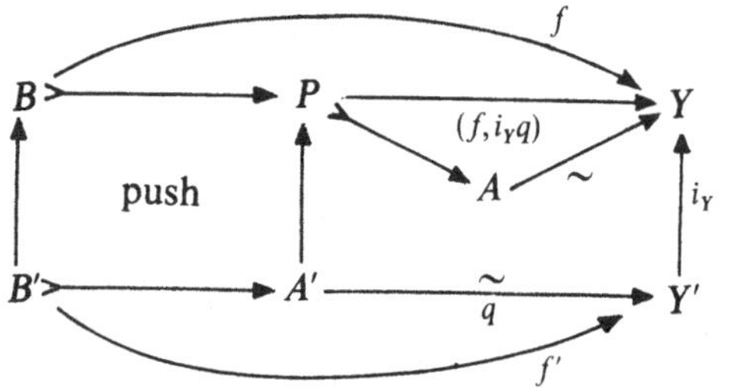

$$\tag{3}$$

where A' and A are given by factorization in **C**, (C3). This shows that (A, A') is a factorization for (f, f').

Also the axiom on fibrant objects (C4) is satisfied in **Pair (C)**: In the diagram

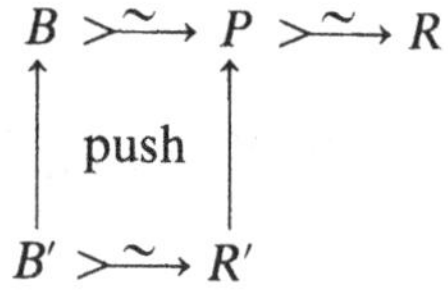

$$
\begin{array}{ccc}
B \overset{\sim}{\rightarrowtail} P \overset{\sim}{\rightarrowtail} R \\
\uparrow \quad \text{push} \quad \uparrow \\
B' \overset{\sim}{\rightarrowtail} R'
\end{array}
\qquad , \qquad (4)
$$

we choose fibrant models R' and R in **C**. Then the pair (R, R') is a fibrant model of (B, B'). We have to check that (R, R') is fibrant in **Pair (C)**. To this end we prove the proposition on fibrant objects in (1.5): Let A and B be fibrant in **C** and let

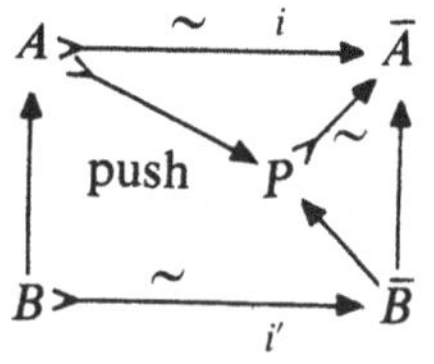

$$
\begin{array}{ccc}
A \overset{\sim}{\underset{i}{\rightarrowtail}} \bar{A} \\
\uparrow \quad \text{push} \quad P \nearrow \\
B \overset{\sim}{\underset{i'}{\rightarrowtail}} \bar{B}
\end{array}
\qquad (5)
$$

be a trivial cofibration $i = (i, i')$ in **Pair (C)**. We have to show that there is a retraction $r = (r, r') : (\bar{A}, \bar{B}) \rightarrow (A, B)$ of i.

First we choose a retraction r' of i' in **C**. This is possible since B is fibrant in C, see the definition in (I.1.1). Then we obtain the following commutative diagram

$$\begin{array}{ccccc} & & \bar{A} & & \\ & {\scriptstyle\sim}\nearrow & & \searrow{\scriptstyle r} & \\ A & \rightarrowtail & P & \xrightarrow{q} & A \\ \uparrow & \text{push} & \uparrow & & \uparrow{\scriptstyle i_A} \\ B & \overset{i'}{\rightarrowtail} & \bar{B} & \xrightarrow{r'} & B \end{array},$$

where $q=(1_A, i_A r')$ and where $r=q^{\sim}$ is an extension of q by (1.6). Thus (r,r') is a retraction for $i=(i,i')$.

Next assume (A,B) is fibrant in **Pair (C)** and let $\alpha: A \overset{\sim}{\rightarrowtail} \bar{A}$ and $\beta: B \overset{\sim}{\rightarrowtail} \bar{B}$ be trivial cofibrations in **C**. We obtain retractions for α and β in **C** since we have retractions for the trivial cofibrations

$$(\bar{\beta},\beta):(A,B) \overset{\sim}{\rightarrowtail} \left(A \bigcup_B \bar{B}, \bar{B}\right),$$

$$(\alpha,1):(A,B) \overset{\sim}{\rightarrowtail} (\bar{A},B),$$

in **Pair (C)**. This shows that A and B are fibrant in **C**. Now the proof of (1.5) is complete. □

In *addition to the factorization axiom* (C3) we get

(1.8) **Lemma.** *Let Y be fibrant. Then for a map $f:B \to Y$ there is a factorization $f:B \rightarrowtail A \xrightarrow{\sim} Y$ of f where A is fibrant.*

In particular, for a cofibration $Y \subset X$ we can choose a cylinder $I_Y X$ *which is fibrant if* X is fibrant.

Proof. By (C3) we have a factorization $\bar{A}$ of f. By (C4) we have a fibrant model $j_0:\bar{A} \overset{\sim}{\rightarrowtail} A$ and by (1.6) we obtain the commutative diagram

$$\begin{array}{cccc} & A & & \\ & {\scriptstyle\sim}\uparrow & \searrow{\scriptstyle p} & \\ f:B \rightarrowtail & \bar{A} & \xrightarrow{\sim} & Y \end{array},$$

where p is a weak equivalence by (C1). □

(1.9) **Lemma.** *Let $u,v:X \to U$ be maps and let $u \simeq v$ rel Y. If u is a weak equivalence so is v.*

Proof. We apply (C1) twice to the commutative diagram

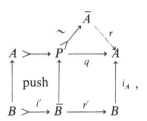

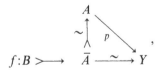

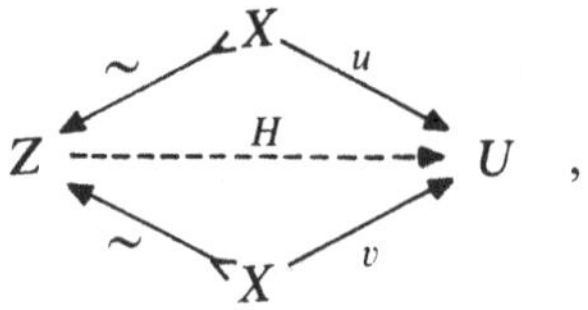

where we use (I.1.5).   □

(1.10) **Weak lifting lemma.** *Any commutative diagram*

$$\begin{array}{ccc} B & \longrightarrow & X \\ \downarrow & & \downarrow \sim \\ A & \longrightarrow & Y \end{array}$$ (*)

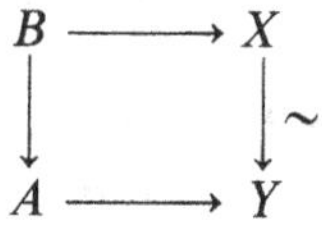

can be embedded in a commutative diagram

(**)

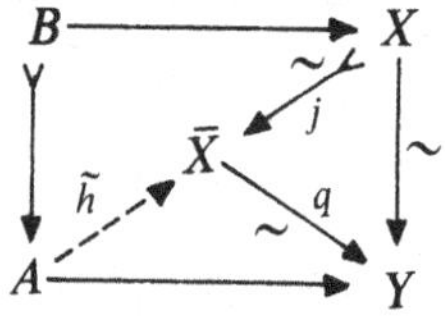

We call the pair $L = (\tilde{h}, j)$ in (**) a **weak lifting** for diagram (*). The map $\tilde{h}$ is a cofibration provided $B \to X$ is a cofibration.

Proof. Apply (C3) to the map $A \bigcup_B X \to Y$ which is defined by (*) and use (C1).
                                              □

(1.11) **Lifting lemma.** *Consider the commutative diagram of unbroken arrows:*

$$\begin{array}{ccc} B & \xrightarrow{\ f\ } & X \\ \downarrow{\scriptstyle i} & {\scriptstyle h}\nearrow & \downarrow{\scriptstyle \sim}\ p \\ A & \xrightarrow{\ g\ } & Y \end{array} \ .$$

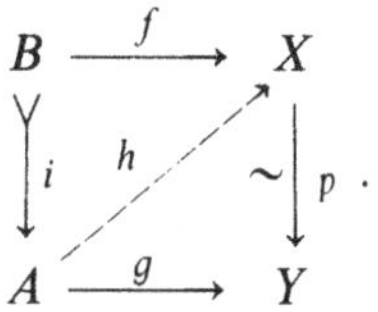

(a) *If X is fibrant there is a map h for which the upper triangle commutes.*

(b) *If X and Y are fibrant there is a map h for which the upper triangle commutes and for which ph is homotopic to g rel B. We call a map h with these properties a **lifting** for the diagram.*

(c) *If X and Y are fibrant a lifting of the diagram is unique up to homotopy rel B.*

Proof. Since X is fibrant there is a retraction $r: \bar{X} \xrightarrow{\sim} X$ for $j: X \underset{\sim}{\rightarrowtail} \bar{X}$ in (**)

of (1.10). Let $h = r\tilde{h}$. This proves (*a*). Now assume also that X and Y are fibrant. With the notation in (1.10) we have to prove

$$ph = pr\tilde{h} = qjr\tilde{h} \simeq q\tilde{h} = g \text{ rel } B. \tag{1}$$

We consider the commutative diagram of unbroken arrows:

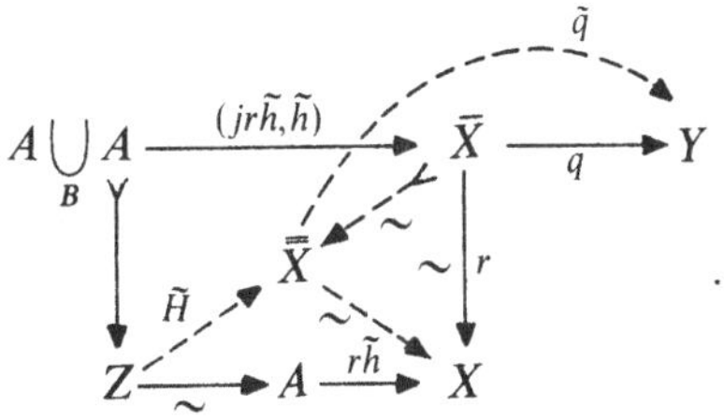

(2)

Here Z denotes a cylinder on $B \subset A$. As in (1.10) we obtain the weak lifting $\tilde{H}$ and by (1.6) we have the map $\tilde{q}$. Thus $H = \tilde{q}\tilde{H}$ is a homotopy as in (1). This proves (*b*). We prove (*c*) in (2.6). □

The following result is an easy consequence of (1.11):

(1.12) **Corollary.** *Let $i: Y \rightarrowtail X$ be a cofibration and a weak equivalence between fibrant objects. Then Y is a* **deformation retract** *of X. That is, there is a retraction $r: X \to Y$ with $r | Y = 1_Y$ and with $ir \simeq 1_X$* rel Y.

Proof. Consider the diagram

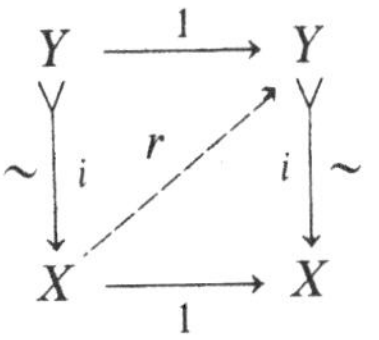

and apply (1.11). □

(1.13) **Corollary.** *Let Y be fibrant and let*

$$B \underset{i}{\rightarrowtail} A_1 \underset{p}{\overset{\sim}{\twoheadrightarrow}} Y, \quad B \underset{j}{\rightarrowtail} A_2 \underset{q}{\overset{\sim}{\twoheadrightarrow}} Y$$

be factorizations of a given $f: B \to Y$ such that A_1 and A_2 are fibrant, see (1.8). *Then there is up to homotopy* rel B *a unique weak equivalence $\alpha: A_1 \to A_2$ with $\alpha i = j$ and $q\alpha \simeq p$* rel B.

By (2.12) below we know that α in (1.13) is also a homotopy equivalence under B.

(1.14) **Corollary.** *A retract of a fibrant object is fibrant.*

Proof. Consider a push out as in the proof of (I.2.6) and use the same argument as in (I.2.6) or use (1.6). □

§2 Sets of homotopy classes

For a cofibration $Y \subset X$ and a map $u: Y \to U$ let

$$\text{Hom}(X, U)^u \tag{2.1}$$

be the set of all maps $f: X \to U$ in **C** for which $f | Y = u$. We say f is an **extension** of u. On this set of extensions of u we have as in I, §1 the homotopy relation relative Y which we denote by '$\simeq$ rel Y'.

(2.2) **Proposition.** *Let U be fibrant. Then all cylinders on $Y \subset X$ define the same homotopy relation relative Y on the set* (2.1). *Moreover, the homotopy relation relative Y is an equivalence relation.*

Thus, if U is fibrant, we have the set

(2.3)(a) $$[X, U]^Y = [X, U]^u = \text{Hom}(X, U)^u / \simeq \text{rel } Y$$

of homotopy classes. We write $\{f\}$ for the homotopy class of f and we write $[X, U]^Y$ if the choice of u is clear from the context.

For an initial object ϕ of **C** let

(b) $$[X, U] = [X, U]^\phi = \text{Hom}(X, U) / \simeq \text{rel } \phi$$

be the set of all homotopy classes of maps from X to U. Here we assume that X is cofibrant and that U is fibrant. For the objects $i_X: Y \rightarrowtail X$ and $u: Y \to U$ in $\mathbf{C}^Y$ we have by (1.4)

(c) $$[X, U]^Y = [i_X, u].$$

Proof of (2.2). Let Z_1 and Z_2 be cylinders and assume that there is a homotopy $H_1: \bar{u} \simeq \bar{v}$ rel Y which is defined on Z_1. Then we obtain a homotopy $H_2: \bar{u} \simeq \bar{v}$ rel Y defined on Z_2, as follows: We apply (1.10) and (1.6) to the diagram

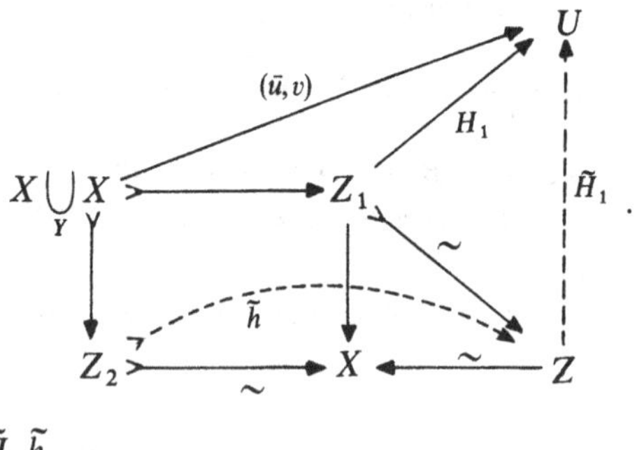

(*)

and we set $H_2 = \tilde{H}_1 \tilde{h}$.

We now show that $(\simeq \operatorname{rel} Y)$ is an equivalence relation. Clearly, $\bar{u} \simeq \bar{u} \operatorname{rel} Y$ since $\bar{u}p: Z \to X \to U$ is a homotopy $\bar{u} \simeq \bar{u} \operatorname{rel} Y$. Moreover, we apply (1.10) and (1.6) to the diagram

(2.4)

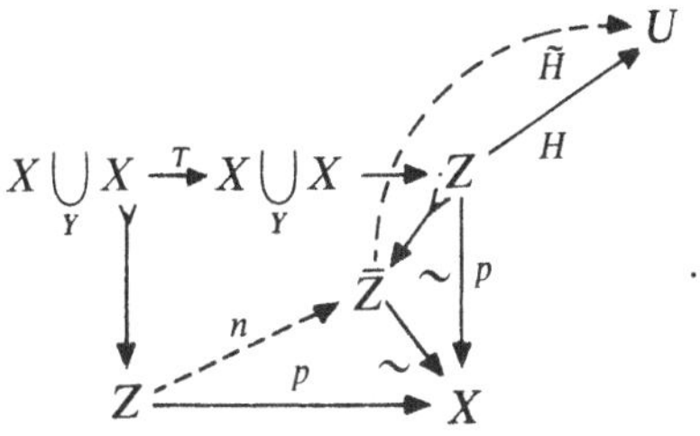

Here T is the interchange map of the two factors X. For a homotopy $H: \bar{u} \simeq \bar{v} \operatorname{rel} Y$ the composition $\tilde{H}n$ is a homotopy $\bar{v} \simeq \bar{u} \operatorname{rel} Y$. We call $-H = \tilde{H}n$ *a negative* of the homotopy H.

We now consider the push out

$$\begin{array}{ccccc} Z & \xrightarrow[\sim]{\bar{i}_0} & Z \bigcup_X Z & \xrightarrow[\sim]{(p,p)} & X \\ \uparrow i_1 & & \uparrow & & \\ X & \xrightarrow[i_0]{\sim} & Z & & \end{array} .$$

Since i_0 is a weak equivalence also $\bar{i}_0$ is one by (C2). Thus by (C1) the map (p, p) is a weak equivalence since $p = (p, p)\bar{i}_0$. For homotopies $H: \bar{u} \simeq \bar{v}$ and $G: \bar{v} \simeq \bar{w}$ we obtain the next diagram in which we apply again (1.10) and (1.6):

(2.5)

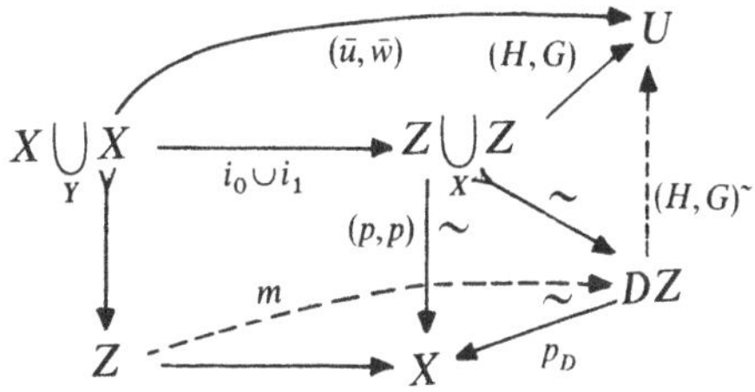

$H + G = (H, G)^{\sim} m$ is a homotopy $\bar{u} \simeq \bar{w} \operatorname{rel} Y$. This proves the proposition. □

(2.6) *Proof of* (1.11) (*c*). Let h and $\bar{h}$ be liftings and let H and G be homotopies relative B from ph to g and from g to $p\bar{h}$ respectively. We apply (1.11)(a) to the diagram

$$\begin{array}{ccc} A\bigcup_B A & \xrightarrow{(h,\bar h)} & X \\ \downarrow{\scriptstyle i_0\cup i_1} & \overset{F}{\nearrow} & \downarrow{\scriptstyle \sim} \\ Z\bigcup_A Z & \xrightarrow[(H,G)]{} & Y \end{array}.$$

Here $i_0 \cup i_1$ is a cofibration by (1.2). Now F is a homotopy from h to $\bar h$ relative B since we have:

(2.7) ***Remark.*** The sequence

$$A\bigcup_B A \underset{i_0\cup i_1}{\rightarrowtail} Z\bigcup_A Z \underset{(p,p)}{\xrightarrow{\sim}} A$$

is a cylinder for $B \rightarrowtail A$ in the sense of (I.1.5). In fact $(p,p)(i_0 \cup i_1) = (1,1)$ is the folding map. (Using this cylinder we see that (2.5) follows from $(*)$ in the proof of (2.2).) □

Next we study induced functions on homotopy sets. Let U and V be fibrant. A map $g: U \to V$ and a pair map $(f, f'):(A, B) \to (X, Y)$ induce functions

(2.8) $$g_*:[X,U]^u \longrightarrow [X,V]^{gu},$$ $$f^*:[X,U]^u \longrightarrow [A,U]^{uf'},$$

where $u: Y \to U$. We set $g_*\{x\} = \{gx\}$ and $f^*\{x\} = \{xf\}$. Clearly, if $f \simeq f_1$ rel B then $f^* = f_1^*$.

(2.9) **Lemma.** *g_* and f^* in (2.8) are well defined.*

Proof. Let $H: x \simeq y$ rel Y be a homotopy. Then $gH: gx \simeq gy$ rel Y and thus g_* is well defined. Moreover, f^* is well defined since

$$H^{\sim}(If): xf \simeq yf \text{ rel } B.$$

Here $H^{\sim}$ and If are given by the following diagram where we apply (1.6) and (1.10): The pair map f yields the commutative diagram of unbroken arrows

(2.10)

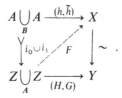

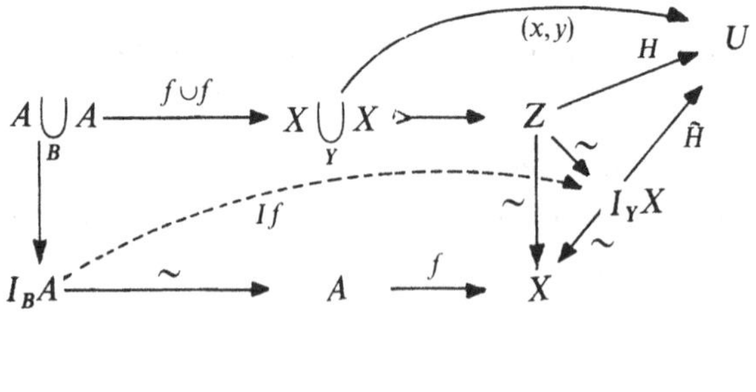

$I_B A$ denotes a cylinder on $B \subset A$. Z and $I_Y X$ are cylinders on $Y \subset X$, compare (I.1.5). $\qquad\square$

(2.11) Proposition.

   (a) *If g is a weak equivalence, then g_* is a bijection.*

   (b) *If f and f' are weak equivalences or if (f,f') is a push out then f^* is a bijection.*

Proof. g_* is surjective since for $\{v\} \in [X, V]^{gu}$ we can apply (1.11)(b) to the diagram

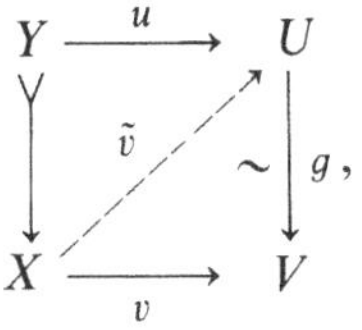

$$\begin{array}{ccc}
Y & \overset{u}{\longrightarrow} & U \\
\Big\downarrow & \overset{\tilde{v}}{\nearrow} & \sim\Big\downarrow g \\
X & \underset{v}{\longrightarrow} & V
\end{array} \qquad (1)$$

and we obtain $\tilde{v}$ with $g_*\{\tilde{v}\} = \{v\}$. Next g_* is injective since we can apply (1.11)(c).

Now assume that (f,f') is a push out. For $\{a\} \in [A, U]^{uf'}$ we have

$$f^*\{(a,u)\} = \{(a,u)f\} = \{a\}. \qquad (2)$$

Therefore f^* is surjective. Now let

$$H : xf \simeq yf \,\mathrm{rel}\, B \qquad (3)$$

be a homotopy. We consider the push out diagram

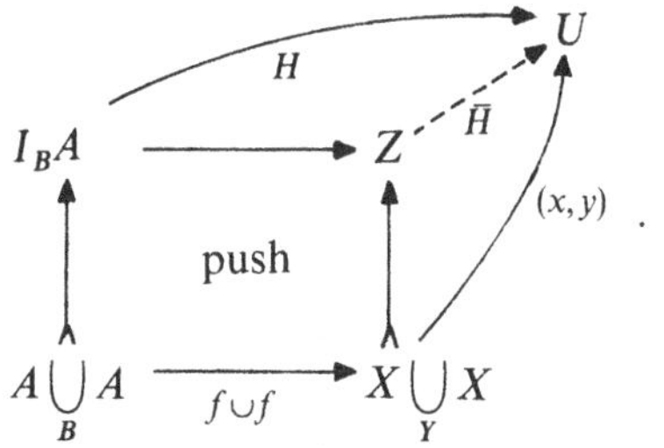

$$(4)$$

One can check that Z is a cylinder on $Y \subset X$. Hence we obtain the homotopy $\bar{H} : x \simeq y \,\mathrm{rel}\, Y$. This proves that f^* is injective.

Next assume that f and f' are weak equivalences. Then (f,f') is the composition of pair maps

$$\begin{array}{ccccc}
A & \overset{f''}{\underset{\sim}{\longrightarrow}} & P & \overset{g}{\underset{\sim}{\longrightarrow}} & X \\
\Big\uparrow & \text{push} & \Big\uparrow & & \Big\uparrow i_X \\
B & \overset{f'}{\underset{\sim}{\longrightarrow}} & Y & \underset{1}{\longrightarrow} & Y
\end{array} \qquad (5)$$

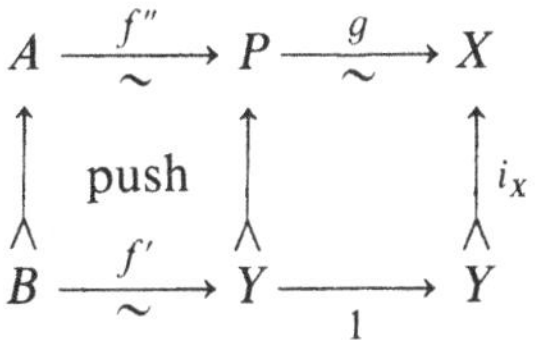

where $g = (f, i_X)$ is a weak equivalence by (C1) and (C2). Since we have seen that the push out (f'', f') induces a bijection it remains to show that g^* is a bijection. To this end we prove the following special case: Let $j: X \rightarrowtail^\sim RX$ be a fibrant model of X by (C4). Then

$$j^* : [RX, U]^Y \cong [X, U]^Y \tag{6}$$

is a bijection. Clearly, j^* is surjective by (1.6). Moreover, j^* is injective since the push out

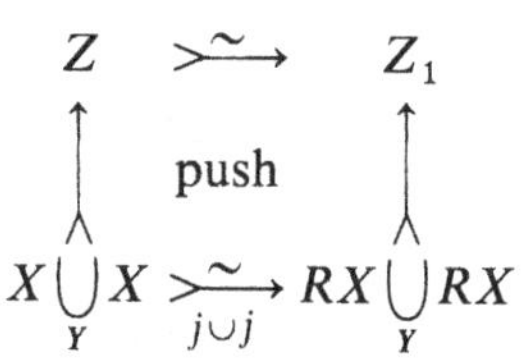

$$
\begin{array}{ccc}
Z & \rightarrowtail^\sim & Z_1 \\
\uparrow & & \uparrow \\
\wedge & \text{push} & \wedge \\
X \underset{Y}{\cup} X & \underset{j \cup j}{\rightarrowtail^\sim} & RX \underset{Y}{\cup} RX
\end{array}
\tag{7}
$$

yields a cylinder Z_1 on $Y \rightarrowtail RX$. Thus, we can use the same argument as in (4). Now g in (5) extends to a commutative diagram (see (1.6)).

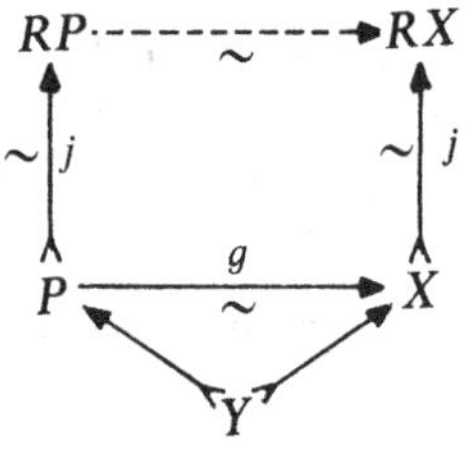

$$\tag{8}$$

By (2.11)(a) we know that $\tilde{g}$ is a homotopy equivalence under Y, see (2.12) below. Thus $\tilde{g}^* : [RX, U]^Y \to [RP, U]^Y$ is a bijection. We deduce from (6) and (8) that g^* is a bijection. $\qquad\square$

We derive from (2.11) (a) the following general form of a *theorem of Dold* (compare Dold (1963), (1966) and 2.18, 6.21 in tom Dieck–Kamps–Puppe). Clearly, the dual of the following result is also true in a fibration category.

(2.12) **Corollary.** *Consider the commutative diagram*

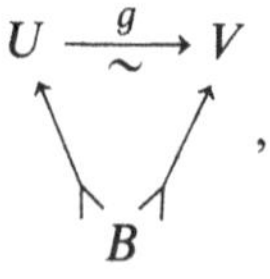

where U and V are fibrant. If g is a weak equivalence, then g is a homotopy *equivalence under B. That is, there is $f: V \to U$ under B with $gf \simeq 1_V$ rel B and $fg \simeq 1_U$ rel B.*

Proof. By (2.11)(*a*) there is $\{f\} \in [V, U]^B$ with $g_*\{f\} = \{1_V\}$. On the other hand for $\{fg\} \in [U, U]^B$ we have $g_*\{fg\} = \{gfg\} = \{g\} = g_*\{1_U\}$. Since g_* is injective we get $\{fg\} = \{1_U\}$. This proves (2.12). □

Next let RU and $R'U$ be two fibrant models of the object U as in (C4). Then we obtain by (1.6) the commutative diagram

(2.13)
$$\begin{array}{ccc} RU & \xrightarrow{\alpha} & R'U \\ & {\scriptstyle\sim}\nwarrow \quad \nearrow{\scriptstyle\sim} & \\ & U & \end{array}$$

The map α is well defined up to homotopy rel U and by (2.12) α is a homotopy equivalence under U. This shows that fibrant models are essentially unique.

Next we describe the relative cylinders and the homotopy extension property.

For a cofibration $(Y, B) \rightarrowtail (X, A)$ in **Pair (C)** we have the folding map $(X \bigcup_Y X, A \bigcup_B A) \to (X, A)$, a factorization of which is a cylinder $I_{(Y,B)}(X, A)$ in **Pair (C)**. We obtain this cylinder by (1.7) (3) together with the commutative diagram

(2.14)
$$\begin{array}{ccccccc} X \bigcup_Y X & \rightarrowtail & X \cup I_B A \cup X & \rightarrowtail & I_Y X & \xrightarrow{\sim} & X \\ \uparrow & \text{push} & \uparrow & & & & \uparrow \\ A \bigcup_B A & \rightarrowtail & I_B A & & \xrightarrow{\sim} & & A \end{array}$$

where the horizontal maps compose to the folding (1,1). Thus $I_B A$ and $I_Y X$ are in fact also cylinders in **C**; $I_{(Y,B)}(X, A) = (I_Y X, I_B A)$.

(2.15) ***Definition.*** A **triple** $(\bar{A}, A, B)$ in **C** is a sequence of cofibrations $B \rightarrowtail A \rightarrowtail \bar{A}$. *A* map $(\bar{f}, f, f'): (\bar{A}, A, B) \to (\bar{X}, X, Y)$ between triples corresponds to a pair of pair maps $(\bar{f}, f)$ and (f, f'). ||

A triple $(\bar{A}, A, B)$ gives us the cofibration $(B, B) \overset{i}{\rightarrowtail} (\bar{A}, A)$ in **Pair (C)**, see (1.3). Thus we have by (2.14) the cylinder on i which is a pair $(I_B\bar{A}, I_B A)$ with

(2.16) $$I_B A \rightarrowtail \bar{A} \cup I_B A \cup \bar{A} \overset{j}{\rightarrowtail} I_B \bar{A}.$$

This, we say, is the **relative cylinder** of the triple $(\bar{A}, A, B)$. Compare axiom (I4) in (I.§ 3).

For the push out diagram

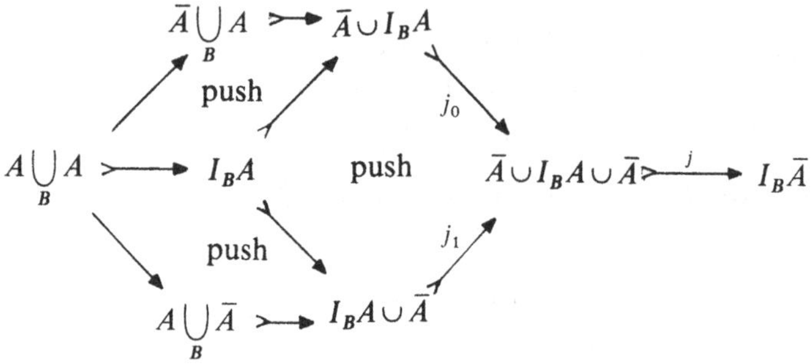

the maps jj_0 and jj_1 are weak equivalences and cofibrations. This yields by (1.6) the following **homotopy extension property** of cofibrations (compare (I3) in (I.§ 3)). Consider for $\tau = 0$ or $\tau = 1$ the diagram

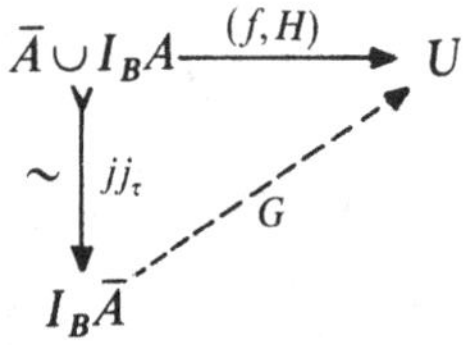

where $f: \bar{A} \to U$ and $H: I_B A \to U$ are given with $f|A = Hi_\tau$.

(2.17) **Proposition.** *If U is fibrant there exists a homotopy G which extends (f, H). G is unique up to homotopy* rel $\bar{A} \cup I_B A$.

Now assume U is fibrant and let $u: A \to U$ be given. The pair maps $i:(A, B) \rightarrowtail (\bar{A}, B)$ and $j:(\bar{A}, B) \to (\bar{A}, A)$ induce

$$[\bar{A}, U]^u \underset{j^*}{\longrightarrow} [\bar{A}, U]^{u|B} \underset{i^*}{\longrightarrow} [A, U]^{u|B}.$$

From (2.17) we easily derive the exactness of this sequence, that is

(2.18) **Corollary.**

$$\operatorname{Im} j^* = i^{*^{-1}}\{u\}.$$

The prolongation of this sequence will be discussed in (5.17) and in § 10 below.

Cylinders in an arbitrary cofibration category have a similar property as we described in the push out axiom for cylinders in (I2), see (I.§ 3). To see this we consider the map $(\bar{f}, f, f'):(\bar{A}, A, B) \to (\bar{X}, X, Y)$ between triples. Then $((\bar{f}, f), (f', f'))$ is a map between pairs in **Pair (C)**, compare (2.15). From (1.5) and (2.10) we deduce the pair map

$$(2.19) \qquad (I_B\bar{A}, I_BA) \xrightarrow{(I\bar{f}, If)} (I_Y\bar{X}, I_YX)$$

between relative cylinders.

(2.20) **Lemma.** *If* $(\bar{f}, f):(\bar{A}, A) \to (\bar{X}, X)$ *is a push out (see (1.3)) we can assume that also* $(I\bar{f}, If)$ *and* $(I\bar{f}, \bar{f} \cup If \cup \bar{f})$ *are push outs.*

We leave the proof as an exercise.

§3 The homotopy category of fibrant and cofibrant objects

Let **C** be a cofibration category with an initial object ϕ. Then we have the full subcategories

$$(3.1) \qquad \mathbf{C}_{cf} \subset \mathbf{C}_c \subset \mathbf{C},$$

where $\mathbf{C}_{cf}$ consists of objects which are both fibrant and cofibrant, see (I.1.2). The category $\mathbf{C}_c$ of cofibrant objects is a cofibration category, see (I.1.3), but $\mathbf{C}_{cf}$ in general is not, see (I.1.2). From (2.2) and (2.9) we derive:

(3.2) **Lemma.** *Homotopy relative* ϕ *is a natural equivalence relation on the morphism sets of* $\mathbf{C}_{cf}$.

We thus have the **homotopy category**

$$(3.3) \qquad \mathbf{C}_{cf}/\simeq \; = \mathbf{C}_{cf}/(\simeq \text{rel}\, \phi).$$

The morphism sets in this category are the sets $[X, Y]$ in (2.3)(b) which are well defined since X, Y are cofibrant and fibrant. We consider the quotient functor

$$(3.4) \qquad q:\mathbf{C}_{cf} \longrightarrow \mathbf{C}_{cf}/\simeq,$$

which carries the morphism f to its homotopy class rel ϕ. This functor has the following universal property:

(3.5) ***Definition.*** Let **C** be an arbitrary category and let S be a subclass of the class of morphisms in **C**. By the **localization** of **C** with respect to S we mean the category $S^{-1}\mathbf{C}$ together with a functor $q:\mathbf{C} \to S^{-1}\mathbf{C}$ having the following universal property: For every $s \in S$, $q(s)$ is an isomorphism; given any functor $F:\mathbf{C} \to \mathbf{B}$ with $F(s)$ an isomorpshim for all $s \in S$, there is a unique functor $\theta:S^{-1}\mathbf{C} \to \mathbf{B}$ such that $\theta q = F$. Except for set-theoretic difficulties the category $S^{-1}\mathbf{C}$ exists, see Gabriel–Zisman (1967). Let $Ho(\mathbf{C})$ be the localization of **C** with respect to the given class of weak equivalences in **C**.

We show that the functor q in (3.4) has the universal property in (3.5). Therefore the localization $Ho\mathbf{C}_{cf} = \mathbf{C}_{cf}/\simeq$ exists. Moreover, the inclusions of categories in (3.1) induce equivalences of categories

(3.6) **Proposition.**

$$Ho\mathbf{C}_{cf} \xrightarrow[i]{\sim} Ho\mathbf{C}_c \xrightarrow[j]{\sim} Ho\mathbf{C}.$$

Proof of (3.6). We first show that q in (3.4) has the universal property in (3.5). Let $f:A\to X$ be a weak equivalence in $\mathbf{C}_{cf}$, then the theorem of Dold (2.12) shows that f is a homotopy equivalence. Hence $q(f)$ is an isomorphism in $\mathbf{C}_{cf}/\simeq$.

Next let $F:\mathbf{C}_{cf}\to B$ be a functor which carries weak equivalences to isomorphisms. We have to show that F factors uniquely over q. Let $H:f\simeq g$ be a homotopy rel ϕ. Then we get (with the notation in (I.1.5)):

$$F(f) = F(Hi_0) = F(H)F(i_0) \quad \text{and} \quad F(g) = F(Hi_1) = F(H)F(i_1).$$

Now $F(i_0) = F(i_1)$ since $F(i_0)F(p) = 1 = F(i_1)F(p)$ where $F(p)$ is an isomorphism, hence $F(f) = F(g)$.

It remains to show that $Ho\mathbf{C}_c$ and $Ho\mathbf{C}$ exist and that i and j in (3.6) are equivalences of categories.

For each object X in $\mathbf{C}_c$ we choose a fibrant model RX in $\mathbf{C}_{cf}$. We set $RX = X$ if X is an object in $\mathbf{C}_{cf}$. Let $Ho\mathbf{C}_c$ be the category having the same objects as $\mathbf{C}_c$ and with the homotopy set $[RU, RV]$ as set of morphisms $U\to V$. Let $q:\mathbf{C}_c\to Ho\mathbf{C}_c$ be the functor which is the identity on objects and which carries g to the extension Rg,

(3.7)
$$\begin{array}{ccc} RU & \xrightarrow{Rg} & RV \\ \uparrow\wr & & \uparrow\wr \\ U & \xrightarrow{g} & V \end{array},$$

which we get by (1.6). One checks that q is a well defined functor which carries weak equivalences to isomorphisms and that q is universal with respect to this property. This proves that $Ho\mathbf{C}_c$ exists and by construction the inclusion i in (3.6) is an equivalence of categories. Indeed, i is full and faithful and satisfies the realizability condition; hence i is an equivalence of categories. By (3.7) we actually obtain a functor

(3.8) $$R:\mathbf{C}_c\to \mathbf{C}_{cf}/\simeq,$$

which induces the inverse of the equivalence i. The functor R depends on the choice of models but different choices yield canonically isomorphic functors.

Similarly, we prove the result for $Ho\mathbf{C}$, compare the proof of theorem 1 in Quillen (1967). We define $Ho\mathbf{C}$ as follows: objects are the same as in $\mathbf{C}$. We choose by (C3) a factorization

$$\phi \rightarrowtail MX \underset{m}{\xrightarrow{\sim}} X, \tag{3.9}$$

for each object X in $\mathbf{C}$. Thus MX is cofibrant. Let the homotopy set $[RMX, RMY]$ be the set of morphisms $X \to Y$ in $Ho\mathbf{C}$. We define a functor $q: \mathbf{C} \to Ho\mathbf{C}$ which carries weak equivalences to isomorphisms and which is universal with respect to this property: Consider the diagram:

(3.10)

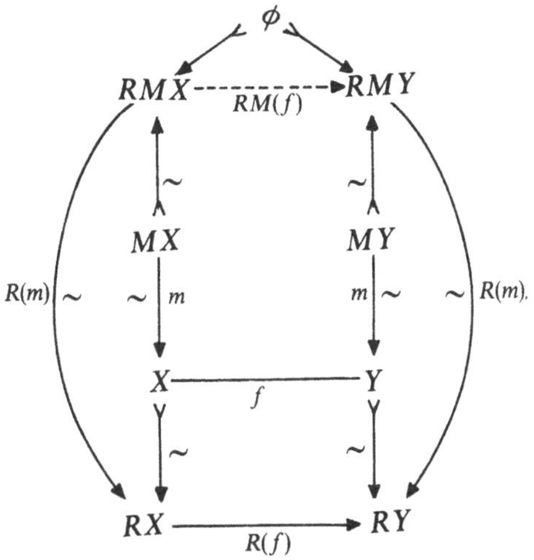

Here $R(m)$ and $R(f)$ are given as in (3.7). By (1.11) there is a lifting $RM(f)$. Let $q(f)$ be the homotopy class of $RM(f)\,\mathrm{rel}\,\phi$. One checks that $q(f)$ is well defined, see (2.11), and that q is a functor with the universal property. This proves that $Ho\mathbf{C}$ exists and that j is an equivalence of categories. By (3.10) we obtain a functor

$$RM: \mathbf{C} \longrightarrow \mathbf{C}_{cf}/\simeq, \tag{3.11}$$

which depends on the choice of models. Different choices yield canonically isomorphic functors. The functor RM induces the inverse of the equivalence ji in (3.6). Now the proof of (3.6) is complete. □

(3.12) ***Lemma.*** *Let* $(i: Y \to X)$ *and* $(u: Y \to U)$ *be objects in* $\mathbf{C}^Y$. *Moreover, let* $Y \rightarrowtail MX \xrightarrow{\sim} X$ *be a factorization of* i *and let* $U \overset{\sim}{\rightarrowtail} RU$ *be a fibrant model* *of U in* **C**. *Then the set* $[MX, RU]^Y$ *can be identified with the set of morphisms from i to u in* $Ho(\mathbf{C}^Y)$.

Compare (2.3)(c).

Proof of (3.12). Let $Y \rightarrowtail MU \xrightarrow{\sim} U$ be a factorization of u and let RMU and RMX be fibrant models. Then we have a weak equivalence $RMU \xrightarrow{\sim} RU$ induced by $m: MU \xrightarrow{\sim} U$, see (3.7). Therefore

$$[RMX, RMU]^Y \xrightarrow{\approx} [RMX, RU]^Y \xrightarrow{\approx} [MX, RU]^Y. \qquad \square$$

By (3.12) we may define

(3.13) $$[X, U]^Y = [MX, RU]^Y,$$

provided maps $i: Y \to X$ and $u: Y \to U$ are given. An arbitrary element $\{f\} \in [X, U]^Y$ is this represented by a commutative diagram

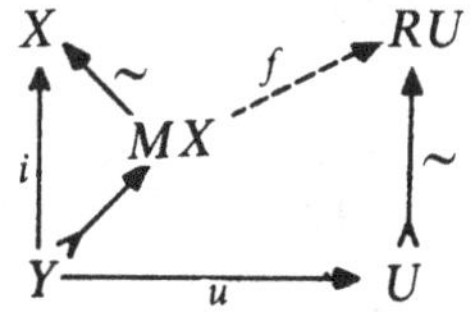

in **C**. The set $[MX, RU]^Y$ is an 'honest' homotopy set as defined in (2.3).

§ 4 Functors between cofibration categories

Each functor $\alpha: \mathbf{C} \to \mathbf{K}$ which preserves weak equivalences induces by the universal property in (3.5) a functor $Ho\alpha: Ho\mathbf{C} \to Ho\mathbf{K}$, compare the notation in (I.1.10).

(4.1) **Definition.** Assume $\alpha, \beta: \mathbf{C} \to \mathbf{K}$ are functors which preserve weak equivalences. A **natural weak equivalence** $\tau: \alpha \xrightarrow{\sim} \beta$ is a natural transformation such that $\tau: \alpha(X) \xrightarrow{\sim} \beta(X)$ is a weak equivalence for all X. Moreover, we say α and β are **natural weak equivalent** if there is a finite chain $\alpha \sim \alpha_1 \cdots \sim \beta$ of natural weak equivalences. ||

Clearly, if α and β are natural weak equivalent we get a natural equivalence

(4.2) $$Ho(\alpha) \sim Ho(\beta)$$

of the induced functors on homotopy categories.

(4.3) **Proposition.** *Let* $\alpha, \beta: \mathbf{C} \to \mathbf{K}$ *be functors between cofibration categories which are natural weak equivalent. If* α *preserves weak equivalences then also*

β preserves weak equivalences. If α is compatible with a homotopy push out then so is β. Thus if α is a model functor then so is β, see (I.1.10).

Since the identical functor of a cofibration category **C** is a model functor we get:

(4.4) **Corollory.** *Let **C** be a cofibration category and let $\alpha:\mathbf{C}\to\mathbf{C}$ be a functor naturally weak equivalent to the identical functor of **C**. Then α is a model functor.*

Proof of (4.3). We have the commutative diagram

$$\begin{array}{ccc} \alpha X & \xrightarrow[\sim]{\tau} & \beta X \\ {\scriptstyle \alpha f}\downarrow & & \downarrow{\scriptstyle \beta f,} \\ \alpha Y & \xrightarrow[\sim]{\tau} & \beta Y \end{array}$$

which shows that βf is a weak equivalence if f is one. We now check that β is compatible with push outs, see (I.1.10). We obtain a factorization $M\beta A$ of $\beta(i)$ by the commutative diagram

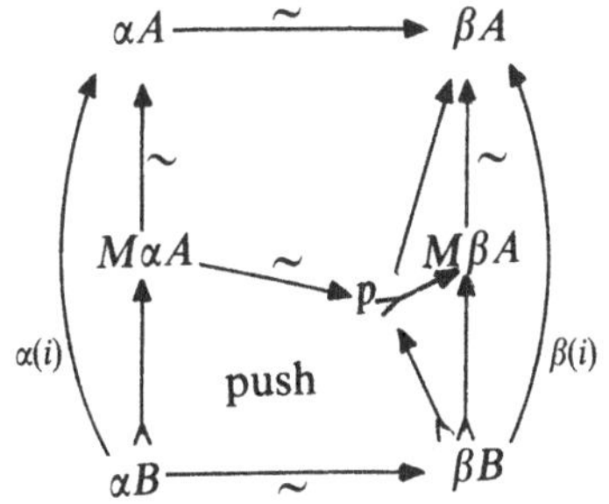

This induces the commutative diagram, see (I.1.10)(3):

$$\begin{array}{ccc} \alpha Y & \xrightarrow{\sim} & \beta Y \\ {\scriptstyle q}\uparrow{\scriptstyle \wr} & & \uparrow{\scriptstyle q} \\ M\alpha Y & \xrightarrow[\sim]{} & M\beta Y, \\ \uparrow & & \uparrow \\ \alpha X & \xrightarrow[\sim]{} & \beta X \end{array}$$

where the induced map $M\alpha Y\to M\beta Y$ is a weak equivalence by (1.2)(b). This shows that $q:M\beta Y\to\beta Y$ is a weak equivalence. □

(1) ***Example.*** Let **Top** be the cofibration category of topological spaces with the CW-structure in (I.5.6). Then the *realization of the singular set* yields a functor

$$|S|:\mathbf{Top}\to\mathbf{Top},$$

which carries a space X to the CW-complex $|SX|$, see (I.5.8.). The functor $|S|$ is natural weak equivalent to the identical functor via the natural map $|SX|\xrightarrow{\sim}X$. Therefore (4.4) above shows that $|S|$ is a model functor which carries objects to cofibrant objects.

(2) ***Example.*** Consider the cofibration category **C** in (I.5.10). From 3.3 in Bousfield (1975) we derive a '*localization functor*' $\mathbf{C}\to\mathbf{C}$ which carries objects to fibrant objects and which is natural weak equivalent to the identical functor of **C**. Thus by (4.4) this localization functor is a model functor.

(4.5) **Proposition.** *Let $f:B\to X$ be a map in the cofibration category* **C**. *Then the* **cobase change functor**

$$f_*:(\mathbf{C}^B)_c\to(\mathbf{C}^X)_c$$

(*which carries $B\rightarrowtail A$ to $X\rightarrowtail A\cup_B X$) is a based model functor which carries cofibrations to cofibrations, see* (1.4).

We leave the proof as an exercise. The cobase change functor induces the functor

(4.6) $$Ho(f_*):Ho(\mathbf{C}^B)_c\to Ho(\mathbf{C}^X)_c$$

which is an equivalence of categories provided f is a weak equivalence. For this we show that each object $i:X\rightarrowtail A$ is realizable by Hof_*:Consider the diagram

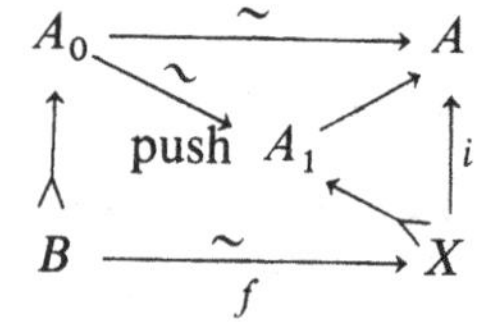

where A_0 is a factorization of if. Now $f_*A_0=A_1$ is isomorphic to A in $Ho(\mathbf{C}^X)_c$ by $A_1\xrightarrow{\sim}A$.

§5 The groupoid of homotopies

For a cofibration $Y\subset X$ we choose a cylinder I_YX. Let $x,y:X\to U$ be extensions of $u:Y\to U$ where U is fibrant. A homotopy $x\simeq y$ rel Y is a map $G:I_YX\to U$, which extends $(x,y):X\cup_Y X\to U$. We consider the set of homotopy classes relative (x, y) of such homotopies:

$$H_Y(x,y) = [I_YX, U]^{(x,y)}. \tag{5.1}$$

The elements of this set are called **tracks** from x to y relative Y. The set (5.1) depends only up to canonical bijection on the choice of the cylinder I_YX, compare diagram (*) in the proof of (2.2). If U is not fibrant we define the set (5.1) by (3.13).

A map $g: U \to V$ and a pair map $f: (A, B) \to (X, Y)$ induce functions as in (2.8)

$$\begin{cases} f^* = (If)^*: H_Y(x,y) \longrightarrow H_B(xf, yf) \\ g_* \qquad\quad : H_Y(x,y) \to H_Y(gx, gy) \end{cases}. \tag{5.2}$$

For the definition of f^* we use If in (2.10). The function f^* does not depend on the choice of If.

We show that we have the structure $(+, -, 0)$ of a groupoid for the sets in (5.1). Let $p: I_YX \to X$ be the projection of the cylinder. We call

$$0 = \{xp\} \in H_Y(x,x) \tag{5.3}$$

the **trivial track**. Let

$$- = n^*: H_Y(x,y) \longrightarrow H_Y(y,x) \tag{5.4}$$

be defined by n in (2.4). We call $-G$ the **negative** of the track G. Moreover, we define the addition

$$+: H_Y(x,y) \times H_Y(y,z) \longrightarrow H_Y(x,z). \tag{5.5}$$

Here we set $\{H\} + \{G\} = \{(H, G)\tilde{}\, m\}$, see (2.5). The addition is also called **track addition.**

One can check that the functions in (5.4) and (5.5) are well defined.

(5.6) **Proposition.** *For $H \in H_Y(w,x)$, $G \in H_Y(x,y)$ and $F \in H_Y(y,z)$ we have the following equations* (1), ..., (9).

(1) $H + (G + F) = (H + G) + F$,
(2) $H + 0 = 0 + H = H$,
(3) $H + (-H) = 0, (-H) + H = 0$,
(4) $f^*(H + G) = f^*H + f^*G$,
(5) $f^*(-H) = -f^*H$,
(6) $g_*(H + G) = g_*H + g_*G$,
(7) $g_*(-H) = -g_*H$.

Now consider the following commutative diagram

$$\begin{array}{ccccccc} A' & \xrightarrow{f',g'} & A & \xrightarrow{f,g} & X & \xrightarrow{x,y} & U \\ \uparrow & & \uparrow & & \uparrow & \nearrow_{u} & \\ B' & \longrightarrow & B & \longrightarrow & Y & & \end{array}$$

and let $H \in H_Y(x, y)$, $G \in H_B(f, g)$ *and* $G' \in H_{B'}(f', g')$ *be tracks*

(8) $H * G = f^*H + y_*G = x_*G + g^*H$

(9) $(H * G) * G' = H * (G * G')$

The equations (1)···(9) correspond to the equations which define a **2-category**. Therefore the proposition implies:

Corollary. *Let* **C** *be a cofibration category and let* Y *be an object in* **C**. *Then the category* $(\mathbf{C}^Y)_C$ *of cofibrant objects in* $\mathbf{C}^Y$ *is a 2-category in which the 2-morphisms are tracks.*

Compare Kamps, Marcum (1976) and Kelley–Street.

Proof of (5.6). We leave (1)···(7) as an exercise to the reader. For (8) and (9) representatives of the tracks yield maps

$$I_B(I_B A) \xrightarrow{IG} I_Y X \xrightarrow{H} U,$$

$$I_{B'}(I_{B'}(I_{B'}A')) \xrightarrow{IIG'} I_B(I_B A) \xrightarrow{H(IG)} U.$$

Now we can use an argument as in the proof of (5.15) below. □

Let $Y \subset X \subset \bar{X}$ be cofibrations and for $x, y: X \to U$ let $G \in H_Y(x, y)$ be a track. Then G induces a bijection of homotopy sets

(5.7) $$G^{\#}: [\bar{X}, U]^x \longrightarrow [\bar{X}, U]^y$$

with the following properties:

(5.8) $$\begin{cases} 0^{\#} = \text{identity}, \\ (G_1 + G_2)^{\#} = G_2^{\#} \circ G_1^{\#}. \end{cases}$$

We define $G^{\#}$ by the homotopy extension property (2.17). That is, we choose for $\{\alpha\} \in [\bar{X}, U]^x$ and for $G = \{H\}$ a homotopy $\bar{H}$ on $I_Y \bar{X}$ with $\bar{H} | I_Y X = H$, $\bar{H} i_0 = \alpha$, and we set: $G^{\#}\{\alpha\} = \{\bar{H} i_1\}$.

(5.9) **Lemma.** *$G^{\#}$ is well defined and satisfies the equation in* (5.8).

Proof. We choose the relative cylinder for the triple $X \bigcup_Y X \subset \bar{X} \cup I_Y X \cup \bar{X} \subset I_Y \bar{X}$, as in (2.16). Here we assume the cylinder $I_{X \cup_Y X}(\bar{X} \cup I_Y X \cup \bar{X})$ to be the push out of cylinders. By the homotopy extension property of this relative cylinder we see that $G^{\#}\{\alpha\}$ does not depend on the choice of H and α. Clearly,

$0^{\#} = \text{identity}$. Moreover, we derive formula (5.8) from the commutative diagram: $(Z = I_Y X, \bar{Z} = I_Y \bar{X})$

$$\begin{array}{ccccc} Z & \xrightarrow{m} & DZ & \overset{\sim}{\longleftarrow\!\!<} & Z \bigcup_X Z \\ \downarrow & & \downarrow & \text{push} & \downarrow \\ \bar{Z} & \xrightarrow{m} & D\bar{Z} & \longleftarrow\!\!< & \bar{Z} \bigcup_{\bar{X}} \bar{Z} \end{array},$$

which we deduce from (1.5). □

Let $g: U \to V$ and that let $(\bar{f}, f, f_0): (\bar{A}, A, B) \to (\bar{X}, X, Y)$ be a map between triples. Then we have the commutative diagrams, compare (5.2),

(5.10)
$$\begin{array}{ccc} [\bar{X}, U]^x & \xrightarrow{G^{\#}} & [\bar{X}, \bar{U}]^x \\ \bar{f}^* \downarrow & & \downarrow \bar{f}^* \\ [\bar{A}, U]^{xf} & \xrightarrow{(f^*G)^{\#}} & [\bar{A}, U]^{yf} \end{array}, \qquad \begin{array}{ccc} [\bar{X}, U]^x & \xrightarrow{G^{\#}} & [\bar{X}, U]^y \\ g_* \downarrow & & \downarrow g_* \\ [\bar{X}, V]^{gx} & \xrightarrow{(g_*G)^{\#}} & [\bar{X}, V]^{gy} \end{array}.$$

Proof. For $\bar{f}^*$ this follows from (2.19). In fact

$$\bar{f}^* G^{\#}\{\alpha\} = \bar{f}^*\{\bar{H} i_1\} = \{\bar{H} i_1 \bar{f}\} = \{\bar{H}(I\bar{f}) i_1\}.$$

Here $\bar{H}(I\bar{f})$ extends $H(If) \in f^*G$ by (2.19). □

We define the **torus** $\Sigma_Y X$ on $Y \subset X$ by the push out diagram:

(5.11)
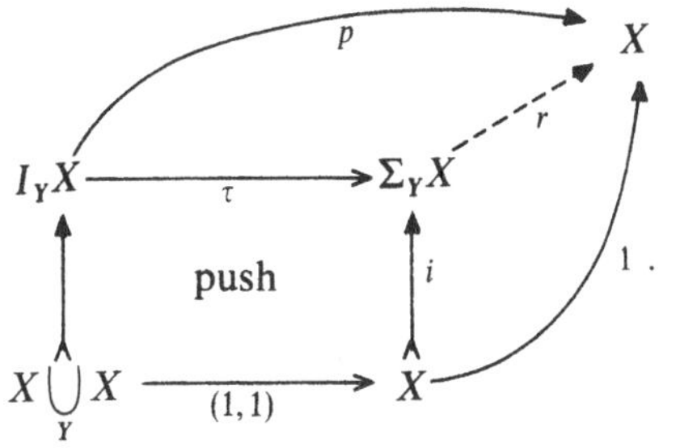

A pair map $f: (A, B) \to (X, Y)$ induces $\Sigma \cdot f: \Sigma_B A \to \Sigma_Y X$ with $\Sigma \cdot f = If \cup f$, see (2.10).

Let $u: X \to U$ be a map where U is fibrant. By (2.11) the push out τ induces the bijection of sets

(5.12)
$$[\Sigma_Y X, U]^u \underset{\approx}{\xrightarrow{\tau^*}} [I_Y X, U]^{(u,u)} = H_Y(u, u).$$

Here $H_Y(u, u)$ is a group by (5.6) which via τ^* induces a group structure on the set $[\Sigma_Y X, U]^u$. We denote this group by

(5.13)
$$\pi_1(U^{X|Y}, u) = ([\Sigma_Y X, U]^u, +, -, 0), \text{ see § 10.}$$

(5.14) ***Remarks.*** Let $Y \rightarrowtail X$ be a cofibration in the category of topological spaces **Top** and let $u: X \to U$ be a map in **Top**. Then we can define the subspace $U^{X|Y} = \{f \in U^X; f|Y = u|Y\}$ of the function space $U^X = \{f; f: X \to U\}$ with the compact open topology. The map u is the basepoint of $U^{X|Y}$. The **fundamental group** $\pi_1(U^{X|Y}, u) = [\Sigma_Y X, U]^u$ coincides with the group in (5.13) up to the canonical isomorphism which depends on the choice of the cylinder in (5.11).

For a track $G \in [I_Y X, U]^{(u,v)} = H_Y(u, v)$ we obtain the diagram

$$(5.15)\qquad \begin{array}{ccccc} [\Sigma_Y X, U]^u & \xrightarrow[\approx]{\tau^*} & H_Y(u,u) & & H \\ \downarrow{\scriptstyle G^\#} & & \downarrow & & \downarrow \\ [\Sigma_Y X, U]^v & \xrightarrow[\approx]{\tau^*} & H_Y(v,v) & & -G+H+G \end{array},$$

where we use the triple $(\Sigma_Y X, X, Y)$, see (5.7).

Lemma. *Diagram* (5.15) *commutes.*

This shows by (5.6) that $G^\#$ in (5.15) is an isomorphism of groups which is an inner automorphism if $u = v$.

Proof. By (5.10) we see $\tau^* G^\# (\tau^*)^{-1} H = ((1,1)^* G)^\# H$.

We know that $Z = I_Y X \bigcup_Y I_Y X$ is a cylinder on $X \bigcup_Y X$. Thus we have $(1,1)^* G = (G, G): Z \to U$.

By (2.16) we have a relative cylinder $(\bar{Z}, Z)$ on $X \bigcup_Y X \subset I_Y X$ such that in the diagram

$$\begin{array}{ccc} I_Y X \cup Z & \xrightarrow{H,(G,G)} & U \\ {\scriptstyle jj_0}\downarrow & \nearrow{\scriptstyle \Gamma} & \\ \bar{Z} & & \end{array}$$

jj_0 is a cofibration and a weak equivalence. Therefore the extension Γ gives us $(G, G)^\# H = \{\Gamma i_1\}$, see (5.7). We now consider the following commutative diagram of unbroken arrows

$$\begin{array}{ccc} I_Y X \bigcup I_Y X & \xrightarrow{K, i_1} & \bar{Z} \\ {\scriptstyle \partial}\,\big\downarrow\,{\scriptstyle (i_0, i_1)} & \nearrow{\scriptstyle \bar{\Lambda}} & {\scriptstyle p}\downarrow{\scriptstyle \sim} \\ & & I_Y X \\ & & {\scriptstyle p}\downarrow{\scriptstyle \sim} \\ I_\partial(I_Y X) & \xrightarrow[\Lambda]{} & X \end{array}.$$

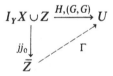

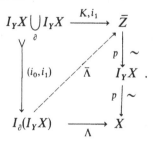

Here we use an appropriate cylinder for $\partial = X\bigcup_Y X \subset I_Y X$. The map K represents $-(Ii_0)+i_0+(Ii_1)$, where $(Ii_0, Ii_1): Z \subset \bar{Z}$ is the inclusion. Since by (5.6) we know $-0+0+0=0$, there is a homotopy Λ which makes the diagram above commute. By (1.10) we choose the lifting $\bar{\Lambda}$ which gives the homotopy $\Gamma\bar{\Lambda}:\Gamma i_1 \simeq -G+H+G \operatorname{rel} \partial$. This proves the commutativity of (5.15). □

Let $(\bar{A}, A, B)$ be a triple and let $v: A \to U$ be given. We derive from (5.7), (5.8) the **action of the fundamental group**

(5.16) $$[\bar{A}, U]^A \times [\Sigma_B A, U]^A \xrightarrow{+} [\bar{A}, U]^A,$$

which carries (u, H) to $u+H = H^{\#}(u)$. This action leads to the **exact sequence** of sets (see § 10).

(5.17) $$[\Sigma_B A, U]^A \xrightarrow{u^+} [\bar{A}, U]^A \xrightarrow{j} [\bar{A}, U]^B \xrightarrow{i^*} [A, U]^B,$$

where j and i^* are defined as in (2.18) and where $u^+(H) = u+H$. One can check that $\operatorname{image}(u^+) = j^{-1}j(u)$. In (2.18) we have seen that image $(j) = (i^*)^{-1}(\{v\})$. The exact sequence (5.17) is of great help for the comparison of the categories $Ho(\mathbf{C}^B)$ and $Ho(\mathbf{C}^A)$ if $B \rightarrowtail A$. For example, we have the following useful results in topology:

(5.18) ***Example.*** Consider the cofibration category **Top** in (5.1) and the triple $(A, *, \phi)$ where ϕ is the empty set and where $*$ is a basepoint of A such that $* \rightarrowtail A$ is a cofibration in **Top**. For this triple the exact sequence (5.17) yields the exact sequence

$$\begin{array}{ccccccc} [\Sigma_\phi *, U]^* & \xrightarrow{+} & [A, U]^* & \to & [A, U]^\phi & \longrightarrow & [*, U]^\phi \\ \| & & & & & & \| \\ \pi_1(U) & & & & & & \pi_0(U) \end{array} .$$

This shows that for a path connected space U we have the equations

$$[A, U]^*/\pi_1(U) = [A, U]^\phi,$$

where the fundamental group π_1 acts by

$$X \longmapsto \alpha^{\#} X, \quad \alpha \in \pi_1, \quad X \in [A, U]^*.$$

(5.19) ***Example.*** Let $p: U \twoheadrightarrow D$ be a fibration in **Top** with fiber $F = p^{-1}(*)$ and with basepoint $* \in F \subset U$. Moreover, let $A \to D$ be a basepoint preserving map. If $* \rightarrowtail A$ is a closed cofibration in **Top** there is the exact sequence.

$$\begin{array}{ccccccc} [\Sigma_\phi *, U]^*_D & \xrightarrow{+} & [A, U]^*_D & \longrightarrow & [A, U]^\phi_D & \longrightarrow & [*, U]^\phi_D. \\ \| & & & & & & \| \\ \pi_1(F) & & & & & & \pi_0(F) \end{array}$$

This as well is an example of the exact sequence (5.17). In fact, let $\overline{\mathbf{Top}_D}$ be the cofibration category of spaces over D obtained by Strøm's model category (I.4a.4) via (I.4a.5). Then $(A, *, \phi)$ is a triple in $\overline{\mathbf{Top}_D}$ and U is a fibrant object in $\mathbf{Top}_D$ so that we can apply (5.17). In particular we get

$$[A, U]^*_D/\pi_1 F = [A, U]^\phi_D,$$

provided the fiber F is path connected.

(5.20) ***Example.*** Let $\mathbf{F} = \mathbf{Top}$ be the fibration category of topological spaces in (5.2). For a triple $\bar{X} \twoheadrightarrow X \twoheadrightarrow Y$ in $\mathbf{F}$ we have the exact sequence

$$[U, \Omega_Y X]_X \xrightarrow{+} [U, \bar{X}]_X \to [U, \bar{X}]_Y \to [U, X]_Y$$

which is dual to (5.17). We consider the special case with $Y = *$ and $U = p^{-1}(*) = F \to * \in X$, $p: \bar{X} \twoheadrightarrow X$. Thus we get for $1 = 1_F$

$$\begin{array}{ccl} [F, \Omega_* X]_X & \xrightarrow{1^+} & [F, \bar{X}]_X \to [F, \bar{X}] \to [F, X] \\ \| & & \quad \| \\ [F \times S^1/F \times *, X]^* & & [F, F] \\ \cup & & \quad \cup \\ [S^1, X]^* = \pi_1(X) & \longrightarrow & \mathrm{Aut}(F), \ \alpha \longmapsto \alpha^\# \end{array}$$

Here $\mathrm{Aut}(F)$ is the group of homotopy equivalences of F in $\mathbf{Top}/\simeq$. The operator $\alpha \longmapsto \alpha^\#$ satisfies $(\alpha + \beta)^\# = \beta^\# \circ \alpha^\#$. This operator induces the action of $\pi_1 X$ from the right on the homology of the fiber by $x^\alpha = (\alpha^\#)_*(x)$ for $x \in H_* F$. If F is simply connected (or simple) we obtain by $\pi_1 X \to \mathrm{Aut}(F)$ the action of $\pi_1 X$ on the homotopy groups $\pi_n F = [S^n, F]$, $n \geqq 2$.

§ 5a Appendix: homotopies and functors

Let $\mathbf{C}$ and $\mathbf{K}$ be cofibration categories and let $\alpha: \mathbf{C} \to \mathbf{K}$ be a functor which *preserves weak equivalences.* We here show that α carries tracks in $\mathbf{C}$ to tracks in $\mathbf{K}$.

First we consider the induced functions on homotopy sets. Let Y be an object in $\mathbf{C}$. Then α induces the functor

(5a.1) $$\alpha: \mathbf{C}^Y \longrightarrow \mathbf{K}^{\alpha Y},$$

which carries the object $i:Y \to X$ in $\mathbf{C}^Y$ to the object $\alpha(i):\alpha Y \to \alpha X$ in $\mathbf{K}^{\alpha Y}$. Clearly, this functor as well preserves weak equivalences and thus we get the induced functor on homotopy categories

(5a.2) $$Ho\alpha: Ho\mathbf{C}^Y \longrightarrow Ho\mathbf{K}^{\alpha Y}.$$

Now assume that $i: Y \rightarrowtail X$ is a cofibration in $\mathbf{C}$ and that $u: Y \to U$ is given where U is fibrant in $\mathbf{C}$. For the homotopy set $[X, U]^Y$ we obtain by $Ho\alpha$ in (5a.2) the function (compare (3.13))

(5a.3) $$\alpha = Ho\alpha: [X, U]^Y \longrightarrow [\alpha X, \alpha U]^{\alpha Y} = [M\alpha X, R\alpha U]^{\alpha Y}.$$

Usually α alone is sufficient notation for this function since it will be clear from the context whether we apply α to a map in $\mathbf{C}$ or to a homotopy class in $Ho\mathbf{C}^Y$. In (5a.3) we choose for $\alpha(i):\alpha Y \to \alpha X$ the factorization

$$\alpha Y \rightarrowtail M\alpha X \xrightarrow[q]{\sim} \alpha X, \tag{1}$$

and we choose for the object αU the fibrant model

$$\alpha U \overset{\sim}{\underset{j}{\rightarrowtail}} R\alpha U. \tag{2}$$

Then the function $\alpha = Ho\alpha$ carries the homotopy class of $f: X \to U$ rel Y to the homotopy class, rel αY, of the composition

$$\bar{f}: M\alpha X \xrightarrow[q]{\sim} \alpha X \xrightarrow[\alpha f]{} \alpha U \overset{\sim}{\underset{j}{\rightarrowtail}} R\alpha U. \tag{3}$$

Thus the construction $f \longmapsto \bar{f}$ depends on the choices in (1) and (2).

Now consider the set of tracks $[I_Y X, U]^{(x,y)}$ where $x, y: X \to U$ are maps which coincide on Y. Then the functor α induces the following function α_L which carries tracks from x to y rel Y to tracks from $\bar{x}$ to $\bar{y}$ rel αY; (here $\bar{x}$ and $\bar{y}$ are given by (3) above).

(5a.4)
$$\begin{array}{ccc} [I_Y X, U]^{(x,y)} & \xrightarrow{\alpha} & [\alpha I_Y X, \alpha U]^{\alpha(x,y)} \\ \Big\downarrow{\scriptstyle \alpha_L} & & \Big\| \\ [I_{\alpha Y} M\alpha X, R\alpha U]^{(\bar{x},\bar{y})} & \xleftarrow{L^*} & [M\alpha I_Y X, R\alpha U]^{j\alpha(x,y)} \end{array}$$

In this diagram the function α is defined by (5a.3) and $L^* = h^*(j^*)^{-1}$ is induced by a weak lifting $L = (h, j)$ in the following commutative diagram with $L' = ((\alpha i_0)q, (\alpha i_1)q)$.

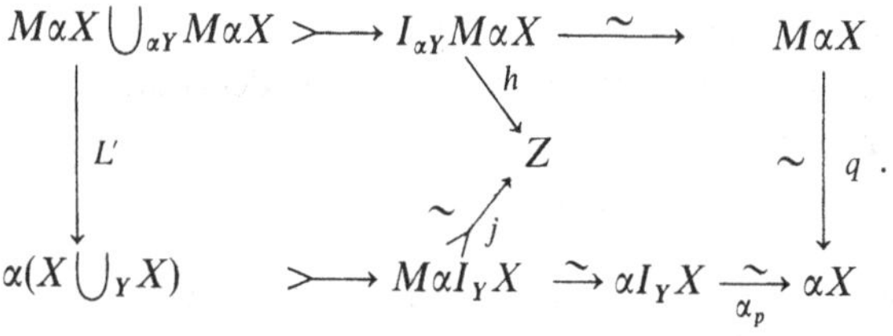

It is clear that L^* in (5a.4) is well defined, compare (1.10) and (1.11). Moreover, L^* is bijection provided α is compatible with the push out $X\bigcup_Y X$, see (I.1.10) and (2.11).

(5a.5) **Theorem.** *Let* $H\in[I_Y X, U]^{w,x}$ *and* $G\in[I_Y X, U]^{x,y}$ *be tracks as in* (5.6). *Then* α_L *in* (5a.4) *satisfies the formulas*:

$$\begin{aligned}\alpha_L(H+G) &= \alpha_L(H)+\alpha_L(G),\\ \alpha_L(-H) &= -\alpha_L(H),\\ \alpha_L(0) &= 0,\\ (R\alpha g)_*\alpha_L(G) &= \alpha_L(g_*G) \quad \textit{where } g:U\to U',\\ (M\alpha f)^*\alpha_L(G) &= \alpha_L(f^*G) \quad \textit{where } f:(X',Y')\to(X,Y).\end{aligned}$$

Here $R\alpha g$ is an extension $R\alpha U\to R\alpha U'$ of $\alpha g:\alpha U\to\alpha U'$. Moreover $M\alpha f = (f_1, i)$ is a weak lifting of the diagram

(5a.6)

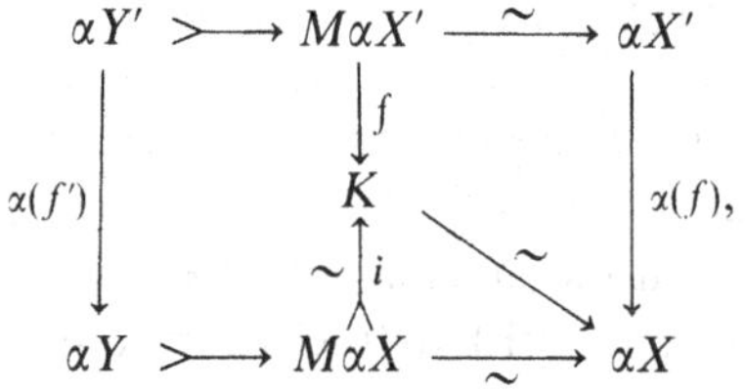

which induces $(M\alpha f)^* = f_1^*(i^*)^{-1}$. We point out that in (5a.5) we only assume that the functor α preserves weak equivalences.

Proof of (5a.5). Consider the diagrams in (2.4), (2.5) and (2.10). These diagrams are commutative diagrams in **C** of the type

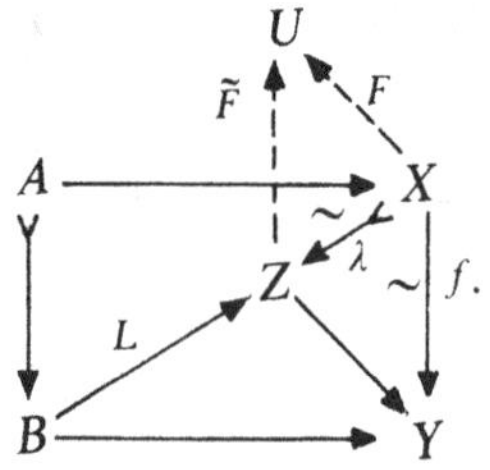

(1)

Here U is a fibrant object and $\tilde{F}$ denotes an extension of F, (L, λ) is a 'weak lifting' as in (1.10). The functor α carries the unbroken arrows of diagram (1) to the subdiagram in the middle of the following commutative diagram of unbroken arrows.

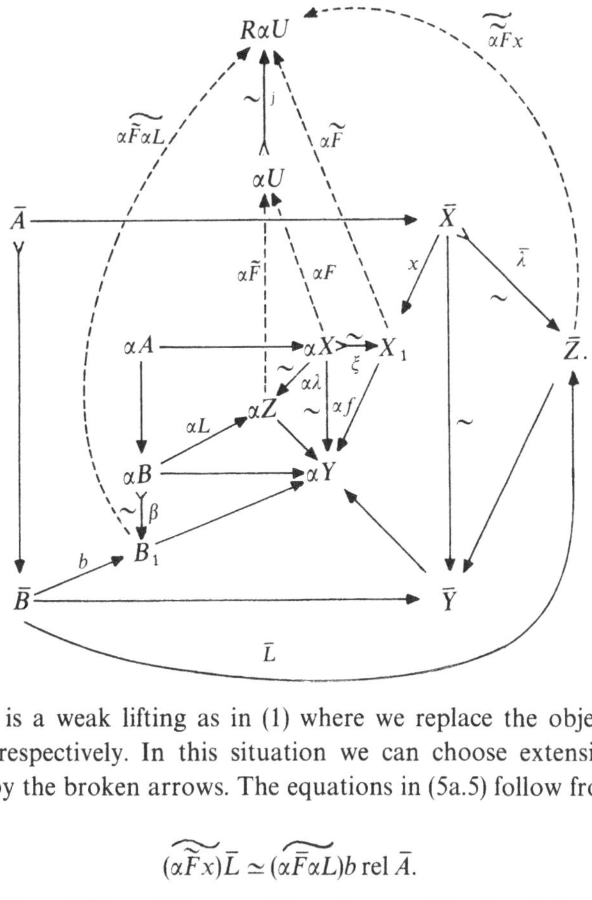

(2)

Here $(\bar{L}, \bar{\lambda})$ is a weak lifting as in (1) where we replace the objects $A, B\ldots$ by $\bar{A}, \bar{B}\ldots$ respectively. In this situation we can choose extensions of αF described by the broken arrows. The equations in (5a.5) follow from the fact that

$$(\widetilde{\alpha\tilde{F}x})\bar{L} \simeq (\widetilde{\alpha\tilde{F}\alpha L})b \operatorname{rel} \bar{A}. \tag{3}$$

We prove this as follows: Diagram (2) can be considered as being a diagram in $Ho(\mathbf{K}^{\bar{A}})$. In this category we get the equation

$$(\alpha\lambda)^{-1}\alpha L\beta^{-1}b = \xi^{-1}x\bar{\lambda}^{-1}\bar{L}. \tag{4}$$

We see this since we can apply the isomorphism αf to both sides. Hence commutativity of (2) yields (4). Now the right-hand side of (3) represents $j\alpha F(\alpha\lambda)^{-1}\alpha L\beta^{-1}b$, the left-hand side represents $j\alpha F\xi^{-1}x\bar{\lambda}^{-1}\bar{L}$. Therefore if we apply $j\alpha F$ to both sides of (4) we get (3); here we use (3.12) in the category $Ho(\mathbf{K}^{\bar{A}})$.

As an example we derive from (3) the equation $\alpha_L(-H) = -H$. In this case (1) is given by (2.4) and $b = x = \bar{q}$ and $\beta = \xi = r$ are given by diagram (5a.4)(1). Moreover, F represents the track H. Now the left-hand side of (3) represents $-\alpha_L H$ and the right-hand side of (3) represents $\alpha_L(-H)$. □

Next we consider the operator $G^{\#}$ in (5.7) and we show that $G^{\#}$ is compatible with the functor α. Let $Y \rightarrowtail X \rightarrowtail \bar{X}$ be a triple in **C**. We choose the commutative diagram

(5a.7)
$$\begin{array}{ccccc} \alpha Y & \longrightarrow & \alpha X & \longrightarrow & \alpha\bar{X} \\ & \nwarrow & \uparrow\sim & & \uparrow\sim \\ & & M\alpha X & \rightarrowtail & \bar{M}\alpha\bar{X}. \end{array}$$

(5a.8) **Proposition.** *Let $G \in [I_Y X, U]^{(x,y)}$ be a track and let $\alpha_L G$ be given by (5a.4). Then we have the commutative diagram*

$$\begin{array}{ccc} [\bar{X}, U]^x & \xrightarrow{\quad G^{\#}\quad} & [\bar{X}, U]^y \\ \downarrow\alpha & & \downarrow\alpha \\ [\bar{M}\alpha\bar{X}, R\alpha U]^{\bar{x}} & \xrightarrow{(\alpha_L G)^{\#}} & [\bar{M}\alpha\bar{X}, R\alpha U]^{\bar{y}} \end{array}$$

Here α carries $\xi: \bar{X} \to U$ to the composition $\bar{\xi}: \bar{M}\alpha\bar{X} \xrightarrow{\sim} \alpha\bar{X} \xrightarrow{\alpha f} \alpha U \rightarrowtail{\sim} R\alpha U$.

Proof. Since α preserves weak equivalences also the induced functor

$$\alpha: \mathbf{Pair\ C} \longrightarrow \mathbf{Pair\ K} \tag{1}$$

preserves weak equivalences. The triple $(\bar{X}, X, Y)$ is the same as a cofibration

$$i:(Y, Y) \rightarrowtail (\bar{X}, X) \quad \text{in } \mathbf{Pair\ C}. \tag{2}$$

Then a factorization

$$\alpha(i):(\alpha Y, \alpha Y) \rightarrowtail (\bar{M}, M) \xrightarrow{\sim} (\alpha\bar{X}, \alpha X) \tag{3}$$

is the same as a choice in (5a.7), $\bar{M} = \bar{M}\alpha\bar{X}, M = M\alpha X$. Now we can apply (5a.4) for the functor α in (1). This yields the function

$$\begin{array}{c} H \in [I_{(Y,Y)}(\bar{X}, X), (U, U]^{(\xi,\eta)} \\ \Big\downarrow \alpha_L \\ \alpha_L H \in [I_{\alpha(Y,Y)}(\bar{M}, M), R\alpha(U, U)]^{(\bar{\xi},\bar{\eta})} \end{array} , \tag{4}$$

For $G = H | I_Y X$ we have $\{\eta\} = G^{\#}\{\xi\}$ and similarly for $\alpha_L G = (\alpha_L H) | I_{\alpha Y} M$ we have $\{\bar{\eta}\} = (\alpha_L G)^{\#}\{\bar{\xi}\}$. This proves (5a.8) since $\alpha G^{\#}\{\xi\} = \alpha\{\eta\} = \{\bar{\eta}\} = (\alpha_L G)^{\#}\{\bar{\xi}\} = (\alpha_L G)^{\#}\alpha\{\xi\}$ by definition of α in (5a.8). □

(5a.9) ***Remark.*** Proposition (5a.8) shows that the functor α carries the exact sequence (5.17) is **C** to the corresponding exact sequence in **K**.

§6 Homotopy groups

Let **C** be a cofibration category with an *initial object* which we denote by $*$.

(6.1) ***Definition.*** A **based object** in **C** is a pair $X = (X, O_X)$ where X is a cofibrant object (that is, $* \to X$ is a cofibration) and where $O = O_X: X \to *$ is a map from X to the initial object. We call $O = O_X$ the **trivial map** on X. A map $f: X \to Y$ between based objects is **based** if $Of = O$ and f is **based up to homotopy** if $Of = O$ in $Ho\mathbf{C}_c$. ||

(6.2) ***Remark.*** If the initial object $*$ is also final object of **C** then each cofibrant object in **C** is based by the unique map $O: X \to *$. In general there might be many maps $X \to *$. For example, let **B** be a cofibration category and let Y be an object in **B**. Then $\mathbf{C} = \mathbf{B}^Y$ is a cofibration category with the initial object $* = (1_Y: Y \to Y)$. A based object in **C** is given by cofibration $Y \rightarrowtail X$ and by a retraction $O_X: X \to Y$. There might be several different retractions from X to Y.

(6.3) ***Example.*** Let B be a cofibration category and let $B \rightarrowtail A$ be a cofibration in **B**. Then we have the cofibration $i_2: A \rightarrowtail A \bigcup_B A$ (inclusion of the second summand) so that $A \bigcup_B A$ is a cofibrant object in $\mathbf{B}^A$. The object $A \bigcup_B A$ in $\mathbf{B}^A$ is based by the folding map $O = (1,1): A \bigcup_B A \longrightarrow A$. We denote this based object in $\mathbf{B}^A$ by $\Sigma_B^o A = (A \underset{i_2}{\rightarrowtail} A \bigcup_B A \underset{O}{\longrightarrow} A)$. We use this example in §10.

(6.4) ***Definition.*** Two based objects X_1, X_2 are **equivalent** if there is a commutative diagram

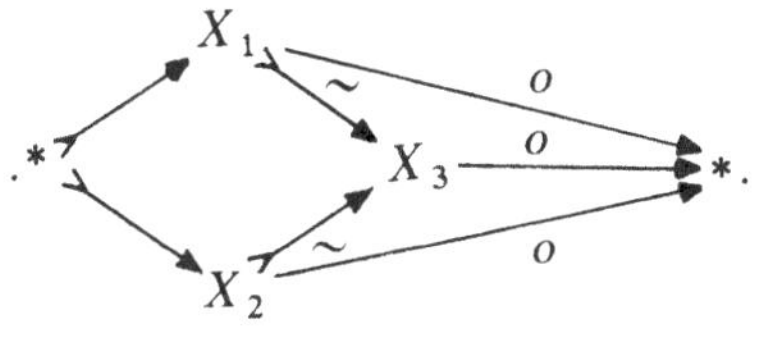

||

Let X be any object in **C** and assume a map $O_X: X \to *$ is given. Then any factorization MX of $* \to X$ is a based object by

$$* \rightarrowtail MX \xrightarrow{\sim} X \longrightarrow * \tag{1}$$

Two such factorizations are equivalent as based objects.

On the other hand a fibrant model $X \overset{\sim}{\rightarrowtail} RX$ of a based object X admits a commutative diagram

$$\begin{array}{ccccc} * & \rightarrowtail & X & \xrightarrow{0_X} & * \\ & & {\scriptstyle\sim}\downarrow & & \downarrow{\scriptstyle\sim} \\ & & RX & \xrightarrow[RO_X]{} & R* \end{array}, \qquad (2)$$

where RO_X is an extension as in (1.6). *In case* $* = R*$ *is fibrant* we can use RO_X as a trivial map of RX; two such choices again are equivalent as based objects. Actually we always may assume that $R* = *$. In case $R* \neq *$ we have $i: * \overset{\sim}{\rightarrowtail} R*$ and hence the functor $i_*: \mathbf{C} = \mathbf{C}^* \to \mathbf{C}^{R*}$ (see (4.5)) induces an equivalence of homotopy theories.

For an object Y and a based object X we always have the **trivial map** $0: X \to * \to Y$. If Y is fibrant this map represents the trivial homotopy class

(6.5) $$0 \in [X, Y] = \pi_0^X(Y).$$

Here $[X, Y]$ denotes the **set of homotopy classes relative** $*$, see (2.3)(b).

For cofibrant objects A and B in $\mathbf{C}$ we get the **sum** $A \vee B$ which is the push out of $A \leftarrowtail * \rightarrowtail B$. If A and B are based then $A \vee B$ is based by $(0, 0): A \vee B \to *$. The inclusions

(6.6) $$i_1: A \rightarrowtail A \vee B, \quad i_2: B \rightarrowtail A \vee B$$

induce the bijection

(6.7) $$[A \vee B, Y] = [A, Y] \times [B, Y].$$

This follows from the fact that $I_*A \vee I_*B$ is a cylinder on $* \rightarrowtail A \vee B$. Here I_*A is a cylinder on $* \rightarrowtail A$.

(6.8) ***Definition.*** For a based object A the **suspension** ΣA is the based object which is defined by the push out diagram

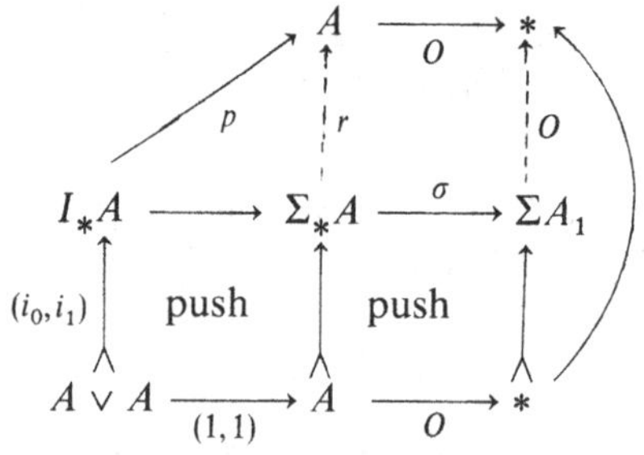

Here Σ_*A is the **torus on** $* \rightarrowtail A$, see (5.11). Clearly, ΣA depends on the choice of the trivial map $0: A \to *$. The composition $\pi_0: I_*A \to \Sigma_*A \to \Sigma A$ is the

canonical homotopy $\pi_0: 0 \simeq 0$ on ΣA. Since ΣA is a based object we can define inductively

(6.9) $$\Sigma^n A = \Sigma(\Sigma^{n-1} A), \quad n \geqq 1, \quad \Sigma^o A = A.$$

The push out σ in the diagram above yields the bijection of sets (see (5.13))

(6.10) $$[\Sigma A, U] \overset{\sigma^*}{=} [\Sigma_* A, U]^0 \text{ with } 0: A \to U.$$

By (5.13) this set has a group structure, via σ^* also $[\Sigma A, U]$ is a group for which 0 is the neutral element. More generally we obtain for $n \geqq 1$ the **homotopy groups**

(6.11) $$\pi_n^A(U) = [\Sigma^n A, U].$$

These groups are only defined with respect to a based object A. A priori, they depend on the choice of cylinders in (6.8), but different choices of cylinders yield *canonically* isomorphic groups. For $n \geqq 2$ the groups $\pi_n^A(U)$ are abelian, see (9.10).

(6.12) ***Example.*** Let $\mathbf{C} = \mathbf{Top}^*$, see (I.5.4). If $A = S^0$ is the *zero sphere*, $S^0 = \{0, 1\}$, then

$$\pi_n X = \pi_n^{S^0}(X) = [\Sigma^n S^0, X] = [S^n, X]$$

is the nth **homotopy group** of the pointed space X. $\pi_1 X$ is the fundamental group of X. The fundamental group $\pi_1 X$ acts on the groups $\pi_n X$. This action is available for the groups in (6.11) as follows:

By use of $\pi_0: I_* A \to \Sigma_* A \to \Sigma A$ we identify a map $\alpha: \Sigma A \to U$ with a homotopy $\alpha\pi_0: 0 \simeq 0, \alpha\pi_0: I_* A \to U$. This leads by (5.7) to the (natural) group action

(6.13) $$\pi_n^A(U) \times \pi_1^A(U) \longrightarrow \pi_n^A(U) \quad (n \geqq 1),$$
$$(\xi, \alpha) \longmapsto \xi^\alpha,$$

with $\xi^\alpha = (\alpha\pi_0)^\#(\xi)$. Here we use (6.10) for $n = 1$ and (10.4)(3) for $n > 1$. Moreover, $\xi \mapsto \xi^\alpha$ is an automorphism of $\pi_n^A(U)$ for each α, in fact, the inner automorphism $\xi \mapsto \xi^\alpha = -\alpha + \xi + \alpha$ for $n = 1$, see (5.15).

(6.14) **Proposition.** *If A is a suspension, $A = \Sigma A'$, then the operation in* (6.13) *is trivial ($\xi^\alpha = \xi$ for all α).*

This is clear for $n = 1$ since $\pi_1^A = \pi_2^{A'}$ is abelian, for $n > 1$ we prove this in (11.14).

For a **based map** $f: A \to B$ we obtain the based map

(6.15) $$\Sigma f: \Sigma A \longrightarrow \Sigma B$$

by using If in (2.10) and by the push out diagram (6.8). More generally we get

$\Sigma^n f:\Sigma^n A \to \Sigma^n B$. These maps induce homomorphisms $f^* = (\Sigma^n f)^*:\pi_n^B(U) \to \pi_n^A(U)$. One can check that $(gf)^* = f^*g^*$ for a based map $g:B \to C$.

(6.16) ***Remark.*** For $X = \Sigma A \vee \Sigma A$ the group structure on $[\Sigma A, X]$, see (3.13), yields maps $m = i_0 + i_1 \in [\Sigma A, \Sigma A \vee \Sigma A]$ and $n = -1 \in [\Sigma A, \Sigma A]$ in the category $Ho\mathbf{C}$, compare (3.6). One easily verifies that $(\Sigma A, m, n, O)$ is a **cogroup** in $Ho\mathbf{C}$, that is: In $Ho\mathbf{C}$ we have the equations $(m \vee 1)m = (1 \vee m)m$, $(1, n)m = (n, 1)m = 0$, and $(1, 0)m = (0, 1)m = 1$. The maps m and n determine the group structure on the functor $\pi_1^A(\ )$ by $+ = m^*$ and $- = n^*$.

§ 6a Homotopy groups and functors

Let $\alpha:\mathbf{C} \to \mathbf{K}$ be a functor between cofibration categories which preserves weak equivalences and which is based, that is α carries the initial object $*$ in $\mathbf{C}$ to the initial object $* = \alpha(*)$ in $\mathbf{K}$, compare (I.1.10); (if α is not based we consider the based functor $\alpha:\mathbf{C} \to \mathbf{K}^{\alpha(*)}$).

A based object $* \rightarrowtail X \xrightarrow{o} *$ in $\mathbf{C}$ yields by the choice of a factorization $M\alpha X$ a based object

(6.a.1) $$* = \alpha(*) \rightarrowtail M\alpha X \xrightarrow{\sim} \alpha X \xrightarrow{\alpha(0)} \alpha * = *$$

in $\mathbf{K}$. Hence the suspension $\Sigma M\alpha X$ is defined. The functor α induces the binatural homomorphism of groups

(6a.2) $$\alpha_L:[\Sigma X, U] \longrightarrow [\Sigma M\alpha X, R\alpha U].$$

This homomorphism is a special case of (5a.4) since we use (6.10) and (5.12). From (5a.5) we deduce the properties of α_L.

There is a map $q_0:\Sigma M\alpha X \to M\alpha\Sigma X$ in $Ho\mathbf{C}$ such that the following diagram commutes

$$\begin{array}{ccc} [\Sigma X, U] & \xrightarrow{\alpha} & [\alpha\Sigma X, \alpha U] \\ \downarrow{\scriptstyle \alpha_L} & & \| \\ [\Sigma M\alpha X, R\alpha U] & \xleftarrow[q_0^*]{} & [M\alpha\Sigma X, R\alpha U] \end{array} .$$

Clearly, q_0 can be obtained by $\alpha_L(\{i\})$ where $i:\Sigma X \rightarrowtail R\Sigma X$ is a fibrant model of ΣX. The function α is given by $Ho\alpha$ is § 4, see also (5a.3). We now define inductively the binatural homomorphism

(6a.3) $$\alpha_L = \alpha_L^n:[\Sigma^n X, U] \longrightarrow [\Sigma^n M\alpha X, R\alpha U] \text{ by } \alpha_L^1 = \alpha_L \quad \text{and}$$
$$\alpha_L^n = (\Sigma^{n-1} q_0)^* \alpha_L^{n-1}.$$

Here the suspension $\Sigma^{n-1} q_0$ is defined as in § 9 below since q_0 is based up to homotopy, $n \geqq 2$. In (8.27)(3) we describe a different way to obtain the map q_0. This shows that q_0^* and $(\Sigma^{n-1}q_0)^*$ are bijections provided α is a model functor.

(6a.4) ***Example.*** Let $\alpha = C_*$ be the functor of (reduced) singular chains from pointed topological spaces to chain complexes, see I.§ 6. Then α_L in (6a.3) yields the classical *Hurewicz homomorphism*

$$\pi_n(U) = [\Sigma^n S^0, U] \longrightarrow [\Sigma^n C_* S^0, C_* U] = H_n U.$$

§7 Relative homotopy groups and the exact homotopy sequence of a pair

Let **C** be a cofibration category with an initial object $*$. Then the category **Pair (C)** of pairs in **C** has the initial object $* = (*, *)$. A **based pair** (A, B), that is, a based object in **Pair (C)**, is given by maps $* \rightarrowtail B \rightarrowtail A \to *$ in **C**. Hence A and B are based objects in **C**. For a based pair we have the homotopy groups in the cofibration category **Pair (C)**, see (6.11),

(7.1) $$[\Sigma^n(A, B), (U, V)] \quad (n \geqq 0),$$

where (U, V) is a fibrant pair. This is a pointed set for $n = 0$, a group for $n = 1$ and an abelian group for $n \geqq 2$. The pair $\Sigma^n(A, B)$ is a pair of suspensions in **C**

(7.2) $$\Sigma^n(A, B) = (\Sigma^n A, \Sigma^n B).$$

This follows from (2.16). Moreover, the comultiplication m on $\Sigma^n(A, B)$ is a map

(7.3) $$m: (\Sigma^n A, \Sigma^n B) \longrightarrow (\Sigma^n A \vee \Sigma^n A, \Sigma^n B \vee \Sigma^n B)$$

in *Ho* **Pair (C)** which restricts to the comultiplication on $\Sigma^n A$ and $\Sigma^n B$ respectively, see (6.16). This shows that the boundary function

(7.4) $$\partial: [\Sigma^n(A, B), (U, V)] \longrightarrow [\Sigma^n B, V]$$

with $\partial\{(f, f')\} = \{f'\}$ is a homomorphism, $n \geqq 1$.

For a based object A in **C** we define the **cone** CA by the push out:

(7.5)

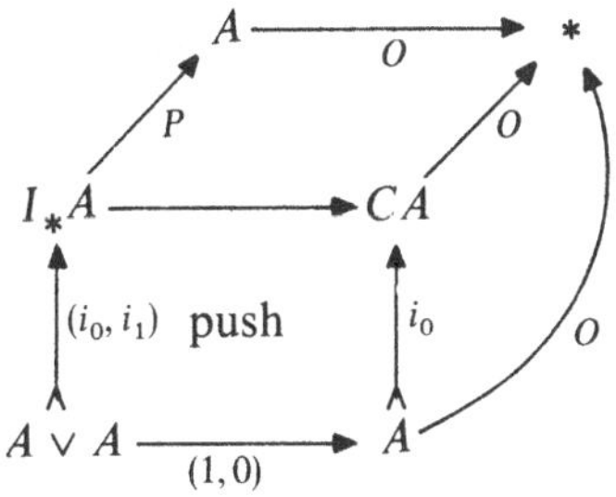

A map $f: A \to U$ is **nullhomotopic** ($f \simeq 0$) if and only if there is an extension

$\bar{f}: CA \to U$ of f. If we replace in (7.5) the map $(1,0)$ by $(0,1)$ we obtain $i_1: A \rightarrowtail C'A$. Since $A \rightarrowtail I_*A$ is a weak equivalence also $* \to CA$ is weak equivalence and thus $CA \to *$ is a weak equivalence.

More generally we say that for a based object A any factorization

$$A \rightarrowtail C \overset{\sim}{\to} * \tag{1}$$

of the trivial map $0: A \to *$ is a **cone on A**; $C = CA$. For two cones C_1, C_2 there is always a commutative diagram

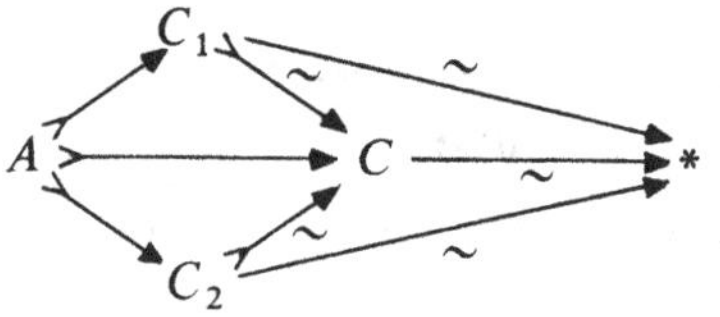

$$\tag{2}$$

where C is a cone on A.

We have the push out map

$$\tag{7.6} \pi: (CA, A) \longrightarrow (\Sigma A, *),$$

compare (6.8) and (1.3). Since the pair (CA, A) in (7.5) is based we obtain $\Sigma^n(CA, A) = (\Sigma^n CA, \Sigma^n A)$, as in (7.2). By (1.10) we can choose a weak lifting T for the diagram

$$
\begin{array}{ccc}
\Sigma^n A & \rightarrowtail & \Sigma^n CA \\
\downarrow{\scriptstyle i_0} & \overset{T}{\nearrow} & \downarrow{\scriptstyle \sim} \\
C\Sigma^n A & \underset{\sim}{\longrightarrow} & *
\end{array}
$$

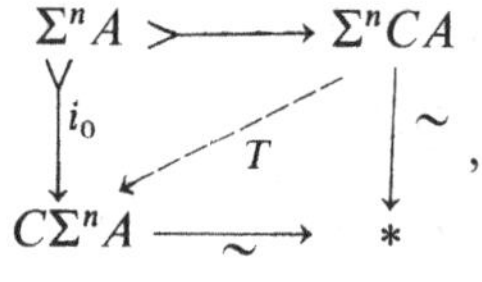

which gives a bijection T^* for the following sets:

$$\tag{7.7} \pi_{n+1}^A(U, V) = [(C\Sigma^n A, \Sigma^n A), (U, V)] \underset{T^*}{\approx} [\Sigma^n(CA, A), (U, V)].$$

We call the set $\pi_{n+1}^A(U, V)$ with the group structure induced by T^* the **relative homotopy groups** of the fibrant pair (U, V). This is an abelian group for $n \geq 2$. The relative homotopy groups are the linking terms in the following *long exact homotopy sequence of the pair* (U, V):

For $n \geq 0$ we consider the sequence

$$\underset{i}{\to} \pi_{n+1}^A(U) \underset{j}{\to} \pi_{n+1}^A(U, V) \underset{\partial}{\to} \pi_n^A(V) \underset{i}{\to} \pi_n^A(U).$$

Here i is induced by $V \to U$, ∂ is given by restriction as in (7.4) and $j = \pi^*$ is induced by the pair map $\pi: (C\Sigma^n A, \Sigma^n A) \to (\Sigma^{n+1} A, *)$ in (7.6).

(7.8) Proposition. *For $n \geq 0$ the sequence is exact, for $n \geq 1$ the functions i, j, ∂*

are homomorphisms of groups. Moreover, the sequence is natural with respect to pair maps $(U, V) \to (U', V')$ in Ho **Pair (C)** *and with respect to based maps $A \to B$ in* **C**.

Remark. (7.8) is as well the cofibration sequence in **Pair(C)** for $(A, *) \to (A, A) \to (CA, A)$, see (8.25) below

Proof. Since we can replace A by $\Sigma^n A$ it is enough to check exactness for $n = 0$. Let $A \xrightarrow{\alpha} V \rightarrowtail U$ be nullhomotopic. Then we have an extension $\bar{\alpha}: CA \to U$ which is a pair map $\bar{\alpha}$: $(CA, A) \to (U, V)$ with $\partial\{\bar{\alpha}\} = \{\alpha\}$. This proves exactness at $\pi_0^A(V)$. Now suppose, we have $(\bar{\alpha}, \alpha):(CA, A) \to (U, V)$ with $H: \alpha \simeq 0$. Then the homotopy extension property of $A \subset CA$ gives us the pair map

$$(G, H):(I_* CA, I_* A) \longrightarrow (U, V).$$

For $(Gi_1, 0):(CA, A) \to (U, V)$ there is $\beta: \Sigma A \to U$ with $j\{\beta\} = \{Gi_1\} = \{Gi_0\} = \{\bar{\alpha}\}$ since $Gi_1 | A = 0$. This proves exactness at $\pi_1^A(U, V)$. Next, suppose $\beta: \Sigma A \to U$ is given with $j\{\beta\} = 0$. Then we have a nullhomotopy

$$(G, H): C(CA, A) = (CCA, CA) \longrightarrow (U, V)$$

with $Gi = G|(CA, A) = \beta\pi$. Since $\beta\pi | A = 0$ we obtain by H a map $\gamma: \Sigma A \to V$ with $\gamma\pi = H$. For the two cofibrations $i_0, Ci_0: CA \rightarrowtail CCA$ (given by the cone on CA and by the pair $C(CA, A)$ respectively) there is a homotopy $Gi_0 \simeq G(Ci_0)$ rel A since $i_0 | A = (Ci_0)| A$ and since CCA is contractible. This proves that $i\{\gamma\} = \{\beta\}$. Thus we have also exactness at $\pi_1^A(U)$.

It remains to show that for $n \geqq 1$ the functions i, j, ∂ are homomorphisms of groups. This is clear for i and ∂. For j we use the following argument:

The based map $\pi:(CA, A) \to (\Sigma A, *)$ and its suspension $\Sigma^n\pi$ in **Pair (C)** induce certainly for $n \geqq 1$ a homomorphism of groups $(\Sigma^n\pi)^*$. Moreover, for the isomorphism $(-1)^n$ the diagram (7.9) commutes. This proves that j is a homomorphism since $n \geqq 1$.

(7.9)
$$\begin{array}{ccc} \pi_{n+1}^A(U) & \xleftarrow[\cong]{(-1)^n} & [\Sigma^{n+1} A, U] \\ {\scriptstyle j}\downarrow & & \downarrow{\scriptstyle (\Sigma^n\pi)^*} \\ \pi_{n+1}^A(U, V) & \xrightarrow[\cong]{T^*} & [\Sigma^n(CA, A), (U, V)] \end{array}.$$
□

Next we consider the action of the fundamental group $\pi_1^A(V)$ on the homotopy sequence (7.8). First, we define the action

(7.10)
$$\pi_{n+1}^A(U, V) \times \pi_1^A(V) \longrightarrow \pi_{n+1}^A(U, V)$$

on the relative homotopy groups as follows: We consider the pair $(n \geqq 1)$.

$Q = (C\Sigma^n A \vee \Sigma A, \Sigma^n A \vee \Sigma A)$. Its long exact homotopy sequence gives us the split short exact sequence

$$0 \longrightarrow \pi^A_{n+1}(Q) \underset{\partial}{\longrightarrow} \pi^A_n(\Sigma^n A \vee \Sigma A) \underset{(0,1)_*}{\longrightarrow} \pi^A_n(\Sigma A) \longrightarrow 0.$$

For the elements $i_0 : \Sigma^n A \to \Sigma^n A \vee \Sigma A, i_1 : \Sigma A \to \Sigma^n A \vee \Sigma A$ we have

(7.11) $$\mu = \mu_n = i_0^{i_1} : \Sigma^n A \to \Sigma^n A \vee \Sigma A \quad \text{in } Ho\mathbf{C},$$

which induces the action in (6.13), that is $\xi^\alpha = \mu^*(\xi, \alpha)$. We call μ the **universal example** of the action.

We now apply the short exact sequence above. Since $(0,1)_*\mu = \mu^*(0,1) = 0^1 = 0$, we know that $\mu \in \text{Image } \partial$. Since ∂ is injective there is a unique map

(7.12) $$(C\Sigma^n A, \Sigma^n A) \xrightarrow{\bar{\mu}} (C\Sigma^n A \vee \Sigma A, \Sigma^n A \vee \Sigma A)$$

in Ho **Pair (C)** with $\partial\bar{\mu} = \mu$. This pair map $\bar{\mu}$ is the universal example for the action (7.10). That is, we set for $\bar{\xi} \in \pi^A_{n+1}(U, V)$ and $\alpha \in \pi^A_1(V)$

$$\bar{\xi}^\alpha = \bar{\mu}^*(\bar{\xi}, \alpha).$$

One easily checks that this is in fact a group action acting via automorphisms. By naturality we have $\partial(\bar{\xi}^\alpha) = (\partial\bar{\xi})^\alpha$, this means, ∂ is an *equivariant* homomorphism. The group $\pi^A_1(V)$ acts on $\pi^A_n(U)$ by (6.13) and by the homomorphism $i : \pi^A_1(V) \to \pi^A_1(U)$.

(7.13) **Proposition.** *All homomorphisms i, j, ∂ of the exact homotopy sequence* (7.8) $(n \geq 1)$ *are equivariant with respect to the action of* $\pi^A_1(V)$.

Proof. We have to check

$$j(\gamma^{i\alpha}) = j(\gamma)^\alpha, \quad \gamma \in \pi^A_{n+1}(U).$$

The universal example for this equation is the following commutative diagram in Ho **Pair (C)**

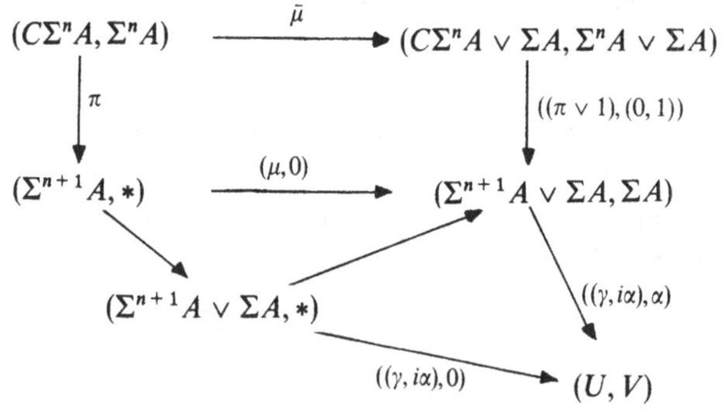

In (11.15) below we show that this diagram commutes. □

(7.14) ***Definition.*** A homomorphism between groups, $\partial:\rho_2 \to \rho_1$, is a **crossed module** if ρ_1 acts from the right on ρ_2 via automorphisms ($\rho_2 \to \rho_2$, $x \longmapsto x^\alpha$, $\alpha \in \rho_1$) such that for $x, y \in \rho_2$

$$\partial(x^\alpha) = -\alpha + \partial x + \alpha. \tag{1}$$

$$-y + x + y = x^{\partial(y)}. \tag{2}$$

A **map** (F, φ) between crossed modules is a commutative diagram

$$\begin{array}{ccc} \rho_2 & \xrightarrow{F} & \rho_2' \\ \Big\downarrow{\scriptstyle \partial} & & \Big\downarrow{\scriptstyle \partial'} \\ \rho_1 & \xrightarrow[\varphi]{} & \rho_1' \end{array} \tag{3}$$

in the category of groups such that F is φ equivariant, that is $F(x^\alpha) = (Fx)^{\varphi\alpha}$. ||

(7.15) ***Remark.*** For a crossed module $\partial:\rho_2 \to \rho_1$ the kernel of ∂ lies in the center of ρ_2, in fact for $y \in \text{kernel}(\partial)$ we have $-y + x + y = x^{\partial y} = x^0 = x$ for all x.

(7.16) **Proposition.** *The homomorphism $\partial:\pi_2^A(U, V) \to \pi_1^A(V)$ is a crossed module for all A, U, and V as in* (7.8).

Proof. (1) in (7.14) is clearly satisfied by (7.13) and (5.15). It remains to check (2) in (7.14). For this we consider the commutative diagram

$$\begin{array}{ccc} \bar{\bar{\mu}} \in \pi_2^A(C(\Sigma A \vee \Sigma A), \Sigma A \vee \Sigma A) & & \\ \Big\uparrow{\scriptstyle i_*} & \searrow{\scriptstyle \partial}\ \cong & \\ & & \pi_1^A(\Sigma A \vee \Sigma A), \\ \bar{\mu} \in \pi_2^A(C\Sigma A \vee \Sigma A, \Sigma A \vee \Sigma A) & \nearrow{\scriptstyle \partial'} & \end{array}$$

where i is the inclusion. For $\bar{\mu}$ in (7.12) we know $\partial'\bar{\mu} = \mu = -i_2 + i_1 + i_2$. On the other hand, for $\bar{\bar{\mu}} = -j_2 + j_1 + j_2$ we get $\partial\bar{\bar{\mu}} = -i_2 + i_1 + i_2$ as well. (Clearly i_1, i_2, j_1, j_2 denote the obvious inclusions). Since $C(\Sigma A \vee \Sigma A)$ is contractible we see that ∂ is an isomorphism. Therefore we get $i_*\bar{\mu} = \bar{\bar{\mu}}$. This, in fact, is the universal example for equation (2) in (7.14). □

(7.17) **Corollary.** *The image of $\pi_2^A(U) \to \pi_2^A(U, V)$ lies in the center of the group $\pi_2^A(U, V)$.*

This is an immediate consequence of (7.15) and of the exactness in (7.8). Finally we consider the end of the exact sequence in (7.8)

$$\pi_1^A(U) \xrightarrow{j} \pi_1^A(U, V) \xrightarrow{\partial} \pi_0^A(V) \xrightarrow{i} \pi_0^A(U)$$

The last three terms are pointed sets. Moreover the sequence comes with a natural action

$$\pi_1^A(U, V) \times \pi_1^A(U) \xrightarrow{+} \pi_1^A(U, V) \tag{7.18}$$

such that j is given by $j(x) = 0 + x$ and such that elements of $\pi_1^A(U, V)$ are in the same orbit if and only if they have the same image in $\pi_0^A(V)$ via ∂. We define $+$ in (7.18) by the map $\mu: CA \to CA \vee \Sigma A$ in $Ho(\mathbf{C}^A)$ which is given by track addition, compare (8.7) below. Now it is easy to check that the action (7.18) has the properties as described.

The homotopy sequence of a pair in (7.8) is compatible with certain functors. Let $\alpha: \mathbf{C} \to \mathbf{K}$ be a functor between cofibration categories which preserves weak equivalences and which is based, (that is, $\alpha(*) = *$). By (7.5) we get

$$M\alpha(CA, A) = (CM\alpha A, M\alpha A). \tag{7.19}$$

The left-hand side is a factorization of $* \to \alpha(CA, A)$ in **Pair (K)**. By (6a.3), applied to the functor α: **Pair (C)** $\to$ **Pair (K)**, we obtain the homomorphism

$$\begin{array}{c} [\Sigma^n(CA, A), (U, V)] = \pi_{n+1}^A(U, V) \\ \downarrow \alpha \\ [\Sigma^n(CM\alpha A, M\alpha A), R\alpha(U, V)] = \pi_{n+1}^{M\alpha A}(R\alpha U, R\alpha V) \end{array} \tag{7.20}$$

between relative homotopy groups. Here $R\alpha(U, V) = (R\alpha U, R\alpha V)$ is a fibrant model of $\alpha(U, V) = (\alpha U, \alpha V)$ is **Pair (K)**.

(7.21) **Theorem.** *With the notation above the functor α induces a commutative diagram*

$$\begin{array}{ccccccc} \pi_{n+1}^A(U) & \to & \pi_{n+1}^A(U, V) & \xrightarrow{\partial} & \pi_n^A(V) & \to & \pi_n^A(u) \\ \downarrow \alpha & & \downarrow \alpha & & \downarrow \alpha & & \downarrow \alpha \\ \pi_{n+1}^{M\alpha A}(R\alpha U) & \to & \pi_{n+1}^{M\alpha A}(R\alpha U, R\alpha V) & \xrightarrow{\partial} & \pi_n^{M\alpha A}(R\alpha V) & \to & \pi_n^{M\alpha A}(R\alpha U) \end{array},$$

the rows of which are the exact sequences in (7.8). *Moreover, α is compatible with the action in* (7.10), *that is,* $\alpha(\xi^\beta) = (\alpha\xi)^{(\alpha\beta)}$.

This is essentially clear by the naturality of α in (6a.3).

(7.22) ***Remark.*** The commutative diagram in (7.21) is a well-known fact for the Hurewicz homomorphism in (6a.4).

§8 Principal cofibrations and the cofiber sequence

Let $f:A\to B$ be a map and let A be a based object. We define the **cofiber (mapping cone)** C_f by the push out diagram in **C**

(8.1)
$$\begin{array}{ccc} CA & \xrightarrow{\pi_f} & C_f \\ {\scriptstyle i_0}\uparrow & \text{push} & \uparrow{\scriptstyle i_f} \\ A & \xrightarrow{f} & B \end{array}$$

Here CA is a cone on A as defined in (7.5). If B is also a based object and if f is a based map then C_f is based by $O=(O_{CA},O_B):C_f\to *$, where O_{CA} is the map in (7.5). In this case i_f is a based map.

(8.2) ***Warning.*** If $B=*$ and if $f=O_A$ then $C_f=\Sigma A$. However, there might be maps $f:A\to *$ with $f\neq O_A$. In this case we have $C_f\neq\Sigma A$. Moreover, C_f is not based in this case.

(8.3) ***Definition.*** We call a cofibration $i:B\rightarrowtail C$ a **principal cofibration** with attaching map $f\in[A,B]$ if there is a map $f_0:A\to RB$, which represents f, together with a weak equivalence $C_{f_0}\xrightarrow{\sim}RC$ in $\mathbf{C}^B$. ||

For example, i_f in (8.1) is a principal cofibration.

A principal cofibration i has the following *characteristic property*. Consider the diagram of unbroken arrows in **C**

(8.4)
$$\begin{array}{ccccc} & & C & & \\ & & {\scriptstyle i}\uparrow & \searrow{\scriptstyle w} & \\ A & \overset{f}{\dashrightarrow} & B & \xrightarrow{u} & U \end{array},$$

where U is fibrant and where $f\in[A,B]$ is the attaching map. The map u can be extended over C (that is, a map w with $w|B=u$ exists) exactly if the element

(8.5)
$$f^*\{u\}\in[A,U]$$

is the trivial element, $f^*\{u\}=0$. For this reason we call $f^*\{u\}$ the **primary obstruction for extending** u.

The obstruction property of $f^*\{u\}$ is a consequence of the fact that the push out (8.1) induces the bijection

(8.6)
$$[I_*A,U]^{uf,0}=[CA,U]^{uf}=[C_f,U]^u$$

of homotopy sets by (2.11). By (8.6) we identify a map $w:C_f\to U$ with a null-homotopy $uf\simeq 0$.

The suspension ΣA **cooperates from the right** on C_f by means of the map

(8.7) $$\mu: C_f \longrightarrow C_f \vee \Sigma A$$

in $Ho\mathbf{C}^B$. By using the identification in (8.6) we define the cooperation μ by track addition $\mu = i_0\pi_f + i_1\pi_0$, where $\pi_0: I_*A \to \Sigma A$ and $\pi_f: I_*A \to CA \to C_f$ are the canonical homotopies and where i_0 and i_1 are the inclusions of C_f and ΣA respectively in $C_f \vee \Sigma A$. Compare (5.5). Let $Y \rightarrowtail B \rightarrowtail C_f$ be cofibrations in $\mathbf{C}$ and let $v: Y \to U$ be given. The cooperation (8.7) induces the function $\mu^* = +$:

(8.8) $$[C_f, U]^Y \times [\Sigma A, U] \xrightarrow{+} [C_f, U]^Y.$$

By (5.6) this is a group action of the group $[\Sigma A, U]$ on the set $[C_f, U]^Y$. If $B \rightarrowtail C$ is a principal cofibration as in (8.3) we have this action as well on the set $[C, U]^v$ by the bijection $[C, U]^v = [C_{f_0}, U]^v$.

If $Y = B$ the action (8.8) is transitive and effective so that for an extension $w: C_f \to U$ of $u: B \to U$ we have the bijection

(8.9) $$w^+: [\Sigma A, U] \xrightarrow{\approx} [C_f, U]^u,$$

defined by $w^+(\alpha) = \{w\} + \alpha$.

If $Y = *$ we obtain by (8.8) the usual action

(8.10) $$[C_f, U] \times [\Sigma A, U] \xrightarrow{+} [C_f, U].$$

The relative cylinder for the triple (C_f, B, Y) gives us the cofibration (see (2.16))

(8.11) $$C_f \cup I_Y B \cup C_f \rightarrowtail I_Y C_f.$$

By track addition we have the map in $Ho\mathbf{C}$

$$\begin{cases} w_f: \Sigma A \longrightarrow C_f \cup I_Y B \cup C_f \\ w_f = -i_0\pi_f + If + i_1\pi_f \end{cases},$$

where $If: I_*A \to I_Y B$ is induced by f, compare (2.10).

(8.12) **Lemma.** *The cofibration* (8.11) *is a principal cofibration with attaching map* w_f.

Proof. By (2.20) we know that $(I_*CA, CA \cup I_*A \cup CA) \to (I_Y C_f, C_f \cup I_Y B \cup C_f)$ is a push out. Since we have the weak equivalence $(C\Sigma A, \Sigma A) \xrightarrow{\sim} (I_*CA, CA \cup I_*A \cup CA)$ of pairs we obtain (8.12) □

Let $u_0, u_1: C_f \to U$ be maps and let $H: u_0|B \simeq u_1|B$ $(H: I_Y B \to U)$ be a homotopy of their restrictions. Then we define the **difference**

(8.13) $$d(u_0, H, u_1) = w_f^*(u_0, H, u_1) \in [\Sigma A, U].$$

We set $d(u_0, u_1) = d(u_0, 0, u_1)$ if $u_0|B = u_1|B$, here 0 is the trivial homotopy. By (8.12) $d(u_0, H, u_1) = 0$ iff the homotopy H is extandable to a homotopy $G: u_0 \simeq u_1$. Clearly, the homotopy $-u_0\pi_f + H(If) + u_1\pi_f: \Sigma A \to U$ represents $d(u_0, H, u_1)$. By (5.6) we see

(8.14) $$d(u_0, H, u_1) + d(u_1, G, u_2) = d(u_0, H + G, u_2).$$

Moreover, for the action $+$ in (8.8) we have the equation

(8.15) $$\{u_0\} + d(u_0, H, u_1) = \{u_1\} \in [C_f, U]^Y,$$

where $u_0|Y = u_1|Y$ and where $H: u_0|B \simeq u_1|B$ is a homotopy rel Y.

We now describe the *exact classification sequence* for a principal cofibration. Suppose we have a commutative diagram in **C**

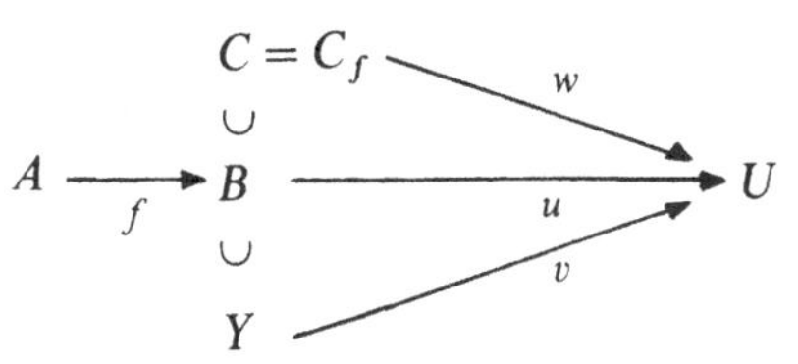

Then we have the following sequence of homotopy sets:

(8.16) $$[\Sigma_Y B, U]^u \xrightarrow[\Sigma(w,f)]{} [\Sigma A, U] \xrightarrow[w^+]{} [C_f, U]^v \xrightarrow[i_f^*]{} [B, U]^v \xrightarrow[f^*]{} [A, U].$$

Here we define w^+ by $w^+(\alpha) = \{w\} + \alpha$ and we set $\Sigma(w, f)(H) = d(w, H, w)$, where we identify the element $H \in [\Sigma_Y B, U]^u$ with a track $H: u \simeq u$. By (8.14) it is easy to see that $\Sigma(w, f)$ is a homomorphism of groups. Moreover,

(8.17) **Proposition.** *The sequence* (8.16) *is exact in the following sense*:

$$\text{Image}\,(i_f^*) = (f^*)^{-1}(0),$$
$$\text{Image}\,(w^+) = (i_f^*)^{-1}\{u\},$$
$$\text{Image}\,\Sigma(w,f) = (w^+)^{-1}(\{w\}).$$

Thus the group 'Image $\Sigma(w,f)$' is the **isotropy group** of the action $+$ in (8.8) in $\{w\}$. If the group $[\Sigma A, U]$ is abelian the isotropy group depends only on the orbit of $\{w\}$ which is characterized by the restriction $\{u\}$ of $\{w\}$. Moreover, if $[\Sigma A, U]$ is abelian the homomorphism $\Sigma(w,f) = \Sigma(u,f)$ depends only on u. In § 10 we discuss the prolongation of (8.16).

Proof of (8.17). The first equation is a consequence of the obstruction property in (8.5). The second equation follows from (8.15). The third equation is obtained as follows: Let $\{w\} + \{\alpha\} = \{w\}$ and let $H: (w, \alpha)\mu \simeq w$ rel Y be a homotopy. Since $(w, \alpha)\mu | B = u$ the restriction of H to B is a self homotopy $H': u \simeq u$ rel Y for which by (8.13) $0 = d(w + \alpha, H', w) = d(w + \alpha, w) +$

$d(w, H', w) = -d(w, w+\alpha) + \Sigma(w,f)(H') = -\alpha + \Sigma(w,f)(H')$. Therefore $\alpha \in \operatorname{Im}\Sigma(w,f)$. On the other hand if $\alpha \in \operatorname{Im}\Sigma(w,f)$ then H' exists such that $0 = -\alpha + \Sigma(w,f)H'$. The equations above then imply $0 = d(w+\alpha, H', w)$. Therefore H' can be extended to a homotopy $w + \alpha \simeq w \operatorname{rel} Y$. This proves the third equation. □

Assume now in (8.16) that $Y = *$, that B is a based object and that f is a based map. Then also C_f is based and we obtain for $w = 0: C_f \to * \to U$ the following special case of the exact sequence (8.16):

(8.18) $$[\Sigma B, U] \xrightarrow{(\Sigma f)^*} [\Sigma A, U] \xrightarrow{0^+} [C_f, U] \xrightarrow{i_f^*} [B, U] \xrightarrow{f^*} [A, U].$$

This is the classical cofibration sequence (Puppe sequence). We now describe a different approach which yields this sequence as well.

Let A be a based object and let $A \rightarrowtail B$ be a cofibration. Then we define the **cofiber** B/A by the push out diagram.

(8.19) $$\begin{array}{ccc} B & \xrightarrow{q} & B/A \\ {\scriptstyle i}\uparrow & \text{push} & \uparrow \\ A & \xrightarrow[0]{} & * \end{array} .$$

We call $A \overset{i}{\rightarrowtail} B \xrightarrow{q} B/A$ a **cofiber sequence**. By (2.18) we see that the induced sequence of homotopy sets

(8.20) $$[B/A, U] \xrightarrow{q^*} [B, U] \xrightarrow{i^*} [A, U]$$

is exact, that is $(i^*)^{-1}0 = \text{Image } q^*$.

If (B, A) is a based pair then B/A is based and also $B \to B/A$ is a based map. For example we have the following cofibers which are based:

(8.21) $$\begin{cases} CA = I_*A/i_1A, & C'A = I_*A/i_0A, \\ \Sigma A = CA/A = C'A/A = I_*A/(A \vee A) \end{cases}$$

If $f: A \rightarrowtail B$ is a cofibration then also $\pi_f: CA \rightarrowtail C_f$ is a cofibration. In this case we have canonical weak equivalences of cofibers

(8.22) $$C_f \xrightarrow{\sim} C_f/CA = B/A,$$

which are based maps if f is based. We see this by the push out diagrams

$$\begin{array}{ccccc} B & \rightarrowtail & C_f & \xrightarrow{\sim} & C_f/CA \\ {\scriptstyle f}\uparrow & \text{push} & \uparrow & \text{push} & \uparrow \\ A & \rightarrowtail & CA & \xrightarrow{\sim} & * \end{array} .$$

Next we define the cofiber of a map $f: A \to B$ where A is a based object. We choose a factorization A_1 by (C3) and we get the diagram

(8.23)
$$\begin{array}{ccc} A & \xrightarrow{\quad f \quad} & B \\ & \searrow \qquad \nearrow{\scriptstyle \sim}\ p & \\ & A_1 & \xrightarrow[f_1 = q]{\qquad} B_1 = A_1/A \end{array} .$$

We call $B_1 = A_1/A$ the **cofiber of** f. For example we can choose for A_1 the mapping cylinder Z_f of f, see (I.1.8). In this case the cofiber of f is the mapping cone C_f of f, see (8.1).

Assume now that f is a based map. Then also A_1 is based by $0 = 0_B p$ and (A_1, A) is a based pair. Thus f_1 is a based map and we can apply an inductive procedure by replacing f in (8.23) by f_1, etc. This yields the following commutative diagram of based objects and based maps:

(8.24)
$$\begin{array}{ccccccc} & B & & B_1 & & B_2 & \\ {\scriptstyle f}\nearrow & & \nwarrow{\scriptstyle\sim}\ \nearrow & & \nwarrow{\scriptstyle\sim}\ \nearrow & & \nwarrow{\scriptstyle\sim} \\ A & \rightarrowtail & A_1 & \rightarrowtail & A_2 & \rightarrowtail & A_3 \rightarrowtail \cdots \end{array} ,$$

where B_{i+1} is the cofiber of $A_i \rightarrowtail A_{i+1}$ for $i \geq 1$ and where A_{i+1} is a factorization of $q: A_i \to B_i$.

On the other hand the cofiber of $B \rightarrowtail C_f$ is $C_f/B = CA/A = \Sigma A$. This yields the top row of the diagram

$$\begin{array}{ccccccccccc} A & \xrightarrow{f} & B & \overset{i}{\rightarrowtail} & C_f & \xrightarrow{(-1)q} & \Sigma A & \xrightarrow{\Sigma f} & \Sigma B & \xrightarrow{\Sigma i} & \\ \| & & \uparrow\sim & & \uparrow\sim & & \uparrow\sim & & \uparrow\sim & \cdots & \\ A & \rightarrowtail & A_1 & \rightarrowtail & A_2 & \rightarrowtail & A_3 & \rightarrowtail & A_4 & \rightarrowtail & \cdots \end{array} . \qquad (*)$$

Here f, i and q are based maps, the suspensions of which form the prolongation $\Sigma^n f, \Sigma^n i, \Sigma^n((\pm 1)q)$ $(n \geq 1)$ of the top row. As in the classical case we see

(8.25) **Proposition.** *There exists a commutative diagram* $(*)$ *in* $Ho\mathbf{C}$ *in which all vertical arrows are isomorphisms in* $Ho\mathbf{C}$.

If we apply the functor $[\ ,U]$ we derive from (8.25) and (8.20) the exact sequence in (8.18) together with the long exact continuation:

$$\cdots \longrightarrow [\Sigma^2 A, U] \xrightarrow{(\Sigma q)^*} [\Sigma C_f, U] \xrightarrow{(\Sigma i)^*} [\Sigma B, U] \xrightarrow{(\Sigma f)^*} \cdots .$$

(8.26) **Proposition.** $(\Sigma q)^*[\Sigma^2 A, U]$ *lies in the center of the group* $[\Sigma C_f, U]$.

We prove (8.26) in the proof of (9.10) below.

Finally, we consider principal cofibrations in connection with functors. Let $\alpha:\mathbf{C}\to\mathbf{K}$ be a *based functor which preserves weak equivalences*, see (I.1.10). For the principal cofibration (8.1) we obtain the following commutative diagram in $\mathbf{K}$ for which the front square is a push out:

(8.27)

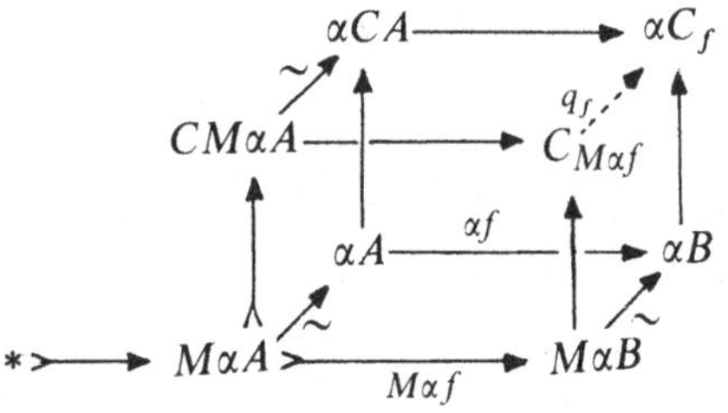

Here $* \rightarrowtail M\alpha f \xrightarrow{\sim} \alpha f$ is a factorization of $* \to \alpha f$ in **Pair (K)** and $CM\alpha A$ is obtained by a factorization of $M\alpha A \to \alpha A \to \alpha CA$. Diagram (8.27) yields the canonical map q_f which is a weak equivalence provided the functor α is compatible with the push out $C_f = CA \bigcup_f B$, see (I.1.10) and (I.4b.4).

Clearly, $M\alpha A$ is a based object by

$$* \rightarrowtail M\alpha A \xrightarrow{\sim} \alpha A \xrightarrow{\alpha 0} \alpha * = *. \tag{1}$$

Moreover, $M\alpha f$ is a based map if f is based. In case $f = 0: A \to *$ is the trivial map we get by q_f the map

$$q_0:\Sigma M\alpha A = C_{M\alpha f}/M\alpha B \longrightarrow \alpha\Sigma A. \tag{2}$$

The map q_f is compatible with the cooperation in the sense that the following diagram commutes in $Ho(\mathbf{K}^{M\alpha B})$.

$$\begin{array}{ccc} C_{M\alpha f} & \xrightarrow{\ \mu\ } & C_{M\alpha f} \vee \Sigma M\alpha A \\ \Big\downarrow{\scriptstyle q_f} & & \Big\downarrow{\scriptstyle ((\alpha i_1)q_f,(\alpha i_2)q_0)} \\ \alpha C_f & \xrightarrow{\ \alpha\mu\ } & \alpha(C_f \vee \Sigma A) \end{array} \quad . \tag{3}$$

Here $i_1: C_f \to C_f \vee \Sigma A$ and $i_2: \Sigma A \to C_f \vee \Sigma A$ are the canonical maps. Since μ is defined by addition of tracks we derive from (5a.5) that diagram (3) commutes.

Now assume that f is a based map between based objects. Then also $M\alpha f$ is a based map between based objects and hence we have cofiber sequences (8.25) for f and $M\alpha f$ respectively. These yield the following commutative diagram in $Ho(\mathbf{K})$

$$\begin{array}{ccccccccc}
\alpha A & \xrightarrow{\alpha f} & \alpha B & \longrightarrow & \alpha C_f & \longrightarrow & \alpha\Sigma A & \xrightarrow{\alpha\Sigma f} & \alpha\Sigma B\cdots \\
\sim\uparrow & & \uparrow\sim & & \uparrow q_f & & \uparrow q_0 & & \uparrow q_0\cdots \\
M\alpha A & \xrightarrow[M\alpha f]{} & M\alpha B & \longrightarrow & C_{M\alpha f} & \longrightarrow & \Sigma M\alpha A & \xrightarrow[\Sigma M\alpha f]{} & \Sigma M\alpha B\cdots
\end{array} \tag{4}$$

where q_f, $q_0\cdots$ are isomorphisms in $Ho(\mathbf{K})$ provided α is a model functor.

§9 Induced maps on cofibers

Let $f:(A,B)\to(X,Y)$ be a pair map. We consider the diagram

(9.1)
$$\begin{array}{ccccccc}
 & \overset{o}{\swarrow} & B & \rightarrowtail & A & \longrightarrow & A/B \\
* & & \downarrow f' & & \downarrow f & & \downarrow f'' \\
 & \underset{o}{\nwarrow} & Y & \rightarrowtail & X & \xrightarrow{q} & X/Y
\end{array}$$

where B and Y are based objects. If f' is a based map, that is $Of'=0$, then the induced map f'' on cofibers is well defined by $f''=f\cup 1$, see (1.2). In this case diagram (9.1) commutes. If (A,B) and (X,Y) are based pairs and if (f,f') is based map then also f'' is a based map.

Next we assume that f' in (9.1) is only *based up to homotopy*, see (6.1). Then we obtain a map f'' in the homotopy category $Ho\mathbf{C}_c$ by lemma (9.4) below.

For a based object B the group $[\Sigma B,*]$ needs not to be trivial. This group acts on the homotopy set $[A/B,U]=[A,U]^B$ of homotopy classes relative $O:B\to *\to U$ as follows. By the canonical homotopy π_0 in (6.8) we associate with $\beta\in[\Sigma B,*]$ the homotopy $\bar\beta:O\simeq O$ with $\bar\beta:I_*B\xrightarrow{\pi_0}\Sigma B\xrightarrow{\beta}R*\to U$. By (5.7) we thus obtain the action

(9.2)
$$\begin{aligned}
[A/B,U]\times[\Sigma B,*] &\longrightarrow [A/B,U], \\
(x,\beta) &\longmapsto \bar\beta^{\#}(x)=x^\beta
\end{aligned}$$

which we call the **$*$-action on the cofiber.**

(9.3) ***Examples for the $*$-action.***

(a) Consider cofiber sequence $A\rightarrowtail\Sigma_*A\to\Sigma A$. From (5.15) we deduce that the action of $[\Sigma A,*]$ on $[\Sigma A,U]$ is given by inner automorphisms, that is: $\xi^\alpha=-\bar\alpha+\xi+\bar\alpha$, where $\bar\alpha:\Sigma A\xrightarrow{\alpha}R*\to U$.

(b) Along the same lines as in (5.15) one checks that the $*$-action for the cofiber sequence $A\rightarrowtail CA\to\Sigma A$ is given by $\xi^\alpha=-\bar\alpha+\xi$.

In particular, we see that the action of $[\Sigma A,*]$ on $[\Sigma A,U]$ depends on how

ΣA is obtained as a quotient. Note that even if $A=\Sigma A'$ is a suspension, the action in (b) in non-trivial if there are non-trivial homotopy classes $\Sigma A \to *$, while the action in (a) is trivial in this case since, as we will see, the group $[\Sigma\Sigma A', U]$ is abelian.

(9.4) **Lemma.** *Let φ: $(A,B)\to(X,Y)$ be a map in Ho* **Pair** **($\mathbf{C}_c$)**, *where (A,B) and (X,Y) are given as in (9.1) (B and Y are based objects). If the restriction $\varphi': B\to Y$ of φ in $Ho\mathbf{C}_c$ is based up to homotopy, that is $O_*\varphi = O$, then there exists a commutative diagram*

$$\begin{array}{ccc} (A,B) & \xrightarrow{\ \varphi\ } & (X,Y) \\ \downarrow{\scriptstyle q} & & \downarrow{\scriptstyle q} \\ (A/B,*) & \xrightarrow{\ \varphi''\ } & (X/Y,*) \end{array}$$

in Ho **Pair ($\mathbf{C}_c$)**. *φ'' is well defined up to the $*$-action on the cofiber by this property.*

We call any φ'' as in (9.4) a **quotient map** for (φ, φ').

Proof of (9.4). We consider the following diagram in which (f,f') represents φ.

$$\begin{array}{ccccccc} & \overset{0}{\swarrow} & B & \rightarrowtail & A & \xrightarrow{\ q\ } & A/B \\ R* & ① & \downarrow{\scriptstyle f'} & ② & \downarrow{\scriptstyle f} & ③ & \downarrow{\scriptstyle f''} \\ & \underset{RO}{\nwarrow} & RY & \rightarrowtail & RX & \xrightarrow[Rq]{} & R(X/Y) \end{array} .$$

Here RX is a fibrant model with $R* \overset{i}{\rightarrowtail} RX$. ② commutes, but ① homotopy commutes (φ' is based up to homotopy). Let $H:(RO)f' \simeq O \text{ rel} *$, $H: I_*B \to R*$ be a homotopy. Then

$$f'' = \bar{H}^{\#}\{(Rq)f\} \in [A, R(X/Y)]^B = [A/B, R(X/Y)] \tag{9.5}$$

where $\bar{H}: I_*B \xrightarrow{H} R* \overset{i}{\rightarrowtail} RX \xrightarrow{Rq} R(X/Y)$. Now f'' represents φ''. Any φ'' for which the diagram in (9.4) commutes can be constructed this way. This implies that the set of quotient maps for (φ, φ') is given by the orbit $(\varphi'')^\alpha$, $\alpha \in [\Sigma B, *]$. □

(9.6) **Lemma.** *Let $\varphi, \varphi', \varphi''$ be given as in (9.4). If φ' and φ: $A\to X$ are isomorphisms in $Ho\mathbf{C}_c$ then also φ'' is an isomorphism in $Ho\mathbf{C}_c$.*

Next we want to derive from (9.4) the suspension operator for homotopy classes. To this end we consider the functor

$$\Sigma. : Ho\mathbf{C}_c \to Ho\,\mathbf{Pair}(\mathbf{C}_c), \tag{9.7}$$

which carries X to the pair $(\Sigma_* X, X)$ and $\varphi: A \to X$ to $(\Sigma.\varphi, \varphi):(\Sigma_* A, A) \to (\Sigma_* X, X)$. Here $\Sigma.\varphi$ is defined as follows: If φ is represented by $f: A \to RX$ then $\Sigma.\varphi$ is represented by $\Sigma.f:\Sigma_* A \to \Sigma_* RX$, compare (5.11). We leave it to the reader to check that $\Sigma.$ is a well defined functor.

Now let A and X be based objects and let $\varphi: A \to X$ be a map in $Ho\mathbf{C}_c$ which is based up to homotopy. Then a **suspension** of φ, $\Sigma\varphi$, is given by a quotient map of $(\Sigma.\varphi, \varphi)$ as in (9.4). That is, $\Sigma\varphi$ is any map for which the diagram

$$\begin{array}{ccc} (\Sigma_* A, A) & \xrightarrow{(\Sigma\varphi, \varphi)} & (\Sigma_* X, X) \\ \downarrow q & & \downarrow q \\ (\Sigma A, *) & \xrightarrow{\Sigma\varphi} & (\Sigma X, *) \end{array}$$

commutes in Ho **Pair** $(\mathbf{C}_c)$. Since $\Sigma\varphi$ is only well defined up the $*$-action we cannot say that Σ is a functor. However, if the grop $[\Sigma A, \Sigma X]$ is *abelian* then the $*$-action is trivial by example (9.3) (a), and in this case the function

(9.8) $$\Sigma:[A, X]_0 \to [\Sigma A, \Sigma X]_0$$

is well defined. Here $[A, X]_0$ denotes the subset of all elements $\varphi \in [A, X]$ with $0_* \varphi = 0$. The suspension of a based map in $\mathbf{C}$ is defined in (6.15). By (9.8) we have the suspension of a map which is based up to homotopy. Using cones we can characterize the suspension Σf as follows.

(9.9) **Lemma.** *Let $f: A \to B$ be based up to homotopy and assume $[\Sigma A, \Sigma B]$ is abelian. Then $\Sigma f: \Sigma A \to \Sigma B$ is the unique map in $Ho\mathbf{C}$ which is a quotient map of $(Cf, f):(CA, A) \to (CB, B)$ and which is based up to homotopy.*

In the lemma $Cf: CA \to CB$ denotes any extension of f; Cf exists since f is based up to homotopy. The map (Cf, f) however is not well defind in Ho **Pair** $(\mathbf{C}_c)$. The map (Cf, f) is a quotient map (as in (9.4)) of $(\mathrm{I}f, f \vee f):(I_* A, A \vee A) \to (I_* B, B \vee B)$. This yields the proposition in (9.9).

(9.10) **Proposition.**

(*a*) *Let $\varphi \in [A, X]_0$ and let U be a fibrant object in $\mathbf{C}$. Then any suspension $\Sigma\varphi$ of φ induces a homomorphism of groups*

$$(\Sigma\varphi)^*: [\Sigma X, U] \to [\Sigma A, U].$$

(*b*) *For $n \geqq 2$ the group $[\Sigma^n A, U]$ is abelian.*

(*c*) *The diagram*

$$\begin{array}{ccc} \Sigma\Sigma A & \xrightarrow{\Sigma m} & \Sigma(\Sigma A \vee \Sigma A) \\ \| & & \uparrow \sim \, (\Sigma i_1, \Sigma i_2) \\ \Sigma^2 A & \xrightarrow{m_\Sigma} & \Sigma^2 A \vee \Sigma^2 A \end{array}$$

commutes in $Ho\mathbf{C}_c$. Here m and m_Σ are the comultiplications on ΣA and $\Sigma^2 A$ respectively, see (6.16).

(*d*) *The function Σ in* (9.8) *is a well defined homomorphism of groups if $A = \Sigma A'$ is a suspension.*

Proof. We first proof (*a*). Let $\varphi \in [A, X]_0$ be represented by $f: A \to RX$. Choose a map $u: R* \to U$. Then we have the following natural isomorphism of groups:

$$[\Sigma X, U] \approx [\Sigma_* X, U]^O \approx [\Sigma_* RX, U]^{u(RO)},\ RO: RX \to R*,$$
$$[\Sigma A, U] \approx [\Sigma_* A, U]^O.$$

By (9.5) and (9.7) we see that the function

$$(\Sigma\varphi)^*: [\Sigma_* RX, U]^{u(RO)} \to [\Sigma_* A, U]^O$$

is given by the formula $(\Sigma\varphi)^*(\xi) = (uH)^\#(\Sigma.f)^*(\xi)$. Since $(uH)^\#$ and $(\Sigma.f)^*$ are homomorphisms of groups, $(\Sigma\varphi)^*$ is also a homomorphism of groups.

We now prove (*b*). It is enough to consider $n = 2$. For $n = 2$ the proposition in (*b*) is a special case of (8.26). In fact, let f in (8.26) be the trivial map $O: A \to *$; then $C_f = \Sigma A$ and $q = 1$. We therefore *prove more generally* **(8.26).** Let $f: A \to B$ be a based map between based objects and let $q: C_f \to C_f/B = \Sigma A$ be the quotient map and let $\mu: C_f \to C_f \vee \Sigma A$ be the coaction in (8.7). Now q is a based map and μ is based up to homotopy. (For $C_f = \Sigma A$ the map μ coincides with m.)

Let $p_1 = (1, 0): C_f \vee \Sigma A \to C_f$ and $p_2 = (0, 1): C_f \vee \Sigma A \to \Sigma A$ be the projections. We have $p_1\mu = 1$ and $p_2\mu = q$. Since μ is based up to homotopy we can choose a suspension $\Sigma\mu$ of μ. Now $\Sigma\mu$ is only well defined up to the $*$-action, so we cannot expect that $\Sigma p_1 \Sigma\mu = 1$ and $\Sigma p_2 \Sigma\mu = \Sigma q$. But there are $\alpha, \delta \in [\Sigma C_f, *]$ such that $\Sigma p_1 \Sigma\mu = 1^\alpha$ and $\Sigma p_2 \Sigma\mu = (\Sigma q)^\delta$. Since μ is well defined in $Ho(\mathbf{C}^B)$ we may assume that $\delta = (\Sigma q)^*(\beta)$ for some $\beta \in [\Sigma^2 A, *]$. The map $(\Sigma i_1, \Sigma i_2): \Sigma C_f \vee \Sigma^2 A \to \Sigma(C_f \vee \Sigma A)$ is an isomorphism in $Ho(\mathbf{C}_c)$. Therefore the homomorphism.

$$\lambda = ((\Sigma i_1)^*, (\Sigma i_2)^*): [\Sigma(C_f \vee \Sigma A), U] \to [\Sigma C_f, U] \times [\Sigma^2 A, U]$$

is bijective. Using (a) we see that the map

$$\tilde{\mu} = (\Sigma\mu)^* \lambda^{-1}: [\Sigma C_f, U] \times [\Sigma^2 A, U] \to [\Sigma C_f, U]$$

is a homomorphism of groups; we define $\xi \oplus \eta = \tilde{\mu}(\xi^{-\alpha}, \eta^{-\beta})$. Since the maps

$\xi \to \xi^{-\alpha}, \eta \to \eta^{-\beta}$ are also homomorphisms, there is the following distributivity law:

$$(\xi + \xi') \oplus (\eta + \eta') = (\xi \oplus \eta) + (\xi' \oplus \eta'). \tag{$*$}$$

We now show that the equations

$$\xi \oplus 0 = \xi \quad \text{and} \quad 0 \oplus \eta = (\Sigma q)^* \eta \tag{$**$}$$

hold. Note that $i_1, i_2, p_1, p_2, 1$ are based maps, so their suspensions are well defined, and we have

$$\Sigma p_j \Sigma i_k = \begin{cases} 1 & \text{if } j = k \\ 0 & \text{if } j \neq k \end{cases}$$

It follows that $\lambda(\Sigma p_1)^*(\xi^{-\alpha}) = (\xi^{-\alpha}, 0)$, hence $\xi \oplus 0 = \tilde{\mu}(\xi^{-\alpha}, 0) = (\Sigma \mu)^*(\Sigma p_1)^* \xi^{-\alpha} = (1^\alpha)^*(\xi^{-\alpha}) = \xi$. Similarly $0 \oplus \eta = (\Sigma q)^* \eta$. By ($*$) and ($**$) we get:

$$\xi + (\Sigma q)^* \eta = (\xi \oplus 0) + (0 \oplus \eta) = (\xi + 0) \oplus (0 + \eta) = \xi \oplus \eta,$$

and

$$(\Sigma q)^* \eta + \xi = (0 \oplus \eta) + (\xi \oplus 0) = (0 + \xi) \oplus (\eta + 0) = \xi \oplus \eta.$$

This proves the proposition in (8.26) and (9.10)(*b*).

Finally, we prove (*c*) and (*d*) in (9.10). For $f = 0$ we see by (9.3)(*a*) and (9.10)(*b*) that the $*$-action is trivial and hence $\xi^{-\alpha} = \xi, \eta^{-\beta} = \eta$. This shows $\xi + \eta = \xi \oplus \eta = \tilde{\mu}(\xi, \eta)$. Now (*c*) is just the universal example for this equation (put $U = \Sigma(\Sigma A \vee \Sigma A)$ and $\xi = \Sigma i_1, \eta = \Sigma i_2$). Finally (9.10)(*d*) is an immediate consequence of (9.10)(*a*), (*b*), (*c*). □

§ 10 Homotopy groups of function spaces

Let **C** be a cofibration category and let (A, B) be a pair in **C**; hence (A, B) denotes a map $i_A: B \to A$ in **C**. A factorization $i_A = qi: B \rightarrowtail MA \xrightarrow{\sim} A$ of this map yields the push out diagram

(10.1)
$$\begin{array}{ccccc} MA & \longrightarrow & MA \bigcup_B A & \xrightarrow{0=(q,1)} & A \\ {\scriptstyle i}\uparrow & \text{push} & {\scriptstyle i}\uparrow & \nearrow {\scriptstyle 1} & \\ B & \xrightarrow[i_A]{} & A & & \end{array} .$$

This shows that $\Sigma_B^0 A = (A \rightarrowtail MA \bigcup_B A \to A)$ is a based object in the cofibration category $\mathbf{B} = \mathbf{C}^A$ of objects under A. In case i_A is a cofibration $\Sigma_B^0 A$ is the example in (6.3). The n-fold suspension $\Sigma^n = \Sigma_{\mathrm{B}}^n$ in the category **B** gives us the based object in **B**

(10.2)
$$\Sigma_{\mathrm{B}}^n(\Sigma_B^0 A) = (\Sigma \rightarrowtail \Sigma_B^n A \xrightarrow{r} A).$$

Here $\Sigma_B^1 A = \Sigma_B A$ is the torus in (5.11) provided i_A is a cofibration in **C**. We see this by the following remark on suspensions in $\mathbf{C}^A$.

(10.3) ***Remark.*** Let $X = (A \rightarrowtail X \xrightarrow{o} A)$ be a based object in $\mathbf{C}^A = \mathbf{B}$. Then the torus $\Sigma_A X$ (in (5.11) yields the suspension $\Sigma_{\mathrm{B}}(X)$ in **B** by the following push out diagram in **C**:

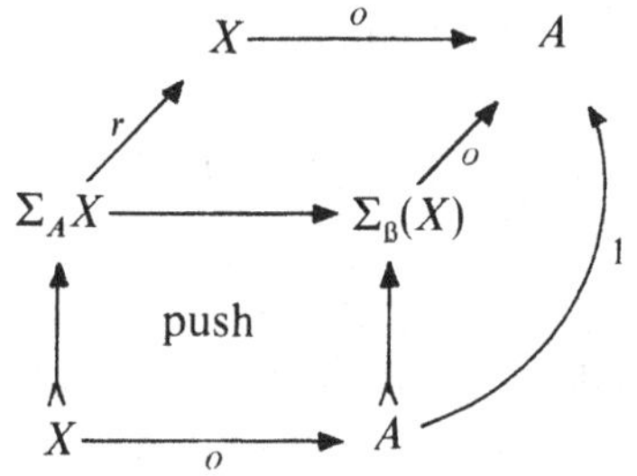

(1)

We have by (2.20) the equation

$$\Sigma_A\left(MA \bigcup_B A\right) = (\Sigma_B MA) \bigcup_B A. \tag{2}$$

This shows that $\Sigma_{\mathrm{B}}(\Sigma_B^0 A) = \Sigma_B MA \bigcup_{MA} A$ where $MA = A$ if i_A is a cofibration. More generally we get

$$\Sigma_{\mathrm{B}}^n(\Sigma_B^0 A) = \Sigma_B^n MA \bigcup_{MA} A, \tag{3}$$

where $\Sigma_B^n MA$ is defined by (10.2) and where $(r, 1_A) = 0$ is the trivial map of the suspension (3). For a cofibration $B \rightarrowtail A$ the object (10.2) can be obtained inductively by the push out diagram ($n \geqq 1$) in **C**

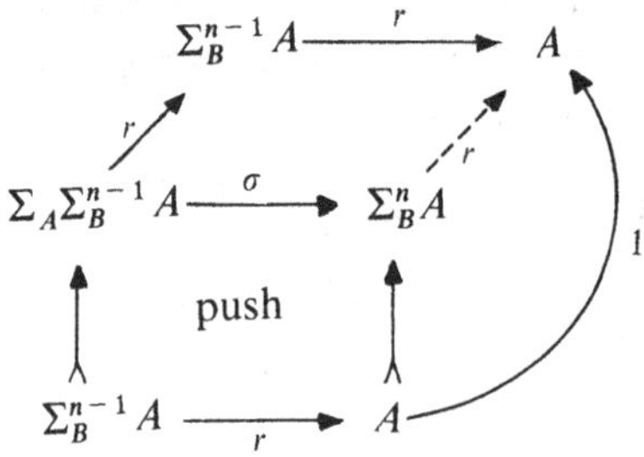

(4)

This follows from (1) and (2).

Now let $u: A \to U$ be a map in **C**; hence u is an object in $\mathbf{B} = \mathbf{C}^A$ which is fibrant in **B** if U is fibrant in **C**. We define the **nth homotopy group of the function space** by

(10.4) $$\pi_n(U^{A|B}, u) = [\Sigma_B^n A, U]^u = [\Sigma_{\mathrm{B}}^n(\Sigma_B^0 A), u]$$

Here the right-hand side is a homotopy group in **B** as defined in (6.11). For $n=1$ the group (10.4) coincides with the group in (5.13). For $n \geqq 2$ the groups (10.4) are abelian by (9.10).

Using the push out (σ, r) in (10.3) (4) one has the isomorphism of groups

$$\sigma^*:[\Sigma_B^n A, U]^u \cong [\Sigma_A \Sigma_B^{n-1} A, U]^{ur}, \tag{1}$$

where the right-hand side is the group defined by (5.13); compare (2.11) (*b*) and (6.10).

For $n=0$ we obtain the bijection of sets; compare (3.13):

$$\pi_0(U^{A|B}, u) = [A, U]^{u|B} = [\Sigma_B^0 A, u]. \tag{2}$$

The map $u|B$ denotes the composition $ui_A: B \to U$.

Finally we get for a based object $A = (* \rightarrowtail A \to *)$ in **C** the equation $\Sigma^n A = \Sigma_*^n A/A$ and hence

$$[\Sigma^n A, U] = \pi_n(U^{A|*}, 0) = [\Sigma_*^n A, U]^0. \tag{3}$$

(10.5) ***Remark.*** In the cofibration category **Top** the homotopy group (10.4) coincides with the nth homotopy group of the function space $U^{A|B}$ with basepoint u, see (5.14). Here we assume that $i_A: B \rightarrowtail A$ is a cofibration in **Top**.

(10.6) ***Remark.*** Different choices of factorizations MA of i_A yield equivalent based objects $\Sigma_B^0 A$ in $\mathbf{C}^A$, see (6.4). Hence also the corresponding suspensions $\Sigma_B^n(\Sigma_B^0 A)$ are equivalent. This shows that the homotopy groups of the function space in (10.4) are well defined up to a *canonical* isomorphism by the pair (A, B) and by $u: A \to U$.

A map $(f, f'):(X, Y) \to (A, B)$ in **Pair (C)** induces a homomorphism

$$f^*: \pi_n(U^{A|B}, u) \to \pi_n(U^{X|Y}, uf), \tag{10.7}$$

as follows. We choose a weak lifting $(Mf, f'):(MX, Y) \to (MA, B)$ for (f, f'). This gives us the map

$$Mf \cup 1: \Sigma_f^0 = MX \bigcup_Y A \to \Sigma_B^0 A = MA \bigcup_B A, \tag{1}$$

which is a based map between based objects in $\mathbf{B} = \mathbf{C}^A$. The push out Σ_f^0 in **C** (given by $i_A f': Y \to A$) is based by the map $0 = (qMf, 1_A)$ where q is the weak equivalence in (10.1). The suspension $\Sigma_B^n(Mf \cup 1)$ is defined in the cofibration category **B** and hence we have with the notation in (10.2), (10.3) the commutative diagram

$$\begin{array}{ccccc} \Sigma_Y^n X & \longrightarrow & \Sigma_B^n(\Sigma_f^0) & \xrightarrow{\Sigma_B^n(Mf \cup 1)} & \Sigma_B^n(\Sigma_B^0 A) = \Sigma_B^n A \\ \uparrow & \text{push} & \uparrow & & \uparrow \\ X & \xrightarrow[f]{} & A & \xrightarrow[1]{} & A \end{array} \qquad (2)$$

One can check that the left-hand side of this diagram is a push out. This diagram induces f^* in (10.7), compare (10.4) and (2.11) (*b*). More generally than (5.15), we obtain by (5.7) the isomorphism $G^\#$ of groups for which the following diagram commutes, see (5.10):

$$\begin{array}{ccc} \pi_n(U^{X|Y}, uf) & \xrightarrow{(f^*G)^\#} & \pi_n(U^{X|Y}, vf) \\ \uparrow f^* & & \uparrow f^* \\ \pi_n(U^{A|B}, u) & \xrightarrow{G^\#} & \pi_n(U^{A|B}, v) \end{array} \qquad (3)$$

As in (5.16) we define the action of the fundamental group $\pi_1(U^{A|B}, u)$ on $\pi_n(U^{A|B}, u)$ by $x + H = H^\#(x)$.

A triple $(\bar{A}, A, B)$ in $\mathbf{C}$ $(B \rightarrowtail A \rightarrowtail \bar{A})$ gives us the pair maps

(10.8) $$(A, B) \overset{i}{\rightarrowtail} (\bar{A}, B) \overset{j}{\longrightarrow} (\bar{A}, A)$$

in **Pair** (**C**) which induce the exact sequence in (2.18) and (5.17). We now describe the prolongation of this sequence. Let $u: \bar{A} \to U$ be a map and let $v = u|A$.

(10.9) **Proposition.** *We have the long exact sequence* ($n \geqq 0$):

$$\xrightarrow[i^*]{} \pi_{n+1}(U^{A|B}, v) \xrightarrow[\partial]{} \pi_n(U^{\bar{A}|A}, u) \xrightarrow[j^*]{} \pi_n(U^{\bar{A}|B}, u) \xrightarrow[i^*]{} \pi_n(U^{A|B}, v).$$

For $n \geqq 1$ this is an exact sequence of homomorphisms of groups (of abelian groups for $n \geqq 2$). For $n = 0$ the sequence coincides with the sequence in (5.17). *For $n = 1$ the image of ∂ lies in the center of $\pi_1(U^{\bar{A}|A}, u)$.*

The sequence in (10.9) is natural, that is, a map $\bar{f}:(\bar{A}, A, B) \to (\bar{X}, X, Y)$ between triples, a map $g: U \to V$, and a homotopy $H: u \simeq w$ rel B respectively induce functions $\bar{f}^*, g_*, H^\#$ as in (10.7) (3). These yield commutative diagrams, the rows of which are exact sequences as in (10.9).

(10.10) ***Remark.*** The result in (10.9) is well known in the category **Top** of topological spaces. Cofibrations $B \rightarrowtail A \rightarrowtail \bar{A}$ induce the following sequence in **Top** of function spaces

$$U^{\bar{A}|A} \xrightarrow{j^*} U^{\bar{A}|B} \xrightarrow{i^*} U^{A|B}.$$

Here i^* is a fibration with fiber j^*. Thus the exact sequence of homotopy groups for a fibration provides us with an exact sequence as in (10.9). In fact, (up to canonical isomorphism) this exact sequence coincides with the one above for $\mathbf{C} = \mathbf{Top}$. Compare (10.5).

Proof of (10.9). Let $\mathbf{B} = \mathbf{C}^{\bar{A}}$ be the cofibration category of objects under $\bar{A}$. The triple $(\bar{A}, A, B)$ yields the following sequence of based objects and based maps in $\mathbf{B}$

(10.11) $$A \bigcup_B \bar{A} \underset{i}{\rightarrowtail} \Sigma_B^0 \bar{A} \underset{q}{\rightarrow} \Sigma_A^0 \bar{A}.$$

The cofibration i is given by $A \rightarrowtail \bar{A}$ and q is defined by the identity on $\bar{A}$, compare (10.1). One easily verifies that (10.11) is in fact a *cofiber sequence* (in the sense of 8.19)) *in the category* $\mathbf{B}$. The long exact sequence (8.26) or (8.18) yields for the cofiber sequence (10.11) the exact sequence in (10.9). Here we apply the functor $[\,.\,, u]$ where u is the object $u: \bar{A} \to U$ in $\mathbf{B}$. □

Next we consider the special case of (10.9) in which $\bar{A}$ is a mapping cone. Let $f: A \to B$ be a map where A is a based object in $\mathbf{C}$ and let $Y \rightarrowtail B$ be a cofibration. Then we have the triple $(C_f = C, B, Y)$ to which we apply (10.9). Then we get the following prolongation of the exact classification sequence in (8.16):

(10.12) **Corollary.** *We have the commutative diagram of groups and group homomorphisms* $(w: C_f \to U, u = w|B)$:

$$\xrightarrow{i^*} \pi_{n+1}(U^{B|Y}, u) \xrightarrow{\bar{\partial}} [\Sigma^{n+1}A, U] \xrightarrow{w^+} \pi_n(U^{C|Y}, w) \xrightarrow{i^*} \pi_n(U^{B|Y}, u).$$

$$\pi_{n+1}(U^{B|Y}, u) \xrightarrow{\partial} \pi_n(U^{C|B}, w), \qquad [\Sigma^{n+1}A, U] \underset{\cong}{\xrightarrow{\sigma_w}} \pi_n(U^{C|B}, w), \qquad \pi_n(U^{C|B}, w) \xrightarrow{j^*} \pi_n(U^{C|Y}, w)$$

Here the row is exact since we have the isomorphism σ_w *which gives us* $\bar{\partial} = (\sigma_w)^{-1}\partial$ *and* $w^+ = j^*\sigma_w$. *For* $n = 1$ *we have image* $i^* = $ *kernel* $\Sigma(w, f)$, *where* $\Sigma(w, f)$ *is defined in* (8.16).

Proof. Let $\mathbf{B} = \mathbf{C}^C$ where $C = C_f$. For the pair (C_f, B) we have the canonical equivalence (in $Ho\mathbf{B}$):

(10.13) $$\Sigma_B^0 C_f = C_f \bigcup_B C_f \xrightarrow{\bar{\mu}} \Sigma A \vee C_f.$$

We obtain $\bar{\mu}$ since $f:A\to B\rightarrowtail C_f$ is null homotopic. More explicitly let $(i_2+i_1:C_f\to \Sigma A\vee C_f\in[C_f,\Sigma A\vee C_f]^B$ be given by the cooperation μ in (8.7). Then we set $\bar{\mu}=(i_2+i_1,i_2)$. Here $i_1:\Sigma A\to\Sigma A\vee C_f$ and $i_2:C_f\to \Sigma A\vee C_f$ are the inclusions. The object $\Sigma_B^0 C_f$ is a based object in **B**, also $\Sigma A\vee C_f$ is a based object in **B** with $o=(0,1):\Sigma A\vee C_f\to C_f$. In fact, $\Sigma A\vee C_f$ is a *suspension in* **B** with

(10.14) $$\Sigma_{\mathrm{B}}(C_f\rightarrowtail A\vee C_f\xrightarrow{o}C_f)=\Sigma A\vee C_f.$$

One easily checks that $\bar{\mu}$ in (10.13) is an equivalence in *Ho***B**. Moreover, $\bar{\mu}$ is *based up to homotopy.*

(10.15) ***Remark.*** In the topological case we never can assume that $\bar{\mu}$ is a based map in **B**. This, in fact, is the reason why we study in §9 maps which are only based up to homotopy.

Since $\Sigma A\vee C_f$ is a suspension in **B** we know by (9.11) that the n-fold suspension $\Sigma^n=\Sigma_{\mathrm{B}}^n$ of $\bar{\mu}$ in the category **B** is well defined:

$$\Sigma_{\mathrm{B}}^n\bar{\mu}:\Sigma_{\mathrm{B}}^n(\Sigma_B^0 C_f)\to\Sigma_{\mathrm{B}}^n(\Sigma A\vee C_f)=\Sigma^{n+1}A\vee C_f.$$

This is an equivalence in *Ho***B** which induces the isomorphism

(10.16) $$\begin{array}{ccc}\pi_n(U^{C|B},w) & = & [\Sigma_{\mathrm{B}}^n(\Sigma_B^0 C),w]\\ \uparrow{\scriptstyle \sigma=\sigma_w} & & \uparrow{\scriptstyle (\Sigma_{\mathrm{B}}^n\bar{\mu})^*}\\ [\Sigma^{n+1}A,U] & = & [\Sigma_{\mathrm{B}}^n(\Sigma A\vee C_f),w]\end{array}$$

of groups ($n\geqq 1$) in (10.12), compare (10.4) and the following section §11.

□

We have the following special case of (10.13) for which $B=*$ and $C_f=\Sigma A$. Then

$$\bar{\mu}=(i_2+i_1,i_2):\Sigma A\vee\Sigma A\to\Sigma A\vee\Sigma A$$

is an equivalence in $Ho(\mathbf{C}^{\Sigma A})$ and is based up to homotopy in $\mathbf{C}^{\Sigma A}$. The left-hand side is based by the folding map $(\Sigma A\xrightarrow{i_2}\Sigma A\vee\Sigma A\xrightarrow{(1,1)}\Sigma A)$. The right-hand side is based by $(\Sigma A\xrightarrow{i_2}\Sigma A\vee\Sigma A\xrightarrow{(0,1)}\Sigma A)$. Now $\bar{\mu}$ induces the equivalence in $Ho\mathbf{C}^{\Sigma A}$

(10.17) $$\Sigma_*(\Sigma A)\to\Sigma^2A\vee\Sigma A$$

This is the suspension of $\bar{\mu}$ in $\mathbf{C}^{\Sigma A}$. Moreover this equivalence induces the isomorphism

(10.18) $$\sigma:[\Sigma^2A,U]\cong[\Sigma_*\Sigma A,U]^w$$

where $w:\Sigma A\to U$ is any map in **C**.

§ 10a Appendix: homotopy groups of function spaces and functors

Let $\alpha: \mathbf{C} \to \mathbf{K}$ be a functor between cofibration categories which preserves weak equivalences and let A be an object in $\mathbf{C}$. This yields the functor $\alpha^A: \mathbf{B} = \mathbf{C}^A \to \mathbf{H} = \mathbf{K}^{\alpha A}$ which preserves weak equivalences and which is based. Let (A, B) be a pair in $\mathbf{C}$ given by a map $i_A: B \to A$ and let $u: A \to U$ be a map in $\mathbf{C}$. The functor α^A induces the binatural homomorphism α_L which is defined by the commutative diagram ($X = \Sigma_B^0 A$):

(10a.1)

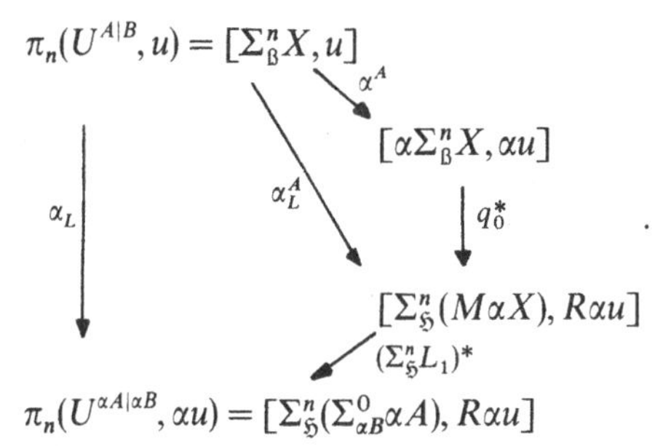

In this diagram the function α^A is given by the functor $Ho(\alpha^A): Ho(\mathbf{B}) \to Ho(\mathbf{H})$. The composition $\alpha_L^A = q_0^* \alpha^A$ is a special case of (6a.3). Moreover, the based map L_1 in $\mathbf{H}$ is given by the commutative diagram

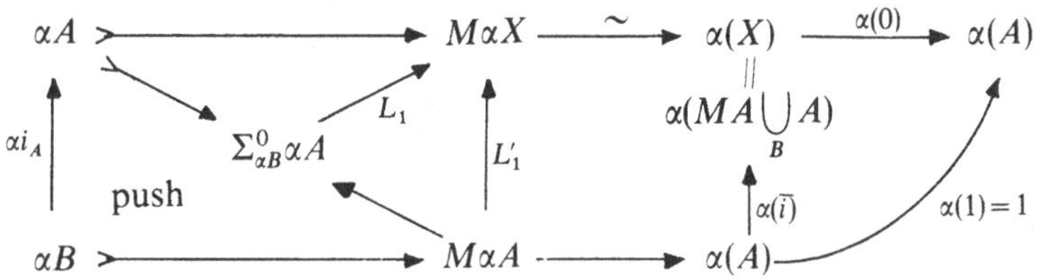

where L_1' is a lifting of $\alpha(\bar{i})$, see (10.1). The map L_1 is a weak equivalence provided α is compatible with the push out $MA \bigcup_B A$, see (I, 10.1) (3). This proves the first part of the following remark:

(10a.2) ***Remark.*** The function $(\Sigma_{\mathfrak{H}}^n L_1)^*$ in (10a.1) is a bijection if α is compatible with $MA \bigcup_B A$. By (6a.4) the function q_0^* in (10a.1) is a bijection if α is compatible with the push out $\Sigma_{\mathrm{B}} X = C_{\mathrm{B}} X / X$ in $\mathbf{B}$.

By naturality of α_L we get:

(10a.3) **Proposition.** *The operator α_L induces a map from the exact sequence for $(\bar{A}, A, B)$ in (10.9) to the exact sequence for $(\overline{M\alpha A}, M\alpha A, \alpha B)$, see (5a.7); in particular $\alpha_L \partial = \partial \alpha_L$.*

(10a.4) **Proposition.** *α_L is compatible with $G^{\#}$ in* (10.7) (3), *that is*, $\alpha_L(G^{\#}\xi) = (\alpha_L G)^{\#}(\alpha_L \xi)$.

For $n = 1$ this is a direct consequence of (5a.5).

§ 11 The partial and functional suspensions

Let **C** be a cofibration category with an initial object $*$. Let B be an object and let A be a based object in **C**. Then $A \vee B$ is the push out of $A \leftarrow\!\!\prec * \rightarrow B$ and $B \rightarrowtail A \vee B$ is a cofibration. We define the retraction

$$r = (0, 1): A \vee B \rightarrow B. \tag{11.1}$$

Thus $(B \rightarrowtail A \vee B \xrightarrow{r} B)$ is a based object in the category $\mathbf{B} = \mathbf{C}^B$. Let Σ_{B} be the suspension in **B**. Then we have

$$\Sigma_{\mathrm{B}}(A \vee B) = \Sigma A \vee B. \tag{11.2}$$

For a fibrant object $u: B \rightarrow U$ in **B** we deduce from (11.1) the isomorphism of groups

$$[\Sigma_{\mathrm{B}}(A \vee B), u] = [\Sigma A \vee B, U]^u = [\Sigma A, U]. \tag{11.3}$$

We derive (11.2) and (11.3) from the fact that $I_* A \vee B$ is a cylinder on $B \rightarrowtail A \vee B$ in **B**.

Let X be a further based object in **C**. We say $\xi \in [X, A \vee B]$ is **trivial on** B if $r_* \xi = 0$ where r is the retraction in (11.1). We write

$$\pi_n^X(A \vee B)_2 = [\Sigma^n X, A \vee B]_2 = \operatorname{kernel} r_* \tag{11.4}$$

for the set of elements which are trivial on B. This set can be interpreted as follows: Let $\mathbf{B}[X \vee B, A \vee B]_0$ be the set of maps

$$(X \vee B \rightarrow A \vee B) \in Ho\mathbf{B}$$

which are *based up to homotopy in* $\mathbf{B} = \mathbf{C}^B$. Then we get

$$[X, A \vee B]_2 = \mathbf{B}[X \vee B, A \vee B]_0 \tag{11.5}$$

For the mapping cone C_g of $g: A \rightarrow B$ we have the following push out diagram, $(\pi_g, 1) = (\pi_g, i_g)$,

$$\begin{array}{ccc} CA \vee B & \xrightarrow{(\pi_g, 1)} & C_g \\ \uparrow & & \uparrow{\scriptstyle i_g} \\ A \vee B & \xrightarrow{(g, 1)} & B \end{array},$$

compare (8.1). This is a pair map which induces the commutative diagram

$$
(11.6)\qquad
\begin{array}{ccccccc}
 & & & & \pi_1^X(CA \vee B, A \vee B) & \overset{\partial}{\cong} & \pi_0^X(A \vee B)_2 \\
 & & & & \big\downarrow (\pi_g, 1)_* & & \big\downarrow (g,1)_* \\
\pi_1^X(B) & \xrightarrow[i]{} & \pi_1^X(C_g) & \xrightarrow[j]{} & \pi_1^X(C_g, B) & \xrightarrow[\partial]{} & \pi_0^X(B)
\end{array} .
$$

Each row is a portion of the exact homotopy sequence of a pair as in (7.8). We assume that X *is a suspension* so that by exactness ∂ in the top row is an isomorphism of groups. By exactness of the lower row we obtain the functional operation

$$
(11.7)\qquad
\begin{cases}
E_g : \text{kernel}(g, 1)_* \to \pi_1^X(C_g)/i\pi_1^X(B), \\
E_g = j^{-1}(\pi_g, 1)_* \partial^{-1},
\end{cases}
$$

which is a homomorphism of groups. We call E_g the **functional suspension.** We also call an element $f \in [\Sigma X, C_g]$ with $f \in E_g(\xi)$, $\xi \in [X, A \vee B]_2$, a *functional suspension of* ξ or equivalently a **twisted map associated to** ξ. If $g = 0 : A \to * \to B$ is the trivial map we have $C_g = \Sigma A \vee B$ and thus (11.6) yields the diagram

$$
\begin{array}{ccc}
 & & \pi_1^X(CA \vee B, A \vee B) \overset{\partial}{\cong} \pi_0^X(A \vee B)_2 \\
 & & \big\downarrow (\pi \vee 1)_* \\
\pi_1^X(\Sigma A \vee B)_2 & \underset{j}{\cong} & \pi_1^X(\Sigma A \vee B, B)
\end{array}
$$

We call the homomorphism

$$
(11.8)\qquad
\begin{cases}
E : \pi_0^X(A \vee B)_2 \to \pi_1^X(\Sigma A \vee B)_2 \\
E = j^{-1}(\pi \vee 1)_* \partial^{-1}
\end{cases}
$$

the **partial suspension.** For the suspension Σ we have the commutative diagram

$$
(11.9)\qquad
\begin{array}{ccc}
\pi_0^X(A \vee B)_2 & \xrightarrow{E} & \pi_1^X(\Sigma A \vee B)_2 \\
\big\uparrow i & & \big\uparrow i \\
\pi_0^X(A)_0 & \xrightarrow[\Sigma]{} & \pi_1^X(\Sigma A)_0
\end{array} ,
$$

where i is the inclusion, see (9.8).

Moreover, by use of (11.5) we have the commutative diagram

$$
(11.10)\qquad
\begin{array}{ccc}
[X, A \vee B]_2 & \xrightarrow{E} & [\Sigma X, \Sigma A \vee B]_2 \\
\| & & \| \\
\mathbf{B}[X \vee B, A \vee B]_0 & \xrightarrow{\Sigma_B} & \mathbf{B}[\Sigma X \vee B, \Sigma A \vee B]_0
\end{array} .
$$

Here Σ_B is the suspension in $\mathbf{B}$ which is defined by (9.8) as well. Therefore, the partial suspension is a particular case of the suspension in the category $\mathbf{B} = \mathbf{C}^B$.

From the definition in (11.8) we obtain the following commutative diagram in $Ho\,\mathbf{Pair}(\mathbf{C})$:

$$\begin{array}{ccc} (CX, X) & \xrightarrow{(\bar{\xi},\xi)} & (CA \vee B, A \vee B) \\ \downarrow \pi & & \downarrow \pi \vee 1 \\ (\Sigma X, *) & \xrightarrow[(E\xi,0)]{} & (\Sigma A \vee B, B) \end{array} \tag{11.11}$$

where $\partial\bar{\xi} = \xi$. In fact, there is a unique $E\xi$, which is trivial on B, for which this diagram commutes, see (9.9). On the other hand we can characterize the partial suspension as follows. The map $\xi: X \to A \vee B$ gives us the pair map $\xi:(X, *) \to (A \vee B, B)$. If we apply (5.11) we obtain

$$\begin{array}{ccc} \Sigma_* X & \xrightarrow{\Sigma \cdot \xi} & \Sigma_B(A \vee B) = \Sigma_* A \vee B \\ \downarrow & & \downarrow \\ \Sigma X & \xrightarrow{E\xi} & \Sigma A \vee B \end{array} \tag{11.12}$$

If X is a suspension there is a unique $E\xi$, trivial on B, which makes the diagram commute in $Ho\mathbf{C}$.

(11.13) **Proposition.** *The action map $\mu_n: \Sigma^n A \to \Sigma^n A \vee \Sigma A$ in* (7.11) *is trivial on ΣA and we have $E\mu_n = \mu_{n+1}$.*

(11.14) **Corollary.** *If A is a suspension then $\mu_1 = i_0$ is the inclusion and thus $E^n \mu_1 = \mu_{n+1} = i_0$ is the inclusion. This implies* (6.14).

(11.15) **Corollary.** *By* (11.11) *the diagram in the proof of* (7.13) *commutes.*

Proof of (11.13). The action map μ corresponds by the equivalence $\Sigma^n_* A \bigcup_A CA \xrightarrow{\sim} \Sigma^n_* A/A = \Sigma^n A$ to the cooperation $\Sigma^n_* A \bigcup_A CA \xrightarrow{\mu} \Sigma^n_* A \bigcup_A (CA \vee \Sigma A)$ in (8.7). We now deduce (11.13) from (11.12). □

Let $k: Y \to B$ be a map in $\mathbf{C}$, let $\eta: Z \to X \vee Y$ be trivial on Y, $Z = \Sigma Z'$, and let $\xi: X \to A \vee B$ be trivial on B, $X = \Sigma X'$. Then $(\xi, i_2 k)\eta: Z \to X \vee Y \to A \vee B$ is trivial on B and we get with $(\xi, k) = (\xi, i_2 k)$

(11.16) **Proposition.**

$$E((\xi, k)\eta) = (E\xi, k)(E\eta).$$

This follows easily from the definition (11.8) together with (11.11) or from (11.10). Also we deduce from (11.10) the

(11.17) **Lemma.** *For* $\xi \in \pi_0^X(A \vee B)_2$, $u \in [B, U]$ *the function*

$$(E\xi)^*(\cdot, u): [\Sigma A, U] \to [\Sigma X, U], \alpha \longmapsto (E\xi)^*(\alpha, u)$$

is a homomorphism of groups.

§12 The difference operator

Let $g: A \to B$ be a map (where A is a based object) and let $i_g: B \rightarrowtail C_g$ be its mapping cone. For $i_1: \Sigma A \to \Sigma A \vee C_g$, $i_2: C_g \rightarrowtail \Sigma A \vee C_g$, we have

(12.1) $$i_2 + i_1: (C_g, B) \to (\Sigma A \vee C_g, B),$$

by the action in (8.8). This is just the cooperation μ in (8.7) except for the order in which the objects appear in the wedge. For $f: \Sigma X \to C_g$ the difference element

$$\nabla f = -f^*(i_2) + f^*(i_2 + i_1): \Sigma X \to \Sigma A \vee C_g$$

is trivial on C_g. This gives us the **difference operator**

(12.2) $$\nabla: \pi_1^X(C_g) \to \pi_1^X(\Sigma A \vee C_g)_2.$$

The difference operator has the following property with respect to the functional suspension E_g in (11.7).

(12.3) **Proposition.** *Let* X *be a suspension. If* $f \in E_g(\xi)$ *for* $\xi \in \pi_0^X(A \vee B)_2$ *then* $\nabla f = (1 \vee i_g)_*(E\xi)$.

We have the following generalization of this result. We introduce the **relative difference operator** ∇ by the commutative diagram

$$\begin{array}{ccccc}
\pi_1^X(C_g) & \xrightarrow{j} & \pi_1^X(C_g, B) & \xrightarrow{(i_2+i_1)_*} & \pi_1^X(\Sigma A \vee C_g, B) \\
& \searrow^{\nabla} & \downarrow{\scriptstyle\nabla} & & \downarrow{\scriptstyle\delta} \\
& & \pi_1^X(\Sigma A \vee C_g)_2 & \xrightarrow[j]{\cong} & \pi_1^X(\Sigma A \vee C_g, C_g)
\end{array},$$

where X is a suspension. Here δ is induced by the identity on $\Sigma A \vee C_g$. Since j is an isomorphism we can define the composition

(12.4) $$\nabla = j^{-1}\delta(i_2 + i_1)_*.$$

One easily checks that for ∇ in (12.2) the diagram commutes. From (11.6) we derive by diagram chase

(12.5) **Proposition.** *For* $\xi \in \pi_0^X(A \vee B)_2$ *the element* $\bar{\xi} = (\pi_g, 1)_* \partial^{-1}\xi$ *satisfies* $\nabla\bar{\xi} = (1 \vee i_g)_*(E\xi)$.

Clearly, (12.3) is a special case of this result. The difference operator ∇ in (12.4) has the following important property:

(12.6) **Proposition.** *Let* $f: X \to Y$ *and let* $F:(C_f, Y) \to (C_g, B)$ *be a pair map. For the element* $\{\pi_f\} \in \pi_1^X(C_f, Y)$, *see* (8.1), *we have*

$$\nabla_F = \nabla F_*\{\pi_f\} = d(i_2 F, (i_2 + i_1)F) \in \pi_1^X(\Sigma A \vee C_g)_2.$$

Here the difference is well defined since $(i_2F)_{|Y} = ((i_2 + i_1)F)_{|Y}$, *see* (8.13).

Proof. The difference is trivial on C_g. Thus we have to show

$$\delta(i_2 + i_1)_* F_* \{\pi_f\} = jd(i_2 F, (i_2 + i_1)F).$$

This is clear by definition of d. □

From (8.15) we deduce the following corollary. Let $F:(C_f, Y, Y_0) \to (C_g, B, B_0)$ be a map of triples and let $u: C_g \to U$ be given with $v = u|B_0$, $\bar{v} = v(F|Y_0)$.

(12.7) **Corollary.** *For* $\alpha \in [\Sigma A, U]$ *we have the equation* $F^*(\{u\} + \alpha) = (F^*\{u\}) + (\nabla_F)^*(\alpha, \{u\})$ *in* $[C_f, U]^{\bar{v}}$, $\{u\} \in [C_g, U]^v$.

Here we use the action in (8.8). Equivalently to (12.7), we can state that the following diagram homotopy commutes rel Y

$$(12.8)\qquad \begin{array}{ccc} C_f & \xrightarrow{F} & C_g \\ \big\downarrow{\scriptstyle i_2+i_1} & & \big\downarrow{\scriptstyle i_2+i_1} \\ \Sigma X \vee C_f & \xrightarrow[(\nabla_F, i_2F)]{} & \Sigma A \vee C_g \end{array}.$$

This is the **left distributivity law** for maps between mapping cones. For $u: C_g \to U$ we consider the diagram ($q \geqq 0$)

$$(12.9)\qquad \begin{array}{ccc} \pi_q(U^{C_g/B}, u) & \xrightarrow{F^*} & \pi_q(U^{C_f/Y}, uF) \\ {\scriptstyle \sigma_u}\big\uparrow \cong & & {\scriptstyle \sigma_{uF}}\big\uparrow \cong \\ [\Sigma^{q+1}A, U] & \xrightarrow[(E^q\nabla_F)^*(\cdot, u)]{} & [\Sigma^{q+1}X, U]. \end{array}$$

with σ in (10.16). By (12.8) this diagram commutes. Compare the definition of $\bar{\mu}$ in (10.13) and use (11.10), see also (11.17).

§ 13 Double mapping cones

We consider a double mapping cone as in the following diagram (A and Q are based objects, U is fibrant)

(13.1)
$$\begin{array}{ccccc} & & C = C_f & \overset{w}{\searrow} & \\ & & \cup & & \\ A & \xrightarrow{f} & B = C_h & \xrightarrow{u} & U, \\ & & \cup & & \\ Q & \xrightarrow{h} & T & \overset{t}{\nearrow} & \\ & & \cup & & \\ & & Y & \overset{v}{\nearrow} & \end{array}$$

compare the diagram in (8.16). We assume that $A = \Sigma A'$ is a *suspension*. For homotopy groups of function spaces we obtain the following diagram ($q \geqq 1$):

(13.2)
$$\begin{array}{ccccc} & & \pi_q(U^{B|Y}, u) & & \\ & \overset{\gamma}{\nearrow} & & \overset{\beta}{\searrow} & \\ \pi_q(U^{B|T}, u) & & \overset{d_1}{\dashrightarrow} & & \pi_{q-1}(U^{C|B}, w) \\ \cong \uparrow \sigma_u & & & & \cong \uparrow \sigma_w \\ [\Sigma^{q+1} Q, U] & & \xrightarrow{(E^q \nabla f)^*(\cdot, u)} & & [\Sigma^q A, U] \end{array}$$

Here $\gamma = j^*$ is induced by $(B, Y) \subset (B, T)$ and $\beta = \partial$ is the boundary in (10.9) for the triple (C, B, Y). We set $d_1 = \beta\gamma$. The isomorphisms σ are defined in (10.16). $E^q \nabla f$ is the q-fold partial suspension of ∇f which gives us for $q \geqq 1$ the homomorphism $\alpha \longmapsto (E^q \nabla f)^*(\alpha, u)$, see (11.17).

(13.3) **Proposition.** *Diagram* (13.2) *commutes.*

The homomorphism d_1 is also the primary differential in a spectral sequence, see (III.4.12).

Proof of (13.3). We have the diagram of based objects in $\mathbf{B} = \mathbf{C}^C$

$$\begin{array}{ccccc} A \vee C & \xrightarrow{f \cup 1} & B \bigcup_Y C & \overset{i}{\rightarrowtail} & C \bigcup_Y C \\ & \overset{\nabla}{\searrow} & \downarrow q & & \\ & & B \bigcup_T C & & \\ & & \bar{\mu} \downarrow \sim & & \\ & & \Sigma Q \vee C & & \end{array}$$

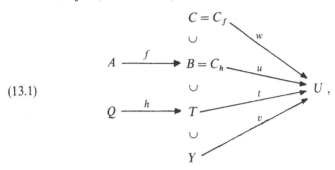

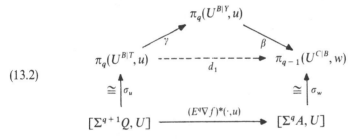

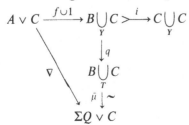

Here the row is a principal cofibration i in **B** with attaching map $f \cup 1$. The map i induces the exact sequence for β in (13.2), compare (10.11). The map q induces γ in (13.2). By (10.13) we have the equivalence $\bar{\mu}$ which induces σ (to the left in diagram (13.2)). Let $\nabla = \bar{\mu}q(f \cup 1)$ be the composition. Then we have $\nabla | A = (1 \vee i_f)(i_1 + i_2)f$, where $i_f : B \rightarrowtail C$. Since $i_f f = 0$ we get

$$\nabla | A = (1 \vee i_f)(-f^*(i_2) + f^*(i_2 + i_1)) = (1 \vee i_f)(\nabla f).$$

This proves (13.3), compare (11.10) and (10.7)(2). □

Next we consider the cylinder $I_Y C$ on the double mapping cone in (13.1). This leads to the following diagram of principal cofibrations and attaching maps (compare (8.11)):

$$\begin{array}{ccc}
 & & I_Y C \\
 & & \uparrow{\scriptstyle i} \\
\Sigma A & \xrightarrow{\ w_f\ } & C \cup I_Y B \cup C = I_Y^{\cdot} C_f \\
 & & \uparrow{\scriptstyle j} \\
\bar{Q} = A \vee \Sigma Q \vee A & \xrightarrow{\ W\ } & B \cup I_Y T \cup B = \bar{Z}
\end{array}$$

where w_f is defined in (8.11) and where

(13.5) $$W = (i_0 f, w_h, i_1 f) : \bar{Q} \to \bar{Z}.$$

As we have seen in (8.12) the cofibration i is a principal cofibration with attaching map w_f. Similarly, we see that j is a principal cofibration with attaching map W since $B \subset C = C_f$ is a principal cofibration, see (13.1).

For the map W in (13.5) we have the functional suspension (see (11.7)): $E_W = j^{-1}(\pi_W, 1)_* \partial^{-1}$. Here we use the homomorphisms of the following diagram

$$\begin{array}{ccccc}
 & & \pi_1^A(C\bar{Q} \vee \bar{Z}, \bar{Q} \vee \bar{Z}) & \underset{\partial}{\xrightarrow{\ \cong\ }} & \pi_0^A(\bar{Q} \vee \bar{Z})_2 \\
 & & {\scriptstyle (\pi_W, 1)_*}\downarrow & & \downarrow{\scriptstyle (W,1)_*} \\
w_f \in \pi_1^A(I_Y^{\cdot} C_f) & \xrightarrow{\ j\ } & \pi_1^A(I_Y^{\cdot} C_f, \bar{Z}) & \longrightarrow & \pi_0^A(\bar{Z})
\end{array}$$

We assume that $A = \Sigma A'$ is a suspension. We define

(13.6) $$\xi \in \pi_0^A(\bar{Q} \vee \bar{Z})_2 \quad \text{with } (W, 1)_*(\xi) = 0$$

as follows:

Let i_0^A, i^Q, i_1^A be inclusions of factors into $\bar{Q}$ and let $i_0^B, i_1^B : B \rightarrowtail \bar{Z}$ be given by the cylinder $I_Y B$. By the cooperation on $C_h = B$ we have

$$i_0^B + w_h - i^Q : C_h \to \bar{Q} \vee \bar{Z}.$$

This gives us the element

$$\xi = -i_0^A + (i_0^B + w_h - i^Q)f - i_1^B f + i_1^A.$$

Since $i_0^B + w_h = i_1^B$ (see (8.15)) we see that $\xi \in \pi_0^A(\bar{Q} \vee \bar{Z})_2$ and

$$(W,1)_*(\xi) = -i_0^B f + (i_0^B + w_h - w_h)f - i_1^B f + i_1^B f = 0.$$

This proves that the functional suspension E_W is defined on ξ. The following result is a crucial fact for the cylinder in (13.4).

(13.7) **Theorem**

$$w_f \in E_W(\xi).$$

In (12.2) we defined the difference ∇f for $f: A \to C_h$. We now define similarly

$$\bar{\nabla} f = f^*(i_2 + i_1) - f^*(i_2) = f^*(i_2) + (\nabla f) - f^*(i_2) \quad \in \pi_0^A(\Sigma Q \vee C_h)_2.$$

If the group is abelian we have $\bar{\nabla} f = \nabla f$. Moreover, we have in $\pi_0^A(\Sigma Q \vee C_f)_2$ the element $\nabla^f = (1 \vee i_f)_* \nabla f = (1 \vee i_f)_* \bar{\nabla} f$. From $i_0^B + w_h = i_1^B$, see (8.15), and from the definition of ξ we deduce

(13.8) $$\xi = -i_0^A + (\bar{\nabla} f)^*(-i^Q, i_1^B) + i_1^A.$$

This leads to the following important formula for the difference in (8.13):

(13.9) **Corollary.** *Let $u_0, u_1 : C \to U$ with $u_0 | Y = u_1 | Y$ and let $H: C_{w_h} \simeq I_Y B \to U$ be a homotopy $H: u_0 | B \simeq u_1 | B$. Then we have the formula*

$$d(u_0 + \alpha_0, H + \beta, u_1 + \alpha_1) = d(u_0, H, u_1) - \alpha_0 + \alpha_1 - (E\nabla^f)^*(\beta, u_1).$$

Here $\alpha_0, \alpha_1 : \Sigma A \to U$, $\beta: \Sigma^2 Q \to U$.

For $H + \beta$ in this formula we use the equivalence $C_{w_h} \simeq I_Y B$. The corollary follows from (13.7) by use of (12.3).

The proof of (13.7) is quite technical, it is somewhat easier to follow the argument if it is assumed that all objects in **C** are fibrant. In this case each map in *Ho***C** which is defined on a cofibrant object can be represented by a map in **C**.

Proof of (13.7). Since the projection $p: I_Y B \to I_T B$ is a weak equivalence it is enough to prove the theorem for $Y = T$. In this case we have

$$C_h \bigcup_T C_h = \bar{Z} = I_T^{\cdot} C_h \rightarrowtail I_T B \quad \text{with } B = C_h. \tag{1}$$

We have the following commutative diagram

$$\begin{array}{ccccccccc}
CA & \xrightarrow{\pi_\tau} & C_\tau & \xrightarrow{\bar{\tau}} & I_*A & \xrightarrow{If} & I_TC_h & \underset{\sim}{\xrightarrow{\bar{\lambda}}} & C\Sigma Q \vee C_h \\
\uparrow & \text{push} & \nwarrow & & \nearrow & & \uparrow & & \uparrow \\
A & \xrightarrow[\tau]{} & & A \vee A & & \xrightarrow[f \cup f]{} & \bar{Z} & \overset{\sim}{\xrightarrow[\lambda]{}} & \Sigma Q \vee C_h \, . \\
& \searrow^{x} & & & \bar{Q} \vee \bar{Z} & & & \swarrow_{(i^0, i_1^B)} &
\end{array} \tag{2}$$

Here we define $\tau = i_0^A - i_1^A$ ($A = \Sigma A'$ is a suspension!) where i_0^A and i_1^A are the inclusions of A into $A \vee A$. We define λ by

$$\lambda = (i^B - i^Q, i^B) \tag{3}$$

with $i^B: C_h \subset \Sigma Q \vee C_h$ and $i^Q: \Sigma Q \subset \Sigma Q \vee C_h$. The difference $i^B - i^Q$ is given by the cooperation on C_h. By definition of λ and τ we see that the composition x in (2) represents the element

$$\{x\} = (\bar{\nabla} f)^*(-i^Q, i_1^B), \tag{4}$$

compare (13.8). There is a canonical null homotopy of $A \xrightarrow{\tau} A \vee A \subset I_*A$ which gives us the map $\bar{\tau}$, see (9) below. Moreover, by use of (6), there is a canonical null homotopy of

$$\Sigma Q \xrightarrow[w_h]{} C_h \bigcup_T C_h \xrightarrow[\lambda]{} \Sigma Q \vee C_h \subset C\Sigma Q \vee C_h.$$

This null homotopy gives us by (8.12) the map $\bar{\lambda}$. The map λ is a weak equivalence. Also $\bar{\lambda}$ is a weak equivalence. We consider the commutative diagram of groups and group homomorphism:

$$\begin{array}{ccccc}
& & \pi_1^A(C\Sigma Q \vee \bar{Z}, \Sigma Q \vee \bar{Z}) & \underset{\cong}{\xrightarrow{\partial}} & \pi_0^A(\Sigma Q \vee \bar{Z})_2 \\
& & \Big\downarrow {\scriptstyle (\pi_{w_h}, 1)^*} & & \Big\downarrow {\scriptstyle (w_h, 1)^*} \\
\pi_1^A(I_*A, A \vee A) & \xrightarrow[(If)_*]{} & \pi_1^A(I_TC_h, \bar{Z}) & \xrightarrow{\partial} & \pi_0^A(\bar{Z}) . \\
& & {\scriptstyle \cong} \Big\downarrow {\scriptstyle \bar{\lambda}_*} & & {\scriptstyle \cong} \Big\downarrow {\scriptstyle \lambda_*} \\
& & \pi_1^A(C\Sigma Q \vee C_h, \Sigma Q \vee C_h) & \hookrightarrow & \pi_0^A(\Sigma Q \vee C_h)
\end{array} \tag{5}$$

Next we observe that the diagram

$$\begin{array}{ccc}
\Sigma Q \vee C_h & \xrightarrow{1} & \Sigma Q \vee C_h \\
\Big\downarrow {\scriptstyle 1 \vee i_1^B} & & \Big\uparrow {\scriptstyle \lambda} \\
\Sigma Q \vee \bar{Z} & \xrightarrow[(w_h, 1)]{} & \bar{Z}
\end{array} \tag{6}$$

homotopy commutes. This follows since

$$\lambda i_1^B = 1_B$$
$$\lambda w_h = d(i_1^B - i^Q, i_1^B) = i^Q + d(i_1^B, i_1^B) = i^Q.$$

By (5) and (6) we see that for $\xi_0 = (1 \vee i_1^B)_* \xi_1$ with $\xi_1 = \lambda_* \partial(If)_*(\{\bar{\tau}\pi_\tau\})$ we have

$$(If)_*(\{\bar{\tau}\pi_\tau\}) = (\pi_{w_h}, 1)^* \partial^{-1} \xi_0. \tag{7}$$

Here, $\{\bar{\tau}\pi_\tau\} \in \pi_1^A(I_* A, A \vee A)$ is given by the pair map in (2) or equivalently by equation (9) below.

By naturality of w_f in (8.12) we have the commutative diagram ($w = w_{1_A}$)

$$\begin{array}{ccccc} & \overset{w}{\nearrow} & CA \cup I_* A \cup CA = I_*^{\cdot} CA & \leftarrow\!\!< & A \vee A \\ \Sigma A & & \downarrow I^{\cdot} f & & \downarrow f \cup f \\ & \underset{w_f}{\searrow} & C_f \cup I_T B \cup C_f = I_T^{\cdot} C_f & \leftarrow\!\!< & \bar{Z} \end{array},$$

where we set $I^{\cdot} f = \pi_f \cup If \cup \pi_f$. This gives us the commutative diagram

$$\begin{array}{ccc} \pi_1^A(I_T^{\cdot} C_f) & \xrightarrow{j} & \pi_1^A(I_T^{\cdot} C_f, \bar{Z}) \\ \uparrow & & \uparrow \\ \pi_1^A(I_*^{\cdot} CA) & \xrightarrow{j} & \pi_1^A(I_*^{\cdot} CA, A \vee A) \end{array}. \tag{8}$$

Since we have by definition of w in (8.12)

$$j\{w\} = -\{i_0^{CA}\} + \{\bar{\tau}\pi_\tau\} + \{i_1^{CA}\} \tag{9}$$

we can deduce the theorem from (7), (8) and (4), see (13.8). □

We now apply the result above to the classification sequence in (8.16) and (10.12).

(13.10) **Proposition.** *Let f and h be given as in* (13.1) *where we set $Y = T$. Then we have the long exact sequence* ($n \geqq 1$):

$$\pi_{n+1}(U^{C|Y}, w) \xrightarrow{\bar{i}} [\Sigma^{n+2} Q, U] \xrightarrow{\nabla^{n+1}(u,f)} [\Sigma^{n+1} A, U] \xrightarrow{w^+} \pi_n(U^{C|Y}, w)$$
$$\xrightarrow{\bar{i}} \cdots \pi_1(U^{C|Y}, w) \xrightarrow{\bar{i}} [\Sigma^2 Q, U] \xrightarrow{\nabla(u,f)} [\Sigma A, U] \xrightarrow{w^+} [C_f, U]^v$$
$$\xrightarrow{i_f^*} [B, U]^v \xrightarrow{f^*} [A, U],$$

the final part of which is isomorphic to (8.16). *Here we set $\nabla = \nabla^1$ and*

$$\nabla^{n+1}(u, f)(\beta) = (E^{n+1} \nabla f)^*(\beta, u),$$

as in (13.2), $n \geqq 0$. *The operator* ω^+ *is the same as in* (10.12) *and* $\bar{\imath}$ *is given by* $\sigma^{-1}i^*$.

Proof. For $Y = T$ the map γ in (13.2) is the identity and therefore β in (13.2) is the same as ∂ in (10.12). Now (10.12), (13.3) and (13.9) yield the result. □

We leave it to the reader to consider in more detail the special case of (13.10) in which $Y = T = *$ and $h = 0$. Then $f: A = \Sigma A' \to B = \Sigma Q$ is a map between suspensions. Compare the examples in § 16 below.

Remark. In topology Barcus–Barratt (1958) and Rutter (1967) investigated the isotropy groups of the $[\Sigma A, U]$-action on $[C_f, U]^*$ (where $T = *$) which we identify with the image of $\nabla(u, f)$ in (13.10). In essence they use the homomorphism $\nabla(u, f)$, which in Rutter's notation is $\Gamma(u, f)$. Above we have shown that this homomorphism can be expressed by the partial suspension of the difference ∇f and that this construction is actually available in any cofibration category. Therefore we immediately obtain the result dual to (13.10) in a fibration category. If we take the fibration category of topological spaces then (13.10) yields theorem 2.2 of James–Thomas (1966). This as well is studied by Nomura (1969). Compare also the discussion in Baues (1977).

§ 14 Homotopy theory in a fibration category

As we pointed out in the appendix of § 1 in Chapter I all results and constructions for a cofibration category are in a dual way available for a fibration category **F**. We fix the notation as follows:

Let **F** be a fibration category with a final object $*$ (**F** needs not to have an initial object). A based object in **F** is a fibrant object X (that is $X \twoheadrightarrow *$ is a fibration) together with a map $o: * \to X$.

The **path object** P is dual to the cylinder and was defined in the appendix of I.§ 1a. The **loop object** Ω is dual to the suspension Σ. The dual of an attaching map is the **classifying map** $f: A \to B$ which maps to a based object B and which yields the principal fibration $P_f \to A$. Here P_f is dual to the mapping cone and is obtained by the pull back diagram

$$(14.1)\qquad \begin{array}{ccccc} P_f & \xrightarrow{\pi_f} & WB & \longrightarrow & PB = B^I \\ {\scriptstyle q}\downarrow & \text{pull} & \downarrow & \text{pull} & \downarrow \\ A & \xrightarrow[f]{} & B & \xrightarrow[(1,0)]{} & B \times B \end{array}$$

$B \times B$ is the pull back of $B \twoheadrightarrow * \twoheadleftarrow B$ and WB denotes the **contractible path**

object which is dual to the cone. For a fibration $p: E \twoheadrightarrow B$ over a based object B we have the pull back diagram

(14.2)
$$\begin{array}{ccc} F & \longrightarrow & E \\ \downarrow & & \downarrow p \\ * & \xrightarrow{o} & B \end{array}$$

F is called the **fiber** of p and $F \to E \twoheadrightarrow B$ is a **fiber sequence**. The fiber of $WB \twoheadrightarrow B$ is the loop object ΩB.

We leave it to the reader to translate all results and constructions of this chapter into the dual language of a fibration category.

As an example we describe the dual of the exact homotopy sequence of a pair in (7.8). We write

(14.3)
$$\pi_B^n(X) = [X, \Omega^n B]$$

for the homotopy groups dual to (6.11). Here X is cofibrant and B is a based object in **F**. A **pair** $(A|B)$ in **F** is a map $A \to B$, see (1.3). The category **Pair (F)** is defined dually to (1.3) and is a fibration category by (1.5). Dually to (7.7) we obtain the relative homotopy groups in **F** by

(14.4)
$$\pi_B^{n+1}(U|V) \approx [(U|V), \Omega^n(WB|B)].$$

Here $(WB|B)$ is the pair in (14.1) which is a based object in **Pair (F)**. Now (7.8) states that we have the exact sequence

(14.5)
$$\to \pi_B^{n+1}(V) \xrightarrow{p^*} \pi_B^{n+1}(U) \xrightarrow{j} \pi_B^{n+1}(U|V) \xrightarrow{\partial} \pi_B^n(V) \to .$$

Here p^* is induced by the fibration $p: U \twoheadrightarrow V$. Which is given by the pair $(U|V)$. Moreover j is induced by $\Omega^{n+1}B \to W\Omega^n B \xrightarrow{\simeq} \Omega^n WB$. Again, ∂ is the restriction.

Next we consider the partial and the functional loop operations which are dual to the partial and the functional suspensions respectively in § 11.

We have the pull back $B \times A$ of $B \twoheadrightarrow * \leftarrow A$ and $B \times A \twoheadrightarrow A$ is a fibration which has the section $s = (0, 1): A \to B \times A$ compare (11.1). Now let X be a further based object in **F**. We say that $\xi \in [B \times A, \Omega^n X]$ is **trivial on** A if $s^*\xi = 0$. We write

(14.6)
$$\pi_X^n(B \times A)_2 = [B \times A, \Omega^n X]_2 = \text{kernel } s^*$$

for the subset of elements which are trivial on B. This is the notation dual to (11.4). For the principal fibration $P_f \twoheadrightarrow A$ in (14.1) we have the following pull back diagram

$$(14.7)\qquad \begin{array}{ccc} P_f & \xrightarrow{(\pi_f,q)} & WB \times A \\ {\scriptstyle q}\downarrow & \text{pull} & \downarrow{\scriptstyle q\times 1} \\ A & \xrightarrow[(f,1)]{} & B \times A \end{array},$$

which is a map $(\pi_f, 1):(P_f|A)\to(WB \times A|B \times A)$ in the category **Pair(F)**. Therefore we obtain by (14.5) the following commutative diagram which is dual to (11.6).

$$(14.8)\qquad \begin{array}{ccccccc} & & & & \pi_X^1(WB\times A|B\times A) & \overset{\partial}{\cong} & \pi_X^0(B\times A)_2 \\ & & & & \downarrow{\scriptstyle(\pi_f,1)^*} & & \downarrow{\scriptstyle(f,1)^*} \\ \pi_X^1(A) & \xrightarrow{q^*} & \pi_X^1(P_f) & \xrightarrow{j} & \pi_X^1(P_f|A) & \xrightarrow{\partial} & \pi_X^0(A) \end{array}.$$

Here we assume that X is a loop object, so that by exactness ∂ in the top row is an isomorphisms of groups. Now

(14.9) $\qquad L_g$:kernel $(f, 1)^* \to \pi_X^1(P_f)/\text{im}\, q^*,\quad L_g = j^{-1}(\pi_f, 1)^*\partial^{-1}$,

is the **functional loop operation** dual to E_g in (11.7). If $f = 0: A \twoheadrightarrow * \xrightarrow{0} B$ we have $P_f = \Omega B \times A$ and j in (14.8) yield the isomorphism

$$\pi_X^1(\Omega B \times A)_2 \overset{j}{\cong} \pi_X^1(\Omega B \times A|A).$$

Thus we have the **partial loop operation**

(14.10) $\qquad L: \pi_X^0(B \times A)_2 \to \pi_X^1(\Omega B \times A)_2,\quad L = j^{-1}(\pi_0, 1)^*\partial^{-1}$.

Now it is easy to translate the important results in § 12 and § 13 into the dual language of a fibration category.

For the category **F** of pointed topological spaces, see (I.5.5), the partial and the functional loop operators are studied in Chapter 6 of Baues (1977).

§ 15 Whitehead products

Let **C** be a cofibration category with an initial object $*$. We first obtain the following 'splitting' result which is well known in topology.

(15.1) **Proposition.** *Let $i: A \rightarrowtail B$ be a based cofibration between based objects in* **C**. *If i admits a retraction $r \in [B, A]$ in Ho*(**C**), *$ri = 1_A$, which is based up to homotopy, then there is an isomorphism in Ho*(**C**)

$$\Sigma B \simeq \Sigma(B/A) \vee \Sigma A.$$

Proof. Let U be a fibrant object and let $q: B \to B/A = F$ be the projection for the cofiber. By exactness of the cofiber sequence (8.24) we obtain the short exact sequence of groups

$$0 \to [\Sigma F, U] \xrightarrow{(\Sigma q)^*} [\Sigma B, U] \underset{(\Sigma r)^*}{\overset{(\Sigma i)^*}{\rightleftarrows}} [\Sigma A, U] \to 0. \tag{1}$$

In fact, the homomorphism $(\Sigma i)^*$ is surjective since a suspension Σr, see (9.8), yields the isomorphism $(\Sigma i)^*(\Sigma r)^*$ which differs only by a $*$-action from the identity. The same argument shows that $(\Sigma^2 i)^*$ is surjective and whence $(\Sigma q)^*$ in (1) is injective by exactness. Now (1) shows that

$$\xi = (\Sigma q) + (\Sigma r): \Sigma B = X \to \Sigma(B/A) \vee \Sigma A = Y \tag{2}$$

induces a bijection of sets $\xi^*: [Y, U] \approx [X, U]$ for all fibrant U and thus for all U in **C**. By specializing U the proposition follows. □

(15.2) ***Definition.*** A **fat sum** $A \curlyvee B$ in **C** is a commutative diagram

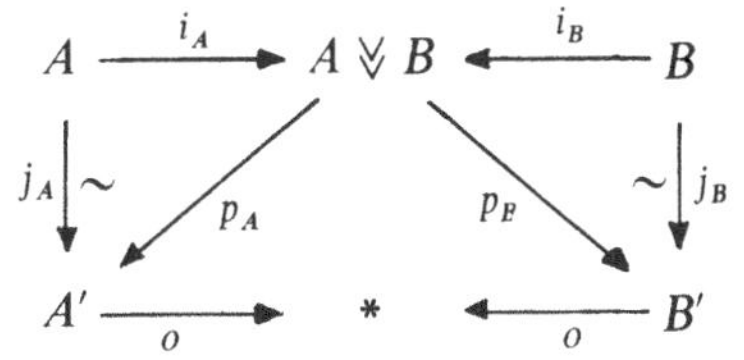

in **C** such that $p_B i_A = o_1$ and $p_A i_B = o_2$ in $Ho(\mathbf{C})$. Where $o_1: A \to * \to B'$ and $o_2: B \to * \to A'$ are given by oj_A and oj_B respectively. ||

In many examples the weak equivalences j_A and j_B are actually the identities of A and B respectively. If $*$ is the final object of **C** it is enough to assume that the diagram in (15.2) is commutative in $Ho(\mathbf{C})$ since then suspensions of homotopy classes are well defined.

We say that a fat sum $A \curlyvee B$ is **based** if all objects in (15.2) are cofibrant and if

(15.3) $$i = (i_A, i_B): A \vee B \rightarrowtail A \curlyvee B$$

is a cofibration. In this case all objects of (15.2) are based objects in **C** and all maps are based maps. Moreover, i in (15.3) is a based map which yields a cofiber sequence as in (8.24). We call the cofiber

(15.4) $$A \wedge B = (A \curlyvee B)/(A \vee B),$$

the **smash product** associated to the based fat sum $A \curlyvee B$. Clearly, the smash product $A \wedge B$ again is a based object in **C**.

The dual of a fat sum in a fibration category is called a **fat product** and the dual of a smash product is a **cosmash product** associated to a fat product.

(15.5) ***Remark.*** For a fat sum $A \curlyvee B$, as in (15.2), we obtain a based fat sum $MA \curlyvee MB$ as follows. As in (6.4)(1) we first choose factorizations $* \rightarrowtail MA \xrightarrow{\sim} A$ and $* \rightarrowtail MB \xrightarrow{\sim} B$ and then we choose a factorization

$$MA \vee MB \rightarrowtail MA \curlyvee MB \xrightarrow{\sim} A \curlyvee B$$

of the obvious map $MA \vee MB \to A \curlyvee B$ given by i_A and i_B. Next we choose a weak lifting $(MA)'$ as in the diagram

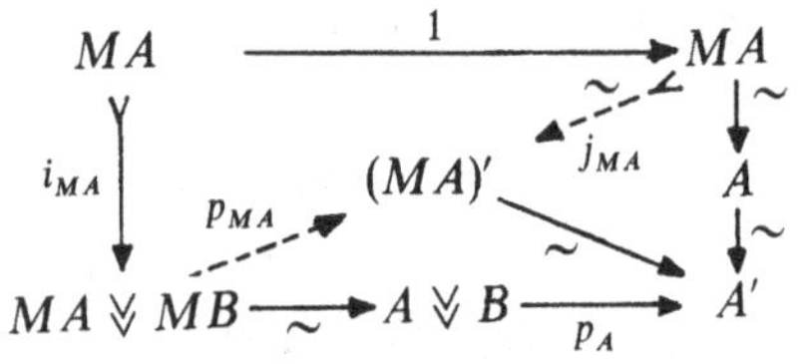

Similarly we choose a weak lifting $(MB)'$. Clearly $(MA)'$ and $(MB)'$ are cofibrant. One readily checks that $MA \curlyvee MB$ is a well defined based fat sum. Hence we have the **smash product** $MA \wedge MB$ associated to $MA \curlyvee MB$ by (15.4).

(15.6) ***Remark.*** Let $\tau: \mathbf{C} \to \mathbf{K}$ be a based functor between cofibration categories which preserves weak equivalences. Then τ carries a fat sum $A \curlyvee B$ in $\mathbf{C}$ to a fat sum $\tau A \curlyvee \tau B = \tau(A \curlyvee B)$ in $\mathbf{K}$. If $A \curlyvee B$ is based then $\tau A \curlyvee \tau B$ needs not to be a based fat sum. In this case we can apply the remark in (15.5) and we get $M\tau A \curlyvee M\tau B$ and $M\tau A \wedge M\tau B$.

(15.7) ***Example.*** Let $\mathbf{C}$ be a cofibration category and let $\mathbf{F}$ be a fibration category and assume $\mathbf{C}$ and $\mathbf{F}$ have an object $*$ which is initial and final. Then a product in $\mathbf{C}$ is a fat sum in $\mathbf{C}$ and dually a sum in $\mathbf{F}$ is a fat product in $\mathbf{F}$. In particular, the product $A \times B$ in $\mathbf{Top}^*$ is a fat sum and the sum $A \vee B$ of spaces in $\mathbf{Top}^*$ is a fat product, see (I.5.4) and (I.5.5).

(15.8) ***Example.*** Consider the cofibration category $\mathbf{C} = \mathbf{DA}_*$ (flat) in (I.7.10). Then the tensor product $A \otimes B$ in (I.7.31) is a fat sum in $\mathbf{C}$ with projections $p_A = 1_A \otimes \varepsilon$, $p_B = \varepsilon \otimes 1_B$ where ε is the augmentation.

(15.9) **Proposition.** *Let $A \curlyvee B$ be a based fat sum. Then there is an isomorphism in $Ho(\mathbf{C})$*

$$r: \Sigma(A \curlyvee B) \simeq \Sigma A \vee \Sigma B \vee \Sigma(A \wedge B).$$

For products in $\mathbf{Top}^*$ as in (15.7) the proposition yields a well-known result due to D. Puppe. We prove (15.9) by use of the cofiber sequence for the cofibration $i = (i_A, i_B)$ in (15.3). This cofiber sequence yields the short exact sequence of groups

(15.10)
$$\begin{array}{ccccc} 0\to[\Sigma(A\wedge B),U] & \xrightarrow{(\Sigma q)^*} & [\Sigma(A \mathbin{\underline{\vee}} B),U] & \xrightarrow{(\Sigma i)^*} & [\Sigma(A\vee B),U]\to 0 \\ & & & \nwarrow^{s} & \cong\uparrow{\scriptstyle j} \\ & & & & [\Sigma A,U]\times[\Sigma B,U]\end{array}$$

Here q is the quotient map for the cofiber (15.4) and the isomorphism j of groups carries (α, β) to the map $\Sigma(A \vee B) \simeq \Sigma A \vee \Sigma B \to U$ given by (α, β). We define the function s by the sum of elements

$$s(\alpha, \beta) = q_A^*(\alpha) + q_B^*(\beta) \tag{1}$$

in the group $[\Sigma(A \mathbin{\underline{\vee}} B), U]$. Here $q_A = (\Sigma j_A)^{-1}(\Sigma p_A)$ and $q_B = (\Sigma j_B)^{-1}(\Sigma p_B)$ are well-defined maps in $Ho(\mathbf{C})$ which induce homomorphisms of groups q_A^* and q_B^* respectively. Now the definition in (15.2) shows by (9.3)(a)

$$(\Sigma i)^* s = j. \tag{2}$$

Therefore $(\Sigma i)^*$ is actually surjective. Similarly we see that $(\Sigma^2 i)^*$ is surjective and therefore $(\Sigma q)^*$ in (15.10) is injective by exactness.

Proof of (15.9). We use a similar argument as in the proof of (15.1) relying on the exact sequence (15.10). The isomorphism r is the sum (3) $r = i_1 q_A + i_2 q_B + i_3(\Sigma q)$ where i_1, i_2, i_3 are the inclusions. □

By use of the exact sequence (15.10) we define the **Whitehead product**

(15.11) $$[\ ,\ ]:[\Sigma A, U] \times [\Sigma B, U] \to [\Sigma(A \wedge B), U]$$

which is **associated to the fat sum** $A \mathbin{\underline{\vee}} B$. For $\alpha \in [\Sigma A, U]$, $\beta \in [\Sigma B, U]$ the product $[\alpha, \beta]$ is given by a commutator in the group $[\Sigma(A \mathbin{\underline{\vee}} B), U]$, namely

$$[\alpha, \beta] = (\Sigma q)^{*-1}(-q_A^*(\alpha) - q_B^*(\beta) + q_A^*(\alpha) + q_B^*(\beta)). \tag{1}$$

This element is well defined since the homomorphism $(\Sigma i)^*$ in (15.10) carries the commutator to the trivial element and whence this commutator is in the image of the injective map $(\Sigma q)^*$ by exactness. For the inclusions $i_1:\Sigma A \to \Sigma A \vee \Sigma B$ and $i_2:\Sigma B \to \Sigma A \vee \Sigma B$ we get the Whitehead product map

$$w = [i_1, i_2]:\Sigma(A \wedge B) \to \Sigma A \vee \Sigma B, \tag{2}$$

which is well defined in $Ho(\mathbf{C})$ and for which

$$[\alpha, \beta] = w^*(\alpha, \beta). \tag{3}$$

Clearly, the Whitehead product is natural with respect to maps $v:U \to U'$, $v_*[\alpha, \beta] = [v_*\alpha, v_*\beta]$. We also use the exact sequence (15.10) for the definition of the group actions

$$(15.12)\quad \begin{cases}[\Sigma(A\wedge B),U]\times[\Sigma A,U]\to[\Sigma(A\wedge B),U],(\xi,\alpha)\longmapsto\xi^{\alpha},\\ [\Sigma(A\wedge B),U]\times[\Sigma B,U]\to[\Sigma(A\wedge B),U],(\xi,\beta)\longmapsto\xi_{\beta},\end{cases}$$

which are defined by

$$\xi^{\alpha}=(\Sigma q)^{*-1}(-q_A^*(\alpha)+(\Sigma q)^*\xi+q_A^*(\alpha)),$$
$$\xi_{\beta}=(\Sigma q)^{*-1}(-q_B^*(\beta)+(\Sigma q)^*\xi+q_B^*(\beta)).$$

(15.13) *Remark.* Let $\mathbf{C}=\mathbf{Top}^*$ be the cofibration category of topological spaces and let the fat sum $A \mathbin{\underline{\vee}} B=A\times B$ be given by the product of topological spaces A and B. (This is a based fat sum if A and B are very well pointed). The Whitehead product associated to $A\times B$ is the classical Whitehead product, originally defined by J.H.C. Whitehead for spheres A and B. One can show that the second action ξ_{β} (15.12) associated to $A\times B$ coincides with the action (6.13) provided $A=S^{n-1}$ is a sphere (in (6.13) we replace A by B). If A is a suspension then the first action is trivial, $\xi^{\alpha}=\xi$, similarly $\xi_{\beta}=\xi$ if B is a suspension. Compare Baues (1977).

From (15.11) and (15.12) we derive the

(15.14) **Lemma.** $[\alpha,0]=[0,\beta]=0$, $\quad[\alpha'+\alpha,\beta]=[\alpha',\beta]^{\alpha}+[\alpha,\beta]$ and $[\alpha,\beta'+\beta]=[\alpha,\beta]+[\alpha,\beta']_{\beta}$.

These formulas are deduced from the corresponding equations for commutators which are valid in any group.

Addendum. *Assume U is a based object and assume α or β are based up to homotopy. Then $[\alpha,\beta]$ is based up to homotopy and $\Sigma[\alpha,\beta]=0$,* see (9.8).

This follows readily from the fact that the group $[\Sigma^2(A \mathbin{\underline{\vee}} B),U]$ is abelian and that Σ in (9.8) is a homomorphism, see (9.10)(*d*).

We also use the exact sequence (15.10) for the

(15.15) ***Definition of the Hopf-construction.*** Let $\mu: A \mathbin{\underline{\vee}} B\to X$ be a based map between based objects with $\alpha=\mu i_A$ and $\beta=\mu i_B$. Then the element

$$\begin{cases}H(\mu)\in[\Sigma(A\wedge B),\Sigma X],\\ H(\mu)=(\Sigma q)^{*-1}(-s(\Sigma\alpha,\Sigma\beta)+\Sigma\mu),\end{cases}$$

is the Hopf-construction of μ. Here the function s is defined as in (15.10)(1). ||

(15.16) ***Example.*** Let $\mathbf{C}=\mathbf{Top}^*$ and let $\mu: S^1\times S^1\to S^1$ be the multiplication on the unit sphere. Then the **Hopf-map** $\eta = H(\mu) \in [\Sigma(S^1 \wedge S^1), \Sigma S^1] = \pi_3(S^2)$ is a generator of the homotopy group $\pi_3(S^2) = [S^3, S^2] \cong \mathbb{Z}$.

(15.17) **Theorem.** *Let $\tau: \mathbf{C} \to \mathbf{K}$ be a based model functor between cofibration categories. Then* Whitehead *products and* Hopf *constructions are compatible with τ. This means*

$$\tau_L[\alpha, \beta] = [\tau_L\alpha, \tau_L\beta] \in [\Sigma(M\tau A \wedge M\tau B), \tau U],$$
$$\tau_L H(\mu) = H(\tau_M \mu) \in [\Sigma(M\tau A \wedge M\tau B), \Sigma M\tau X].$$

Here we use the based fat sum $M\tau A \underset{\vee}{\vee} M\tau B$ in **K** given as in (15.6) and we use the homomorphism τ_L defined in (6a.2). The map $\tau_M \mu = M\tau\mu$ in **K** is a model of $\tau\mu$ as in (8.27) with

$$\begin{array}{ccc} \tau(A \underset{\vee}{\vee} B) & \xrightarrow{\tau\mu} & \tau X \\ \uparrow\sim & & \uparrow\sim \\ M\tau A \underset{\vee}{\vee} M\tau B & \underset{\tau_M\mu}{\rightarrowtail} & M\tau X \end{array}.$$

Proof of (15.17). As in (8.27)(4) we have the commutative diagram in $Ho(\mathbf{K})$

$$\begin{array}{ccccc} \tau\Sigma(A \vee B) & \xrightarrow{\tau\Sigma i} & \tau\Sigma(A \underset{\vee}{\vee} B) & \xrightarrow{\tau\Sigma q} & \tau\Sigma(A \wedge B) \\ \uparrow\sim & & \uparrow\sim & & \uparrow\sim \\ \Sigma(M\tau A \vee M\tau B) & \xrightarrow{\Sigma i} & \Sigma(M\tau A \underset{\vee}{\vee} M\tau B) & \xrightarrow{\Sigma q} & \Sigma(M\tau A \wedge M\tau B) \end{array},$$

where the vertical arrows are isomorphisms in $Ho(\mathbf{K})$ since τ is compatible with homotopy push outs. □

(15.18) ***Remark.*** For the functor $\tau = SC_*\Omega$ in (I.7.29) the first equation of (15.17) yields a classical result of Samelson on Whitehead products. We will show this in (17.24) below.

The Whitehead products associated to a product $S^n \times S^n$ of spheres in **Top*** are the most important examples. They have the following nice properties:

(15.19) ***Example.*** Let $\mathbf{C} = \mathbf{Top}^*$ and let $A \wedge B = (A \times B)/(A \vee B)$ where A and B are very well pointed spaces in **Top***. We identify the n-sphere S^n with the n-fold smash product

$$S^n = S^1 \wedge \cdots \wedge S^1 = S^1 \wedge S^{n-1} = \Sigma S^{n-1}. \tag{1}$$

This shows $S^{n+m+1} = \Sigma S^n \wedge S^m$. For the homotopy groups

$$\pi_{n+1}(U) = [\Sigma S^n, U] = [S^n, \Omega U] = \pi_n(\Omega U) \tag{2}$$

we have the **Whitehead product associated to** $S^n \times S^m$

$$[\ ,\ ]:\pi_n(\Omega U)\times\pi_m(\Omega U)\to\pi_{n+m}(\Omega U). \tag{3}$$

Now assume that $\pi_1(U)=\pi_0(\Omega U)=0$. Then $(\pi_*(\Omega U),[\ ,\ ])$ is a **graded Lie algebra**, see (I.9.1). The bilinearity of (3) is a consequence of (15.14).

Further properties of Whitehead products in **Top*** are discussed in Baues (1981).

§ 15a Appendix: Whitehead products, co-Whitehead products and cup products in topology

Let $\mathbf{C}=\mathbf{Top}^*$ be the cofibration category in (I.5.4) and let $\mathbf{F}=\mathbf{Top}^*$ be the fibration category in (I.5.5). Suppose that A and B are very well pointed CW-spaces, hence A and B are fibrant and cofibrant in **C** and in **F**. We have the cofibration, resp. fibration sequence

(15a.1)
$$\begin{cases} A\vee B \overset{i}{\rightarrowtail} A\times B \overset{q}{\longrightarrow} A\wedge B \text{ in } \mathbf{C}, \text{ and} \\ A \curlywedge B \overset{p}{\longrightarrow} A\vee B \overset{i}{\longrightarrow} A\times B \text{ in } \mathbf{F}. \end{cases}$$

Here $A\wedge B=(A\times B)/(A\vee B)$ is the smash product associated to the product $A\times B$ in **C**, and $A \curlywedge B$ is the cosmash product in **F** associated to the sum $A\vee B$ (this is the homotopy theoretic fiber of the inclusion i, see (14.1)), compare also (15.7). The product $A\times B$ is a based fat sum in **C**; the sum $A\vee B$, however, is a fat product in **F** which is *not* based since i is not a fibration (here we use (15.5)).

By (15.11) we have the Whitehead product $[\ ,\ ]$ associated to $A\times B$ in **C** and we have dually the co-Whitehead product $[\ ,\ ]^{op}$ associated to $A\vee B$ in **F**:

(15a.2)
$$\begin{cases} [\ ,\ ]:[\Sigma A,U]\times[\Sigma B,U]\to[\Sigma(A\wedge B),U] \text{ in } \mathbf{C}, \\ [\ ,\ ]^{op}:[U,\Omega A]\times[U,\Omega B]\to[U,\Omega(A\curlywedge B)] \text{ in } \mathbf{F}. \end{cases}$$

By the addendum of (15.14) we know that $\Sigma[\alpha,\beta]=0$ and that $\Omega[\alpha,\beta]^{op}=0$.

We now assume that U is a very well pointed CW-space. The next three propositions are due to Arkowitz.

(15a.3) **Proposition.** *The inclusion* $i:\Sigma A\vee\Sigma B\rightarrowtail\Sigma A\times\Sigma B\simeq C_w$ *is a principal cofibration with the attaching map* $w=[i_1,i_2]:\Sigma(A\wedge B)\to\Sigma A\vee\Sigma B$ *given by the Whitehead product map.*

Proof. Let $CA\dot{\times}CB=CA\times B\cup A\times CB$ be the indicated subspace of the product $CA \times CB$ of cones and let $\pi_0:(CA, A) \to (\Sigma A, *)$ be the quotient map. Consider the following commutative diagram where Q is the push out

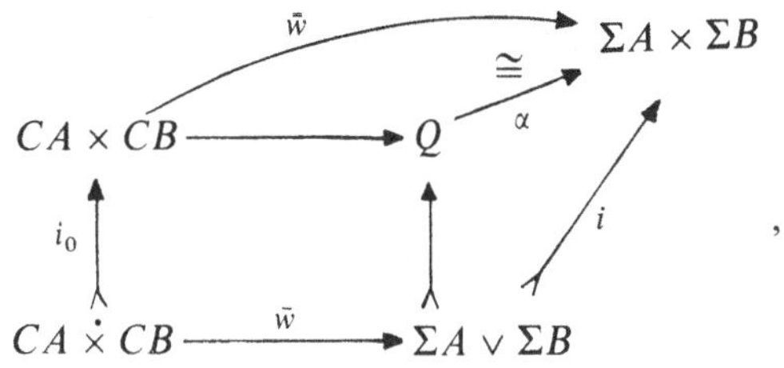

,

where $\bar{\bar{w}} = \pi_0 \times \pi_0$ and where $\bar{w}$ is the restriction of $\bar{\bar{w}}$. The induced map α is a homotopy equivalence and i_0 is a cone since $CA \times CB$ is contractible. Now (15a.3) follows since there is a homotopy equivalence h for which the composition

$$w:\Sigma(A \wedge B) \xrightarrow[\simeq]{h} CA \dot{\times} CB \xrightarrow{\bar{w}} \Sigma A \vee \Sigma B$$

is the Whitehead product map. □

Now let $\tilde{\Delta}$,

(15a.4) $$\tilde{\Delta} = q\Delta : U \to U \times U \to U \wedge U,$$

be the **reduced diagonal** of U, $\Delta(x) = (x, x)$ for $x \in U$.

(15a.5) **Proposition.** *The co-Whitehead product* $[\alpha, \beta]^{op}$ *above satisfies the equation*

$$[\alpha, \beta]^{op} = (\Omega t) j(\alpha \wedge \beta)\tilde{\Delta},$$

where $t:\Sigma(\Omega A \wedge \Omega B) \simeq A \mathbin{\hat{\hat{\wedge}}} B$ *is a natural homotopy equivalence and where* $j:\Omega A \wedge \Omega BL \to \Omega\Sigma(\Omega A \wedge \Omega B)$ *is the adjoint to the identity of* $\Sigma(\Omega A \wedge \Omega B)$.

The proof uses arguments invented by Ganea.

Proof of (15a.5). For the proof it is convenient to use the following result of Strøm (1968):

Lemma. *Let* $p:E \twoheadrightarrow B$ *be a fibration and let* $A \rightarrowtail B$ *be a cofibration in* **Top** *then* $p^{-1}(A) \rightarrowtail E$ *is a cofibration in* **Top**. (1)

It is easy to see that for W in (14.1) the diagram

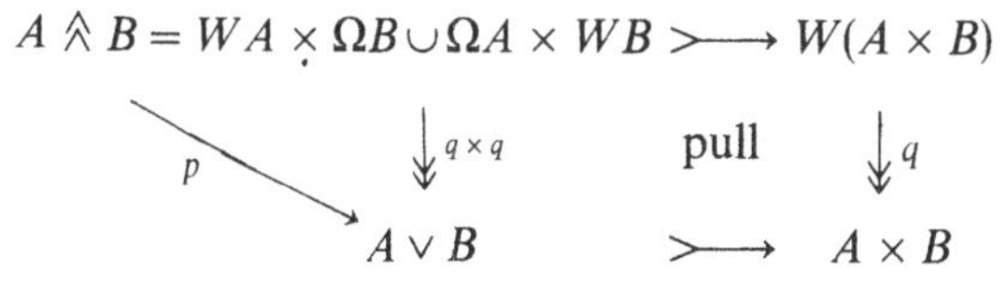

(2)

is a pull back diagram. Here $\Omega A \rightarrowtail WA$ is a cone by (1) since WA is contractible, see (7.5)(1). There is a commutative diagram

$$\begin{array}{ccc} (WA, \Omega A) & \xleftarrow{\sim} & (C\Omega A, \Omega A) \\ \downarrow q & & \downarrow \pi_0 \\ (A, *) & \xleftarrow[R_A]{} & (\Sigma\Omega A, *) \end{array} \quad . \tag{3}$$

in **Pair (Top*)**. Here R_A is the **evaluation map** on $\Sigma\Omega A$ which is adjoint to the identity on ΩA.

Now (3) yields the homotopy commutative diagram

$$\begin{array}{ccc} A \mathbin{\hat{\wedge}} B & \xrightarrow{p} & A \vee B \\ \uparrow \simeq & & \uparrow R_A \vee R_B \\ C\Omega A \mathbin{\dot{\times}} C\Omega B & \xrightarrow{\bar{w}} & \Sigma\Omega A \vee \Sigma\Omega B \, , \\ \wr\| h & \nearrow w & \\ \Sigma\Omega A \wedge \Omega B & & \end{array} \tag{4}$$

with $t: \Sigma\Omega A \wedge \Omega B \to A \mathbin{\hat{\wedge}} B$,

where w is the Whitehead product map; compare the proof of lemma (15a.3) above. Next consider the commutator in

$$[\Sigma U, A \vee B] = [U, \Omega(A \vee B)], \tag{5}$$

which is given by

$$\begin{aligned} (\Omega p)_* [\alpha, \beta]^{op} &= -i_1\alpha - i_2\beta + i_1\alpha + i_2\beta \\ &= ([i_1\bar{\alpha}_1 i_2\bar{\beta}]\Sigma\tilde{\Delta})^- \end{aligned} \tag{6}$$

Here we use (15a.4). The element $\bar{\alpha} \in [\Sigma U, A]$ is the adjoint of α and satisfies

$$\bar{\alpha} = R_A(\Sigma\alpha). \tag{7}$$

Thus we get

$$\begin{aligned} [i_1\bar{\alpha}, i_2\bar{\beta}] &= [i_1 R_A(\Sigma\alpha), i_2 R_B(\Sigma\beta)] \\ &= [i_1 R_A, i_2 R_B](\Sigma(\alpha \wedge \beta)), \end{aligned} \tag{8}$$

and hence the result follows by (4) and (6). □

We now assume that A and B are Eilenberg–Mac Lane spaces as defined in (III.6.1) below. Let $\bar{A}$ and $\bar{B}$ be abelian groups and let

$$A = K(\bar{A}, n+1), \quad B = K(\bar{B}, m+1).$$

Then $\Omega A = K(\bar{A}, n)$ and

(15a.6) $$\tilde{H}^n(U, \bar{A}) = [U, K(\bar{A}, n)] = [U, \Omega A]$$

is the reduced cohomology of U. Moreover, we have the well-known **cup product**

(15a.7) $$\cup:\tilde{H}^n(U,\bar{A})\otimes\tilde{H}^m(U,\bar{B})\to\tilde{H}^{n+m}(U,\bar{A}\otimes\bar{B}),$$

which carries $\alpha\otimes\beta$ to $\alpha\cup\beta$. This cup product can be described by the co-Whitehead product.

(15a.8) **Proposition.** $\alpha\cup\beta=(\Omega k)_*[\alpha,\beta]^{op}$.

Here $k:A \mathbin{\hat{\wedge}} B\to K(\bar{A}\otimes\bar{B},n+m+1)$ is the first k-invariant of the space $A \mathbin{\hat{\wedge}} B\simeq\Sigma(\Omega K(\bar{A},n+1)\wedge\Omega K(\bar{B},m+1))$, see (III.7.4) below. The proposition shows us the precise meaning of the statement that 'cup products and Whitehead products are dual to each other', compare Eckmann (1962). From (15.14) we derive the well-known fact $\Omega(\alpha\cup\beta)=0$.

For the proof of (15a.8) recall that the cup product (15a.7) is induced by the **cup product map**

$$\bigcup_{n,m}:K(\bar{A},n)\times K(\bar{B},m)\xrightarrow{q}K(\bar{A},n)\wedge K(\bar{B},m)\xrightarrow{\tilde{\bigcup}_{n,m}}K(\bar{A}\otimes\bar{B},n+m),$$

where q is the quotient map and where $\tilde{\bigcup}_{n,m}$ is the first k-invariant. We have

(15a.9) $$\alpha\cup\beta=\bigcup_{n,m}(\alpha\times\beta)\Delta=\tilde{\bigcup_{n,m}}(\alpha\wedge\beta)\tilde{\Delta},$$

whence (15a.8) follows from (15a.5).

Finally, we describe an important property of the Whitehead product map and of the cup-product map respectively. We know that the Whitehead product map

$$w=w_{A,B}:\Sigma A\wedge B\to\Sigma A\vee\Sigma B$$

is trivial on ΣB. Therefore the partial suspension

$$E(w_{A,B}):\Sigma^2 A\wedge B\to\Sigma^2 A\vee\Sigma B$$

is defined by (11.10). On the other hand, the cup product map above is trivial on $K(\bar{B},m)$ and thus the partial loop operation L in (14.10) yields the map

$$L(\bigcup_{n,m}):\Omega K(\bar{A},n)\times K(\bar{B},m)\to\Omega K(\bar{A}\otimes\bar{B},n+m).$$

Here we identify $\Omega K(\bar{A},n)=K(\bar{A},n-1)$.

(15a.10) **Proposition.**

$$E(w_{A,B})=w_{\Sigma A,B},$$

and

$$L\left(\bigcup_{n,m}\right)=\bigcup_{n-1,m}$$

Compare 3.1.11 and 6.1.12 in Baues (1977). The proposition will be used in the next section.

§ 16 Examples on the classification of maps in topology

In this section we illustrate the use of the exact classification sequence in (13.10) where we set $T = Y = *$. We work in the category **Top*** of topological spaces with basepoint which is a cofibration category **C** and a fibration category **F** by (I.5.4) and (I.5.5) respectively.

We consider maps in **Top***

(16.1)
$$\begin{cases} f:\Sigma A_1 \vee \cdots \vee \Sigma A_r \to \Sigma A_{r+1} \vee \cdots \vee \Sigma A_k, \\ g:K_1 \times \cdots \times K_r \to K_{r+1} \times \cdots \times K_k, \end{cases}$$

where for simplicity we assume that all A_i are suspensions (for examples spheres of dimension $\geqq 1$) and that all K_i are Eilenberg–Mac Lane spaces of abelian groups. For example the Whitehead-product map $w_{A,B}$ and the cup product map $\bigcup_{m,n}$ in (15a.10) are such maps f and g respectively.

Let U be a very well pointed CW-space. We can apply the classification sequence in (13.10) for the computation of the sets of homotopy classes

(16.2)
$$[C_f, U] \quad \text{and} \quad [U, P_g].$$

respectively. Here C_f is the mapping cone of f and P_g is the homotopy theoretic fiber of g. We will describe various examples for which we obtain explicit descriptions of the sets in (16.2).

(16.3) ***Remark.*** The general method of computation for the sets (16.2) proceeds as follows. Using the Hilton–Milnor theorem and the Künneth theorem respectively it is always possible to decompose the maps f and g in (16.1) into a sum of maps f_j and g_j respectively. For these summands it is easy to compute the differences ∇f_j and ∇g_j. In most cases (15a.10) as well allows the computation of the partial suspension $E^n\nabla f_j$ and of the partial loop operation $L^n\nabla g_j$. Then we can apply the classification sequence (13.10) which yields a description of the sets $[C_f, U]$ or $[U, P_g]$ in terms of exact sequences. Such applications were obtained with different methods by Rutter (1967), Nomura (1969) and originally by Barcus–Barratt (1958), compare also Baues (1977).

We now consider the example where $f = w_{A,B} = w$ is the Whitehead product map and where $g = \bigcup_{n,m}$ is the cup product map. By (15a.3) we have the principal cofibration

(16.4)
$$\Sigma A \wedge B \xrightarrow{f} \Sigma A \vee \Sigma B \rightarrowtail \Sigma A \times \Sigma B \simeq C_f.$$

Dually (in the Eckmann–Hilton sense) we have the principal fibration

(16.5) $P_g = K(\bar{A},n) \cup K(\bar{B},m) \twoheadrightarrow K(\bar{A},n) \times K(\bar{B},m) \xrightarrow{g} K(\bar{A} \otimes \bar{B}, n+m)$.

In both cases we can apply the classification sequence (13.10) and we can use (15a.10). This yields the following two results.

(16.6) **Theorem.** *Let A and B be suspensions. Then there is a bijection of sets*

$$[\Sigma A \times \Sigma B, U] \approx \bigcup_{(\alpha,\beta)} [\Sigma^2 A \wedge B, U]/I_{\alpha,\beta},$$

where the disjoint union is taken over all pairs $(\alpha,\beta) \in [\Sigma A, U] \times [\Sigma B, U]$ *with* $[\alpha,\beta] = 0$ *and where* $I_{\alpha,\beta}$ *in the subgroup of all elements*

$$[x,\beta] + \tau^*[\alpha,y] \in [\Sigma^2 A \wedge B, U]$$

with $x \in [\Sigma^2 A, U]$ *and* $y \in [\Sigma^2 B, U]$.

The map τ is the obvious interchange map $\Sigma A \wedge \Sigma B \approx \Sigma^2 A \wedge B$. Here $[\alpha,\beta]$, $[x,\beta]$ and $[\alpha,y]$ denote the Whitehead products in (15a.2).

Addendum: *Let* $u: T = \Sigma A \times \Sigma B \to U$ *be a basepoint preserving map and let* $U^{T|*}$ *be the function space of all such maps. Let* $(\alpha,\beta) = ui$ *be the restriction of u to* $\Sigma A \vee \Sigma B$. *Then for* $n \geqq 1$ *there is a short exact sequence of groups.*

$$0 \to \operatorname{coker} \nabla^{n+1} \to \pi_n(U^{T|*}, u) \to \ker \nabla^n \to 0.$$

$$\begin{cases} \nabla^n : \pi^A_{n+1}(U) \times \pi^B_{n+1}(U) \to \pi^{A \wedge B}_{n+1}(U), \\ \nabla^n(x,y) = [x,\beta] + \tau^*[\alpha,y]. \end{cases}$$

(16.7) **Theorem.** *There is a bijection of sets*

$$[U, K(\bar{A},n) \cup K(\bar{B},m)] \approx \bigcup_{(\alpha,\beta)} H^{n+m-1}(U, \bar{A} \otimes \bar{B})/I_{\alpha,\beta},$$

where the disjoint union is taken over all pairs $(\alpha,\beta) \in \tilde{H}^n(U,\bar{A}) \times \tilde{H}^m(U,\bar{B})$ *with* $\alpha \cup \beta = 0$ *and where* $I_{\alpha,\beta}$ *is the subgroup of all elements*

$$x \cup \beta + \alpha \cup y \in H^{n+m-1}(U, \bar{A} \otimes \bar{B})$$

with $x \in \tilde{H}^{n-1}(U,\bar{A})$ *and* $y \in \tilde{H}^{m-1}(U,\bar{B})$.

Here $\alpha \cup \beta$, $x \cup \beta$ and $\alpha \cup y$ denote the cup products in the cohomology of U, see (15a.7).

Addendum. *Let* $u: U \to P = K(\bar{A},n) \cup K(\bar{B},m)$ *be a basepoint preserving map and let* $P^{U|*}$ *be the function space of all such maps. Let* $(\alpha,\beta) = pu$ *be the projection of u to* $K(\bar{A},n) \times K(\bar{B},m)$. *Then for* $k \geqq 1$ *there is the short exact* *sequence of groups*

$$0 \to \operatorname{coker} \nabla_{k+1} \to \pi_k(P^{U|*}, u) \to \ker \nabla_k \to 0,$$

$$\begin{cases} \nabla_k : \tilde{H}^{n-k}(U, \bar{A}) \times \tilde{H}^{m-k}(U, \bar{B}) \to \tilde{H}^{n+m-k}(U, \bar{A} \otimes \bar{B}), \\ \nabla_k(x, y) = x \cup \beta + \alpha \cup y. \end{cases}$$

Proof of (16.6) *and* (16.7). We apply (13.10) with $T = Y = *$ for the computation of (16.2). For $f = w_{A,B} = w$ the difference ∇w is the map

$$\nabla w = -i_2\omega + (i_2 + i_1)w : \Sigma A \wedge B \to \Sigma(A_1 \vee B_1) \vee \Sigma(A_2 \vee B_2),$$

where $A_1 = A_2 = A$, $B_1 = B_2 = B$. For the obvious inclusions a_i, b_i $(i = 1, 2)$ of A_i and B_i respectively we get

$$\begin{aligned} \nabla w &= -w^*(a_2, b_2) + w^*(a_2 + a_1, b_2 + b_1) \\ &= -[a_2, b_2] + [a_2 + a_1, b_2 + b_1] \\ &= [a_1, b_1] + [a_1, b_2] - (\Sigma T)^*[b_1, a_2], \end{aligned}$$

where we use the bilinearity of the Whitehead product. Since the partial suspension E is a homomorphism and since $\Sigma[a_1, b_1] = 0$ we derive from (11.9) and (15a.10)

$$E^n \nabla w = [\bar{a}_1, b_2] - (\Sigma^{n+1} T)^*[\bar{b}_1, a_2].$$

Here $\bar{a}_1$ and $\bar{b}_1$ are the inclusions of $\Sigma^{n+1} A_1$ and $\Sigma^{n+1} B_1$ respectively.

Since $\nabla^n(u, w) = (E^n \nabla w)^* (\cdot, u)$ we get the result in (16.6) by (13.10). We leave the proof of (16.7) as an exercise. The arguments are of the same type as above, they use the bilinearity of the cup product and (15a.10) and clearly the sequence dual to (13.10) in the fibration category $\mathbf{F} = \mathbf{Top}^*$. □

We next describe further examples where we use the method in (16.3). We say that a homomorphism $\alpha : \pi_2(M) \to \pi_2(U)$ between homotopy groups is **realizable** if there is a map $u : M \to U$ with $\alpha = u_*$.

(16.8) **Theorem.** *Let M be a closed 4-dimensional manifold which is simply connected. Then there is a bijection*

$$[M, U] \approx \bigcup_{\alpha} H^4(M, \pi_4 U)/I_\alpha,$$

where the disjoint union is taken over all realizable $\alpha \in \operatorname{Hom}(\pi_2 M, \pi_2 U) = H^2(M, \pi_2 U)$. The subgroup I_α is the image of the homomorphism

$$\begin{cases} \nabla_\alpha : H^2(M, \pi_3 U) \to H^4(M, \pi_4 U), \\ \nabla_\alpha(x) = (\eta^*)_* Sq^2(x) + [\ ,\ ]_*(x \cup \alpha). \end{cases}$$

Here $\eta^* : \pi_3(U) \otimes \mathbb{Z}/2 \to \pi_4(U)$ is induced by the suspension of the Hopf map (15.16) and Sq^2 is the Steenrod square. Moreover $x \cup \alpha$ is the cup product and $[\ ,\ ]: \pi_3 U \otimes \pi_2 U \to \pi_4 U$ is the Whitehead product.

Addendum. *Let $u: M \to U$ be a basepoint preserving map and let $U^{M|*}$ be the function space of all such maps. Assume that u induces $\alpha \in H^2(M, \pi_2 U)$ as above. Then for $n \geqq 1$ there is a short exact sequence of groups*

$$0 \to \operatorname{coker} \nabla_\alpha^{n+1} \to \pi_n(U^{M|*}, u) \to \ker \nabla_\alpha^n \to 0,$$

$$\begin{cases} \nabla_\alpha^n: H^2(X, \pi_{n+2} U) \to H^4(X, \pi_{n+3} U), \\ \nabla_\alpha^n(x) = (\eta^*)_* Sq^2(x) + [\ ,\ ]_*(x \cup \alpha). \end{cases}$$

Here ∇_α^n is defined in the same way as $\nabla_\alpha = \nabla_\alpha^1$ above.

Proof. Let B be a basis of the free abelian group $H^2 M = \bigoplus_B \mathbb{Z}$ and choose an ordering $<$ of B. Then the Eckmann–Hilton homology decomposition shows that M is homotopy equivalent to the mapping cone C_f with

$$f: S^3 \longrightarrow \bigvee_B S^2.$$

$$f = \Sigma_{b \in B} (b \cup b) \cdot (i_b \eta) + \Sigma_{a < b} (a \cup b) \cdot [i_a, i_b].$$

Here $a \cup b \in H^4(M) = \mathbb{Z}$ denotes the cup product and i_b is the inclusion of S^2 for $b \in B$. We compute ∇f by the bilinearity of the Whitehead product and by the distributivity law for the Hopf map

$$\eta^*(\alpha + \beta) = \eta^*(\alpha) + \eta^*(\beta) + [\alpha, \beta],$$

where $\alpha, \beta \in \pi_2(U)$. Since $Sq^2(x) = x \cup x \bmod 2$ we obtain the result (16.8) in the same way as in the proof of (16.6). □

Next we describe classical examples which can be derived as well from the classification sequence (13.10) in a fibration category similarly as (16.7).

(16.9) **Theorem** (Pontrjagin, Steenrod). *Let X be a CW-complex with dim $(X) \leqq n + 1$ and let $u \in H^n(S^n, \mathbb{Z})$ be a generator. Then the degree map $\deg: [X, S^n] \to H^n(X, \mathbb{Z})$ with $\deg(F) = F^*(u)$ is surjective and*

$$\deg^{-1}(\xi) \approx \begin{cases} H^3(X, \mathbb{Z})/2\xi \cup H^1(X, \mathbb{Z}), & n = 2 \\ H^{n+1}(X, \mathbb{Z}/2)/Sq_{\mathbb{Z}}^2 H^{n-1}(X, \mathbb{Z}), & n \geqq 3 \end{cases}.$$

Here Sq^2 is the Steenrod square. For $n = 1$ the theorem is a special case of the following result, compare Spanier. $\mathbb{C}P_n$ denotes the **complex projective space** with $\mathbb{C}P_1 \approx S^2$.

(16.10) **Theorem.** *Let X be a CW-complex with* $\dim(X) \leqq 2n + 1$ *and let $u \in H^2(\mathbb{C}P_n, \mathbb{Z})$ be a generator. Then the degree map* $\deg: [X, \mathbb{C}P_n] \to H^2(X, \mathbb{Z})$

with $\deg(F) = F^*(u)$ *is surjective and*

$$\deg^{-1}(\xi) \approx H^{2n+1}(X, \mathbb{Z})/(n+1)\xi^n \cup H^1(X, \mathbb{Z}).$$

Here ξ^n denots the n-fold cup product of ξ.

Proof of (16.10). The first k-invariant of $\mathbb{C}P_n$ is the map $k: \mathbb{C}P_\infty = K(\mathbb{Z}, 2) \to K(\mathbb{Z}, 2n+2)$ which is represented by the cup product power u^{n+1}. Therefore we have by $\dim(X) \leqq 2n+1$

$$[X, P_k] = [X, \mathbb{C}P_n].$$

Compare (III.7.2). We can compute $\nabla k: K(\mathbb{Z}, 2) \times K(\mathbb{Z}, 2) \to K(\mathbb{Z}, 2n+2)$ as follows

$$\begin{aligned}\nabla k &= -kp_2 + k(p_2 + p_1) \\ &= -p_2^* u^{n+1} + (p_2 + p_1)^*(u^{n+1}) \\ &= -p_2^* u^{n+1} + ((p_2 + p_1)^* u)^{n+1} \\ &= -p_2^* u^{n+1} + (p_2^* u + p_1^* u)^{n+1} \\ &= \sum_{k=0}^{n} \binom{n+1}{k} (p_2^* u^k) \cup (p_1^* u^{n+1-k}).\end{aligned}$$

From (15a.10) we derive

$$L(p_1^* u^{n+1-k} \cup p_2^* u^k) = (\Omega p_1^* u^{n+1-k}) \cup p_2^* u^k,$$

where $\Omega p_1^* u^{n+1-k} = 0$ for $n+1-k > 1$ since $\Omega(\alpha \cup \beta) = 0$. Moreover, we have

$$\Omega p_1^* u = p_1 : \Omega K(\mathbb{Z}, 2) \times K(\mathbb{Z}, 2) \to \Omega K(\mathbb{Z}, 2).$$

Thus the map

$$L\nabla k: \Omega K(\mathbb{Z}, 2) \times K(\mathbb{Z}, 2) \to \Omega K(\mathbb{Z}, 2n+2)$$

is given by $L(\nabla k) = (n+1) \cdot (p_1 \cup p_2^* u^n)$. Now the result (16.10) is a consequence of (13.10), compare the proof of (16.7). We leave it to the reader to formulate an addendum of (16.10) which corresponds to the addendum of (16.7). □

Also an old result of Pontrjagin and Dold–Whitney is an illustration of the classification sequence (13.10). Let $BSO(n)$ be the classifying space of the special orthogonal group. The homotopy set $[X, BSO(n)]$ determines the oriented $(n-1)$ sphere bundles with structure group $SO(n)$ over X.

(16.11) **Theorem** (Dold–Whitney). *Let* X *be a connected* 4*-dimensional polyhedron. Then there is a bijection* ($n \geqq 3$)

$$[X, BSO(n)] = \bigcup_{\alpha} ((H^4(X, \mathbb{Z})/I_\alpha^n) \oplus A^n),$$

where the disjoint union is taken over all $\alpha \in H^2(X, \mathbb{Z}/2)$. *We have* $A^4 = H^4(X, \mathbb{Z})$ *and* $A^n = 0$ *otherwise. Moreover,* $I^n_\alpha = 0$ *for* $n \geqq 5$. *For* $n = 3$ *and* $n = 4$ *the subgroup* I^n_α *is the image of*

$$\begin{cases} \nabla_\alpha : H^1(X, \mathbb{Z}/2) \to H^4(X, \mathbb{Z}), \\ \nabla_\alpha(x) = (\beta x) \cup (\beta x) + \beta(x \cup \alpha). \end{cases}$$

Here β *is the* Bockstein *homomorphism for* $\mathbb{Z} \to \mathbb{Z} \to \mathbb{Z}/2$.

Proof. We only describe the connection of the result with the classification sequence (13.10): In low dimensions the space $BSO(n)$, $n \geqq 3$, has the homotopy groups $\pi_1 = 0 = \pi_3$ and $\pi_2 = \mathbb{Z}/2$ and $\pi_4 = \mathbb{Z}$ for $n \neq 4$ and $\pi_4 = \mathbb{Z} \oplus \mathbb{Z}$ for $n = 4$. Thus the first k-invariant of the Postnikov tower (II.7.2) is

$$k : K(\pi_2, 2) \to K(\pi_4, 5).$$

Moreover, the principal fibration $P_k \twoheadrightarrow K(\pi_2, 2)$ yields for $\dim X \leqq 4$ a bijection

$$[X, BSO(n)] = [X, P_k].$$

Thus we can apply (13.10). In fact, we have $[K(\mathbb{Z}/2, 2), K(\mathbb{Z}, 5)] = \mathbb{Z}/4$ and k is a generator for $n = 3$. For $n = 4$ the first coordinate of k is a generator and the second coordinate is trivial. Moreover, k is twice a generator for $n > 4$. This is a result of F. Peterson, compare formula 22 in Dold–Whitney. By use of the explicit description of the generator one obtains ∇k and $L(\nabla k)$ with

$$L(\nabla k)_*(x, \alpha) = \nabla_\alpha(x). \qquad \square$$

§ 17 Homotopy groups in the category of chain algebras

We describe explicitly homotopy groups (and homotopy groups of function spaces) in the category of chain algebras **DA**; in particular, we compute the homotopy addition map. Similar results can be obtained in the category of commutative cochain algebras $\mathbf{CDA}^0_*$ and in the category of chain Lie algebras **DL**. The author, however, is not aware of an explicit formula for the homotopy addition map on the cylinders in (I.8.19) and (I.9.18) respectively. Such a formula would be an interesting analogue of the Baker–Campbell–Hausdorff formula. Moreover, by the universal enveloping functor U one can compare the 'rational' homotopy addition map in **DL** with the 'integral' one in $\mathbf{DA}_*$ which we describe in (17.4) below.

Let R be a commutative ring of coefficients and let **DA** be the category of chain algebras over R with the structure in (I.§7). Recall that $\mathbf{DA}^B_c$ is an I-category. *A* homotopy equivalence in $\mathbf{DA}^B_c$ is also a weak equivalence.

The converse is true if R is a principal ideal domain and if B is R-flat. Moreover, we know that **DA**(flat) and **DA**$_*$(flat) are cofibration categories provided R is a principal ideal domain, see (I.7.10).

Let $B \rightarrowtail A = (B \amalg T(W), d)$ be a cofibration in **DA** and let $u\colon B \to U$ be a map in **DA**. Then we have the set of homotopy classes

(17.1) $$[A, U]^u = \mathbf{DA}(A, U)^u / \simeq \operatorname{rel} B,$$

where $\mathbf{DA}(A, U)^u = \{f\colon A \to U \mid f|B = u\}$ and where the homotopy relation is defined in (I.7.13). Clearly, this is an equivalence relation by (2.2). In case $B = *$ is the initial object we set

(17.2) $$[A, U] = \{f\colon A \to U\} / \simeq \operatorname{rel} *,$$

Here $A = (T(W), d)$ is a free chain algebra.

We denote by $\{x\}$ the homotopy class represented by $x\colon A \to U$. If A is a **based object** in **DA** (or, equivalently, if A is an augmented free chain algebra) we have the o-map $\varepsilon = 0\colon A \to *$ which is the augmentation of A; this map yields the **zero element** in the homotopy set $[A, U]$.

Recall that we have an explicit **cylinder** $I_B A = (B \amalg T(W' \oplus W'' \oplus sW), d)$ for the cofibration $B \rightarrowtail A$, see (I.7.11). This yields the problem of computing an explicit **homotopy addition map**

(17.3) $$m\colon I_B A \to I_B A \bigcup_A I_B A = DZ,$$

which is defined in (2.5). Here we have

$$DZ = (B \amalg T(W' \oplus W'' \oplus W''' \oplus sW_0 \oplus sW_1), d)$$

with $W' = W'' = W''' = W_0 = W_1 = W$.

(17.4) ***Computation of m.*** The map m is a homotopy $i' \simeq i'''$ of the inclusions $i', i'''\colon A \to DZ$. Therefore m is given by the homotopy (see (I.7.12)):

$$M\colon A \to DZ$$

with

$$M(b) = 0 \qquad \text{for } b \in B, \tag{1}$$
$$m(sw) = M(w) \qquad \text{for } w \in W, \tag{2}$$
$$Md + dM = i''' - i', \tag{3}$$
$$M(xy) = (Mx)y''' + (-1)^{|x|} x'(My) \quad \text{for } x, y \in A. \tag{4}$$

On the other hand, we obtain from the canonical inclusions

$$\begin{cases} j_0, j_1\colon I_B A \to DZ, \\ j_0 w' = w', j_0 w'' = w'', j_0 sw = sw_0, \\ j_1 w' = w'', j_1 w'' = w''', j_1 sw = sw_1, \end{cases}$$

the homotopies $S_0 = j_0 S$, $S_1 = j_1 S: A \to DZ$, where S is defined in (I.7.11). These homotopies satisfy

$$\left.\begin{aligned} S_0 d + dS_0 &= i'' - i', \\ S_1 d + dS_1 &= i''' - i''. \end{aligned}\right\} \tag{5}$$

Therefore we have the operator

$$\hat{M} = M - S_0 - S_1, \tag{6}$$

which satisfies the following equations:

$$d\hat{M} + \hat{M}d = 0, \tag{7}$$

$$\hat{M}(b) = 0 \quad \text{for } b \in B, \tag{8}$$

$$\hat{M}(xy) = (\hat{M}x)y''' + (-1)^{|x|}x'(\hat{M}y) + (S_0 x)(y''' - y'') - (-1)^{|x|}(x'' - x')(S_1 y). \tag{9}$$

We now construct $\hat{M}$ inductively: Let $v \in W$ be given with $|v| = n$ and assume we know $\hat{M}(w)$ for all $w \in W$ with $|w| < n$. Then by (9) we have a formula for $\hat{M}dv$. By (7) this element is a boundary $\hat{M}dv = d\xi$. We choose such a ξ and we set

$$\hat{M}v = -\xi, \quad Mv = S_0 v + S_1 v - \xi. \tag{10}$$

There is a canonical way to define $\hat{M}$ in (6) as follows. We define a map of *degree* $+2$

$$\begin{array}{ccc} S_m : A & \longrightarrow & DZ \\ \| & & \| \\ B \coprod T(W) & & B \coprod T(W' \oplus W'' \oplus W''' \oplus sW_0 \oplus sW_1), \end{array} \tag{11}$$

$$\begin{cases} S_m(b) = 0 & \text{for } b \in B, \\ S_m(w) = 0 & \text{for } w \in W, \text{ and} \\ S_m(xy) = (-1)^{|x|} S_0 x \cdot S_1 y + S_m(x) \cdot y''' + x' \cdot S_m(y) \end{cases}$$

for $x, y \in A$. As in (I.7.11) one can check that S_m is well defined. Moreover, the map

$$\hat{M} = S_m d - dS_m : A \to DZ \tag{12}$$

satisfies the equations (7), (8) and (9). Therefore (12) yields a canonical choice in (10), compare Helling.

We describe an example for the homotopy addition map in § 2. More generally then (17.1), we have the homotopy groups of function spaces ($n \geqq 1$) in **DA**,

(17.5) $$\pi_n(U^{A|B}, u) = [\Sigma_B^n A, U]^u,$$

compare (§ 6, § 10). We know that π_1 is a group and that π_n, for $n \geqslant 2$, is an abelian group. For $n=1$ the group structure on π_1 is induced by the comultiplication m on $\Sigma_B A$ which makes the diagram

$$\begin{array}{ccc} I_BA & \xrightarrow{m} & I_BA\bigcup_A I_BA \\ \downarrow{\scriptstyle p} & & \downarrow{\scriptstyle p\cup p} \\ \Sigma_BA & \xrightarrow[m]{} & \Sigma_BA\bigcup_A \Sigma_BA \\ \| & & \| \\ A\coprod T(sW) & & A\coprod T(sW_0\oplus sW_1) \end{array}$$

commutative. Here p is the canonical identification map with $pw'=pw''=w$, $psw=sw$ for $w\in W$.

(17.6) **Theorem.** *Let $B\rightarrowtail A=B\coprod T(W)$ be a cofibration and assume W is finite dimensional (that is, $W_n=0$ for $n>N$). Then $\pi_1(U^{A|B},u)$ is a nilpotent group.*

Proof. This is an easy consequence of (III.4.14) since $B\rightarrowtail A$ is a complex of finite length in $\mathbf{DA}_c^B$, see (VII.§4). A less simple proof can be deduced from the explicit formula for m in (17.4). □

From the definition of $\Sigma_B^n A$ in (10.3)(4) we derive

(17.7) $$\Sigma_B^n A=B\coprod T(W\oplus s^nW)=A\coprod T(s^nW).$$

The differential d on $\Sigma_B^n A$ ($n\geqq 1$) is given as follows. Let $S^n:A\to\Sigma_B^n A$ be the map of degree n with

$$\begin{cases} S^nb=0 & \text{for } b\in B,\\ S^nw=s^nw & \text{for } w\in W,\\ S^n(xy)=(S^nx)\cdot y+(-1)^{n|x|}x\cdot(S^ny) & \text{for } x,y\in A.\end{cases}$$

Then d is defined on W by d_A and on s^nW by $ds^nw=(-1)^nS^ndw$.

For a free augmented chain algebra $A=(T(W),d)$ we have the **suspension** $\Sigma A=\Sigma_*A/A$ which is a quotient in **DA**. From (17.7) we derive

(17.8) $$\Sigma A=(T(sW),d)\quad\text{with}\quad d(sw)=-sd_Qw\quad\text{for } w\in W.$$

Here d_Q is the differential on $QA=W$, see (I.7.2). Clearly, the **n-fold suspension** is

(17.9) $$\Sigma^nA=\Sigma_*^nA/A=(T(s^nW),d)\quad\text{with } ds^nw=(-1)^ns^nd_Qw.$$

The homotopy set $[\Sigma^nA,U]$ has a group structure which is abelian for $n\geqq 2$. We leave it to the reader to describe a formula for the comultiplication

(17.10) $$m:\Sigma A=T(sW)\to\Sigma A\vee\Sigma A=T(sW_0\oplus sW_1),$$

which is obtained by dividing out A in the bottom row of diagram (17.5). In general m induces a non abelian group structure for $[\Sigma A, U]$, compare the example in (17.22) below. By use of derivations there is an altenative description of the homotopy groups in (17.5).

(17.11) ***Definition.*** Let $B \subset A$ be a cofibration and let $u: A \to U$ be a map in **DA**. An $(A|B, u)$**-derivation of degree** n $(n \in \mathbb{Z})$ is a map $F: A \to U$ of degree n of the underlying graded modules such that

$$F(b) = 0 \qquad \text{for } b \in B, \tag{1}$$

$$F(xy) = (Fx)(uy) + (-1)^{n|x|}(ux)(Fy) \qquad \text{for } x, y \in A. \tag{2}$$

The set of all such derivations:

$$\mathrm{Der}_n = \mathrm{Der}_n(U^{A|B}, u) \tag{3}$$

is a module by $(F+G)(x) = (Fx) + (Gx)$. We define a boundary operator

$$\begin{aligned} &\partial: \mathrm{Der}_n \to \mathrm{Der}_{n-1} \\ &\partial(F) = F \circ d - (-1)^n d \circ F \end{aligned} \tag{4}$$

where d denotes the differential in A and U respectively. One easily checks that $\partial(F)$ is an element of Der_{n-1} and that $\partial\partial = 0$. Thus we have for all $n \in \mathbb{Z}$ the homology

$$H_n \mathrm{Der}_*(U^{A|B}, u) = \ker(\partial)/\mathrm{im}(\partial) \tag{5}$$

of the chain complex of derivations. ||

(17.12) **Proposition.** *Let $B \subset A = B \coprod T(W)$ be a cofibration in* **DA**. *Then we have for $n \geq 1$ a bijection*

$$\pi_n(U^{A|B}, u) = H_n \mathrm{Der}_*(U^{A|B}, u),$$

which depends on the choice of the generators W. For $n \geq 2$ this is an isomorphism of abelian groups.

Proof. Let $\mathbf{A}(\Sigma_B^n A, U)^u$ be the set of all algebra maps $\bar{F}: \Sigma_B^n A \to U$ between the underlying graded algebras with $\bar{F}|A = u$. We have the bijection

$$\mathbf{A}(\Sigma_B^n A, U)^u = \mathrm{Der}_n, \quad \bar{F} \longmapsto \bar{F}S^n = F, \tag{1}$$

with S^n in (17.7). Now $\bar{F}$ is a chain algebra map $(d\bar{F} = \bar{F}d)$ if and only if $\partial F = 0$. Therefore (1) gives us the bijection (see (17.1))

$$\mathbf{DA}(\Sigma_B^n A, U)^u = \mathrm{kernel}(\partial). \tag{2}$$

It remains to show that for $\bar{F}_1, \bar{F}_2 \in \mathbf{DA}(\Sigma_B^n A, U)^u$ we have

$$\bar{F}_1 \simeq \bar{F}_2 \text{ rel } A \Leftrightarrow \exists G \in \mathrm{Der}_{n+1} \quad \text{with} \quad \partial G = F_2 - F_1. \tag{*}$$

Then, clearly, (2) gives us the bijection in (17.12). Now a **DA**-homotopy $\bar{G}$ from $\bar{F}_1$ to $\bar{F}_2(\text{rel}\, A)$ is given by a map $\bar{G}:\Sigma_B^n A \to U$ of degree $+1$ with

$$\bar{G}x = 0 \quad \text{for } x \in A, \tag{3}$$

$$\bar{G}d + d\bar{G} = \bar{F}_2 - \bar{F}_1, \tag{4}$$

and

$$\bar{G}(xy) = (\bar{G}x)(\bar{F}_2 y) + (-1)^{|x|}(\bar{F}_1 x)(\bar{G}y) \tag{5}$$

for $x, y \in \Sigma_B^n A$. Therefore the map

$$G = \bar{G}S^n \tag{6}$$

is of degree $n+1$ and satisfies by (17.7)

$$Gb = 0 \quad \text{for } b \in B, \tag{7}$$

$$\begin{aligned} G(xy) &= \bar{G}((S^n x)y + (-1)^{n|x|}x(S^n y)) \\ &= (Gx)(uy) + (-1)^{(n+1)|x|}(ux)(Gy) \end{aligned} \tag{8}$$

for $x, y \in A$. Here we use (3) and (5). By (7) and (8) we know that $G \in \text{Der}_{n+1}$. Moreover, by (4) and (17.7) we have for $v \in W$

$$\begin{aligned} \partial G(v) &= \bar{G}S^n dv - (-1)^{n+1} d\bar{G}S^n v \\ &= (-1)^n \bar{G} ds^n v + (-1)^n d\bar{G}s^n v \\ &= (-1)^n (\bar{F}_2 - \bar{F}_1) s^n v \\ &= (-1)^n (F_2 - F_1)(v). \end{aligned} \tag{9}$$

Therefore $\partial(-1)^n G = F_2 - F_1$. The other direction of $(*)$ can be proved in a similar way. It follows from (17.17) and (17.18) below that the bijection is actually an isomorphism of groups for $n \geqq 2$. □

(17.13) ***Definition.*** For chain complexes V, W let

$$\text{Hom}_n = \text{Hom}_n(V, W) \quad (n \in \mathbb{Z})$$

be the **module of linear maps of degree** n from V to W. Let $\partial: \text{Hom}_n \to \text{Hom}_{n-1}$ be defined by $\partial f = fd - (-1)^n df$. Then, since $\partial\partial = 0$, we have the homology $H_n \text{Hom}_*(V, W)$ of the chain complex (Hom_*, ∂). Let $s^n V$ be the chain complex with $ds^n v = (-1)^n s^n dv$. Then $H_n \text{Hom}_*(V, W)$ is just the set of homotopy classes of chain maps $s^n V \to W$. ||

From (17.12) we derive the following special case:

(17.14) **Corollary.** *Let A be an augmented free chain algebra and let QA be the chain complex of indecomposables of A. Then we have for $n \geqq 1$ a bijection $[\Sigma^n A, U] = H_n \text{Hom}_*(QA, U)$, which depends on the choice of generators W for $A = T(W)$, $W \cong QA$. For $n \geqq 2$ this is an isomorphism of abelian groups.*

(17.15) **Addendum.** *If the differential on QA is trivial we have*

$$H_n \operatorname{Hom}_*(QA, U) = \operatorname{Hom}_n(QA, H_*U).$$

Proof of (17.14) *and* (17.15). We have the isomorphism of chain complexes $\operatorname{Der}_*(U^{A|*}, 0) = \operatorname{Hom}_*(QA, U)$. Since QA is a free module we obtain (17.15), compare also III.4.3 in Mac Lane (1967). □

(17.16) ***Definition.*** A cofibration $B \rightarrowtail A$ in **DA** with $A = B \coprod T(W) = \bigoplus_{j \geq 0} B \otimes (W \otimes B)^{\otimes j}$ is **of filtration** $\leq n$ (with respect to W) if

$$dW \subset \bigoplus_{j=0}^{n} B \otimes (W \otimes B)^{\otimes j}. \qquad \|$$

The *elementary* cofibrations are just those of filtration 0. We now consider cofibrations of filtration 1 with $dW \subset B \oplus B \otimes W \otimes B$.

(17.17) ***Example.*** Let $Y \subset X = Y \coprod T(V)$ be a cofibration in **DA**. Then for $n \geq 1$

$$X \subset \Sigma_Y^n X = X \coprod T(s^n V)$$

is a cofibration of filtration ≤ 1 with respect to $s^n V$. In fact, by (17.7) we see $ds^n V \subset X \otimes s^n V \otimes X$.

(17.18) **Proposition.** *Let $B \subset A = B \coprod T(W)$ be a cofibration of filtration ≤ 1. Then the comultiplication m on $\Sigma_B A$ in* (17.5) *is given by $m(\alpha) = \alpha$ for $\alpha \in A$, and $m(sw) = sw_0 + sw_1$ for $w \in W$.*

(17.19) **Corollary.** *If $B \subset A$ is of filtration ≤ 1 then the bijection in* (17.12), *$\pi_1(U^{A|B}, u) = H_1 \operatorname{Der}_*(U^{A|B}, u)$, is an isomorphism of abelian groups.*

Proof of (17.18). For $w \in W$ we have

$$dw = d_1 w + d_2 w \tag{1}$$

with $d_1 w \in B$ and $d_2 w \in B \otimes W \otimes B$. This shows that for $\hat{M}$ in (6) of (17.4) we have

$$\hat{M}(dv) = \hat{M}(d_2 w), \tag{2}$$

see (8) in (17.4). Now (8) and (9) in (17.4) show us that for $\alpha \otimes v \otimes \beta$, $(\alpha, \beta \in B, v \in W)$, we have

$$\hat{M}(\alpha \otimes v \otimes \beta) = (-1)^k \alpha' \otimes (\hat{M} v) \otimes \beta''', k = |\alpha|. \tag{3}$$

Assume now we have constructed $\hat{M}$ with $\hat{M} v = 0$ for $|v| < n$. Then for w, $|w| = n + 1$, the equations (1), (2) and (3) above show $\hat{M}(dw) = 0$. Therefore we can choose $\xi = 0$ in (10) of (17.4). Thus for all $v \in W$ we have $\hat{M}v = 0$ or equivalently $m(sv) = Mv = S_0 v + S_1 v = sv_0 + sv_1$. □

(17.20) ***Definition.*** Let T be the free augmented algebra generated by the generator t with degree $|t| = 0$, clearly $d(t) = 0$. Then $\Sigma^n T = T(s^n t)$ has trivial differential and the **homotopy groups** are

$$\pi_n^T(A) = [\Sigma^n T, A] = H_n(A) \tag{1}$$

for $n \geqq 1$, see § 6 and (17.14). Moreover, the **relative homotopy groups of a pair** (A, B) are

$$\pi_n^T(A, B) = H_n(A, B) = H_n(A/B). \tag{2}$$

Here A/B is the quotient chain complex of the underlying chain complexes of A and B, see (7.7). ||

The **exact homotopy sequence** (7.8) for the functor π_n^T is just the long exact homology sequence

(17.21) $$\cdots \xrightarrow[\partial]{} H_n B \xrightarrow[1]{} H_n A \xrightarrow[j]{} H_n(A, B) \xrightarrow[\partial]{} H_{n-1} B \to \cdots.$$

Remark. T corresponds in topology to the 1-sphere S^1 so that $\pi_n^T(A)$ corresponds to the homotopy groups $\pi_{n+1}(X)$, $n \geqq 1$, of a space X. Clearly, the **homotopy groups of spheres** $\pi_n^T(\Sigma^m T)$ can be easily computed in the category **DA** of chain algebras.

(17.22) ***Example.*** We consider the example which in topology corresponds to a product $S^{n+1} \times S^{m+1}$ of two spheres $(n, m \geqq 1)$. Let v, w and $v \times w$ be elements of degree n, m and $n + m + 1$ respectively and let $A = T(v, w, v \times w)$ be the free chain algebra generated by these elements with the differential

$$\left.\begin{aligned} &dv = dw = 0 \text{ and for } [v, w] = vw - (-1)^{nm} wv \text{ let} \\ &d(v \times w) = (-1)^n [v, w]. \end{aligned}\right\} \tag{1}$$

We construct $\hat{M}$ on $A(X)$. First we set $\hat{M}v = \hat{M}w = 0$. Since $dS_0 v = v'' - v'$ and $dS_1 w = w''' - w''$ we obtain from (9) in (17.4) the equation

$$\begin{aligned} \hat{M}(vw) &= (S_0 v)(w''' - w'') - (-1)^n (v'' - v')(S_1 w) \\ &= (-1)^{n+1} d((S_0 v)\cdot(S_1 w)). \end{aligned}$$

Similarly we get $\hat{M}(wv) = (-1)^{m+1} d((S_0 w)\cdot(S_1 v))$. Therefore we see that $\hat{M}d(v \times w) = d\xi$ with $-\xi = (S_0 v)(S_1 w) + (-1)^{(n+1)(m+1)}(S_0 w)(S_1 v)$. Now we can define M by

$$\left.\begin{aligned} &Mv = S_0 v + S_1 v, Mw = S_0 w + S_1 w, \\ &M(v \times w) = S_0(v \times w) + S_1(v \times w) + (S_0 v)(S_1 w) + (-1)^{(n+1)(m+1)}(S_0 w)(S_1 v). \end{aligned}\right\} \tag{2}$$

We have for $x = sv$, $y = sw$, $z = s(v \times w)$ the degrees $|z| = |x| + |y|$.

Moreover, $\Sigma A = T(x, y, z)$ has trivial differential and the comultiplication

$$m: \Sigma A \to \Sigma A \coprod \Sigma A = T(x_0, x_1, y_0, y_1, z_0, z_1) \tag{3}$$

is given by

$$\begin{cases} mx = x_0 + x_1, \\ my = y_0 + y_1, \\ mz = z_0 + z_1 + x_0 y_1 + (-1)^{|x||y|} y_0 x_1. \end{cases}$$

For the homotopy set (see (17.14) and (17.15))

$$[\Sigma A, U] = H_{|x|}(U) \times H_{|y|}(U) \times H_{|x|+|y|}(U) \tag{4}$$

we obtain by (3) the multiplication $+$:

$$(\alpha_0, \beta_0, \gamma_0) + (\alpha_1, \beta_1, \gamma_1) = (\alpha_0 + \alpha_1, \beta_0 + \beta_1, \gamma_0 + \gamma_1 + \alpha_0 \beta_1 + (-1)^{|x||y|} \beta_0 \alpha_1). \tag{5}$$

Here $\alpha_0 \beta_1$ and $\beta_0 \alpha_1$ are products in the algebra $H_* U$. The commutator of the elements $\bar{\alpha} = (\alpha, 0, 0)$, $\bar{\beta} = (0, \beta, 0) \in [\Sigma A, U]$ is by (5) the element

$$-\bar{\alpha} - \bar{\beta} + \bar{\alpha} + \bar{\beta} = (0, 0, [\alpha, \beta]). \tag{6}$$

Thus, if the Lie bracket

$$[\alpha, \beta] = \alpha \cdot \beta - (-1)^{|\alpha||\beta|} \beta \cdot \alpha \tag{7}$$

of the algebra $H_* U$ is not trivial, the group $[\Sigma A, U]$ is not abelian.

Recall that we have the functor $\tau: \mathbf{Top}_1^* \to \mathbf{DA}_*$ given by $\tau = SC_*\Omega$ in (I.7.29). We know that τ is a model functor, actually τ is natural weak equivalent to a based model functor. We therefore can apply (15.17) for the computation of $\tau[\alpha, \beta]$ where $[\alpha, \beta]$ is a Whitehead product, $\alpha \in \pi_{n+1}(U)$ and $\beta \in \pi_{m+1}(U)$. By the Bott–Samelson theorem we have a canonical weak equivalence σ,

(17.23) $$* \rightarrowtail T(v) \xrightarrow[\sigma]{\sim} \tau(S^{n+1})$$

(where $|v| = n$, $T(v) = \Sigma^n T$), such that the composition

$$h: \pi_n(\Omega U) = [S^{n+1}, U] \xrightarrow{\tau} [\tau S^{n+1}, \tau U] \overset{\sigma^*}{=} [T(v), \tau U] = H_n(\Omega U)$$

is the Hurewicz-map, see (17.20). Now one readily checks that there is a commutative diagram in $\mathbf{DA}_*$

$$\begin{array}{ccccc} T(v) \vee T(w) & & \xrightarrow{((\tau i_1)\sigma, (\tau i_2)\sigma)} & & \\ \downarrow i & & & & \\ (T(v, w, v \times w), d) & \xrightarrow[j_1]{\sim} & T(v) \otimes T(w) & \xrightarrow[j_2]{\sim} & \tau(S^{n+1} \times S^{m+1}). \end{array}$$

Here $j_2 j_1$ is given by the isomorphism in (I.7.32). The map $j_2 j_1$ is compatible with the projections p_1, p_2 on $S^{n+1} \times S^{m+1}$ and on $T(v, w, v \times w)$ respectively. We define $p_1 : T(v, w, v \times w) \to T(v)$ by $p_1(v) = v$ and $p_1(w) = 0 = p_1(v \times w)$. Similarly, we define p_2. The diagram above shows that $(T(v, w, v \times w), i, p_1, p_2)$ is a fat wedge $M\tau S^{n+1} \mathbin{\underset{\sim}{\vee}} M\tau S^{m+1}$ for $\tau(S^{n+1} \times S^{m+1})$ as in (15.5). Hence we can apply (15.17) and we get by (17.22)(7) the Samelson

(17.24) **Theorem.** *For* $\alpha \in \pi_n(\Omega U)$ *and* $\beta \in \pi_m(\Omega U)$ *and for the* Whitehead *product* $[\alpha, \beta] \in \pi_{n+m}(\Omega U)$ *the* Hurewicz-*homomorphism*

$$h : \pi_*(\Omega U) \to H_*(\Omega U)$$

satisfies the formula

$$h([\alpha, \beta]) = [h\alpha, h\beta].$$

Here the bracket at the right-hand side is the Lie bracket associated to the multiplication in the algebra $H_*(\Omega U)$, *compare* (I.9.3).

Compare also (I.9.23). We need no sign in the formula of (17.24) since we define the Whitehead product by the commutator in (15.11).

III

The homotopy spectral sequences in a cofibration category

In Chapter II we obtained the fundamental exact sequences of homotopy theory, namely

(A) the cofiber sequence,
(B) the exact sequence for homotopy groups of function spaces, and
(C) the exact sequence for relative homotopy groups.

In this chapter we derive from these exact sequences the corresponding homotopy spectral sequences. In case (A) this leads to the general form of both the Atiyah–Hirzebruch spectral sequence and the Bousfield–Kan spectral sequence, compare also Eckmann–Hilton (1966). In case (B) we obtain a far reaching generalization of the Federer spectral sequence for homotopy groups of function spaces in topology. In case (C) the homotopy spectral sequence is the general form of the homotopy exact couple in topology (considered by Massey) from which we deduce the 'certain exact sequence of J.H.C. Whitehead' in a cofibration category. All these results on the homotopy spectral sequences are available in any cofibration category. Moreover, a functor between cofibration categories (which carries weak equivalences to weak equivalences) induces a map between these homotopy spectral sequences compatible with the differentials. Various properties of the spectral sequences are proved in this chapter, some of them seem to be new (even for the classical topological spectral sequences).

In all cases (A), (B) and (C) we study the E_2-term of the spectral sequences. For this we introduce complexes, chain complexes, and twisted chain complexes in a cofibration category. Complexes are iterated mapping cones which are obtained by a succession of attaching cones. For example in topology CW-complexes and dually Postnikov-towers are complexes. We define a functor which carries complexes (and filtration preserving maps) to (twisted) chain complexes and chain maps. The (twisted) cohomology, defined in terms of this chain functor, yields the E_2-term of the spectral sequence in case (A) (and in case (B)). In topology the chain complex corresponds to the cellular chain complex of a CW-complex, and the twisted chain complex corresponds to the cellular chain complex of the universal covering of a CW-complex. Therefore the twisted cohomology yields the cohomology with local coefficients of a CW-complex as an example. The spectral sequence for case (B) can also be used for the enumeration of the set π_0 of a function space which is a set of homotopy classes of maps.

A surprising application of the spectral sequence in case (B) is the result that in any cofibration category the group π_1 acts nilpotently on π_n $(n \geqq 1)$ provided that we consider a complex X of finite length. In particular, the homotopy group $[\Sigma X, U]$ is nilpotent; in topology this is a classical result of G.W. Whitehead (1954).

In the first section of the chapter we consider filtered objects X in a cofibration category and we show that the homotopy group $\pi_n(U^X, u)$ is embedded in a short exact $\lim^1$-sequence.

§ 1 Homotopy groups of homotopy limits

Let **C** be a cofibration category. We first generalize the category **Pair(C)** in (II.1.3) by considering 'filtered objects' in **C**.

(1.1) ***Definition.*** Let **Fil(C)** be the following category. Objects are diagrams

$$A = (A_0 \underset{i_0}{\longrightarrow} A_1 \underset{i_1}{\longrightarrow} A_2 \underset{i_2}{\longrightarrow} A_3 \rightarrow \cdots).$$

A morphism $f: A \rightarrow B$ in **Fil(C)** is a sequence of maps $f_n: A_n \rightarrow B_n$ with $i_n f_n = f_{n+1} i_n$, $n \geqq 0$. We say that f is a weak equivalence if each f_n, $n \geqq 0$, is a weak equivalence in **C**. Moreover, f is a cofibration if each map

$$(f_{n+1}, f_n): (A_{n+1}, A_n) \rightarrow (B_{n+1}, B_n)$$

is a cofibration in **Pair(C)**. We call the object $A = \{A_n\} = \{A_n, i_n\}$ a **filtered object** in **C**. We say that A is **constant** if $A_n = U$ and $i_n = 1_U$ for all n; in this case we write $A = \{U\}$. Moreover, A is a **skeleton**, $A = A_{\leq n}$, if $i_m: A_m \rightarrow A_{m+1}$ is the identity for $m \geqq n$. ||

If **C** has an initial object $*$ then also **Fil(C)** has the initial object $* = \{*\}$. An object $A = \{A_n\}$ is cofibrant in **Fil(C)** if all maps

(1.1)(a) $$* \rightarrowtail A_0 \rightarrowtail A_1 \rightarrowtail \cdots \rightarrowtail A_n \rightarrowtail$$

are cofibrations in **C**. Recall that $\mathbf{Fil(C)}_c$ denotes the full subcategory of cofibrant objects in **Fil(C)**.

For a fibration category **F** we define **Fil(F)** dually by diagrams

(1.1)(b) $$A = (A_0 \underset{p_0}{\leftarrow} A_1 \underset{p_1}{\leftarrow} A_2 \leftarrow \cdots)$$

in **F**. This is a fibrant object in **Fil(F)** if all p_i are fibrations and if $A_0 \twoheadrightarrow *$ is a fibration. We call such a fibrant object a *tower of fibrations.* As in (II.1.5) we obtain

(1.2) **Lemma.** *The category* **Fil(C)** *with cofibrations and weak equivalences in* (1.1) *is a cofibration category.*

We only check (C3) and (C4) in **Fil(C)**. For a map $f:\{A_n\} \to \{B_n\}$ we obtain a factorization $f:\{A_n\} \rightarrowtail \{C_n\} \xrightarrow{\sim} \{B_n\}$ by the following commutative diagram.

(1.3) $$\begin{array}{ccccccc}
\uparrow & \cdots\cdots\cdots & & & & & \\
A_2 & \rightarrowtail & P_2 & \rightarrowtail & C_2 & \xrightarrow{\sim} & B_2 \\
\uparrow & & \text{push} & & \uparrow & \nearrow & \\
A_1 & \rightarrowtail & P_1 & \rightarrowtail & C_1 & \xrightarrow{\sim} & B_1 \\
\uparrow & \text{push} & \uparrow & & \nearrow & & \\
A_0 & \rightarrowtail & C_0 & \xrightarrow{\sim} & B_0 & &
\end{array}$$

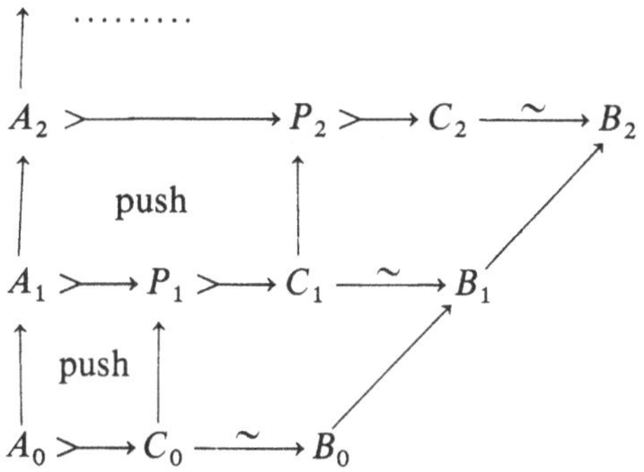

Here we first choose a factorization C_0 of $A_0 \to B_0$ in **C**. Then the push out P_1 and the map $P_1 \to B_1$ is defined. We choose a factorization C_1 of this map. Then the push out P_2 and the map $P_2 \to B_2$ is defined. There is a factorization C_2 of this map. Inductively we obtain the filtered object C with $A \rightarrowtail C \xrightarrow{\sim} B$.

Next we get a fibrant model $\{A_n\} \overset{\sim}{\rightarrowtail} \{RA_n\}$ of $\{A_n\}$ by choosing inductively fibrant models in **C** as in the diagram

(1.4) $$\begin{array}{ccccc}
\uparrow & \cdots & & & \uparrow \\
A_2 & \xrightarrow{\sim} & P_1 & \overset{\sim}{\rightarrowtail} & RA_2 \\
\uparrow & & \text{push} & & \uparrow \\
A_1 & \overset{\sim}{\rightarrowtail} & P_0 & \overset{\sim}{\rightarrowtail} & RA_1 \\
\uparrow & \text{push} & \uparrow & & \\
A_0 & \underset{\sim}{\rightarrowtail} & RA_0 & &
\end{array}$$

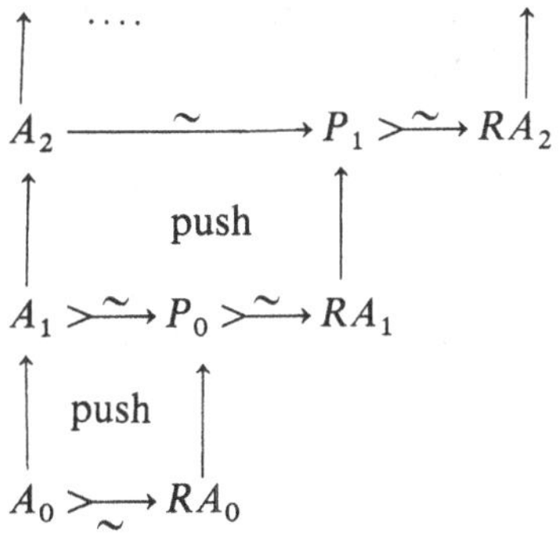

Here RA_{n+1} is the fibrant model in **C** of the push out P_n. An object $A = \{A_n\}$ in **Fil(C)** is *fibrant* if and only if all objects A_n are fibrant in **C**, see (II.1.5).

We now consider cylinders and homotopies in **Fil(C)**. For a cofibrant filtered object A we have the cylinder $I_*A = \{I_*A_n\}$ which is a sequence of cylinders in **C**. We call a map $H: I_*A \to U$ in **Fil(C)** a 0-homotopy or a filtration preserving homotopy. The example of CW-complexes in topology shows that also homotopies are of interest which are not 0-homotopies. We say H is a **k-homotopy**, $k \geqq 0$, between f and g if we have homotopies $H_n: i^k f_n \simeq i^k g_n \operatorname{rel} *$ with $i^k: U_n \to U_{n+k}$ such that the diagram

(1.5)
$$\begin{array}{ccc} I_*A_{n+1} & \xrightarrow{\;H_{n+1}\;} & U_{n+k+1} \\ \uparrow{\scriptstyle j} & & \uparrow \\ I_*A_n & \xrightarrow{\;H_n\;} & U_{n+k} \end{array}$$

commutes, $n \geqq 0$. We write $H: f \overset{k}{\simeq} g$. We are mainly interested in 0-homotopies and 1-homotopies. If U is fibrant, then $\overset{k}{\simeq}$ is an equivalence relation on the set of all filtration preserving maps $A \to U$. Clearly, $f \overset{k}{\simeq} g$ implies $f \overset{k+1}{\simeq} g$. In case U is not fibrant we use a fibrant model RU of U in **Fil(C)** for the definition of $\overset{k}{\simeq}$. In particular, $Ho\,\mathbf{Fil(C)} \cong \mathbf{Fil(C)}_{cf}/\overset{0}{\simeq}$.

In order to obtain homotopy limits in **C** we assume the following

(1.6) **Continuity axiom.** *For each cofibrant object* $A = \{A_n\}$ *in* **Fil(C)** *there exists the colimit* $\lim A_n$ *in* **C**. *Moreover, the functor* $\lim: \mathbf{Fil(C)}_c \to \mathbf{C}$ *preserves weak equivalences and cofibrations.*

The axiom is also used in Anderson (1978). We denote a **colimit** (direct limit) by lim and we denote a **limit** (inverse limit) by Lim.

(1.7) ***Examples.***

(a) The cofibration category of topological spaces in (I.5.1) satisfies the continuity axiom. This follows from the appendix in Milnor (1963), compare also Vogt (1963). This implies that the cofibration category of simplicial sets given by (I.2.12) and the cofibration category **Top** with the CW-structure (I.5.6) both satisfy the continuity axiom.

(b) The fibration category of simplicial sets (and of simplicial sets with base point respectively) given by (I.2.12) satisfies the dual of the continuity axiom. Compare Part II in Bousfield–Kan.

(c) The cofibration category of positive chain complexes with the structure in (I.§6) satisfies the continuity axiom since in $\mathbf{Chain}_R$ we have

$\lim HA_n = H(\lim A_n)$, compare VIII.5.20 in Dold (1972); this also implies that the cofibration categories of chain algebras (I.7.10) and of commutative cochain algebras (I.8.17) satisfy the continuity axiom.

(1.8) ***Remark.*** If the cofibration category **C** satisfies the continuity axiom then for any cofibrant object X in **C** also the cofibration category, $\mathbf{C}^X$, of objects under X satisfies the continuity axiom.

(1.9) ***Definition.*** Let **C** be a cofibration category with an initial object $*$ and assume the continuity axiom is satisfied. For each filtered object $A = \{A_n\}$ in **C** we define the **homotopy colimit** by

$$\operatorname{holim}\{A_n\} = \lim\{MA_n\}.$$

Here $\{MA_n\} = MA$ is given by a factorization $* \rightarrowtail MA \xrightarrow{\sim} A$ of $* \to A$ in **Fil (C)**. Hence holim is a functor

$$\operatorname{holim}: Ho(\mathbf{Fil\,C}) \underset{\sim}{\xrightarrow{M}} Ho(\mathbf{Fil\,C})_c \xrightarrow{\lim} Ho\mathbf{C},$$

compare (II.§3). In case the colimit $\lim\{A_n\}$ exists, the canonical map $\operatorname{holim}\{A_n\} \to \lim\{A_n\}$ needs not to be a weak equivalence. For a cofibrant filtered object A we have the canonical weak equivalence $\operatorname{holim}\{A\} \xrightarrow{\sim} \lim\{A_n\}$, in this case we write $A_\infty = \lim\{A_n\}$. ||

An object A is *based in* **Fil (C)** if $* = A_{-1} \rightarrowtail A_0$ is a cofibration and if all maps $A_n \rightarrowtail A_{n+1}$ are based in **C**. If axiom (1.6) is satisfied the limit A_∞ of a based filtered object A is clearly based in **C**. For the based filtered object A we have the function

(1.10) $$[\Sigma^n A_\infty, U] \xrightarrow[p]{} \operatorname{Lim}[\Sigma^n A_i, U] \quad (n \geqq 0)$$

for homotopy groups. Here U is a fibrant object in **C** and A yields the inverse system of sets (groups for $n \geqq 1$), $\{[\Sigma^n A_i, U]\}$, and p is the obvious map. The based filtered object $\Sigma^n A = \{\Sigma^n A_i\}$ is the n-fold suspension in **Fil(C)**. We derive from the homotopy extension property in (II.2.17) that p is a surjective function for $n \geqq 0$, here we use the cylinder in **Fil(C)**. Clearly, for $n \geqq 1$ the function p is a homomorphism of groups. For the study of the kernel $p^{-1}(0)$ of (1.10) we follow the idea of Milnor (1962), compare also K.S. Brown (1973) and Bousfield–Kan. To this end we introduce

(1.11) ***Definition of*** Lim ***and*** Lim^1 ***for groups.*** A tower of (possibly non-abelian) groups and homomorphisms

$$\cdots \to G_n \xrightarrow{j} G_{n-1} \to \cdots \to G_{-1} = *$$

gives rise to a **right action** $\oplus$ of the product group ΠG_n on the product set ΠG_n given by $(x_0, \ldots, x_i, \ldots) \oplus (g_0, \ldots, g_i, \ldots) = (-g_0 + x_0 + jg_1, \ldots, -g_i + x_i + jg_{i+1}, \ldots)$. Here we write the group structure of G_n additively. Clearly, $\operatorname{Lim}\{G_n\} = \{g \mid 0 \oplus g = 0\}$ is the isotropy group of this action in 0. We define $\operatorname{Lim}^1\{G_n\}$ as the orbit set of the action; this is the set of equivalence classes $[\{x_i\}]$ of ΠG_n under the equivalence relation given by $x \sim y \Leftrightarrow \exists g$ with $y = x \oplus g$. In general $\operatorname{Lim}^1\{G_n\}$ is only a pointed set, but if the G_n are abelian, then $\operatorname{Lim}^1\{G_n\}$ inherits the usual abelian group structure. In fact, in this case $\operatorname{Lim}^1\{G_n\}$ is the cokernel of the homomorphism $d: \Pi G_n \to \Pi G_n$, $d(g) = 0 \oplus g$, and $\operatorname{Lim}\{G_n\}$ is the kernel of d. ||

(1.12) ***Remark.*** Let $\{G_n\}$ be a tower of groups such that $j: G_n \to G_{n-1}$ is surjective for all n. Then $\operatorname{Lim}^1\{G_n\} = 0$. Compare IX.§2 in Bousfield–Kan, where one can find as well further properties of Lim and Lim^1.

(1.13) **Theorem.** *Let* **C** *be a cofibration category which satisfies the continuity axiom* (1.6). *Let U be a fibrant object in* **C** *and let A be a based object in* **Fil(C)**. *Then one has the natural short exact sequence* ($n \geq 0$):

$$0 \to \operatorname{Lim}^1[\Sigma^{n+1}A_i, U] \xrightarrow[j]{} [\Sigma^n A_\infty, U] \xrightarrow[p]{} \operatorname{Lim}[\Sigma^n A_i, U] \to 0.$$

For $n = 0$ this is an exact sequence of sets, (i.e. j injective, p surjective and $p^{-1}(0) = \text{image } j$). *For $n \geq 1$ this is a short exact sequence of groups (of abelian groups for $n \geq 2$). The sequence is natural for based maps $A \to A'$ between based objects in* **Fil(C)** *and the sequence is natural for maps $U \to U'$ in* **C**.

For $n = 1$ the sequence in (1.13) is an extension E of a group G by an abelian group M. Such an extension

(1.14)
$$0 \to M \xrightarrow[i]{} E \xrightarrow[p]{} G \to 0$$

gives *M the structure of a G-module* by defining $a^g = g_0^{-1} \cdot i(a) \cdot g_0$ for $g_0 \in p^{-1}(g)$, $a \in M$, $g \in G$. We also call $(a, g) \longmapsto a^g$ the **action** of G on M **associated to the extension** (1.14). It is well known that the extension (1.14) is classified by an element $\{E\} \in H^2(G, M)$, compare (IV.3.7) and (IV.6.2).

(1.15) **Proposition.** *The action of* $g = \{g_i\} \in \operatorname{Lim}[\Sigma A_i, U]$ *on* $a = [\{a_i\}] \in \operatorname{Lim}^1[\Sigma^2 A_i, U]$ *associated to the extension in* (1.13) ($n = 1$) *is given by the formula*

$$a^g = [\{a_i^{g_i}\}]$$

where $a_i^{q_i}$ *is defined by the action of* $[\Sigma A_i, U]$ *on* $[\Sigma^2 A_i, U]$ *in* (II.6.13).

We prove (1.13) and (1.15) in (1.23) below. Special cases of (1.13) are well known, compare Milnor (1962), Bousfield–Kan, Vogt (1973) and Huber–Meyer (1978). The result in (1.15) seems to be new also in topology.

Example. Using the exact sequence (1.13) in the cofibration category of topological spaces Gray showed that there is an essential map $f:\mathbb{C}P^\infty \to S^3$($\mathbb{C}P^\infty = \infty$-dimensional complex projective space) such that the restrictions $f|\mathbb{C}P^n$ are null homotopic for all n.

Now let $\alpha:\mathbf{C}\to\mathbf{K}$ be a functor between cofibration categories which both satisfy the continuity axiom. Assume that α is based ($\alpha(*)=*$) and that α preserves weak equivalences. Clearly, α induces a functor

(1.16) $$\alpha:\mathbf{Fil(C)}\to\mathbf{Fil(K)}.$$

For a cofibrant object $A=\{A_n\}$ in **Fil(C)** we get $* \rightarrowtail B = M\alpha A \xrightarrow{\sim} \alpha A$ in **Fil(K)** as in (1.9). This gives us the canonical map

(1.17) $$* \rightarrowtail B_\infty = \operatorname{holim} \alpha A \to \alpha \lim A = \alpha A_\infty.$$

Moreover, a factorization of this map yields

$$* \rightarrowtail B_\infty \overset{q_\infty}{\rightarrowtail} M\alpha A_\infty \xrightarrow{\sim} \alpha A_\infty.$$

If A is a based object in **Fil(C)** then B is a based object in **Fil(K)**, hence we obtain by (II.6a.2) the maps

$$\begin{cases} \alpha_i = \alpha_L:[\Sigma^n A_i, U]\to[\Sigma^n B_i, R\alpha U] \\ \alpha_\infty = (\Sigma^n q_\infty)^*\alpha_L:[\Sigma^n A_\infty, U]\to[\Sigma^n B_\infty, R\alpha U] \end{cases}$$

which are homomorphisms of groups for $n \geqq 1$.

(1.18) **Proposition.** *The map* α_∞ *is compatible with the exact sequence in* (1.13), *that is:*

$$(\operatorname{Lim}\{\alpha_i\})\circ p = p\circ\alpha_\infty \text{ and}$$
$$j\circ(\operatorname{Lim}^1\{\alpha_i\}) = \alpha_\infty\circ p.$$

This result is proved in (1.23) below. Next we derive from (1.13) the following corollary on homotopy groups of function spaces in **C**.

(1.19) **Corollary.** *Let* **C** *be a cofibration category and assume* **C** *satisfies axiom* (1.6). *Let* X *be a filtered object in* $\mathbf{Fil(C)}_c$ *and let* $u:X_\infty\to U$ *be a map in* **C** *into a fibrant object* U. *We denote by* u_n *the composition* $X_n\to X_\infty \xrightarrow{u} U$. *Then we have for* $n\geqq 0$ *the short exact sequence*

$$0\to \operatorname{Lim}^1\pi_{n+1}(U^{X_n|X_0}, u_n)\to\pi_n(U^{X_\infty|X_0}, u)\to \operatorname{Lim}\pi_n(U^{X_n|X_0}, u_n)\to 0.$$

The sequence is natural with respect to maps in **Fil(C)**. *Moreover, the sequence has properties as described in* (1.13) *and* (1.15), *in particular, for* $n=1$ *the action associated to the extension is given by the formula in* (1.15) *via the action in* (II.5.16) *and* (II.10.7) (3).

We also point out that the sequence in (1.17) is compatible with functors as in (1.16), compare (10a.1). The following proof of (1.17) is an example for the great advantages of an axiomatic approach.

Proof of (1.19). By (1.8) the cofibration category $\mathbf{B}=\mathbf{C}^{X_\infty}$ satisfies axiom (1.6). The filtered object X in $\mathbf{C}$ yields the following filtered object A in $\mathbf{B}$ which is based. Let

$$A_n = \left(X_\infty \rightarrowtail X_n \bigcup_{X_0} X_\infty \xrightarrow{p} X_\infty \right)$$

where $p=(j_n, 1)$, $j_n: X_n \to X_\infty$ the canonical map. The map $u: X_\infty \to U$ gives us the fibrant object u in $\mathbf{B}$. We obtain by (1.13) the exact sequence for $[\Sigma^n A_\infty, u]$ in $\mathbf{B}$. This sequence is isomorphic to the one in (1.19) by use of (II.10.4). □

For the proof of (1.13) we use the **mapping torus** $Z_{f,g}$ which is given by the push out in $\mathbf{C}$

(1.20)
$$\begin{array}{ccc} I_*X & \xrightarrow{\pi} & Z_{f,g} \\ \uparrow & \text{push} & \uparrow i \\ X \vee X & \xrightarrow[(f,g)]{} & Y \end{array}$$

Here X is a based object. Clearly, if f and g are based then $Z_{f,g}$ is based. The inclusion i also called the **'homotopy equalizer'** of f and g since clearly $if \simeq ig$.

(1.21) **Lemma.** *Assume f and g are based maps. Then $Z_{f,g}$ is embedded in the cofiber sequence of based maps*

(*a*) $Y \underset{i}{\rightarrowtail} Z_{f,g} \underset{q}{\to} \Sigma X \xrightarrow[-\Sigma f+\Sigma g]{} \Sigma Y \underset{\Sigma i}{\to} \Sigma Z_{f,g} \to \cdots.$

Moreover, the diagram

(*b*)
$$\begin{array}{ccc} C_i & \xrightarrow{\mu} & C_i \vee \Sigma Y \\ \sim\downarrow & & \downarrow\sim \\ \Sigma X & \xrightarrow{\bar\mu} & \Sigma X \vee \Sigma Y \end{array}$$

with $\bar{\mu} = -i_2(\Sigma f) + i_1 + i_2(\Sigma g)$ *commutes in* $Ho(\mathbf{C})$. *Here* μ *is the coaction in* (II.8.7). *Also the following diagram commutes in* $Ho(\mathbf{C})$

$$(c)\qquad \begin{array}{ccc} \Sigma^2 X & \xrightarrow{\mu_2} & \Sigma^2 X \vee \Sigma X \\ {\scriptstyle \Sigma q}\uparrow & & \downarrow{\scriptstyle 1 \vee \Sigma(if)} \\ \Sigma Z_{f,g} & \xrightarrow{\tilde{\mu}} & \Sigma^2 X \vee \Sigma Z_{f,g} \end{array}$$

Here we set $\tilde{\mu} = -i_2 + i_1(\Sigma q) + i_2$ *and we define* μ_2 *by the universal example of the coaction in* (II.7.11).

By (*a*) in the lemma we have the exact sequence of groups

$$[\Sigma^2 X, U] \xrightarrow[(\Sigma q)^*]{} [\Sigma Z_{f,g}, U] \xrightarrow[(\Sigma i)^*]{} [\Sigma Y, U].$$

Here the image of $(\Sigma i)^*$ is the group $G = \text{kernel}\,(-\Sigma f + \Sigma g)^*$ and the kernel of $(\Sigma i)^*$ is the abelian group $M = \text{image}\,(\Sigma q)^* = \text{cokernel}\,(-\Sigma^2 f + \Sigma^2 g)^*$. Hence we have the extension

$$0 \to M \to [\Sigma Z_{f,g}, U] \to G \to 0.$$

The associated action of G on M, see (1.14), is computed in (1.21) (*c*); namely, for $\xi \in G \subset [\Sigma Y, U]$ and for $[a] \in M$ with $a \in [\Sigma^2 X\ U]$ we have

$$(1.22)\qquad [a]^{\xi} = [a^{(\Sigma f)^*\xi}] = [a^{(\Sigma g)^*\xi}]$$

where we use the action of $[\Sigma X, U]$ on $[\Sigma^2 X, U]$ defined by μ_2, $\mu_2^*(a, \eta) = a^{\eta}$.

Proof of (1.21). Clearly, $Z_{f,g}/Y = \Sigma X$ by (1.20). On the other hand we obtain an isomorphism in $Ho(\mathbf{C})$

$$\bar{q}: \Sigma X \to C_i = Z_{f,g} \bigcup_{Y} CY \tag{1}$$

by track addition

$$\bar{q} = -Cf + \pi + Cg \tag{2}$$

where π is the track in (1.20). Now (*b*) follows immediately from (2) since μ in (*b*) as well is defined by track addition. Clearly (*b*) implies that $-\Sigma f + \Sigma g$ is part the cofiber sequence in (*a*). Next we prove (*c*). For $\bar{\mu}$ in (*b*) we obtain the principal cofibration $C\bar{\mu}$ and the homotopy push out

$$\begin{array}{ccc} \Sigma X \vee \Sigma Y & \xrightarrow{(0,1)} & \Sigma Y \\ \downarrow & & \downarrow{\scriptstyle \Sigma i} \\ C\bar{\mu} & \xrightarrow{u} & \Sigma Z_{f,g} \end{array} \tag{3}$$

This is clear by the cofiber sequence (*a*) since $(0,1)_* \bar{\mu} = -\Sigma f + \Sigma g$. We have the homotopy

$$\left.\begin{aligned} o - i_1 &: I_*(\Sigma X \vee \Sigma Y) \to \Sigma^2 X \vee \Sigma Z_{f,g} \\ o - i_1 &: i_2(\Sigma i)(0,1) \simeq i_2(\Sigma i)(0,1) \end{aligned}\right\} \tag{4}$$

Which is obtained from the trivial homotopy o by adding $-i_1$ with $i_1 : \Sigma^2 X \subset \Sigma^2 X \vee \Sigma Z_{f,g}$, see (II.13.9). We observe that the difference

$$d(u, o - i_1, u) \in [\Sigma^2 X, \Sigma^2 X \vee \Sigma Z_{f,g}], \tag{5}$$

see (II.8.13), represents $\tilde{\mu}$ in (*c*), that is,

$$\tilde{\mu} = (\Sigma q)^* d(u, o - i_1, u). \tag{6}$$

For this we use the canonical homotopy $(0,1)\bar{\mu}q \simeq (-\Sigma f + \Sigma g)q \simeq 0$. Now the proposition in (*c*) is a consequence of (II.13.9) and (II.11.13). We leave the details as an exercise. □

(1.23) *Proof of* (1.13), (1.15), *and* (1.18). Let $W = \bigvee_m A_m$ be the coproduct of all objects A_m, $m \geqq 0$. This coproduct exists by (1.6). Let f be the identity of W and let g be the map

$$W = \bigvee_{m \geqq -1} A_m \xrightarrow{g} \bigvee_{m \geqq -1} A_{m+1} = W$$

with $g = \bigvee_{m \geqq -1} i_m$, $i_m : A_m \rightarrowtail A_{m+1}$. The double mapping cylinder $Z_{f,g}$ is called the **telescope** for A. For $Z_{f,g}$ we use the cylinder $I_* W = \bigvee_{m \geqq 0} I_* A_m$ on W. This shows that $Z_{f,g}$ is the colimit of $Z = \{Z_n\}$ with

$$Z_n = I_* A_0 \bigcup_{A_0} I_* A_1 \bigcup_{A_1} I_* A_2 \bigcup \cdots \bigcup_{A_{n-1}} I_* A_n.$$

We sketch the telescope of A by:

(1.24) 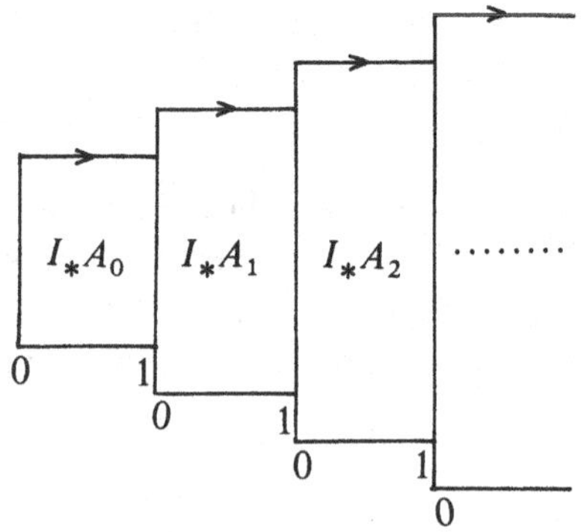

Clearly, the obvious projection $Z_n \to A_n$ is a weak equivalence. Therefore, by (1.6) also $h : \lim Z_n = Z_{f,g} \xrightarrow{\sim} \lim A_n = A_\infty$ is a weak equivalence. We now can use (1.21) for the group $[\Sigma A_\infty, U] = [\Sigma Z_{f,g}, U]$. This yields the result in

(1.13). Moreover (1.15) is a consequence of (1.22). Finally we get (1.18) by (II.8.27)(4). □

§2 The homotopy spectral sequence

We describe an 'extended' homotopy spectral sequence for a filtered object in a cofibration category **C** which is based on the exact cofiber sequences in **C**. This spectral sequence consists in dimension 1 of possibly non-abelian groups, and in dimension 0 of pointed sets, acted on by the groups in dimension 1. We discuss two versions of the spectral sequence, one for homotopy groups and one for homotopy groups of function spaces.

Case (A). Let **C** be a cofibration category with an initial object $*$ and let $X = \{X_n\}$ be a based object in **Fil(C)**, $* \rightarrowtail X \xrightarrow{0} *$. Then we set for $n \geqq 0$

$$U_n = (U^{X_n|*}, 0) \quad \text{and } F_n = (U^{X_n|X_{n-1}}, 0) \tag{2.1}$$

so that $\pi_r U_n = [\Sigma^r X_n, U]$ and $\pi_r F_n = [\Sigma^r (X_n/X_{n-1}), U]$ are homotopy groups in **C**.

Case (B). Let **C** be a cofibration category (which not necessarily has an initial object) and let $X = \{X_n\}$ be an object in **Fil(C)** such that $i: X_n \rightarrowtail X_{n+1}$ is a cofibration in **C**, $n \geqq 0$. Moreover, let $u_n: X_n \to U$ be a map in **C** with $u_{n+1} i = u_n$, hence $u = \{u_n\}: X \to \{U\}$. Then we set

$$U_n = (U^{X_n|X_0}, u_n) \quad \text{and } F_n = (U^{X_n|X_{n-1}}, u_n) \tag{2.2}$$

so that $\pi_r U_n$ and $\pi_r F_n$ are homotopy groups of function spaces in **C**.

In case (A) and in case (B) we can form the homotopy sequences

$$\cdots \to \pi_2 U_{n-1} \to \pi_1 F_n \to \pi_1 U_n \xrightarrow{i} \pi_1 U_{n-1} \to \pi_0 F_n \to \pi_0 U_n \to \pi_0 U_{n-1}. \tag{2.3}$$

This is the cofiber sequence for $X_{n-1} \rightarrowtail X_n$ in case (A), see (II.8.25), and in case (B) this is the exact sequence for homotopy groups of function spaces given by the triple (X_n, X_{n-1}, X_0), see (II.10.9). We have shown in Chapter II that the sequence (2.3) is *exact* in the following sense (this is true for case (A) and case (B)):

(i) The last three objects are sets with base-point 0, all the others are groups, and the image of $\pi_2 U_{n-1}$ lies in the center of $\pi_1 F_n$,
(ii) everywhere 'kernel = image', and
(iii) the sequence comes with a natural action, $+$, of $\pi_1 U_{n-1}$ on $\pi_0 F_n$ such

that $\pi_1 U_{n-1} \to \pi_0 F_n$ is given by $\alpha \longmapsto 0 + \alpha$, and such that 'elements of $\pi_0 F_n$ are in the same orbit if and only if they have the same image in $\pi_0 U_n$'.

From this it follows that one can form the rth **derived homotopy sequence** ($r \geqq 0$):

$$\cdots \to \pi_2 U_{n-2r-1}^{(r)} \to \pi_1 F_{n-r}^{(r)} \to \pi_1 U_{n-r}^{(r)} \to \pi_1 U_{n-r-1}^{(r)}$$
$$\to \pi_0 F_n^{(r)} \to \pi_0 U_n^{(r)} \to \pi_0 U_{n-1}^{(r)}$$

Here we set

$$\pi_i U_n^{(r)} = \text{image}(\pi_i U_{n+r} \to \pi_i U_n) \subset \pi_i U_n,$$
$$\pi_i F_n^{(r)} = \frac{\text{kernel}(\pi_i F_n \to \pi_i U_n/\pi_i U_n^{(r)})}{\text{action of kernel } (\pi_{i+1} U_{n-1} \to \pi_{i+1} U_{n-r-1})}$$

(for $i > 0$ the group $\pi_i F_n^{(r)}$ is the cokernel of the boundary homomorphism between the indicated kernels, for $i = 0$ the set $\pi_0 U_n/\pi_0 U_n^{(r)}$ is the quotient of sets). It is not hard to see that the derived homotopy sequences in (2.4) are also exact in the above sense. This leads to the

(2.5) ***Definition of the homotopy spectral sequence*** $\{E_r^{s,t}\{U_n\}\}$. Let $E_r^{s,t} = \pi_{t-s} F_s^{(r-1)}$ for $t \geqq s \geqq 0$, $r \geqq 1$ and let the differential $d_r: E_r^{s,t} \to E_r^{s+r,t+r+1}$ be the composite map

$$\pi_{t-s} F_s^{(r-1)} \to \pi_{t-s} U_s^{(r-1)} \to \pi_{t-s-1} F_{s+r}^{(r-1)}. \qquad \|$$

This spectral sequence has the following properties:

(i) $E_r^{s,t}$ is a group if $t - s \geqq 1$, which is abelian if $t - s \geqq 2$,
(ii) $E_r^{s,t}$ is a pointed set with basepoint 0 if $t - s = 0$,
(iii) the differential $d_r: E_r^{s,t} \to E_r^{s+r,t+r-1}$ is a homomorphism if $t - s \geqq 2$, and its image is a subgroup of the center if $t - s = 2$; moreover for $t - s \geqq 1$

$$E_{r+1}^{s,t} = \frac{E_r^{s,t} \cap \text{kernel}\, d_r}{E_r^{s,t} \cap \text{image}\, d_r},$$

(iv) there is an action, $+$, of $E_r^{s-r,s-r+1}$ on the set $E_r^{s,s}$ such that d_r: $E_r^{s-r,s-r-1} \to E_r^{s,s}$ is given by $d_r(x) = o + x$, and such that

$$E_{r+1}^{s,s} \subset E_r^{s,s}/\text{action of } E_r^{s-r,s-r+1}.$$

(2.6) ***Convergence for finite skeleta.*** Assume that $X = X_{\leqq N}$ is a finite skeleton, see (1.1). Then $\lim X = X_\infty = X_N$ clearly exists. We have $\pi_i F_n = 0$ for $n > N$, $i \geqq 0$. Therefore we get

$$E_{N+1}^{s,t} = E_{N+2}^{s,t} = \cdots = E_\infty^{s,t}. \qquad (1)$$

(In case (B) we also have $\pi_i F_0 = 0$, this implies $E_N^{s,t} = E_{N+1}^{s,t}$.) We picture the spectral sequence as follows.

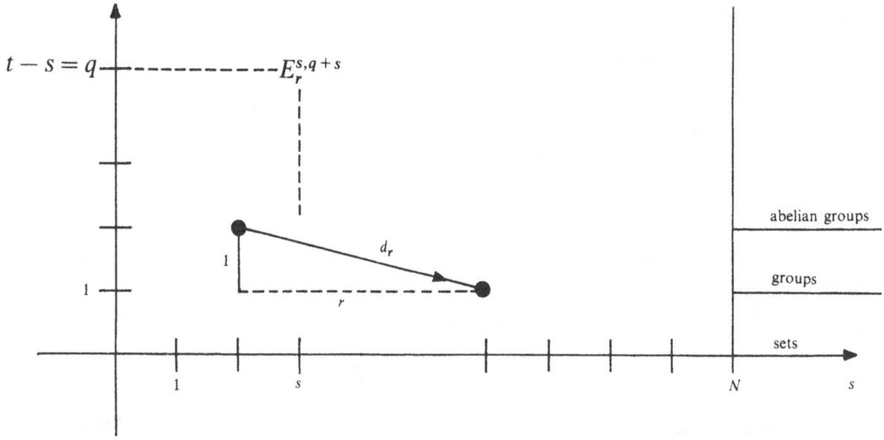

Elements in the column of degree s are represented by elements in $\pi_* F_s$. The row of degree q contributes towards the homotopy group

$$\pi_q U_\infty = \begin{cases} [\Sigma_q X_\infty, U] \text{ for case (A)} \\ \pi_q(U^{X_\infty | X_0}, u_\infty) \text{ for case (B)} \end{cases} \tag{2}$$

We now form the **filtration quotients**

$$Q_s \pi_q = \text{image}(\pi_q U_\infty \to \pi_q U_s). \tag{3}$$

This gives us the tower of groups ($q \geqq 1$)

$$Q_N \pi_q \to Q_{N-1} \pi_q \to \cdots \to Q_0 \pi_q \to o. \tag{4}$$

For $X_\infty = X_N$ we get ($q \geqq 1$)

$$\left.\begin{aligned} &\pi_q U_\infty = Q_N \pi_q, \text{ and} \\ &E_\infty^{s,q+s} = \text{kernel } (Q_s \pi_q \to Q_{s-1} \pi_q) \end{aligned}\right\}. \tag{5}$$

We define a *filtration of* $\pi_q U_\infty (q \geqq 1)$

$$\left.\begin{aligned} &K_{s,q} = \text{kernel } (\pi_q U_\infty \to \pi_q U_s), \\ &\cdots \subset K_{s,q} \subset \cdots \subset K_{0,q} = \pi_q U_\infty \end{aligned}\right\}. \tag{6}$$

The associated graded group of this filtration is by (5) the group

$$G^{s,q} = K_{s-1,q}/K_{s,q} = E_\infty^{s,q+s}. \tag{7}$$

(2.7) ***Complete convergence.*** Let **C** be a cofibration category which satisfies the continuity axioms (1.6). Then $X_\infty = \lim \{X_n\}$ is defined (provided X_0 is cofibrant in case (B)). Assume that for all $s \geqq 0$

(a) $\operatorname{Lim}_r^1(E_r^{s,s+q}) = 0 = \operatorname{Lim}_r^1(E_r^{s,s+q+1})$
then we have for $q \geqq 1$

(b) $\begin{cases} \pi_q U_\infty = \operatorname{Lim}_s(Q_s\pi_q) = \operatorname{Lim}_s(\pi_q U_s), \\ E_\infty^{s,q+s} = \operatorname{kernel}(Q_s\pi_q \to Q_{s-1}\pi_q), \end{cases}$

where we define

(c) $E_\infty^{s,t} = \operatorname{Lim}_r(E_r^{s,t}) = \bigcap_{r \geqq s} E_r^{s,t}$.

Clearly, (a) is satisfied in the finite case by (v)(1); therefore (2.6)(5) is a special case of (b). For a proof of (b) compare IX.5.4 in Bousfield–Kan. The Mittag-Leffler convergence as well is available; for this we refer the reader to IX.5.5 in Bousfield–Kan.

(2.8) ***Naturality.*** A map $g: U \to U'$ in **C** and a map $f: X'' \to X$ in **Fil(C)** (which in case (A) is based) induce functions

$$g_*: E_r^{s,t}\{U_n\} \to E_r^{s,t}\{U'_n\}, \text{ and}$$
$$f^*: E_r^{s,t}\{U_n\} \to E_r^{s,t}\{U''_n\}$$

respectively. Here we have in case (A) resp. in case (B)

$$U'_n = ((U')^{X_n|*}, o) \quad \text{resp.} = ((U')^{X_n|X_0}, gu_n),$$
$$U''_n = (U^{X''_n|*}, o) \quad \text{resp.} = (U^{X''_n|X''_0}, u_n f_n).$$

The functions g_*, f^* are compatible with all the structure of the spectral sequence described in (i)$\cdots$(iv) above. This follows from the corresponding naturality of the sequence (2.3), see (II.8.25) and (II.10.9).

(2.9) ***Invariance in case (A).*** The map g_* in (2.8) depends only on the homotopy class of g in Ho(**C**) Moreover, if $f, f_1: X'' \to X$ are maps which are k-homotopic by a homotopy which is a based map (see (1.5)) than f^* and f_1^* coincide on $E_r^{s,t}$ for $r > k$. This follows by a diagram chase in the same way as the corresponding result in (10.9) below. Further 'invariance' in case (A) and in case (B) can be derived from (3.15) and (4.12) respectively where we compute the E_2-term.

(2.10) ***Naturality with respect to functors.*** Let $\alpha: \mathbf{C} \to \mathbf{K}$ be a functor between cofibration categories which carries weak equivalences to weak equivalences.

Case(A). Assume that α is based, $\alpha * = *$. For $X = \{X_n\}$ we choose $* \rightarrowtail M\alpha X \xrightarrow{\sim} \alpha X \to *$ in **Fil(K)** and we define U'_n by $M\alpha X = B$ as in (1.17).

The maps α_i in (1.17) induce maps

$$\alpha_L: E_r^{s,t}(\{U_n\}) \to E_r^{s,t}(\{U'_n\}),$$

which are compatible with all the above structure (i)...(iv) of the spectral sequence. In particular, $d_r\alpha_L = \alpha_L d_r$.

Case (B). For $X = \{X_n\}$ we choose $* \rightarrowtail M\alpha X \xrightarrow{\sim} \alpha X$ in $\mathbf{Fil}(\mathbf{C}^{\alpha X_0})$ and we define U'_n by $M\alpha X$ and by $\alpha(u_n)$ as in (2.2). Then α_L is defined similarly as in case (A) above.

(2.11) ***Naturality with respect to homotopies in case (B).*** Let $u, u': X = \{X_n\} \to \{U\}$ be maps as in (2.2) and let $I_{X_0}X$ be a cylinder of $\{X_0\} \rightarrowtail X$ in $\mathbf{Fil}(\mathbf{C})$. A homotopy $H: u \simeq u'$ rel X_0, $H: I_{X_0}X \to \{U\}$, induces maps

$$H^{\#}: E_r^{s,t}(\{U_n\}) \to E_r^{s,t}\{U'_n\}),$$

which are compatible with all the above structure (i)...(iv). Here U'_n is defined by u' as in (2.2).

(2.12) ***Compatibility with suspension (case A).*** For the based object X in $\mathbf{Fil}(\mathbf{C})$ we have the suspension ΣX which again is a based object in $\mathbf{Fil}(\mathbf{C})$. Therefore the spectral sequence $\{E_r^{s,t}\{U_n^\Sigma\}\}$ is defined with

$$U_n^\Sigma = (U^{\Sigma X_n|*}, 0).$$

We have for $t - s \geqq 1$

$$E_r^{s,t}\{U_n^\Sigma\} = E_r^{s,t+1}\{U_n\}.$$

Moreover, in the range where this equality holds the differentials coincide up to sign.

(2.13) ***Compatibility with suspension (case B).*** For the filtered object X in $\mathbf{Fil}(\mathbf{C})$ of case (B) we get the filtered object $\Sigma_{X_0}X$ in $\mathbf{Fil}(\mathbf{C})$ which is the torus of $\{X_0\} \rightarrowtail X$. Now the spectral sequence $\{E_r^{s,t}\{U_n^\Sigma\}\}$ is defined with

$$U_n^\Sigma = (U^{\Sigma_{X_0}X_n|X_0}, ur),$$

where $r: \Sigma_{X_0}X \to X$ is the retraction. For $t - s \geqq 1$ one gets

$$E_r^{s,t}\{U_n^\Sigma\} = E_r^{s,t+1}\{U_n\},$$

and in the range where this equality holds the differentials coincide up to sign.

(2.14) ***Remark.*** Chapter IX of Bousfield–Kan deals with a special case of the spectral sequence above. They consider the fibration category $\mathbf{F}$ of pointed simplicial sets. A '*tower of fibrations X in* $\mathbf{F}$' in the sense of Bousfield–Kan is exactly a based object in **Fil(F)**$_f$. The dual homotopy spectral sequence (case (A)) for X, with $U = S^0 =$ 0-sphere, in the *spectral sequence of Bousfield–Kan* for a tower of fibrations.

§3 Based complexes and cohomology

Let **C** be a cofibration category with an initial object $*$. The following notion of a complex is the generalization of a 'CW-complex' available in any cofibration category.

(3.1) ***Definition.*** **A complex** $X = \{X_n, f_n\}$ in **C** is a filtered object $\{X_n\}$ in **C** for which all maps $X_{n-1} \rightarrowtail X_n (n \geqq 1)$ are principal cofibrations (see (II.8.3)) with **attaching maps** $f_n \in [A_n, X_{n-1}], n \geqq 1$. Here all A_n are based objects and we assume that A_n is a suspension for $n \geqq 2$. A_1 needs not to be a suspension. We call X_n the **n-skeleton** of X. Let **Complex** be the full subcategory **Fil(C)** consisting of complexes. ||

Remark. If X is a complex, and if $Y \xrightarrow{\sim} X$ or $X \xrightarrow{\sim} Y$ are weak equivalences in **Fil(C)**, then Y is a complex the attaching maps of which are induced by those of X. In particular, the fibrant model RX in **Fil(C)** is a complex.

(3.2) ***Warning.*** If we have a complex X with $X_0 = *$ then X_1 needs not to be the suspension ΣA_1, compare (II.8.2).

(3.3) ***Definition.*** A complex X as in (3.1) is a **based complex** if $X_0 = *$ and if X is a based filtered object such that all attaching maps are based up to homotopy. Let **Complex**$_0$ be the subcategory of **Complex** consisting of based complexes and of maps $F: X \to Y$ for which each $F_n: X_n \to Y_n$ is based up to homotopy. ||

(3.4) ***Example.*** Let X be a CW-complex with $X^0 = *$. Then $X = \{X^n, f_n\}$ is a complex in **Top*** with attaching maps $f_n \in [A_n, X^{n-1}]$ where

$$A_n = \bigvee_{Z_n} S^{n-1}$$

is a one point union of spheres with $Z_n =$ set of n-cells of X. In particular A_1 is not a suspension, the spaces A_i $(i > 1)$, however, are suspensions. Compare (I.5.7).

We now introduce the following commutative diagram of functors (which is well known for CW-complexes):

$$
(3.5) \qquad \begin{array}{ccc}
\mathbf{Complex}_0/\overset{0}{\simeq} & \xrightarrow{K} & \mathbf{Chain} \\
\big\downarrow p & & \big\downarrow p \\
\mathbf{Complex}_0/\overset{1}{\simeq} & \xrightarrow{K} & \mathbf{Chain}/\simeq
\end{array}
$$

Here $\overset{k}{\simeq}$ is the equivalence relation on the category $\mathbf{Complex}_0$ induced by k-homotopies, see (1.5), and p is the quotient functor.

For the definition of **Chain** we use a subcategory $\mathbf{K}$ of $Ho(\mathbf{C})$ on which the **suspension functor**

$$
(3.6) \qquad \Sigma : \mathbf{K} \to \mathbf{K}
$$

is well defined by (II.9.8). Objects of $\mathbf{K}$ are based objects in $\mathbf{C}$ and morphisms in $\mathbf{K}$ are maps $f: A \to B$ which are based up to homotopy and for which A is a suspension if $B \neq *$.

(3.7) ***Definition.*** A **chain complex** $A = \{A_i, d_i\}$ is a sequence of maps $d_i: A_i \to \Sigma A_{i-1}$ $(i \in \mathbb{Z})$ in $\mathbf{K}$ with $(\Sigma d_{i-1}) d_i = 0$. Actually, we require the slightly stronger condition

$$
d_i^*(i_1 + i_2(\Sigma d_{i-1})) = d_i^*(i_1) \in [A_i, \Sigma A_{i-1} \vee \Sigma^2 A_{i-2}]. \qquad (1)
$$

A **chain map** $f: A \to B$ is a sequence of maps $f_i: A_i \to B_i$ in $\mathbf{K}$ such that $(\Sigma f_{i-1}) d_i = d_i f_i$. Two chain maps $f, g: A \to B$ are **homotopic** if there exist elements $\alpha_n \in [\Sigma B_{n-1}, A_n]_0$, $n \in \mathbb{Z}$, such that in the abelian group $[\Sigma B_{n-1}, \Sigma A_{n-1}]_0$ we have the equation

$$
(\Sigma g_{n-1}) - (\Sigma f_{n-1}) = (\Sigma \alpha_{n-1})(\Sigma d_{n-1}) + d_n \alpha_n, \qquad (2)
$$

$$
\begin{array}{ccccc}
A_n & \xrightarrow{d_n} & \Sigma A_{n-1} & \xrightarrow{\Sigma d_{n-1}} & \Sigma^2 A_{n-2} \\
{\scriptstyle f_n}\big\uparrow{\scriptstyle g_n} & \nwarrow^{\alpha_n} & {\scriptstyle \Sigma f_{n-1}}\big\uparrow{\scriptstyle \Sigma g_{n-1}} & \nwarrow^{\Sigma \alpha_{n-1}} & {\scriptstyle \Sigma^2 f_{n-2}}\big\uparrow{\scriptstyle \Sigma^2 g_{n-2}} \\
B_n & \xrightarrow{d_n} & \Sigma B_{n-1} & \xrightarrow{\Sigma d_{n-1}} & \Sigma^2 B_{n-2}
\end{array} \qquad (3)
$$

We call $\alpha: f \simeq g$ a **chain homotopy**. One readily verifies that homotopy is a natural equivalence relation on the category **Chain** of chain maps. ||

For the definition of the functor K in (3.5) we use a slight

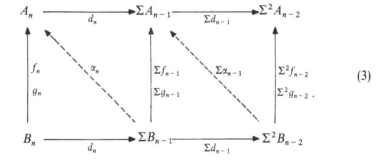

(3.8) ***Generalization of the cofiber sequence.*** Consider a map $f:A=\Sigma A'\to B$ which is based up to homotopy and suppose $B\subset C_f$ is a based pair. Then there is a unique equivalence

$$\Sigma A \simeq C_f/B \tag{1}$$

in $Ho(\mathbf{C})$ which is a quotient map of $\pi_f:(CA,A)\to(C_f,B)$ and which is based up to homotopy. Moreover, for the cofiber sequence of $B\subset C_f$ we obtain the commutative diagram in $Ho(\mathbf{C})$

$$\begin{array}{ccccccc} & & & \Sigma A & & & \\ & & {}^{q}\nearrow & \nwarrow{\sim} & \searrow{}^{\Sigma f} & & \\ B & \rightarrowtail & C_f & \xrightarrow[q]{} & C_f/B & \to & \Sigma B \end{array} \tag{2}$$

and thus $(\Sigma f)q=0$. This follows from (II.9.9). Next let $g:X=\Sigma X'\to Y$ and let (C_g,Y) be a based pair. A pair map $F:(C_f,B)\to(C_g,Y)$ for which $C_f\to C_g$ is based up to homotopy yields a unique quotient map

$$\bar{F}:\Sigma A\simeq C_f/B\to C_g/Y\simeq\Sigma X, \tag{3}$$

which is based up to homotopy.

Proof of (3). Let $\bar{\bar{F}}$ be any quotient map of F (see (II.9.4). Then we know $O_*\bar{\bar{F}}q=O_*qF=O_*F=0$. Therefore by (2) there is $\beta\in[\Sigma B,*]$ with $O_*\bar{\bar{F}}=\beta(\Sigma f)$. Now $\bar{F}=\bar{\bar{F}}^{\beta}$ is a quotient map which is based up to homotopy. □

We now define K in (3.5). A based complex $X=\{X_n,f_n\}$ yields by (3.8) the composite map $(n\in\mathbb{Z})$

(3.9) $$d_{n+1}:A_{n+1}\xrightarrow[f_{n+1}]{}X_n\xrightarrow[q]{}X_n/X_{n-1}\sim\Sigma A_n$$

where we set $A_i=*=X_i$ for $i\leqq 0$. The **boundary map** d_{n+1} of X satisfies $(\Sigma d_n)d_{n+1}=0$ since $(\Sigma f_n)q=0$ by (3.8). This shows that $k(X)=\{A_n,d_n\}$ is a chain complex $\mathbf{C}$. We leave it as an exercise to check condition (1) in (3.7) for $k(X)$. Now let K in (3.5) be given by

(3.10) $$K(X)=\{\Sigma A_n,\Sigma d_n\}=\Sigma k(X).$$

For $F:X\to Y$ in $\mathbf{Complex}_0$ the induced map $KF:KX\to KY$ in **Chain** is obtained by (3.8) (3):

$$\begin{array}{ccccccc} B_n & \xrightarrow{g_n} & Y_{n-1} & \rightarrowtail & Y_n & \longrightarrow & Y_n/Y_{n-1}\sim\Sigma B_n \\ & {}^{0}\swarrow & & & & & \\ * & & \Big\downarrow F_{n-1} & & \Big\downarrow F_n & & \Big\downarrow (KF)_n \\ & {}_{0}\nwarrow & & & & & \\ A_n & \xrightarrow[f_n]{} & X_{n-1} & \rightarrowtail & X_n & \longrightarrow & X_n/X_{n-1}\sim\Sigma A_n \end{array}.$$

By naturality of the cofiber sequence we see that KF is a map in **Chain** so that K is a well-defined functor. In general $(KF)_n$ is not desuspendable.

(3.11) **Lemma.** *K induces a well-defined functor on homotopy categories (see (3.5)), that is: $F \overset{1}{\simeq} G \Rightarrow KF \simeq KG$.*

Proof. For the complex Y we choose a cylinder $I_* Y = \{I_* Y_n\}$ in **Fil(C)**. This cylinder yields the **complex** $ZY = \{(ZY)_n, W_n\}$ with $(ZY)_n = Y_n \cup I_* Y_{n-1} \cup Y_n$ and with attaching maps

$$\begin{Bmatrix} W_n : B_n \vee \Sigma B_{n-1} \vee B_n \to (ZY)_{n-1}, \\ W_n = (i_0 g_n, w_{g_{n-1}}, i_1 g_n), \end{Bmatrix} \tag{1}$$

compare (II.13.5). ZY is a based complex since Y is based. *A* 1-homotopy $H: F \overset{1}{\simeq} G$ is a map $H: ZY \to X$ in **Fil(C)**. Moreover, $H_n:(ZY)_n \to X_n$ is based up to homotopy since F_n and G_n are based up to homotopy. (This fact is one of the reasons why we work with maps which are based up to homotopy.) Since H is a map in **Complex**$_0$ we can apply the chain functor K to H. We obtain the chain map $KH: K(ZY) \to K(X)$, as in (3.10). This chain map gives us the commutative diagram

$$\begin{array}{ccc} \Sigma B_n \vee \Sigma^2 B_{n-1} \vee \Sigma B_n & \xrightarrow[(KH)_n]{} & \Sigma A_n \\ \downarrow \Sigma \bar{d}_n & & \downarrow \Sigma d_n \\ \Sigma^2 B_{n-1} \vee \Sigma^3 B_{n-2} \vee \Sigma^2 B_{n-1} & \xrightarrow[(KH)_{n-1}]{} & \Sigma^2 A_{n-1} \end{array} . \tag{2}$$

Here $\bar{d}_n$ is the boundary in $K(ZY)$. We derive from the definition of W_n in (1) and (II.13.5)

$$(\Sigma \bar{d}_n) | j_0 \Sigma B_n = j_0 (\Sigma d'_n) \tag{3}$$

$$(\Sigma \bar{d}_n) | j_1 \Sigma B_n = j_1 (\Sigma d'_n), \tag{4}$$

$$(\Sigma \bar{d}_n) | \Sigma^2 B_{n-1} = -j_0 - j(\Sigma^2 d'_{n-1}) + j_1. \tag{5}$$

We denote by d'_n the boundary in $K(Y)$. The maps j_0 and j_1 are the first and the second inclusion of ΣB or $\Sigma^2 B$ respectively, j is the inclusion of $\Sigma^3 B_{n-2}$. Equation (5) follows from the definition of w_f in (II.8.11). On the other hand, we have $(KH)_n | j_0 \Sigma B_n = (KF)_n$, and $(KH)_n | j_1 \Sigma B_n = (KG)_n$. We define α by $\alpha_n = (KH)_n | \Sigma^2 B_{n-1}$. The commutativity of (2) shows that $\alpha: KF \simeq KG$ is a homotopy in the sense of (3.7). Here we use (3), (4) and (5). This proves the proposition. □

We use chain complexes for the definition of cohomology as follows: Let U be a fibrant object in **C** and let A be a chain complex in **C** with boundary maps $d_n: A_n \to \Sigma A_n$. Then we get the induced maps ($k \geqq 1$)

$$(3.12) \qquad [\Sigma^{k+1} A_{n-1}, U] \xrightarrow{\delta_{k+1}^{n-1}} [\Sigma^k A_n, U] \xrightarrow{\delta_k^n} [\Sigma^{k-1} A_{n+1}, U]$$

with $\delta_k^n(\alpha) = (\Sigma^{k-1} d_{n+1})^*(\alpha)$. Thus δ_k^n is a homomorphism between abelian groups for $k \geqq 2$. Since A satisfies $(\Sigma d_{n-1})d_n = 0$ we have $\delta_k^n \delta_{k+1}^{n-1} = 0$. Moreover, by conditions (1) in (3.7) the abelian group image (δ_2^{n-1}) acts from the right on the subset kernel (δ_1^n) of $[\Sigma A_n, U]$ by addition in the group $[\Sigma A_n, U]$. This leads to the definition of **cohomology** ($k \geqq 1$)

$$(3.13) \qquad H_k^n(A, U) = \frac{\text{kernel}(\delta_k^n)}{\text{action of image}(\delta_{k+1}^{n-1})}.$$

This is an abelian group for $k \geqq 2$ and this is a set for $k = 1$. Moreover, the cohomology $H_1^n(A, U)$ is an abelian group provided kernel (δ_1^n) is an abelian subgroup of $[\Sigma A_n, U]$. For $k \geqq 0$ we set

$$(3.14) \qquad H_k^n(\Sigma A, U) = H_{k+1}^n(A, U).$$

Here $\Sigma A = \{\Sigma A_n, \Sigma d_n\}$ denotes the suspension of the chain complex A. Clearly the cohomology is a functor on the category $\mathbf{Chain}^{op} \times Ho(\mathbf{C})$. Moreover, a chain homotopy $f \simeq g$ implies $f^* = g^*$; for this consider diagram (3.7)(3). Hence $H_k^n(-, U)$ is a functor on $\mathbf{Chain}/\simeq$.

The crucial property of the cohomology above is the following result on the homotopy spectral sequence (case (A)):

(3.15) ***Proposition.*** *Let X be a based complex in $\mathbf{C}$ and let $E_r^{s,t}\{U_n\}$ be the homotopy spectral sequence with $U_n = (U^{X_n|*}, 0)$, see (2.1). Then one has the natural isomorphism*

$$E_2^{s,q+s}\{U_n\} = H_q^s(K(X), U)$$

for all s and q. Here K is the chain functor in (3.10) with $K(X) = \Sigma k(X)$. In particular, $E_2^{s,q+s}$ is an abelian group for $q \geqq 1$ and 1-homotopic maps $F \overset{1}{\simeq} G$ on X induce $F^* = G^*$ on $E_2^{s,q+s}$.

Proof of (3.15). We consider only the case $q = 0$. Then we get the commutative diagram

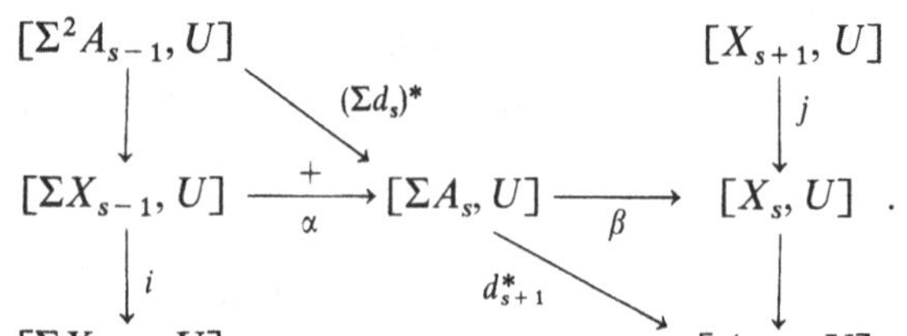

The row and the columns of this diagram are exact cofiber sequences. The definition in (2.5) yields

$$E_2^{s,s} = \frac{\beta^{-1}(\text{image}(j))}{\text{action of } \alpha(\text{kernel}(i))}$$
$$= \frac{\text{kernel}(d_{s+1}^*)}{\text{action of image}(\Sigma d_s)^*}. \qquad \square$$

(3.16) ***Example.*** Let X be a CW-complex with $X^0 = *$ as in (3.4). Then we have the **cellular chain complex** of X in $\mathbf{Chain}_{\mathbb{Z}}$ with

$$C_n X = H_n(X^n, X^{n-1}) \cong \bigoplus_{Z_n} \mathbb{Z}. \tag{1}$$

It is well known that for $n \geqq 1$ the homotopy class of Σd_n,

$$d_n = q f_n : A_n \to X_{n-1} \to \Sigma A_{n-1}, \tag{2}$$

can be identified with the boundary map $\partial_n : C_n X \to C_{n-1} X$ of the cellular chain complex. Hence $d_n^* : [\Sigma^{k+2} A_{n-1}, U] \to [\Sigma^{k+1} A_n, U]$ is the same as

$$\partial_n^* : \text{Hom}(C_{n-1}X, \pi_{n+k}, U) \to \text{Hom}(C_n X, \pi_{n+k} U).$$

This shows for $n + k \geqq 2$ and $k \geqq 0$

$$H_k^n(KX; U) = H^n(X, *; \pi_{n+k}(U)). \tag{3}$$

The right-hand side is the singular cohomology of the pair $(X, *)$. The spectral sequence for X corresponds to the (unstable) Atiyah–Hirzebruch spectral sequences, see Hilton (1971) for the stable version of this spectral sequence. ||

§ 4 Complexes and twisted cohomology

Let $\mathfrak{X}$ be a class of complexes as defined in (3.1) and let **Complex**($\mathfrak{X}$) be the full subcategory of **Complex** consisting of complexes in $\mathfrak{X}$. We assume that for each $X \in \mathfrak{X}$ the colimit $X_\infty = \lim \{X_n\}$ with $X_n \rightarrowtail X_\infty$ exists and that lim yields a functor

(4.1) $$\lim : \mathbf{Complex}(\mathfrak{X}) \to Ho(\mathbf{C}).$$

For example, if each complex in $\mathfrak{X}$ is a finite skeleton then this functor exists. On the other hand lim in (4.1) exists provided the cofibration category **C** satisfies the continuity axiom (1.6) and each X in $\mathfrak{X}$ is cofibrant.

We now introduce the commutative diagram of functors

(4.2)
$$\begin{array}{ccc} \mathbf{Complex}\,(\mathfrak{X})/\overset{0}{\simeq} & \xrightarrow{\check{K}} & \mathbf{Chain}^{\vee} \\ \downarrow p & & \downarrow p \\ \mathbf{Complex}\,(\mathfrak{X})/\overset{1}{\simeq} & \xrightarrow{\check{K}} & \mathbf{Chain}^{\vee}/\simeq \end{array},$$

where the equivalence relation $\overset{k}{\simeq}$ is induced by k-homotopies, see (1.5), and where p is the quotient functor. This diagram is the 'twisted' analogue of diagram (3.5). We proceed in the same way as in § 3 and, in fact, we will see that (4.2) can be considered as being a special case of diagram (3.5).

For the definition of the category **Chain**$^{\vee}$ we use categories **Coef** and **Wedge**:

(4.3) ***Definition of the coefficient category*** **Coef.** Objects are pairs (X, X_1), $X_1 \rightarrowtail X$, in **C**. We denote a pair $\bar{X} = (X, X_1)$ by X if the choice of X_1 is clear from the context. A morphism $\varphi: \bar{X} = (X, X_1) \to \bar{Y} = (Y, Y_1)$ in **Coef** is a map $\varphi: X_1 \to Y$ in $Ho(\mathbf{C})$ for which there is a commutative diagram in $Ho(\mathbf{C})$

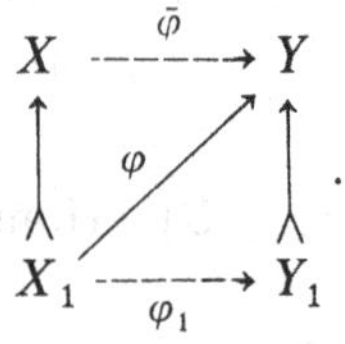

Composition is defined by $\varphi\psi = \varphi\psi_1 = \bar{\varphi}\psi$. Occasionally also the map φ_1 will be denoted by φ. ||

Next we define the twisted analogue **Wedge** of the category **K** in (3.6). For the category **Wedge** the **partial suspension functor**

(4.4) $$E: \mathbf{Wedge} \to \mathbf{Wedge}$$

is well defined by (II.11.8) and by (II.11.16). An object in **Wedge** denoted by $A \vee \bar{X}$, is given by a based object A in **C** and by a pair $\bar{X}$ in **Coef**. A map $f \odot \varphi: A \vee \bar{X} \to B \vee \bar{Y}$ in **Wedge** is a morphism $\varphi: \bar{X} \to \bar{Y}$ in **Coef** together with a map $f \in [A, B \vee Y]_2$ for which there is a commutative diagram in $Ho(\mathbf{C})$

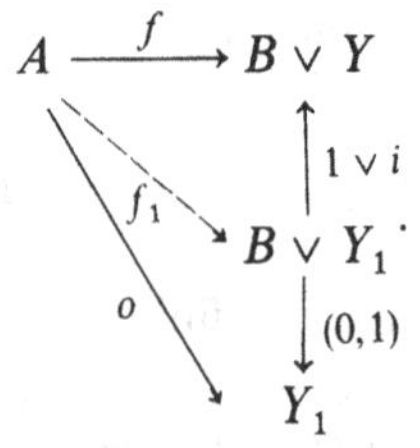

(1)

Here we assume that $A = \Sigma A'$ is a suspension if $B \neq *$. Composition is defined by the formula

$$(f \odot \varphi)(g \odot \psi) = (f, i_Y \bar{\varphi}) g \odot (\varphi \psi). \tag{2}$$

The map $i_Y : Y \to B \vee Y$ is the inclusion. We call $f \odot \varphi$ a φ-**map** in **Wedge**. We also write $f \odot \varphi$ or more precisely $f \odot \bar{\varphi}$ for the map $(f, i_Y \bar{\varphi}) : A \vee X \to B \vee Y$ in $Ho(\mathbf{C})$. Moreover, the map f_1 in (1) occasionally is denoted by f. Finally, the partial suspension functor E is given by

$$\left. \begin{array}{l} E(A \vee \bar{X}) = \Sigma A \vee \bar{X}, \\ E(f \odot \varphi) = (Ef) \odot \varphi. \end{array} \right\} \tag{3}$$

(4.5) *Definition.* A **twisted chain complex** $K = \{\bar{X}, A_n, d_n\}$ is given by a pair $\bar{X} = (X, X_1)$ in **Coef**, by a sequence A_n, $n \in \mathbb{Z}$, of based objects in $\mathbf{C}$ and by a sequence of maps $d_n \odot 1 : A_n \vee \bar{X} \to \Sigma A_{n-1} \vee \bar{X}$ in **Wedge** such that $((Ed_{n-1}) \odot 1) d_n = 0$. Actually we require the slightly stronger condition

$$d_n^*((i_1 + i_2 Ed_{n-1}) \odot 1) = d_n^*(i_1 \odot 1) \in [A_n, (\Sigma A_{n-1} \vee \Sigma^2 A_{n-2}) \vee X]_2. \tag{1}$$

Here i_1 is the inclusion of ΣA_{n-1}, and i_2 is the inclusion of $\Sigma A_{n-1} \vee X$, into $\Sigma A_{n-1} \vee \Sigma^2 A_{n-2} \vee X$. A **twisted chain map** or a φ-**chain map** $(\varphi, f) : K' \to K$ ($K' = \{\bar{Y}, B_n, d_n\}$) is given by a map $\varphi : \bar{Y} \to \bar{X}$ in **Coef** and by a sequence $f = \{f_n\}$ such that the following diagram commutes in **Wedge**

$$
\begin{array}{ccc}
A_n \vee \bar{X} & \xrightarrow{d_n \odot 1} & \Sigma A_{n-1} \vee \bar{X} \\
{\scriptstyle f_n \odot \varphi} \big\uparrow & & \big\uparrow {\scriptstyle Ef_{n-1} \odot \varphi} \\
B_n \vee \bar{Y} & \xrightarrow{d_n \odot 1} & \Sigma B_{n-1} \vee \bar{Y}
\end{array}
\quad . \tag{2}
$$

Clearly composition of twisted chain maps is defined by composition in **Wedge**. Two twisted chain maps (φ, f), (ψ, g) from K' to K are **homotopic** if $\varphi = \psi$ and if there are elements $\alpha_n \in [\Sigma B_{n-1}, A_n \vee X_1]_2$ ($n \in \mathbb{Z}$) such that in the abelian group $[\Sigma B_{n-1}, A_{n-1} \vee X]_2$ we have the equation:

$$(Eg_{n-1}) - (Ef_{n-1}) = (E\alpha_{n-1} \odot \varphi)(Ed_{n-1}) + (d_n \odot 1) \alpha_n, \tag{3}$$

$$
\begin{array}{ccccc}
A_n \vee X & \xrightarrow{\hspace{2cm}} & \Sigma A_{n-1} \vee X & \xrightarrow{\hspace{2cm}} & \Sigma^2 A_{n-2} \vee X \\
& {\scriptstyle d_n \odot 1} & & {\scriptstyle Ed_{n-1} \odot 1} & \\
{\scriptstyle g_n \odot \varphi} \big\uparrow {\scriptstyle f_n \odot \varphi} & {\scriptstyle \alpha_n \odot \varphi} \nwarrow \; \big\uparrow {\scriptstyle \substack{Eg_{n-1} \odot \varphi \\ Ef_{n-1} \odot \varphi}} & & {\scriptstyle E\alpha_{n-1} \odot \varphi} \nwarrow \; \big\uparrow {\scriptstyle \substack{E^2 g_{n-2} \odot \varphi \\ E^2 f_{n-2} \odot \varphi}} & \\
B_n \vee Y & \xrightarrow[{\scriptstyle d_n \odot 1}]{\hspace{2cm}} & \Sigma B_{n-1} \vee Y & \xrightarrow[{\scriptstyle Ed_{n-1} \odot 1}]{\hspace{2cm}} & \Sigma^2 B_{n-2} \vee Y
\end{array}
$$

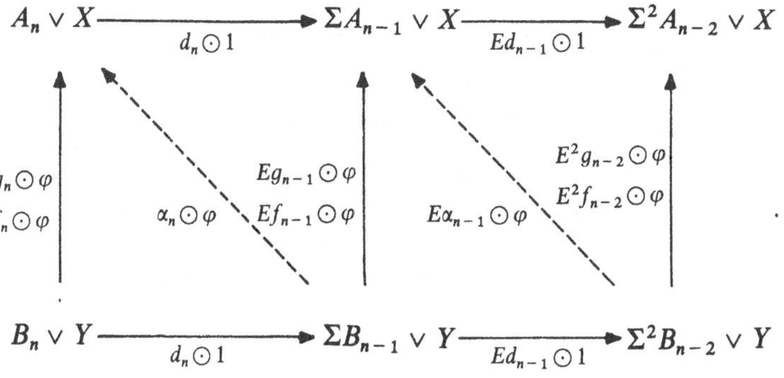

We call $\alpha = \{\alpha_n\}:(\varphi, f) \simeq (\psi, g)$ a **twisted chain homotopy**. It is easy to see that this homotopy is a natural equivalence relation on the category $\mathbf{Chain}^\vee$ of twisted chain maps. Hence the quotient category $\mathbf{Chain}^\vee/\simeq$ is well defined.

We have the forgetful functor

$$c:\mathbf{Chain}^\vee/\simeq \to \mathbf{Coef}, \tag{5}$$

which carries the homotopy class of a φ-chain map to φ. We call c the **coefficient functor**. ||

We now define $\check{K}$ in (4.2). A complex $X = \{X_n, f_n\}$, $X \in \mathfrak{X}$, yields the composite map

(4.6) $$d_{n+1}:A_{n+1} \xrightarrow{f_{n+1}} X_n \xrightarrow{\mu} \Sigma A_n \vee X_n \overset{j}{\rightarrowtail} \Sigma A_n \vee X_\infty.$$

Here $\mu = (i_2 + i_1)$ is the cooperation and f_n is the attaching map of the principal cofibration $X_{n-1} \rightarrowtail X_n$. We set $A_i = *$ for $i \leqq 0$ and $d_i = 0$ for $i \leqq 1$. The **boundary map** d_{n+1} satisfies

$$d_{n+1} = j\mu f_{n+1} = j(\nabla f_{n+1}) \quad (n \geqq 1) \tag{1}$$

where ∇f_{n+1} is the difference construction. We claim that

$$\check{k}(X) = \{X_\infty, A_n, d_n\} \tag{2}$$

is a well-defined twisted chain complex with trivial pair $X_\infty = (X_\infty, X_\infty)$. We define the functor $\check{K}$ in (4.2) by

$$\check{K}(X) = \{X_\infty, \Sigma A_n, Ed_n\} = E\check{k}(X) \tag{3}$$

For $F: Y \to X$ in $\mathbf{Complex}(\mathfrak{X})$ with $F_\infty = \lim F$ the induced map $\check{K}F:\check{K}Y \to \check{K}X$ is the F_∞-chain map given by the elements

$$(\check{K}F)_n = j_* \nabla_{(F_n, F_{n-1})} \in [\Sigma B_n, \Sigma A_n \vee X_\infty]_2 \tag{4}$$

where $\nabla_{(F_n,F_{n-1})}$ is defined by the pair map $(F_n, F_{n-1}):(Y_n, Y_{n-1}) \to (X_n, X_{n-1})$ as in (II.12.6). The map j is defined as in (4.6).

(4.7) Lemma. *The functor $\check{K}$ is well defined and a 1-homotopy $F \overset{1}{\simeq} G$ yields a twisted chain homotopy $KF \simeq KG$.*

This completes the definition of diagram (4.2).

Proof of (4.7). We consider the cofibration category $\mathbf{B} = \mathbf{C}^{X_\infty}$ of objects in $\mathbf{C}$ under X_∞. The complex X in $\mathbf{C}$ gives us the based complex $\bar{X} = (\bar{X}_n, \bar{f}_n)$ in $\mathbf{B}$ as follows. Let

$$\bar{X}_n = \left(X_\infty \underset{i_2}{\rightarrowtail} X_n \bigcup_{X_0} X_\infty \underset{p}{\rightarrow} X_\infty \right) \tag{1}$$

with $p=(i,1)$ and let the attaching map $\bar{f}_n:A_n \vee X_\infty \to X_{n-1} \bigcup_{X_0} X_\infty$ in $Ho(\mathbf{B})$ be given by $\bar{f}_n=(i_1 f_n, i_2)$ where $i_1:X_{n-1} \rightarrowtail X_{n-1} \bigcup_{X_0} X_\infty$. Clearly, the attaching map $\bar{f}_n$ is based up to homotopy. Therefore $\bar{X}$ is a well defined based complex in $\mathbf{B}$ for which $k\bar{X}$ is defined in (3.9). As in the proof of (II.13.3) we see that $\check{k}X = k\bar{X}$. Hence $\check{k}X$ is well defined, in particular, we have $((Ed_n) \odot 1)d_{n+1}$ for d_{n+1} in (4.6).

Next we show that the functor $\check{K}$ can be described in terms of the functor K in (3.10). Let the map $F:Y \to X$ in $\mathbf{Complex}(\mathfrak{X})$ be given as in (4.6) (4). We define a complex $\bar{Y}_F$ in $\mathbf{B}$ by

$$(\bar{Y}_F)_n = (X_\infty \rightarrowtail Y_n \bigcup_{Y_0} X_\infty \xrightarrow[p]{} X_\infty) \tag{2}$$

where $p=(iF_n, 1)$. The attaching maps are $\bar{g}_n:B_n \vee X_\infty \to Y_{n-1} \bigcup_{Y_0} X_\infty$ with $\bar{g}_n=(i_1 g_n, i_2)$. We obtain a map

$$\bar{F}:\bar{Y}_F \to \bar{X} \quad \text{by} \quad \bar{F}_n = F_n \cup 1_{X_\infty}. \tag{3}$$

This map is a based map between based complexes in $\mathbf{B}$. Therefore we obtain by (3.10) the map $K\bar{F}:K\bar{Y}_F \to K\bar{X}$ between chain complexes on $\mathbf{B}$. This map can be identified with the map $\check{K}F$. Compare (II.12.9). Moreover, a 1-homotopy $F \overset{1}{\simeq} G$ yields a 1-homotopy $\bar{F} \overset{1}{\simeq} \bar{G}$ in $\mathbf{B}$ and therefore by (3.11) we get a chain homotopy $K\bar{F} \simeq K\bar{G}$. This completes the proof of (4.7). □

Now we use twisted chain complexes for the definition of twisted cohomology as follows: Let U be a fibrant object in $\mathbf{C}$ and let $u:\bar{X} \to (U, U)$ be a map in $\mathbf{Coef}$. For a twisted chain complex $K = \{\bar{X}, A_n, d_n\}$ the boundary maps d_n induce coboundary functions

$$[\Sigma^{k+1} A_{n-1}, U] \xrightarrow[\delta_{k+1}^{n-1}]{} [\Sigma^k A_n, U] \xrightarrow[\delta_k^n]{} [\Sigma^{k-1} A_{n+1}, U]$$

with $\delta_k^n(\alpha) = (E^{k-1} d_{n+1})^*(\alpha, u)$. For $k \geqq 2$ the coboundary δ_k^n is a homomorphism between abelian groups, see (II.11.17). Since K satisfies $((Ed_{n-1}) \odot 1)d_n = 0$ we get $\delta_k^n \delta_{k+1}^{n-1} = 0$. Moreover, by condition (1) in (4.5) the abelian group image (δ_2^{n-1}) acts from the right on the subset kernel (δ_1^n) of $[\Sigma A_n, U]$ by addition in the group $[\Sigma A_n, U]$. This leads to the definition of **twisted cohomology**, $k \geqq 1$,

(4.8)
$$H_k^n(K,u) = \frac{\text{kernel } (\delta_k^n)}{\text{action of image } (\delta_{k+1}^{n-1})}.$$

This is an abelian group for $k \geqq 2$ and this is a set for $k=1$. Moreover, the cohomology $H_1^n(K,u)$ is an abelian group provided kernel (δ_1^n) is an abelian subgroup of $[\Sigma A_n, U]$. (This, for example, is the case if the attaching map f_n is a functional suspension, see (II.11.7), $n \geqq 2$.) For $k \geqq 0$ we set

(4.9)
$$H_k^n(EK,u) = H_{k+1}^n(K,u).$$

Here $EK = \{\bar{X}, \Sigma A_n, Ed_n\}$ denotes the partial suspension of the twisted chain complex K. A φ-chain map $f: K' \to K$ induces a function ($k \geqq 1$)

(4.10) $$f^*: H^n_k(K, u) \to H^n_k(K', u\varphi), \quad f^*\{\alpha\} = \{(E^k f_n)^*(\alpha, u)\}.$$

Here $\{\alpha\}$ denotes the cohomology class of $\alpha \in [\Sigma^k A_n, U]$. Moreover, a twisted chain homotopy $(\varphi, f) \simeq (\psi, g)$ implies $f^* = g^*$. A map $g: U \to V$ in $Ho(\mathbf{C})$ induces the homomorphism

(4.11) $$g_*: H^n_k(K, u) \to H^n_k(K, gu), \quad g_*\{\alpha\} = \{g_*\alpha\}.$$

The crucial property of the twisted cohomology above is the following result on the homotopy spectral sequence (case (B)) which is the analogue of the result in (3.15).

(4.12) **Theorem.** *Let X be a complex in $\mathbf{C}$, $X \in \mathfrak{X}$, and let $u: X \to \{U\}$ be a map in* **Fil(C)** *which yields $u_\infty: X_\infty \to U$. Consider the homotopy spectral sequence $E^{s,t}_r\{U_n\}$ with $U_n = (U^{X_n|X_0}, u_n)$, see (2.2). Then one has the natural isomorphism*

$$E^{s,q+s}_2\{U_n\} = H^s_q(\check{K}(X), u_\infty)$$

for all s and q. Here $\check{K}$ is the twisted chain functor in (4.6)(3) with $\check{K}(X) = E\check{k}(X)$. In particular, $E^{s,q+s}_2$ is an abelian group for $q \geqq 1$ and 1-homotopic maps $F \overset{1}{\simeq} G$ on X induce $F^* = G^*$ on $E^{s,q+s}_2$.

Proof. The proof of (4.12) is similar to the one in (3.15); the crucial point is the commutativity of diagram (II.13.2). One can also use the construction in the proof of (4.7) so that (4.12) can be considered as being a special case of (3.15). □

(4.13) **Corollary.** *For maps $u, v: X_\infty \to U$ with $u \simeq v$ rel X_0 the groups $E^{p,q}_r(X, u)$ and $E^{p,q}_r(X, v)$ $(r \geqq 2)$ are canonically isomorphic. Thus $E^{p,q}_r(X, u)$ is well defined for $u \in [X_\infty, U]^{X_0}$.*

Proof. Let $H: u \simeq v$ be a homotopy rel X_0. By (2.10) we obtain the isomorphism $H^\#$ of spectral sequences. Since the E_2-term depends only on the homotopy class of $u = u_\infty$, see (4.12), the result follows. □

On the other hand a homotopy $H: u \simeq v$ rel X_0 induces a map $H^\#$ on homotopy groups of function spaces

$$H^\#: \pi_q(U^{X_\infty|X_0}, u) \to \pi_q(U^{X_\infty|X_0}, v),$$

which actually depends on the choice of the track class of H. For example, for $q = 1$ and for a self homotopy $H: u \simeq u$ rel X_0 the map $H^\#$ is an inner automorphism of the group, see (II.5.15). This fact yields the following result:

(4.14) **Corollary.** *Assume the complex $X_0 \subset \cdots \subset X_N = X$ is of finite length and let $u: X \to U$ be a map where U is fibrant. Then the group $\pi_1(U^{X|X_0}, u)$ is nilpotent. In fact, all iterated commutators of $N+1$ elements are trivial.*

Proof. We identify $H \in \pi_1(U^{X|X_0}, u)$ with a self-homotopy $H: u \simeq u$ rel X_0. Then $H^\#$ is the inner automorphism on $\pi_1(U^{X|X_0}, u)$ by H, see (II.5.15). Since $H^\#$ induces the identity on the associated graded group $G^{p,q}$ in (2.6) (7) we obtain the result by the following remark. □

(4.15) ***Remark.*** Assume the group π acts on the group G via automorphisms. **The action is nilpotent** iff there are subgroups ($N < \infty$):

$$0 = K_N \subset \cdots \subset K_p \subset K_{p-1} \subset \cdots \subset K_0 = G$$

with the following properties: For all p the group K_p is normal in K_{p-1} and the quotient group K_p/K_{p-1} is abelian. Moreover, K_p is π-invariant and the π-action induced on K_p/K_{p-1} is trivial. The **group G is nilpotent** if the action of $\pi = G$ on G via inner automorphisms is nilpotent.

More generally than (4.14) we have

(4.16) **Corollary.** *Assume the complex $X_0 \subset \cdots \subset X_N = X$ is of finite length and let $u: X \to U$ be a map into a fibrant object U. Then the group action (defined by* (II.5.16)*) of $\pi_1(U^{X|X_0}, u)$ on $\pi_q(U^{X|X_0}, u)$ is nilpotent ($q \geqq 1$).*

This is proved by (4.13) in the same way as (4.14). By choosing for u the trivial map $u = 0$ we obtain the following special case of (4.16):

(4.17) **Corollary.** *Assume $* \subset X_1 \subset \cdots \subset X_N = X$ is a based complex of finite length ($N < \infty$). Then the group $\pi_1^X(U)$ is nilpotent, in fact all iterated commutators of $N+1$ elements are trivial. Moreover, the action of $\pi_1^X(U)$ on $\pi_q^X(U)$ is nilpotent for $q \geqq 1$, see* (II.6.13).

Remark. For a CW-complex X with $X_0 = *$ and $X_N = X$ it is a classical result of G.W. Whitehead that the group $[\Sigma X, U]$ is nilpotent. It is surprising that this fact by (4.17) is available in any cofibration category.

Finally, we point out that the homotopy spectral sequence in (4.12) is an important tool for the

(4.18) ***Computation of isotropy groups.*** Let $X_0 \subset X_1 \subset \cdots \subset X_N = X$ be a complex with attaching maps $f_n \in [A_n, X_{n-1}]$ and let $u: X_N \to U$ be a map in **C**. Then the action in (II.8.8) gives us the action

$$[X_N, U]^{X_0} \times [\Sigma A_N, U] \xrightarrow{+} [X_N, U]^{X_0}. \tag{1}$$

The isotropy group of this action in $\{u\}\in[X_N, U]^{X_0}$ is the subgroup

$$I_N(\{u\}) = \{\alpha\in[\Sigma A_N, U]:\{u\} + \alpha = \{u\}\}. \tag{2}$$

From (2.6) we derive that the sequence ($N \geqq 1$)

$$0 \to I_N(\{u\}) \hookrightarrow [\Sigma A_N, U] \xrightarrow{p} E_N^{N,N} \to 0 \tag{3}$$

is a short exact sequence of groups. Here $E_N^{N,N}$ is given by the spectral sequence $E_r^{s,t}\{U_n\}$ with $U_n = (U^{X_n|X_0}, u_n)$, $n \leqq N$. For $N = 1$ the exact sequence in (3) coincides with (II.8.9) since p is a bijection in this case.

§5 Cohomology with local coefficients of CW-complexes

We show that cohomology with local coefficients of a CW-complex is a special case of twisted cohomology as defined in section §4.

First we fix some notation on group rings. The fundamental group $\pi = (\pi_1(X), +, -, 0)$ will usually be written additively. The **group ring** $\mathbb{Z}[\pi]$ is the free abelian group generated by the set of elements $[\alpha]$, $\alpha\in\pi$. A typical element is $\Sigma_{\alpha\in\pi} n_\alpha[\alpha]\in Z[\pi]$ where only a finite number of elements $n_\alpha\in\mathbb{Z}$, $\alpha\in\pi$, is non-trivial. The multiplication of the ring $\mathbb{Z}[\pi]$ is defined by the group structure $+$ in π, namely $(\Sigma_\alpha n_\alpha[\alpha])(\Sigma_\beta m_\beta[\beta]) = \Sigma_{\alpha,\beta} n_\alpha m_\beta[\alpha+\beta]$. We have the **augmentation** $\varepsilon:\mathbb{Z}[\pi]\to\mathbb{Z}$, $\varepsilon(\Sigma_\alpha n_\alpha[\alpha]) = \Sigma_\alpha n_\alpha$. $1 = [0]$ is the unit in $\mathbb{Z}[\pi]$ and ε is the ring homomorphism with $\varepsilon[\alpha] = 1$ for $\alpha\in\pi$. A homomorphism $\varphi:\pi\to G$ between groups induces the ring homomorphism $\varphi_\#:\mathbb{Z}[\pi]\to\mathbb{Z}[G]$ with $\varphi_\#[\alpha] = [\varphi\alpha]$. We shall use modules, $M = (\pi, M)$, over the group ring $\mathbb{Z}[\pi]$ which are also called π-**modules**. If not otherwise stated these are **right** π-modules. The action of $\alpha\in\pi$ on $x\in M$ is denoted by $x\cdot\alpha$ or by x^α.

(5.1) ***Definition***. Let $\mathbf{Mod}_{\mathbb{Z}}^{\wedge}$ be the following category. Objects are modules (π, M) over group rings and morphisms are pairs $(\varphi, F):(\pi, M)\to(G, N)$ where $\varphi:\pi\to G$ is a homomorphism between groups and where $F:M\to N$ *is a* φ-**equivariant** homomorphism, that is, $F(x\cdot\alpha) = F(x)\cdot\varphi(\alpha)$. The homomorphism φ induces on N the structure of a right $\mathbb{Z}[\pi]$-module by setting $x\cdot\alpha = x\cdot(\varphi\alpha)$ for $x\in N$, $\alpha\in\pi$. We denote this $\mathbb{Z}[\pi]$-module by φ^*N. Then

$$\mathrm{Hom}_\varphi(M, N) = \mathrm{Hom}_{\mathbb{Z}[\pi]}(M, \varphi^*N)$$

is the abelian group of all φ-equivariant maps from M to N. Similarly, we define the *category* $\mathbf{Chain}_{\mathbb{Z}}^{\wedge}$ *of chain complexes over group rings.* Objects are pairs (π, C) where C is a $\mathbb{Z}[\pi]$-chain complex, see (I.6.1). Morphisms are pairs $(\varphi, F):(\pi, C)\to(G, K)$ where $F:C\to K$ is a **φ-equivariant chain map**. Two such maps are **homotopic**, $(\varphi, F)\simeq(\psi, G)$, if $\varphi = \psi$ and if there exists a φ-equivariant map $\alpha:C\to K$ of degree $+1$ with $d\alpha + \alpha d = -F + G$. ||

We define homology and cohomology for objects in $\mathbf{Chain}_{\mathbb{Z}}^{\wedge}$.

(5.2) ***Definition.*** Let Γ be a *left* $\mathbb{Z}[\pi]$-module. Then the tensor product $C \otimes_{\mathbb{Z}[\pi]} \Gamma$ is a chain complex of abelian groups the **homology** of which is denoted by

$$\hat{H}_*(C, \Gamma) = H_*\left(C \underset{\mathbb{Z}[\pi]}{\otimes} \Gamma \right). \tag{1}$$

Next let Γ be a **right** $\mathbb{Z}[\pi]$-module. Then $\mathrm{Hom}_{\mathbb{Z}[\pi]}(C, \Gamma)$ is a cochain complex of abelian groups with **cohomology** groups

$$\hat{H}^*(C, \Gamma) = H^*(\mathrm{Hom}_{\mathbb{Z}[\pi]}(C, \Gamma)). \tag{2}$$

A map (φ, F) in $\mathbf{Chain}_{\mathbb{Z}}^{\wedge}$ induces homomorphisms

$$\left.\begin{aligned} F_* &= (F \otimes 1_\Gamma)_* : \hat{H}_*(C, \varphi^*\Gamma) \to \hat{H}_*(K, \Gamma) \\ F^* &= \mathrm{Hom}\,(F, 1_\Gamma)_* : \hat{H}^*(K, \Gamma) \to \hat{H}^*(C, \varphi^*\Gamma) \end{aligned}\right\}. \tag{3}$$

These maps depend only on the homotopy class of the chain map F. ||

Let $\mathbf{C} = \mathbf{Top}^*$ be the cofibration category of topological spaces with basepoint, see (II.5.4). Let (X, D) be a **relative CW-complex** as defined in (II.5.7) with skeleta X^n and $* \in D$. We assume that D is path connected and that $D = X^0$ so that X is path connected too. It is easy to see that X is a complex in $\mathbf{C}$ in the sense of (3.1) with skeleta X^n and with attaching maps $(n \geqq 1)$

(5.3) $$f_n : \bigvee_{Z_n} S^{n-1} = A_n \to X^{n-1} \text{ in } Ho(\mathbf{C}).$$

Here Z_n is the set of n-cells in $X - D$. Now let $\mathfrak{X}$ *be a class of such relative CW-complexes in* $\mathbf{C}$. We assume that for each $X \in \mathfrak{X}$ the **universal covering** $\hat{X}$ exists. Let

(5.4) $$p : \hat{X} \to X$$

be the covering projection. We fix for each $X \in \mathfrak{X}$ a basepoint $* \in \hat{X}$ with $p(*) = *$. For each map $F : Y \to X$ in $\mathbf{Complex}(\mathfrak{X})$ there is a unique basepoint preserving map $\hat{F} : \hat{Y} \to \hat{X}$ with $p\hat{F} = Fp$. Here we use the assumption that Y and X are path connected. By **covering transformations** the fundamental group $\pi_1(X)$ acts from the right on $\hat{X}$

(5.5) $$\hat{X} \times \pi_1(X) \to \hat{X},\ (x, \alpha) \longmapsto x \cdot \alpha,$$

and $\hat{F}$ is a φ-equivariant map, that is, $\hat{F}(x \cdot \alpha) = \hat{F}(x) \cdot \varphi(\alpha)$ where $\varphi = \pi_1(F) : \pi_1(Y) \to \pi_1(X)$. We define the relative n-skeleton $\hat{X}^n$ of $\hat{X}$ by $\hat{X}^n = p^{-1}(X^n)$. Next we define the functor

(5.6) $$\hat{C}_* : \mathbf{Complex}(\mathfrak{X}) \to \mathbf{Chain}_{\mathbb{Z}}^{\wedge}$$

which carries the relative CW-complex (X, D) to the relative cellular chain complex $\hat{C}_*X = \hat{C}_*(X, D)$ of the universal covering of X. Let $(n \in \mathbb{Z})$

$$\hat{C}_n(X) = \hat{C}_n(X, D) = H_n(\hat{X}^n, \hat{X}^{n-1}) \tag{1}$$

be the relative singular homology of the pair of skeleta $(\hat{X}^n, \hat{X}^{n-1})$. By covering transformations this is a π-module. The boundary map $d_n: \hat{C}_n(X) \to \hat{C}_{n-1}(X)$ is given by the composition

$$d_n: H_n(\hat{X}^n, \hat{X}^{n-1}) \xrightarrow{\partial} H_{n-1}(\hat{X}^{n-1}) \xrightarrow{j} H_{n-1}(\hat{X}^{n-1}, \hat{X}^{n-2}), \tag{2}$$

where j and ∂ are operators of the long exact homology sequences for pairs. A map $F: X \to Y$ induces

$$\hat{C}_*(F) = \hat{F}_*: \hat{C}_*X \to \hat{C}_*Y. \tag{3}$$

This is a φ-equivariant chain map with $\varphi = \pi_1(F)$ as in (5.5).

(5.7) ***Remark.*** By definition in (II.5.7) we know that $X^n - X^{n-1}$ is a disjoint union of open cells $e^n = D^n - \partial D^n$, namely

$$X^n - X^{n-1} = \bigcup_{e \in Z_n} e. \tag{1}$$

The subset $p^{-1}(e)$ of $\hat{X}$ for $e \in Z_n$ is again a disjoint union of n-cells, homeomorphic to $e \times \pi$, $\pi = \pi_1(X)$. This shows that $\hat{C}_nX$ is the *free* $\mathbb{Z}[\pi]$-module generated by Z_n,

$$\hat{C}_n(X) \cong \bigoplus_{Z_n} \mathbb{Z}[\pi]. \tag{2}$$

Using (5.2) we get the following (relative) **homology** and **cohomology groups with local coefficients**:

(5.8)
$$\begin{cases} \hat{H}_*(X, D; \Gamma) = \hat{H}_*(\hat{C}_*(X, D); \Gamma), \\ \hat{H}^*(X, D; \Gamma) = \hat{H}^*(\hat{C}_*(X, D); \Gamma). \end{cases}$$

The groups (5.8) are also defined by (5.6)(1)(2) in case $D = \varnothing$ is the empty set; then we get $\hat{H}_*(X, \Gamma)$ and $\hat{H}^*(X, \Gamma)$ respectively by (5.8) provided X is path connected. There is a generalization of the cohomology (5.8) in the non-path connected case by use of 'local coefficient systems', see Spanier.

The next result generalizes the well-known isomorphism in (3.16) (3).

(5.9) **Theorem.** *For a relative CW-complex $X = (X, D)$ as above $(X \in \mathfrak{X})$ and for a map $u: X \to U$ in* **Top*** *there is a natural isomorphism*

$$H_k^n(\check{K}X, u) = \hat{H}^n(X, D; \varphi^*\pi_{n+k}(U))$$

with $n \geqq 1$, $n + k \geqq 2$. The left-hand side is the twisted cohomology in §4. The right-hand side is the cohomology with local coefficients where

$\varphi = \pi_1(u): \pi_1(X) \to \pi_1(U)$ *and where* $\pi_{n+k}(U)$ *is a* $\pi_1(U)$ *module by* (II.6.13) $(A = S^0)$.

(5.10) **Corollary.** *The* E_2*-term of the homotopy spectral sequence for* $(U^{X|D}, u)$ *is given by the cohomology groups with local coefficients* $(t \geqq 2)$ $E_2^{s,t} = \hat{H}^s(X, D; \varphi^*\pi_t(U))$, *see* (4.12).

The corollary is originally due to Federer in case the action of the fundamental group $\pi_1 U$ on $\pi_*(U)$ is trivial. Compare also Switzer (1981) and Legrand.

Proof of (5.9) For a discrete set Z let $\Sigma^n Z^+ = \bigvee_Z S^n$ the corresponding one point union of n-spheres. Let Z_n be the set of n-cells of $X^n - X^{n-1}$ and let $\bar{Z}_n$ be the set of n-cells of $Y^n - Y^{n-1}$. Below we construct a canonical isomorphism τ for which the following two diagrams commute where $F: Y \to X$ is a map in **Complex**($\mathfrak{X}$), $(k \geqq 2, n \geqq 1)$.

$$\begin{array}{ccc} \hat{C}_{n+1}X & \overset{\tau}{=} & \pi_k(\Sigma^k Z_{n+1}^+ \vee X)_2 \\ d \downarrow & & \downarrow (E^{k-n}d_{n+1} \odot 1)_* \\ \hat{C}_n X & \overset{\tau}{=} & \pi_k(\Sigma^k Z_n^+ \vee X)_2 \end{array} \qquad (1)$$

$$\begin{array}{ccc} \hat{C}_n Y & \overset{\tau}{=} & \pi_k(\Sigma^k \bar{Z}_n^+ \vee Y)_2 \\ \downarrow \hat{F}_* & & \downarrow (E^{k-n}(\check{K}F)_n \odot F)_* \\ \hat{C}_n X & \overset{\tau}{=} & \pi_k(\Sigma^k Z_n^+ \vee X)_2 \end{array} \qquad (2)$$

The right-hand side of these diagrams is given by the twisted chain functor $\check{K}X$ in §4. Now the proposition in (5.9) follows readily from (1), (2) and from the definitions of cohomology in (5.8) and in (4.8). For $A_n = \Sigma^{n-1} Z_n^+$ we obtain τ by the following commutative diagram where k is the Hurewicz homomorphism $(n \geqq 2)$.

$$\begin{array}{ccccc} \pi_n(\Sigma A_n \vee X^n, X^n) & \underset{j_0}{\overset{\cong}{\longleftarrow}} & \pi_n(\Sigma A_n \vee X^n)_2 & \underset{j_*}{\overset{\cong}{\longrightarrow}} & \pi_n(\Sigma A_n \vee X)_2 \\ \uparrow j' & & & & \\ \pi_n(\Sigma A_n \vee X^n, X^{n-1}) & & & & \cong \uparrow \tau \\ \uparrow \mu_* & & & & \\ \pi_n(X^n, X^{n-1}) & \underset{\cong}{\overset{p_*}{\longleftarrow}} & \pi_n(\hat{X}^n, \hat{X}^{n-1}) & \overset{h}{\longrightarrow} & H_n(\hat{X}^n, \hat{X}^{n-1}) \\ \uparrow j & & & & \\ \pi_n(X^n) & & & & \end{array} \qquad (3)$$

Here h is an isomorphism for $n \geqq 3$, but the isomorphism τ in (3) is well defined by (3) also for $n=2$. The case $n=1$ will be considered in (VI.1.20) (9) below.

Now let c_e be the characteristic map of the cell $e \in Z_n$, $c_e \in \pi_n(X^n, X^{n-1})$, and let $f_e = \partial c_e$ be the attaching map of e. Moreover, let $\tau_e \in \pi_n(\Sigma A_n \vee X)_2$ be given by the inclusion $S^n \subset \Sigma A_n$ associated to e. Then one can check

$$\tau(\hat{e}) = \tau_e \quad \text{for} \quad \hat{e} = hp_*^{-1}(c_e). \tag{4}$$

For $e \in Z_{n+1}$ it is readily seen by the definition in (5.6)(2)

$$hp_*^{-1} j(f_e) = d_{n+1}(\hat{e}). \tag{5}$$

On the other hand, we derive from the definition of the difference construction ∇f_e that

$$j' \mu_* j(f_e) = j_0(\nabla f_e). \tag{6}$$

Using (4), (5) and (6) we get

$$\tau d_{n+1}(\hat{e}) = j_*(\nabla f_e) = d_{n+1} \circ \tau_e = (d_{n+1} \odot 1)_* \tau(\hat{e}). \tag{7}$$

For the second equation compare (4.6) (1). This completes the proof that (1) commutes for $k = n > 2$. For $k \neq n$ we use the result below on the partial suspension E. In a similar way one can check that (2) commutes. □

In the next remark we describe some properties of CW-complexes which are very special cases of the general suspension theorem (V.§ 7a). We leave it to the reader to give a more direct proof.

(5.11) ***Remark.*** For $\Sigma^n Z^+ = \vee_Z S^n$ let $i_e: S^n \subset \Sigma^n Z^+$ be the inclusion associated to $e \in Z$. Let X, Y be relative CW-complexes as in (5.3) above with attaching map $f = f_n: A_n = \Sigma^{n-1} Z_n^+ \to X_{n-1}$.

(a) The homomorphism

$$(\pi_f, 1)_*: \pi_n(CA_n \vee X^{n-1}, A_n \vee X^{n-1}) \to \pi_n(X^n, X^{n-1})$$

is surjective for $n=2$ and is an isomorphism for $n \geqq 2$, see (II.11.6).

(b) The partial suspension

$$E: \pi_n(\Sigma^n Z^+ \vee Y)_2 \to \pi_{n+1}(\Sigma^{n+1} Z^+ \vee Y)_2$$

is surjective for $n=1$ and is an isomorphism for $n \geqq 2$.

(c) The inclusion $Y^k \subset Y$ of the k-skeleton induces a homomorphism

$$\pi_n(\Sigma^n Z^+ \vee Y^k)_2 \to \pi_n(\Sigma^n Z^+ \vee Y)_2$$

$(n \geqq 1)$ which is surjective for $k=1$ and which is an isomorphism for $k \geqq 2$.

(d) For $\pi = \pi_1 Y$ we have the isomorphism of π-modules $(n \geqq 2)$

$$i_n: \bigoplus_Z \mathbb{Z}[\pi] \cong \pi_n(\Sigma^n Z^+ \vee Y)_2$$

which carries $e \in Z$ to the inclusion i_e and which clearly satisfies $Ei_n = i_{n+1}$. By (II.11.13) we see that the partial suspension E is a homomorphism of π-modules.

§ 5a Appendix: admissible classes of complexes

The definition of an admissible class of complexes in a cofibration category is motivated by the properties of CW-complexes. In fact, the class $\mathfrak{X}$ of CW-complexes in (5.4) is an admissible class. In Chapter VII, § 3, § 4 we will describe further examples of admissible classes of complexes.

Let (X, D) be a relative CW-complex $(X \in \mathfrak{X})$ as in § 5 with attaching maps $f_n: A_n \to X_{n-1} = X^{n-1}$. We derive from the properties of CW-complexes in (5.11) that the differential d_n in $\check{k}(X) = \{X_\infty, A_n, d_n\}$ has a factorization as in the following diagram:

(5a.1)

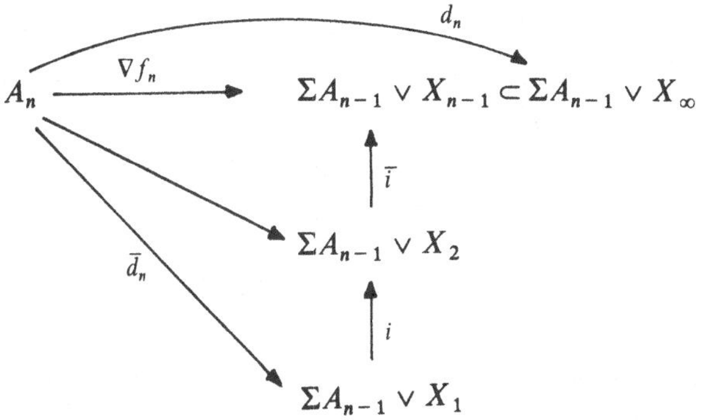

Here $\bar{d}_n$ is trivial on X_1 and the homotopy class of $i\bar{d}_n$ is well defined. This diagram shows that the twisted cohomology $H^n_k(\check{K}X, u)$ in (5.9) actually depends only on the restriction $u_1 = u|X_1$. We derive from (5.11):

(5a.2) **Lemma.** *For a CW-complex* $X = (X, D)$ *as above the twisted chain complex* $\check{k}(X, 2) = \{(X_2, X_1), A_n, \bar{d}_n\}$ *is well defined by the maps* $i\bar{d}_n$ *in* (5a.1). *For* $\check{K}(X, 2) = E\check{k}(X, 2)$ *we get the natural isomorphism*

$$H^n_k(\check{K}(X, \infty), u) = H^n_k(\check{K}(X, 2), u_1),$$

where $u_1:(X^2, X^1) \to (U, U)$ *is the map in* **Coef** *given by u and where* $\check{K}(X, \infty) = \check{K}(X)$ *is the twisted chain complex in* (5.9).

Clearly the homomorphism $\varphi = \pi_1(u)$ in (5.9) as well is determined by $u_1:(X^2, X^1) \to (U, U)$.

Now let **C** be a cofibration category with an initial object $*$. We characterize classes of complexes in **C** which have the properties described in (5a.1) and (5a.2) and we call such classes admissible classes of complexes.

(5a.3) ***Definition.*** Let X be a complex and let A and B be based objects. We say (A, B) is **injective on** X_2 if the inclusion $X_2 \subset X_N (N \geqq 2)$ induces an injective map

$$[A, B \vee X_2]_2 \to [A, B \vee X_N]_2.$$

We say that (A, B) is **surjective on** X_1 if $X_1 \subset X_N (N \geqq 1)$ induces a surjective map

$$[A, B \vee X_1]_2 \to [A, B \vee X_N]_2. \qquad \|$$

(5a.4) ***Notation.*** For an element $\xi \in [A, B \vee X_n]_2$ we frequently denote any composition $(1 \vee j)\xi : A \to B \vee X_n \subset B \vee X_m$, $n \leqq m$, as well by ξ. Here j is the inclusion $X_n \subset X_m$. *We say that* $\xi = \xi'$ *on* X_m if $(1 \vee j)\xi = \xi' \in [A, B \vee X_m]_2$.

(5a.5) ***Definition.*** We say that $\mathfrak{X}$ is an **admissible class** of complexes if for all X, $Y \in \mathfrak{X}$ with attaching maps $A_i \to X_{i-1}$ and $B_i \to Y_{i-1}$, respectively, the following properties (a), (b) are satisfied.

(a) The pairs $(B_{n+1}, \Sigma A_n)$, $(\Sigma B_n, \Sigma A_n)$, and $(\Sigma^2 B_{n-1}, \Sigma A_n)$ are injective on X_2 and surjective on X_1 for $n \geqq 2$.

(b) The pairs $(A_{n+1}, \Sigma^2 A_{n-1})$, $(A_{n+1}, \Sigma A_n \vee \Sigma^2 A_{n-1})$, $(\Sigma B_{n+1}, \Sigma^2 A_n)$, and $(\Sigma^2 B_n, \Sigma^2 A_n)$ are injective on X_2 for $n \geqq 2$.

We also assume that $X \in \mathfrak{X}$ implies that each skeleton X_n is a complex in $\mathfrak{X}$, $n \geqq 1$. $\|$

(5a.6) ***Example.*** It follows readily from (5.11) that the class $\mathfrak{X}$ of relative CW-complexes considered in § 5 is an admissible class of complexes in $\mathbf{C} = \mathbf{Top^*}$.

Now let $\mathfrak{X}$ be an admissible class of complexes as in (5a.5). Then we obtain for each $X \in \mathfrak{X}$ a similar diagram as in (5a.1) (where actually X_∞ is not needed for the construction of $\bar{d}_n$). The conditions in (5a.5) are chosen essentially in such a way that the following lemma holds:

(5a.7) **Lemma.** *Let* $X (X \in \mathfrak{X})$ *be a complex with attaching maps* $f_n: A_n \to X_{n-1}$. *Then the twisted chain complex* $\check{k}(X, 2) = \{(X_2, X_1), A_n, \bar{d}_n\}$ *is well defined by* $\bar{i}\bar{d}_n = \nabla f_n$, *see* (5a.1). *Moreover,* $\check{K}(X, 2) = E\check{k}(X, 2)$ *defines a functor as in* (4.2)

$$\check{K}(-, 2); \mathbf{Complex}(\mathfrak{X})/\overset{0}{\simeq} \to \mathbf{Chain}^\vee$$

which induces a functor

$$\check{K}(-, 2): \mathbf{Complex}(\mathfrak{X})/\overset{1}{\simeq} \to \mathbf{Chain}^\vee/\simeq.$$

Clearly, for the functor $\check{K}(-, \infty) = \check{K}$ in (4.2) the inclusion $X_2 \subset X_\infty$ induces a natural map $\check{K}(X,2) \to \check{K}(X,\infty)$ which induces the isomorphism

$$H^n_k(\check{K}(X,\infty),u) = H^n_k(\check{K}(X,2),u_1) \tag{5a.8}$$

as in (5a.2). We point out that for the definition of $\check{K}(X,2)$ the existence of $X_\infty = \lim X_n$ is not needed.

§ 6 Eilenberg–Mac Lane spaces and cohomology with local coefficients

Recall that an **Eilenberg–Mac Lane space** $K(A,n)$ is a CW-space with basepoint together with an isomorphism

$$\pi_r K(A,n) = \begin{cases} A & r = n \\ 0 & r \neq n \end{cases} \tag{6.1}$$

Here A is a group which is abelian for $n \geqq 2$ and π_r is the rth homotopy group. Such spaces exist and they are unique up to homotopy equivalence in **Top**. For a CW-space X the loop space ΩX is a CW-space too, see Milnor 1959. Hence $\Omega K(A,n)$ is an Eilenberg–Mac Lane space with

$$k: K(A,n-1) \simeq \Omega K(A,n). \tag{6.2}$$

Actually there is a unique homotopy equivalence k in **Top*** for which the composite map $A = \pi_{n-1}K(A,n-1) \to \pi_{n-1}\Omega K(A,n) = \pi_n K(A,n) = A$, induced by k, is the identity of A. Clearly, the loop space in (6.2) is the loop space of a based object in the fibration category **Top** (I.5.2). For a CW-space U we have the well-known natural isomorphism ($n \geq 0$, $k \geqq 0$)

$$\pi^k_{K(A,n+k)}(U) = [U, \Omega^k K(A,n+k)] = [U, K(A,n)] = H^n(U,A). \tag{6.3}$$

Here the left-hand side is a homotopy group in the fibration category **Top**, see (II.14.3), and the right-hand side is the n-th singular cohomology group of U with coefficients in the $\mathbb{Z}$-module A. A similar result as in (6.3) is true for cohomology with local coefficients as defined in § 5 above.

For this we consider the **category $\mathbf{Top}_D$ of spaces over** D which is a fibration category by (I.5.2) and (II.1.4). If $D = *$ is a point we have $\mathbf{Top}_D = \mathbf{Top}$ since $*$ is the final object of **Top**, the spaces with base point are the based objects in this case. In general, a based object in $\mathbf{Top}_D$ is given by a fibration $p: A \twoheadrightarrow D$ in **Top** and by a section $o: D \to A$, $po = 1_D$. For example, if F has a basepoint $*$, then

$$(D \underset{i}{\longrightarrow} F \times D \underset{pr}{\longrightarrow} D) \tag{6.4}$$

with $i(d) = (*, d)$ is a based object in $\mathbf{Top}_D$.

(6.5) ***Definition.*** Let $* \in D$ and let $\pi = \pi_1(D)$ be the fundamental group of the

CW-space D. For a π-module A the (generalized) **Eilenberg–Mac Lane space** $L(A,n)$ in $\mathbf{Top}_D$ is a based object in $\mathbf{Top}_D$ with fiber $K(A,n)$,

$$K(A,n) \xrightarrow{i} L(A,n) \underset{0}{\overset{p}{\rightleftarrows}} D,$$

such that the action of $\pi = \pi_1(D) \subset \pi_1 L(A,n)$ on the group $A = \pi_n(K(A,n)) \subset \pi_n L(A,n)$, given by (II.6.13), coincides with the π-module A. ||

Such objects $L(A,n)$ always exist, compare for example 5.2.6 in Baues (1977), and they are unique up to homotopy equivalence of based objects in $\mathbf{Top}_D$. In particular, we have as in (6.2) the homotopy equivalence in Top_D

(6.6) $$\bar{k}: L(A, n-1) \simeq \Omega L(A,n),$$

which is a based map in Top_D. Here $\Omega L(A,n)$ is the loop space of a based object in the fibration category Top_D. For $n = 1$ we use (6.6) as a definition of $L(A,0)$. On fibers the homotopy equivalence $\bar{k}$ is the homotopy equivalence k in (6.2). By obstruction theory (see for example 5.2.4 in Baues (1977)) we get the following result which generalizes (6.3).

(6.7) **Proposition.** *Let $\alpha: X \to D$ be an object in Top_D where X and D are path connected CW-spaces and let A be a π-module with $\pi = \pi_1 D$. Then there is a natural isomorphism of abelion groups ($k \geqq 0, n \in \mathbb{Z}$)*

$$\pi^k_{L(A,n+k)}(X) = \hat{H}^n(X, \alpha^* A).$$

The left-hand side denotes a homotopy group in the fibration category $\mathbf{Top}_D$, *see* (II.14.3), *the right-hand side is the cohomology with local coefficients which is defined by* (5.8). *The $\pi_1(x)$-module $\alpha^* A$ is induced by $\alpha_*: \pi_1(X) \to \pi_1(D)$, see* (5.1).

Now let $p: U \twoheadrightarrow V$ be a fibration in **Top** between path connected CW-spaces and let $V \to D$ be a map. Then $p = (U \mid V)$ is a fibration in $\mathbf{Top}_D$ for which the exact sequence (II.14.5) is defined. The following result describes a well-known example of this exact sequence.

(6.8) **Theorem.** *Let Z_p be the mapping cylinder of p in* **Top**. *Then we have for $B = L(A, n+k)$ the isomorphism of exact sequences*

$$\begin{array}{ccccccccc}
\pi^k_B(V) & \xrightarrow{p^*} & \pi^k_B(U) & \longrightarrow & \pi^k_B(U \mid V) & \xrightarrow{\partial} & \pi^{k-1}_B(V) & \xrightarrow{p^*} & \\
\wr\| & & \wr\| & & \wr\| & & \wr\| & & . \\
\hat{H}^n(V) & \xrightarrow{p^*} & \hat{H}^n(U) & \xrightarrow{\delta} & \hat{H}^{n+1}(Z_p, U) & \longrightarrow & \hat{H}^{n+1}(V) & \xrightarrow{p^*} &
\end{array}$$

The local coefficients in the cohomology groups of the bottom row are determined by the π-modul A as in (6.7).

The theorem describes the well-known long exact cohomology sequence for the pair (Z_p, U), compare 5.2.4 in Baues (1977).

§ 7 Postnikov towers

We first dualize the notion of a complex in (3.1).

(7.1) ***Definition.*** The notion of a **cocomplex** in a fibration category is obtained by dualizing the notion of a complex in a cofibration category. Thus a cocomplex $E = \{E_1\}$ is a **tower**

$$E_0 \twoheadleftarrow E_1 \twoheadleftarrow E_2 \twoheadleftarrow \cdots$$

of principal fibrations in a fibration category with coattaching maps (= classifying maps) $E_{n-1} \to A_n$ where A_n is a based object (a loop object for $n \geqq 2$). ||

We now show that Postnikov towers yield important examples of cocomplexes in topology. We recall from 5.3.1 in Baues (1977) the following result on the Postnikov decomposition of a fibration.

(7.2) **Theorem.** *Let $p: E \twoheadrightarrow B$ be a fibration in* **Top** *with fiber F and let E, F, and B be path connected CW-spaces. Then there exist fibrations q_n and maps h_n making the diagram of basepoint preserving maps*

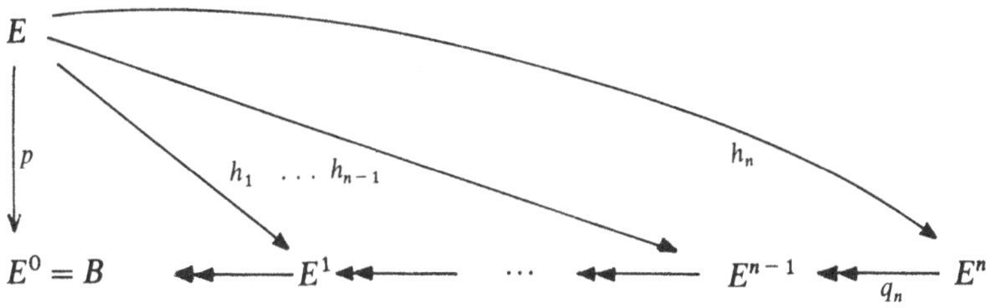

commute, such that for $n \geqq 1$:

(*i*) *q_n is a fibration with fiber $K(\pi_n F, n)$,*
(*ii*) *h_n is $(n+1)$-connected, that is, h_n* induces isomorphisms of homotopy groups *π_r* for *$r \leqq n$ and $\pi_{n+1}(h_n)$ is surjective.*

For $B = *$ we call the tower $\{E^n\}$ the Postnikov-decomposition of the space E. In general the fibrations $q_n: E^n \twoheadrightarrow E^{n-1}$ are not principal fibrations in **Top.** We can, however, apply the following result, see section 5.2 in Baues (1977).

(7.3) **Proposition.** *Let $p: E \twoheadrightarrow B$ be a fibration in* **Top** *between path connected*

CW-spaces with fiber $F = K(A, n)$ where A is abelian, $n \geqq 1$. Let $b: B \to D$ be a map, which induces a surjection $b_*: \pi_1(B) \twoheadrightarrow \pi_1(D) = \pi$, and for which kernel (b_*) acts trivially on A, see (II.5.20). Then A is a π-module and there exists a map $f: B \to L(A, n+1)$ in $\mathbf{Top}_D$ such that $p: E \twoheadrightarrow B$ is a principal fibration in the fibration category $\mathbf{Top}_D$ with classifying map f.

This result gives us many examples for the 'warning' in (II.8.2). The element f,

(7.4) $$f \in [B, L(A, n+1)]_D = \hat{H}^{n+1}(B, b^*A),$$

which is uniquely determined by the fibration $p: E \twoheadrightarrow B$ in (7.3), is called the (twisted) **k-invariant** of the fibration p. Recall that a path connected CW-space F is nilpotent if $\pi_1 F$ acts nilpotently on $\pi_n F$ for $n \geqq 1$, see (I.8.26). We say that F is a **simple** space if $\pi_1 F$ acts trivially on $\pi_n F$, $n \geqq 1$. Moreover, a fibration $p: E \twoheadrightarrow B$ with fiber F is **simple** if F is simple and if $\pi_1(B)$ acts trivially on $\pi_n(F)$, $n \geqq 1$, via (II.5.20). The fibration p is **nilpotent** if F is nilpotent and if $\pi_1(E)$ acts nilpotently on $\pi_n(F) = \pi_{n+1}(Z_p, E)$ via (II.7.10). We derive from (7.2) and (7.3).

(7.5) **Lemma.** *Let $p: E \twoheadrightarrow B$ be a fibration with fiber F and let E, F, and B be path connected CW-spaces.*

(a) *If p is a simple fibration, then $\{E^n\}$ in (7.2) is a cocomplex in* **Top**.

(b) *If F is a simple space, then $\{E^n\}$ in (7.2) is a cocomplex in* $\mathbf{Top}_B$.

(c) *If p is a nilpotent fibration, then each $q_n: E^n \to E^{n-1}$, $n \geqq 1$, in (7.2) is a finite cocomplex in* **Top**, *$E^n = E^n_{N_n} \to \cdots \to E^n_0 = E^{n-1}$, with classifying maps $E^n_k \to K(A^n_k, n+1)$ where A^n_k is a subquotient of $\pi_n F$ with a trivial action of $\pi_1(E)$.*

(d) *If F is a nilpotent space, then each $q_n: E^n \to E^{n-1}$, $n \geqq 1$, in (7.2) is a finite cocomplex in* $\mathbf{Top}_B$, *$E^n = E^n_{N_n} \to \cdots \to E^n_0 = E^{n-1}$, with classifying maps $E^n_k \to L(A^n_k, n+1)$. Here A^n_k is a subquotient of $\pi_n F$ with a trivial action of $\pi_1(F)$ and hence with an induced action of $\pi_1(B)$.*

§8 Nilpotency of function spaces in topology

We consider the commutative diagram in **Top**

(8.1) $$\begin{array}{ccc} Y & \xrightarrow{\;w\;} & E \\ {\scriptstyle i}\downarrow & \overset{f}{\nearrow} & \downarrow {\scriptstyle p}, \\ X & \xrightarrow[\;v\;]{} & B \end{array}$$

where i is a cofibration and p is a fibration in **Top** with fiber F. Let

(8.2) $$\operatorname{Map}(X, E)^Y_B = \{f \in E^X \mid fi = w, pf = v\}$$

be the space of all fillers for diagram (8.1). This is a subspace of the space of all maps, $E^X = \{f, f\colon E \to X\}$, which has the compact open topology.

(8.3) **Theorem.** *Each path component of the function space* $Map(X,E)_B^Y$ *is a nilpotent space provided* (*a*) *or* (*b*) *holds:*

(*a*) $(Y = X_0 \rightarrowtail \cdots \rightarrowtail X_N = X)$ *is a finite complex in* $\mathbf{Top}^Y$.

(*b*) $(E = E_M \twoheadrightarrow \cdots \twoheadrightarrow E_0 = B)$ *is a finite cocomplex in* $\mathbf{Top}_B$.

Clearly (*a*) is satisfied, if (X, Y) is a relative CW-complex for which X and Y are path connected, $* \in Y$, and for which $\dim(X - Y) < \infty$, see (5.3). On the other hand (*b*) is satisfied if E, F and B are path connected CW-spaces and if F is a nilpotent space with only finitely many non-trivial homotopy groups, see (7.4)(*d*). Moreover, we derive the following result.

(8.4) **Corollary.** *Each path component of* $Map(X,E)_B^Y$ *is a nilpotent space if* (X, Y) *is a relative CW-complex with* $\dim(X - Y) < \infty$ *and if* F *is nilpotent.*

In this corollary Y is allowed to be the empty space, $Y = \phi$. We obtain (8.4) since we can replace $E \twoheadrightarrow B$ by an appropriate Postnikov section $E_N \twoheadrightarrow B$, so that we can apply (8.3). The result (8.4) corresponds to a theorem of Scheerer (1980); for $B = *$ the result was obtained by Sullivan (1977) and Hilton–Mislin–Roitberg–Steiner. We give a new proof which is available in many Quillen model categories and which is a simple application of (4.17).

For the proof of (8.3) we use the following lemma:

(8.5) **Lemma.** *Let* $\mathbf{C}$ *be a model category with an initial object* $*$. *Consider maps*

$$Q \rightarrowtail X \rightarrowtail C_f \to D$$

in $\mathbf{C}$ *where* $X \rightarrowtail C_f$ *is a principal cofibration in* $\mathbf{C}$. *Then* $X \rightarrowtail C_f$ *is a principal cofibration in the cofibration category* $\mathbf{C}_D^Q$ *of objects under* Q *and over* D. *Here* $\mathbf{C}_D^Q$ *has the external structure in* (I.4a.5).

Proof. Let $f\colon A \to X$ be the attaching map where A is a based object in $\mathbf{C}$. Let $\mathbf{B} = \mathbf{C}_D^Q$. For the mapping cylinder Z_f of f in $\mathbf{C}$ we obtain the following commutative diagram in $\mathbf{B}$:

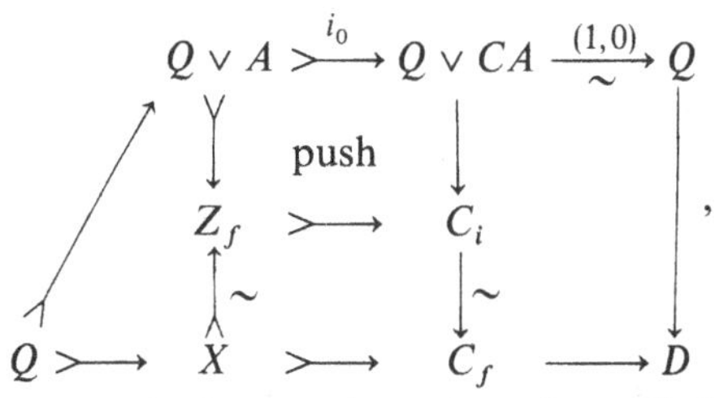

since $C_i/CA = Z_f/A = C_f$ in $\mathbf{C}$. Here the top row is a cylinder in $\mathbf{B}$ of the based object $Q \vee A$ in **B**. Hence $Z_f \rightarrowtail C_i$ is a principal cofibration in **B** which by the diagram above is 'equivalent' to $X \rightarrowtail C_f$. Using the general notion of a principal cofibration in (II.8.3) we obtain the proposition in (8.5). □

Now it is easy to prove (8.3). We use the model category $\overline{\mathbf{Top}}$ of Strøm, see (I.2.10).

(8.6) *Proof of* (8.3). Choose a map $u \in Map(X, E)^Y_B$. First we assume that (*a*) is satisfied. In this case $i_1 : X \rightarrowtail X \bigcup_Y X$ is a based complex in $\mathbf{C} = \overline{\mathbf{Top}}^X$ of finite length. Therefore by (8.5)

$$A = (X \rightarrowtail X \bigcup_Y X \xrightarrow{(1,1)} X \xrightarrow{v} B) \tag{1}$$

is a based complex in $\mathbf{C}^X_B = \mathbf{C}(v)$ of finite length. Moreover,

$$\pi_n(Map(X, E)^Y_B, u) = [\Sigma^n A, B], \tag{2}$$

where $B = (X \xrightarrow{u} E \xrightarrow{p} B)$ is an object in $\mathbf{C}(v)$. Clearly the right-hand side of (2) is defined in the cofibration category $\mathbf{C}(v)$. We can apply (8.3) and the proposition follows immediately. If (*b*) is satisfied we have the dual argument which yields the proof. □

§9 Homotopy groups of function spaces in topology

We consider the homotopy groups

(9.1) $$\pi_n = \pi_n(\mathrm{Map}(X, E)^Y_B, u), \quad n \geqq 1,$$

of the function space in (8.2) where we assume that all spaces in (8.1) are CW-spaces and that X and E, F and B are path connected. It is enough to consider the case where X is a CW-complex with trivial 0-skeleton $X^0 = *$ and where Y is a subcomplex of X. Hence we have the filtration of relative skeleta

(9.2) $$Y \subset X_0 \subset X_1 \subset \cdots \subset X_n \subset \cdots \subset X$$

with $X_n = Y \cup X^n$ where $Y = X_0$ if $* \in Y$, (or equivalently if $Y \neq \varnothing$). On the other hand we have the Postnikov tower

(9.3) $$E \to \cdots \to E^n \twoheadrightarrow \cdots \twoheadrightarrow E^1 \twoheadrightarrow E^0 = B,$$

by (7.2). By naturality of the function space (8.2) we derive from (9.2) and (9.3) the following towers of Serre fibrations in **Top** with

(9.4) $$M_p = \mathrm{Map}(X_p, E)^Y_B \quad \text{and} \quad M^p = \mathrm{Map}(X, E^p)^Y_B,$$

$$\begin{cases} M \to \cdots \to M^p \to M^{p-1} \to \cdots M^0 = *, \\ M \to \cdots \to M_p \to M_{p-1} \to \cdots M_0 = \begin{cases} F & \text{if} \quad Y = \varnothing \\ * & \text{if} \quad Y \neq \varnothing \end{cases} \end{cases}.$$

For the compositions $u_p: X_p \subset X \xrightarrow{u} E$, $u^p: X \xrightarrow{u} E \to E^p$, we have the groups

(9.5) $$(\pi_n)_p = \pi_n(M_p, u_p) \quad \text{and} \quad (\pi_n)^p = \pi_n(M^p, u^p),$$

which are inverse systems by the towers in (9.4). Using (1.13) in the fibration category of Serre fibrations, see (I.2.11) we obtain for $n \geqq 0$ the exact sequences

(9.6) $$\begin{aligned} &0 \to \mathrm{Lim}^1(\pi_{n+1})_p \to \pi_n \xrightarrow{\psi} \mathrm{Lim}(\pi_n)_p \to 0, \\ &0 \to \mathrm{Lim}^1(\pi_{n+1})^p \to \pi_n \xrightarrow{\psi'} \mathrm{Lim}(\pi_n)^p \to 0. \end{aligned}$$

(9.7) ***Remark.*** The exact sequences in (9.6) are isomorphic, that is, kernel ψ = kernel ψ'. We leave the proof as an exercise.

Moreover, we have the following result for the homotopy spectral sequence of Bousfield–Kan, see (2.13).

(9.8) **Theorem.**

(i) *Let Y be path connected with $* \in Y$. Then the homotopy spectral sequence $E_r^{s,t} = E_r^{s,t}\{M_p\}$ satisfies*

$$E_2^{s,t} = \hat{H}^s(X, Y; u^*\pi_t(F))$$

for $t \geqq 2$. For $B = *$ the spectral sequence coincides with the one in (5.10).

(ii) *Let F be a simple space. Then the homotopy spectral sequence $\bar{E}_r^{m,n} = E_r^{m,n}\{M^p\}$ satisfies for $m \geqq 1$*

$$\bar{E}_1^{m,n} = \hat{H}^{2m-n}(X, Y; u^*\pi_m(F)).$$

(iii) *If the assumptions in* (i) *and* (ii) *are satisfied we have an isomorphism of spectral sequences*

$$(E_r^{s,t}, d_r) = (\bar{E}_{r-1}^{t,2t-s}, \bar{d}_{r-1})$$

for $r \geqq 2$, $t - s \geqq 1$.

Recall that we have an action of $\pi_1(E)$ on $\pi_m(F) = \pi_{m+1}(Z_p, E)$ by (II.7.10); this yields an action of $\pi_1(X)$ via $u_*: \pi_1(X) \to \pi_1(E)$ and hence the coefficients in (i) and (ii) above are well defined. A proof of the theorem can be achieved similarly as in (5.10), compare Baues (1977).

§10 The relative homotopy spectral sequence

We now describe an 'extended' spectral sequence for a filtered object in a cofibration category **C** which relies on the exact sequence for relative homotopy groups in (II.§7). We therefore call this spectral sequence the *relative homotopy spectral sequence.*

Let $U = \{U^n\}$ be a cofibrant and fibrant object in **Fil(C)**. If V is any filtered object in **C** we can replace V by its model $RM(V)$ which is cofibrant and fibrant in **Fil(C)**, see (II.§ 3). By (II.§ 7) we can form the exact homotopy sequences of the pairs (U^n, U^{n-1}) where A is a based object in **C**

$$(10.1) \qquad \cdots \to \pi_2^A(U^n) \xrightarrow{j} \pi_2^A(U^n, U^{n-1}) \xrightarrow{\partial} \pi_1^A(U^{n-1}) \xrightarrow{i} \pi_1^A(U^n)$$
$$\xrightarrow{j} \pi_1^A(U^n, U^{n-1}) \xrightarrow{\partial} \pi_0^A(U^{n-1}) \xrightarrow{i} \pi_0^A(U^n).$$

We have shown that the sequence (10.1) is **exact** in the following sense (see II.§ 7):

(i) the last three objects are sets with basepoint 0, all the others are groups, and the image of $\pi_2^A(U^n)$ lies in the center of $\pi_2^A(U^n, U^{n-1})$,
(ii) everywhere 'kernel = image', and
(iii) the sequence comes with a natural action, $+$, of $\pi_1^A(U^n)$ on the set $\pi_1^A(U^n, U^{n-1})$ such that $j: \pi_1^A(U^n) \to \pi_1^A(U^n, U^{n-1})$ is given by $x \mapsto 0 + x$, and such that elements of $\pi_1^A(U^n, U^{n-1})$ are in the same orbit if and only if they have the same image in $\pi_0^A(U^{n-1})$,
(iv) Moreover, the sequence comes with a natural action of $\pi_1^A(U^{n-1})$ on all terms of the top row of (10.1) such that j, ∂, i are equivariant; in addition the action on $\pi_1^A(U^{n-1})$ is given by inner automorphisms, and $\partial: \pi_2^A(U^n, U^{n-1}) \to \pi_1^A(U^{n-1})$ is a crossed module.

The properties (i), (ii), and (iii) above correspond exactly to the properties (i), (ii), and (iii) respectively described in (2.3). We have, however, the fact that i in (10.1) lowers the filtration degree while i in (2.3) raises the filtration degree. We therefore define

$$(10.2) \qquad \begin{array}{ccccc} \pi_i^A(U^m) & \longrightarrow & \pi_i^A(U^{m+1}) & \longrightarrow & \pi_i^A(U^{m+1}, U^m) \\ \| & & \| & & \| \\ \pi_i U_{-m} & \longrightarrow & \pi_i U_{-m-1} & \longrightarrow & \pi_{i-1} F_{-m} \end{array} .$$

Now the bottom row is of the same type as in (2.3). As in (2.3) it follows from (10.1) that one can form the **rth derived homotopy sequences** ($r \geqq 0$, $-m = n \leqq 0$). They are defined in the same way as in (2.4) by the identification in (10.2)

$$(10.3) \qquad \cdots \to \pi_2 U^{(r)}_{-m-2r-1} \to \pi_1 F^{(r)}_{-m-r} \to \pi_1 U^{(r)}_{-m-r} \to \pi_1 U^{(r)}_{-m-r-1}$$
$$\to \pi_0 F^{(r)}_{-m} \to \pi_0 U^{(r)}_{-m} \to \pi_0 U^{(r)}_{-m-1}.$$

Here we set as in (2.4):

$$\pi_i U^{(r)}_{-m} = \text{image}\,(\pi_i^A(U^{m-r}) \to \pi_i^A(U^m)).$$

$$\pi_i F^{(r)}_{-m} = \frac{\text{kernel}\ (\pi_{i+1}^A(U^{m+1}, U^m) \to \pi_i^A(U^m)/\pi_i U^{(r)}_{-m})}{\text{action of kernel}\ (\pi_{i+1}^A(U^{m+1}) \to \pi_{i+1}^A(U^{m+r+1}))}.$$

One can check that the sequence (10.3) is again exact in the sense of (10.1) (i), ..., (iv). Next we obtain as in (2.5).

(10.4) ***Definition of the relative homotopy spectral sequence*** $\{E_r^{s,t}(A, U)\}$. Let $E_r^{s,t} = \pi_{t-s} F_s^{(r-1)}$ for $t \geqq s \leqq 0$, $r \geqq 1$, and let the differential $d_r : E_r^{s,t} \to E_r^{s+r,t+r-1}$ be the composition

$$\pi_{t-s} F_s^{(r-1)} \to \pi_{t-s} U_s^{(r-1)} \to \pi_{t-s-1} F_{s+r}^{(r-1)}. \qquad \|$$

This spectral sequence has again all properties as in (2.5) (i), (ii), (iii), (iv). The main difference is the fact that $E_r^{s,t}$ above is defined for $s \leqq 0$ while in (2.5) the terms $E_r^{s,t}$ are defined for $s \geqq 0$.

(10.5) ***Convergence of the spectral sequence*** (10.4). Assume $U = \{U^n\}$ has the property that the pairs (U^{n+1}, U^n) 'get higher and higher connected', that is $\pi_i^A(U^{n+1}, U^n) = 0$ for $i < N_n$ where $0 \leqq N_0 \leqq N_1 \leqq \cdots$ with $\lim \{N_n\} = \infty$. Then we can find for each $q \geqq 0$ a bound $r = r(q) < \infty$ such that

$$E_r^{s,q+s} = E_{r+1}^{s,q+s} = \cdots = E_\infty^{s,q+s}. \tag{1}$$

We picture the spectral sequence as follows:

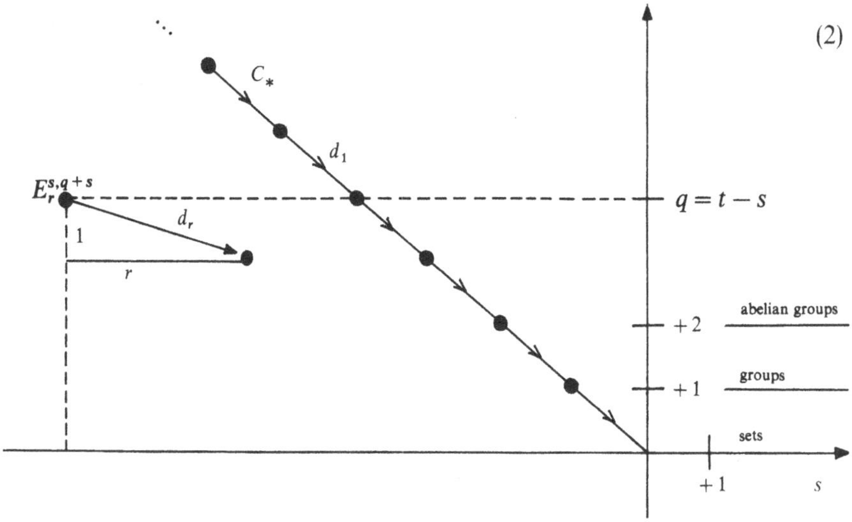

The elements in the column of degree $s \leqq 0$ are represented by elements in $\pi_q F_s = \pi_{q+1}^A(U^{-s+1}, U^{-s})$. The row of degree q contributes towards the homotopy group $\pi_q^A(U^0)$. We define a **filtration** of $\pi_q^A(U^0)(q \geqq 1,\ s \leqq 0)$

$$\left.\begin{aligned} &K_{s,q} = \text{kernel } (\pi_q^A(U^0) \to \pi_q^A(U^{-s}), \\ &\cdots \subset K_{s,q} \subset \cdots \subset K_{-1,q} \subset K_{0,q} = \pi_q^A(U^0). \end{aligned}\right\} \tag{3}$$

By the assumption above we get $\bigcap_{s\leqq 0}K_{s,q}=0$ and the associated graded group of this filtration is

$$G^{s,q}=K_{s+1,q}/K_{s,q}=E_{\infty}^{s,q+s}. \tag{4}$$

Hence the spectral sequence converges to the homotopy groups $\pi_*^A(U^0)$ of the initial term U^0 of the filtered object U.

(10.6) ***The crossed chain complex C_*.*** The indicated diagonal C_* in (10.5) (2), given by $q+s=0$, with all differentials d_1 corresponds to the sequence

$$\cdots\xrightarrow{\partial_4}\pi_3^A(U^3,U^2)\xrightarrow{\partial_3}\pi_2^A(U^2,U^1)\xrightarrow{\partial_2}\pi_1^A(U^1) \tag{1}$$

with $\partial_n=j\partial=d_1$ for $n\geqq 3$ and with $\partial_2=\partial$. Clearly, $\partial_{n-1}\partial_n=0$ for $n\geqq 3$. We call this sequence, together with the action of $\pi_1^A(U^1)$, the **crossed chain complex** $C_*=C_*^A(U)$ where we set

$$C_n^A(U)=\begin{cases}\pi_n^A(U^n,U^{n-1}), & n\geqq 2,\\ \pi_1^A(U^1) & , \quad n=1.\end{cases} \tag{2}$$

The image of ∂_2 is a normal subgroup and the group $\pi=$ cokernel (∂_2) can be identified with the image of $\pi_1^A(U^1)\to\pi_1^A(U^2)$ by exactness. Since ∂_2 is a crossed module we see that kernel (∂_2) is abelian and a π-module. Moreover, the boundaries ∂_n, $n\geqq 4$, and $\partial_3:\pi_3^A(U^3,U^2)\to$ kernel (∂_2) are homomorphisms between π-modules, see (II.7.14), the action is induced by the action of $\pi_1^A(U^1)$. These properties show that $C_*^A(U)$ is a well defined crossed chain complex, compare the definition in (VI.§ 1) below.

We obtain for $n\geqq 2$ the homology of the crossed chain complex

$$H_n^A(U)=H_nC_*^A(U)=\ker\partial_n/\mathrm{im}\,\partial_{n+1}, \tag{3}$$

which is a π-module. Clearly, we have the inclusion $H_2^A(U)\subset E_2^{-1,0}(A,U)$, and we have the equation of $\pi_1^A(U^1)$-modules

$$H_n^A(U)=E_2^{-n+1,0}(A,U)\quad\text{for } n\geqq 3. \tag{4}$$

In fact $E_r^{s,t}$ is a $\pi_1^A(U^1)$-module for $s\geqq 1$ and $t-s=q\geqq 2$ and all differentials are $\pi_1^A(U^1)$-equivariant; (the action of $\pi_1^A(U^1)$ is trivial if A is a suspension, see (II.6.14)).

(10.7) ***Whitehead's certain exact sequence.*** Assume all terms of the spectral sequence are trivial below the diagonal C_* in (10.5)(2), that is, $\pi_q^A(U^n,U^{n-1})=0$ for $q<n$. Then we get for $q\geqq 1$

$$\pi_q^A(U)=\lim\pi_q^A(U)=\pi_q^A(U^{q+1}) \tag{1}$$

and $\pi_q^A(U^q) \twoheadrightarrow \pi_q^A(U)$ is surjective. In particular, the sequence

$$\pi_2^A(U^2, U^1) \xrightarrow{\partial_2} \pi_1^A(U^1) \to \pi_1^A(U) \to 0 \tag{2}$$

is exact and hence kernel (∂_2) is a $\pi_1^A(U)$-module. Moreover, for $n \geqq 2$ also $H_n^A(U)$ is a $\pi_1^A(U)$-module; (the action is trivial if A is a suspension).

Next we define for $q \geqq 2$

$$\Gamma_q^A(U) = \text{image } (\pi_q^A(U^{q-1}) \to \pi_q^A(U^q)). \tag{3}$$

This again is a $\pi_1^A(U)$-module. Now one can form the **long exact sequence of $\pi_1^A(U)$-modules:**

$$\begin{aligned} \cdots \xrightarrow{j} H_4^A(U) \xrightarrow{b} \Gamma_3^A(U) \xrightarrow{i} \pi_3^A(U) \xrightarrow{j} H_3^A(U) \\ \xrightarrow{b} \Gamma_2^A(U) \xrightarrow{i} \pi_2^A(U) \xrightarrow{j} H_2^A(U) \to 0 \end{aligned} \tag{4}$$

Here i is induced by $\pi_q^A(U^q) \to \pi_q^A(U)$ and j is induced by the commutative diagram

$$\begin{array}{ccc} \pi_q^A(U) & \overset{j}{\dashrightarrow} & H_q^A(U) \\ \uparrow\!\!\uparrow & & \uparrow\!\!\uparrow \\ \pi_q^A(U^q) & \xrightarrow{\quad j \quad} & \ker \partial_q \subset \pi_q^A(U^q, U^{q-1}). \\ & {}_{\partial}\nwarrow \quad \nearrow_{\partial_{q+1}} & \\ & \pi_{q+1}^A(U^{q+1}, U^q) & \end{array} \tag{5}$$

We call j the **generalized Hurewicz homomorphism**. Moreover, we obtain the **secondary boundary operator** b by the diagram

$$\begin{array}{ccc} H_{q+1}^A(U) & \overset{b}{\dashrightarrow} & \Gamma_q^A(U) \\ \uparrow\!\!\uparrow & & \downarrow \\ \ker \partial_{q+1} \subset \pi_{q+1}^A(U^{q+1}, U^q) & \xrightarrow{\quad \partial \quad} & \pi_q^A(U^q), \\ & {}_{\partial_{q+1}}\searrow \quad \swarrow_{j} & \\ & \pi_q^A(U^q, U^{q-1}) & \end{array} \tag{6}$$

where kernel $(j) = \Gamma_q^A(U)$. One can check that the exact sequence (4) essentially is a special case of the first derived homotopy sequence in (10.3), $(r = 1)$. For this we use the identification (10.6) (4).

(10.8) ***Naturality.*** A based map $f: A' \to A$ in $\mathbf{C}$ and a map $g: U \to U'$ in $\mathbf{Fil(C)}_{cf}$

induce functions

$$f^*: E_r^{s,t}(A, U) \to E_r^{s,t}(A', U)$$

and

$$g_*: E_r^{s,t}(A, U) \to E_r^{s,t}(A, U')$$

respectively. These maps f^* and g_* are compatible with all the structure of the spectral sequences described above. This follows from the corresponding naturality of the sequences (10.1), see (II.§7).

(10.9) **Invariance.** *Let $f, f_1: A' \to A$ be based maps in* **C** *which are homotopic by a based homotopy then $f^* = f_1^*$. Moreover, if the maps $g, g': U \to U'$ in* **Fil(C)** *are k-homotopic (see (1.5)) then the induced maps g_*, g'_* coincide on $E_r^{s,t}$ for $r > k$.*

Proof of (10.9) The first part is clear. Now assume g, g' are k-homotopic. We consider the following commutative diagram in which all rows are exact sequences:

$$\begin{array}{ccccccccc}
 & & \pi_{i+1}U^{m+1} & \longrightarrow & \pi_{i+1}(U^{m+1}, U^{m-r}) & \longrightarrow & \pi_i U^{m-r} & \longrightarrow & \pi_i U^{m+1} \\
 & & \| & & \downarrow \bar{i}_r & & \downarrow i_r & & \| \\
\pi_{i+1}U^m & \xrightarrow[i]{} & \pi_{i+1}U^{m+1} & \xrightarrow[j]{} & \pi_{i+1}(U^{m+1}, U^m) & \xrightarrow[\partial]{} & \pi_i U^m & \longrightarrow & \pi_i U^{m+1} \\
\| & & \downarrow i_r & & \downarrow \bar{\bar{i}}_r & & \| & & \\
\pi_{i+1}U^m & \xrightarrow[i']{} & \pi_{i+1}U^{m+r+1} & \xrightarrow[j]{} & \pi_{i+1}(U^{m+r+1}, U^m) & \longrightarrow & \pi_i U^m & &
\end{array}$$

Let $x \in \partial^{-1}$ image(i_r) with $\partial x = i_r \bar{x}$. Then a diagram chase shows $x \in$image $(\bar{\bar{i}}_r)$. Let $x = \bar{\bar{i}}_r(\bar{x})$. The whole diagram is natural for g and for g'. Since g and g' are k-homotopic we see

$$\partial g_* x = g_* i_r \bar{x} = g'_* i_r \bar{x} = \partial g'_* x.$$

Hence there is y with $g'_* x = g_* x + jy$. Since g and g' are k-homotopic we know $\bar{i}_r \bar{\bar{i}}_r g_* = \bar{i}_r \bar{\bar{i}}_r g'_*$. This shows $j i_r(y) = 0$. We choose z with $i'(z) = i_r(y)$. Then we get

$$\begin{cases} g'_* x = g_* x + j(y - i(z)) & \text{where} \\ i_r(y - i(z)) = i_r(y) - i'(z) = 0. & \end{cases}$$

Hence $\{g'_* x\}$ and $\{g_* x\}$ coincide in $E_{r+1}^{s,t}(A, U')$. □

(10.10) ***Naturality with respect to functors.*** Let $\alpha: \mathbf{C} \to \mathbf{K}$ be a functor between cofibration categories which carries weak equivalences to weak equivalences and assume that α is based, $\alpha(*) = *$. For U in $\mathbf{Fil(C)}_{cf}$ we choose

$$RM(\alpha U) \quad \text{in} \quad \mathbf{Fil(K)}_{cf} \tag{1}$$

as in (II.3.11). Moreover, for A we choose the based object

$$* \rightarrowtail M\alpha A \to \alpha(A) \to \alpha(*) = *. \tag{2}$$

Then α induces natural functions

$$\alpha: E_r^{s,t}(A, U) \to E_r^{s,t}(M\alpha A, RM\alpha U) \tag{3}$$

which are compatible with all the structure of the spectral sequence described above. This follows from (II.7.21).

(10.11) ***Compatibility with suspension.*** We have the natural isomorphism

$$E_r^{s,t+1}(A, U) = E_r^{s,t}(\Sigma A, U)$$

for $t \geqq s \leqq 0$, $r \geqq 1$, $t - s \geqq 1$, and in this range all differentials of the spectral sequence coincide up to a sign. The equation above is not compatible with the action of $\pi_1^A(U^1)$; this group acts on the right-hand side, yet the corresponding action of $\pi_1^{\Sigma A}(U^1)$ on the left-hand side is trivial.

(10.12) ***Properties of the crossed chain complex.*** A based map $f: A' \to A$ in **C** and a map $g: U \to U'$ in $\mathbf{Fil(C)}_{cf}$ induce homomorphisms f^* and g_* on $C_*^A(U)$. Moreover we obtain induced homomorphisms

$$\left.\begin{aligned} f^*&: H_n^A(U) \to H_n^{A'}(U) \quad \text{and} \\ g_*&: H_n^A(U) \to H_n^A(U') \end{aligned}\right\} \tag{1}$$

which are compatible with the action of $\pi_1^A(U^1)$, that is $f^*(x^\xi) = (f^*x)^{f^*\xi}$, $g_*(x^\xi) = (g_*x)^{g_*\xi}$. If f is homotopic to f_1 by a based homotopy then $f^* = f_1^*$ on $C_*^A(U)$ and on $H_*^A(U)$. Moreover, if g is 1-homotopic to g' then we have $g_* = g'_*$ on $H_*^A(U)$. We have the following compatibility with suspension

$$\left.\begin{aligned} H_3^A(U) &\subset H_2^{\Sigma A}(s^{-1}U) \quad \text{and} \\ H_n^A(U) &= H_{n-1}^{\Sigma A}(s^{-1}U) \quad \text{for } n > 3 \end{aligned}\right\}. \tag{2}$$

Here we define the object $s^{-1}U$ in **Fil(C)** by $(s^{-1}U)^n = U^{n+1}$ for $n \geqq 0$. Finally we remark that a functor α as in (10.10) yields the binatural homomorphisms

$$\left.\begin{aligned} \alpha&: C_*^A(U) \to C_*^{M\alpha A}(RM\alpha U) \\ \alpha&: H_n^A(U) \to H_n^{M\alpha A}(RM\alpha U) \end{aligned}\right\} \tag{3}$$

which are compatible with the action of $\pi_1^A(U^1)$, that is $\alpha(x^\xi) = (\alpha x)^{\alpha(\xi)}$, see (II.7.21)

(10.13) ***Properties of Whitehead's certain exact sequence.*** Clearly, the sequence is natural with respect to a based map $f: A' \to A$ and with respect to a map $g: U \to U'$ in $\mathbf{Fil(C)}_{cf}$. We get the commutative diagram ($n \geq 2$)

$$\begin{array}{ccccccc} H_{n+1}^A(U) & \longrightarrow & \Gamma_n^A(U) & \longrightarrow & \pi_n^A(U) & \longrightarrow & H_n^A(U) \\ \downarrow g_* & & \downarrow g_* & & \downarrow g_* & & \downarrow g_* \\ H_{n+1}^A(U') & \longrightarrow & \Gamma_n^A(U') & \longrightarrow & \pi_n^A(U') & \longrightarrow & H_n^A(U') \end{array} \tag{1}$$

If g is 1-homotopic to g' then all homorphisms g_* in (1) satisfy $g_* = g'_*$. Hence Whitehead's exact sequence is an invariant of the '1-homotopy type' of U. We obtain a similar diagram for f^*. Clearly g_* and f^* are compatible with the action of $\pi_1^A(U)$, that is $g_*(x^\xi) = (g_* x)^{g_* \xi}$ and $f^*(x^\xi) = (f^* x)^{f^* \xi}$. We have the following compatibility with suspension where $s^{-1}U$ is defined as in (10.12).

$$\begin{array}{ccccccccc} \cdots & \longrightarrow & H_4^A(U) & \longrightarrow & \Gamma_3^A(U) & \longrightarrow & \pi_3^A(U) & \longrightarrow & H_3^A(U) \\ & & \| & & \| & & \| & & \cap \\ \cdots & \longrightarrow & H_3^{\Sigma A}(s^{-1}U) & \longrightarrow & \Gamma_2^{\Sigma A}(s^{-1}U) & \longrightarrow & \pi_2^{\Sigma A}(s^{-1}U) & \longrightarrow & H_2^{\Sigma A}(s^{-1}U) \end{array} \tag{2}$$

Here the bottom row is defined since the assumption in (10.7) implies that the spectral sequence $E_r^{s,t}(\Sigma A, s^{-1}U)$ is trivial below the diagonal. Diagram (2) commutes up to a sign. The isomorphisms in (2) are not compatible with the action of $\pi_1^A(U)$, see (10.11). Finally a functor α as in (10.10) yields the commutative diagram

$$\begin{array}{ccccccc} H_{n+1}^A(u) & \longrightarrow & \Gamma_n^A(u) & \longrightarrow & \pi_n^A(u) & \longrightarrow & H_n^A(u) \\ \downarrow \alpha & & \downarrow \alpha & & \downarrow \alpha & & \downarrow \alpha \\ H_{n+1}^{M\alpha a}(U_\alpha) & \longrightarrow & \Gamma_n^{M\alpha A}(U_\alpha) & \longrightarrow & \pi_n^{M\alpha A}(U_\alpha) & \longrightarrow & H_n^{M\alpha A}(U_\alpha) \end{array} \tag{3}$$

where $U_\alpha = RM\alpha(U)$. Here we assume that the spectral sequence $E_r^{s,t}(M\alpha A, U_\alpha)$ is trivial below the diagonal so that the bottom row of (3) is defined by (10.7). All homomorphisms α in (3) are natural in A and U and they are compatible with the action of $\pi_1^A(U)$, that is $\alpha(x^\xi) = (\alpha x)^{\alpha \xi}$.

Remark: J.H.C. Whitehead (1950) introduced the certain exact sequence in the cofibration category **Top*** for a CW-complex U with $U^0 = *$. We study this in more detail in the next section. It is an interesting fact that this exact sequence is available in all cofibration categories for pairs (A, U) which satisfy the assumption in (10.7). For example, in the cofibration category of chain algebras we have this sequence as well (we will describe this in detail elsewhere, compare also (IX.§ 1).

§ 11 The relative homotopy spectral sequence for CW-complexes

Let $\mathbf{C} = \mathbf{Top}^*$ be the cofibration category of topological spaces with basepoint. In this category $*$ is the final and the initial object. Therefore all maps between cofibrant objects are based maps. We apply the results of § 10 to a relative CW-complex $U = (U, D) = \{U^n\}$ with $U^0 = D$. Here D is a path connected CW-space with $* \in D$. We use the based object $A = S^0$ in $\mathbf{Top}^*$ given by the 0-sphere.

We first consider the **crossed chain complex** $C_*^{S^0}(U, D)$ in (10.6) given by

$$(11.1) \qquad \cdots \to \pi_3(U^3, U^2) \to \pi_2(U^2, U^1) \xrightarrow{\partial} \pi_1(U^1).$$

As usual we set $\pi_k^{S^0} = \pi_k$. We will use this crossed chain complex in Chapter VI. It is a result of J.H.C. Whitehead, that ∂ in (11.1) is a **free** crossed module, compare (VI.1.11). The projection $p: \hat{U} \to U$ of the universal covering, see (5.4), induces the isomorphism ($n \geqq 3$)

$$(11.2) \qquad \begin{aligned} C_n^{S^0}(U, D) &= \pi_n^{S^0}(U^n, U^{n-1}) \\ &\cong \pi_n^{S^0}(\hat{U}^n, \hat{U}^{n-1}) \\ &= H_n(\hat{U}^n, \hat{U}^{n-1}) \\ &= \hat{C}_n(U, D). \end{aligned}$$

Compare (5.6)(1). This shows for $\hat{D} = p^{-1}(D)$

$$(11.3) \qquad \begin{cases} H_n^{S^0}(U, D) = H_n(\hat{U}, \hat{D}) & \text{for } n \geqq 4, \text{ and} \\ H_3^{S^0}(U, D) \subset H_3(\hat{U}, \hat{D}). \end{cases}$$

Here we have $H_k(\hat{U}, \hat{D}) = H_k\hat{C}(U, D) = \hat{H}_k(U, D; \mathbb{Z}[\pi])$, $\pi = \pi_1(U)$. In case $D = *$ is a point we get

$$(11.4) \qquad H_n^{S^0}(U, *) = H_n(\hat{U}) \quad \text{for } n \geqq 2, \quad (D = U^0 = *).$$

The cellular approximation theorem shows easely that the assumption in (10.7) is satisfied, namely

$$(11.5) \qquad \pi_q(U^n, U^{n-1}) = 0 \quad \text{for } q < n.$$

For the relative CW-complex $U = (U, D)$ with relative skeleta $D \subset U^n$ we define for $n \geqq 2$

$$(11.6) \qquad \Gamma_n(U, D) = \Gamma_n^{S^0}\{U^n\} = \text{image } (\pi_n U^{n-1} \to \pi_n U^n).$$

By (11.5) and (10.7) (4) we get **Whitehead's exact sequence** of $\pi_1(U)$-modules

$$(11.7) \qquad \cdots \xrightarrow{j} H_4(\hat{U}, \hat{D}) \xrightarrow{b} \Gamma_3(U, D) \xrightarrow{i} \pi_3 U \xrightarrow{j} H_3^{S^0}(U, D)$$

$$\to \Gamma_2(U, D) \to \pi_2 U \to H_2^{S^0}(U, D) \to 0$$

Here the homomorphism j for $n \geqq 4$ is the composite map

$$j: \pi_n(U) \underset{p_*}{\cong} \pi_n(\hat{U}) \xrightarrow[h]{} H_n(\hat{U}) \xrightarrow[j]{} H_n(\hat{U}, \hat{D}),$$

where h is the classical Hurewicz homomorphism. The exact sequence (11.7) is an invariant of the homotopy type under D of the pair (U, D). This follows again from the cellular approximation theorem.

In case $D = *$ is a point the sequence (11.7) gives us the *classical sequence of* J.H.C. Whitehead (1950), $\Gamma_2 U = \Gamma_n(U, *)$,

$$\text{(11.8)} \qquad \xrightarrow{j} H_4\hat{U} \xrightarrow{b} \Gamma_3 U \xrightarrow{i} \pi_3 U \xrightarrow{j} H_3\hat{U} \to 0 \to \pi_2 U \cong H_2\hat{U} \to 0.$$

Here we have $\Gamma_2 U = 0$ since $\pi_2(U^1) = 0$ where U^1 is a one point union of 1-spheres and hence a $K(\pi, 1)$. This gives us the classical result of Hurewicz that for a path connected space we have $\pi_2 U \cong H_2\hat{U}$ and $h: \pi_3 U \to H_3\hat{U}$ is surjective. More generally we have the classical Hurewicz

(11.9) **Theorem.** *Assume $\pi_k(U) = 0$ for $1 < k < n$ so that $\hat{U}$ is $(k-1)$-connected. Then we have $\Gamma_k(U) = 0$ for $1 < k \leqq n$ and therefore $j: \pi_n(U) \cong H_n(\hat{U})$ is an isomorphism and $j: \pi_{n+1}(U) \twoheadrightarrow H_{n+1}(\hat{U})$ is surjective.*

We point out that the isomorphism (11.4) for $n = 2$ is a consequence of the isomorphism in (11.8).

Finally, we consider the relative homotopy spectral sequence $E_r^{s,t}(S^0, U)$ where $U = (U, D)$ is a relative CW-complex with $U^0 = D$ path connected. By (11.5) this sequence is trivial below the diagonal C^* in (10.5) (2). Therefore the spectral sequence converges to $\pi_*(D)$, compare (10.5). For $r \geqq 2$ the spectral sequence is an invariant of the homotopy type of U under D. For $D = *$ this is exactly the '*homotopy exact couple*' of Massey (1952), compare also Hu (1959), p. 252.

(11.10) ***Remark.*** For a finite relative CW-complex (U, D) with $U^N = U$ the spectral sequence $E_r^{s,t}(S^0, U)$ can be considered as being a special case of the spectral sequence of Bousfield–Kan (2.14) since we can choose a commutative diagram in **Top***

$$\begin{array}{ccccccc} U^0 & \subset & U^1 & \subset & \cdots & \subset & U^N \\ \downarrow\simeq & & \downarrow\simeq & & & & \downarrow\simeq \\ U_N & \twoheadrightarrow & U_{N-1} & \twoheadrightarrow & \cdots & \twoheadrightarrow & U_0 \end{array}$$

where the bottom row is a tower of fibrations in **Top***. Then the exact sequences in (2.3) correspond exactly to the exact sequences in (10.1) so that we get

$$E_r^{s,t}(S^0, \{U^n\}) = E_r^{N+s,t}(\{U_n\})$$

for $-N \leqq s \leqq 0$.

IV

Extensions, coverings, and cohomology groups of a category

Here we describe general notions having to do with categories and functors. These notions will be used frequently in the following chapters. We introduce a detecting functor by combining the sufficiency and the realizability conditions used by J.H.C. Whitehead. Examples in Chapter V motivate the notion of an action of abelian groups on a category. We describe various basic properties; linear extensions of categories are special cases of such actions. We classify the equivalence classes of linear extensions by the second cohomology of a small category. This generalizes the Hochschild–Mitchell cohomology. Moreover, we show that the first cohomology classifies linear coverings. The linear extensions and linear coverings lead to the notion of an exact sequence for functors.

§ 1 Detecting functors

Let $\mathbf{K}$ be a category. For objects A, B in $\mathbf{K}$ we denote by $\mathbf{K}(A, B)$ the set of morphisms $A \to B$ of $\mathbf{K}$. $Ob(\mathbf{K})$ denotes the class of objects in $\mathbf{K}$. We write $A \in \mathbf{K}$ or $A \in Ob(\mathbf{K})$ if A is an object in $\mathbf{K}$.

Assume for all objects A, B in $\mathbf{K}$ we have an equivalence relation $\sim$ on $\mathbf{K}(A, B)$. Then $\sim$ is said to be a **natural equivalence relation** on $\mathbf{K}$ if for morphisms

$$A \underset{g}{\overset{f}{\rightrightarrows}} B \underset{b}{\overset{a}{\rightrightarrows}} C$$

in $\mathbf{K}$ we have $(f \sim g \text{ and } a \sim b) \Rightarrow af \sim bg$. In this case we obtain the **quotient category** $\mathbf{K}/\sim$ which has the same objects as $\mathbf{K}$ and for which the set of morphisms is

(1.1) $$(\mathbf{K}/\sim)(A, B) = \mathbf{K}(A, B)/\sim .$$

$\{f\}$ denotes the equivalence class of f. Composition in $\mathbf{K}/\sim$ is defined by $\{a\}\{f\} = \{af\}$. Clearly, the identity of A in $\mathbf{K}/\sim$ is $\{1_A\}$. Each functor $p:\mathbf{A}\to\mathbf{B}$ induces the natural equivalence relation $\sim$ on $\mathbf{A}$ with

(1.2) $$a \sim b \Leftrightarrow pa = pb \quad \text{for } a, b \in \mathbf{A}(A, B).$$

We call the corresponding quotient category $p\mathbf{A} = \mathbf{A}/\sim$ the **image category** of p. The functor p induces the faithful functor $i:p\mathbf{A}\to\mathbf{B}$, which is an inclusion of categories if p is injective on classes of objects. We say p is a **quotient functor** if i is an isomorphism of categories. Following J.H.C. Whitehead we define the following conditions on a functor $p:\mathbf{A}\to\mathbf{B}$:

(1.3) (a) **Sufficiency:** For objects A, A' in $\mathbf{A}$ a morphism $\alpha:A\to A'$ is an equivalence if and only if $p\alpha:pA\to pA'$ is an equivalence in $\mathbf{B}$.

(b) **Realizability:** $p\mathbf{A}\to\mathbf{B}$ is an equivalence of categories. This is equivalent to the following two conditions

(b1) For each object B in $\mathbf{B}$ there is an object A in $\mathbf{A}$ such that pA and B are equivalent in $\mathbf{B}$.

(b2) For objects A, A' in $\mathbf{A}$ and for a morphism $\beta:pA\to pA'$ in $\mathbf{B}$ there is a morphism $\alpha:A\to A'$ with $p\alpha = \beta$.

Compare § 14, theorem 17 in J.H.C. Whitehead (1950). For example, the Whitehead theorem (I.5.9) yields the following well known properties of homotopy groups $\pi_i(X)$ and of homology groups $H_i(X) = H_i(X;\mathbb{Z})$ respectively.

1.4 ***Example***

(A) The functor $\pi_*:\mathbf{Top}_*/\simeq\;\to\mathbf{Gr}_*$ which carries a pointed space X to the graded group $\{\pi_i X, i \geqq 1\}$ satisfies the sufficiency condition on the full subcategory of $\mathbf{Top}/\simeq$ consisting of path connected well pointed CW-spaces.

(B) The functor $H_*:\mathbf{Top}/\simeq\;\to\mathbf{Ab}_*$ which carries a space X to the graded abelian group $\{H_i X, i \geqq 0\}$ satisfies the sufficiency condition on the full subcategory of $\mathbf{Top}/\simeq$ consisting of simply connected CW-spaces.

Clearly the functors in (A) and (B) do not satisfy the realizability condition.

(1.5) ***Definition.*** A functor $p:\mathbf{A}\to\mathbf{B}$ is a **detecting** functor if p satisfies both the sufficiency and the realizability conditions. ||

The problem in many of J.H.C. Whitehead's papers is the construction of such detecting functors for homotopy categories of CW-complexes since this yields the *classification of homotopy types* by the following obvious properties of a detecting functor.

A detecting functor $p:\mathbf{A}\to\mathbf{B}$ induces a 1–1 correspondence between equivalence classes of objects in $\mathbf{A}$ and equivalence classes of objects in $\mathbf{B}$. On morphism sets a detecting functor p induces a surjection $p:\mathbf{A}(A,A')\twoheadrightarrow \mathbf{B}(pA,pA')$. The number of morphisms in $\mathbf{B}(pA,pA')$ can be much smaller than the number of morphisms in $\mathbf{A}(A,A')$. Moreover, a detecting functor induces the surjective homomorphism

$$p:E_{\mathbf{A}}(A)\twoheadrightarrow E_{\mathbf{B}}(pA) \tag{1.6}$$

of groups. Here $E_{\mathbf{A}}(A)$ is the **group of equivalences** $A\xrightarrow{\simeq}A$ in $\mathbf{A}$. We also call $E_{\mathbf{A}}(A)=\mathrm{Aut}_{\mathbf{A}}(\mathbf{A})$ the group of automorphisms of A. In fact, $E_{\mathbf{A}}(A)$ is the subset $p^{-1}E_{\mathbf{B}}(pA)$ in $\mathbf{A}(A,A)$ by sufficiency. An equivalence of categories is a detecting functor. The composition of detecting functors $\mathbf{A}\to\mathbf{B}\to\mathbf{C}$ is again a detecting functor. On the other hand:

(1.7) **Lemma.** *If the composition qp: $\mathbf{A}\to\mathbf{B}\to\mathbf{C}$ is a detecting functor where p is a quotient functor, then q and p are detecting functors.*

§2 Group actions on categories

We say that D is a (natural) action of (abelian) groups on the category $\mathbf{C}$ if for all objects A,B in $\mathbf{C}$ we have an abelian group $D(A,B)$ and a group action

$$\mathbf{C}(A,B)\times D(A,B)\xrightarrow{+}\mathbf{C}(A,B),\ (f,\alpha)\longmapsto f+\alpha, \tag{2.1}$$

such that for $(g,\beta)\in\mathbf{C}(B,C)\times D(B,C)$ there exists $\delta\in D(A,C)$ with $(g+\beta)(f+\alpha)=(gf)+\delta$. This equation is the condition of '*naturality*' for the action. An action D on $\mathbf{C}$ yields the quotient functor $p:\mathbf{C}\to\mathbf{C}/D=\mathbf{C}/\sim$ where for $f,f'\in\mathbf{C}(A,B)$ the equivalence relation $\sim$ (induced by D) is

$$f\sim f'\Leftrightarrow\exists\alpha\in D(A,B)\quad\text{with } f'=f+\alpha. \tag{2.2}$$

Clearly, this is a natural equivalence relation. The morphism set in $\mathbf{C}/D$ is the set of orbits $(\mathbf{C}/D)(A,B)=\mathbf{C}(A,B)/\sim\;=\mathbf{C}(A,B)/D(A,B)$. The isotropy groups of the action are $I_f=\{\alpha\in D(A,B):f+\alpha=f\}$ which depend only on $p(f)$ in $\mathbf{C}/D$. The quotient group $(D/I)_{p(f)}=D(A,B)/I_f$ acts transitively and effectively on the subset $p^{-1}pf$ of $\mathbf{C}(A,B)$. This leads to the following slightly more general notation.

We say that D is a **(natural) action on the functor** $p:\mathbf{C}\to\mathbf{B}$ if for all objects A,B in $\mathbf{C}$ and for all morphisms $\bar{f}\in\mathbf{B}(pA,pB)$ with $\phi\neq p^{-1}(\bar{f})\subset\mathbf{C}(A,B)$ we have an abelian group $d_{\bar{f}}=D(\bar{f})$ together with a transitive action

$$p^{-1}(\bar{f})\times D(\bar{f})\xrightarrow{+}p^{-1}(\bar{f}),\ (f,\alpha)\longmapsto f+\alpha. \tag{2.3}$$

We call $\mathbf{C}$ an **extension of B by the action** D if p satisfies in addition the realizability condition (1.3)(b), (that is $p\mathbf{C} \xrightarrow{\sim} \mathbf{B}$ is an equivalence of categories). Let $I(\bar{f}) = I_f = \{\alpha \in D(\bar{f}) | f + \alpha = f\}$ be the **isotropy group** of the action (2.3). We write $(D/I)_{\bar{f}} = D(\bar{f})/I(\bar{f})$. This group acts transitively and effectively on $p^{-1}(\bar{f})$. We say that the action D is **effective** if $f + \alpha = f$ implies $\alpha = 0$ for all f in $\mathbf{C}$, $\alpha \in D_{p(f)}$. Each action D yields an effective action, D/I, by dividing out the isotropy groups as above. Clearly, D in (2.1) give us an action on the functor $p: \mathbf{C} \to \mathbf{C}/D$ by setting $D_{\bar{f}} = D(A, B)$.

(2.4) ***Definition.*** For an effective action D on $p: \mathbf{C} \to \mathbf{B}$ the formula

$$(g + \beta)(f + \alpha) = (gf) + \Phi_{g,f}(\beta, \alpha) \quad \text{in } \mathbf{C} \tag{1}$$

defines the function

$$\Phi_{g,f}: D_{p(g)} \times D_{p(f)} \longrightarrow D_{p(gf)}. \tag{2}$$

We call (1) and the function Φ the **distributivity law** of the action D. The function $\Phi_{g,f}$ depends on g and f while the group $D_g = D_{p(g)}$ depends only on the morphism $p(g)$ in $\mathbf{B}$. For convenience we often write D_g instead of $D_{p(g)}$. The distributivity law gives us the **induced functions**

$$f^*: D_g \longrightarrow D_{gf}, f^*\beta = \Phi_{g,f}(\beta, 0) \Leftrightarrow (g + \beta)f = gf + f^*\beta,$$

and

$$g_*: D_f \longrightarrow D_{gf}, \; g_*\alpha = \Phi_{g,f}(0, \alpha) \Leftrightarrow g(f + \alpha) = gf + g_*\alpha. \tag{3}$$

||

One easily verifies the following equations

(a) $id_* = id, id^* = id$ for the identity id,
(b) $(hg)_* = h_* g_*: D_f \to D_{hgf}$,
(c) $(gf)^* = f^* g^*: D_h \to D_{hgf}$,
(d) $h_* f^* = f^* h_*: D_g \to D_{hgf}$.

For example (d) follows by $h(g + \beta)f = h(gf + f^*\beta) = hgf + h_* f^*\beta = (hg + h_*\beta)f = hgf + f^* h_*\beta$. The equations (a) $\cdots$ (d) above show that $f \longmapsto D_f$ is a functor on the following category $F(\mathbf{C})$.

(2.5) ***Definition.*** Let $\mathbf{C}$ be a category. Then the **category of factorizations in C**, denoted by $F(\mathbf{C})$, is given as follows. Objects in $F(\mathbf{C})$ are morphisms in $\mathbf{C}$. Morphisms $f \to g$ in $F(\mathbf{C})$ are pairs (α, β) for which

$$\begin{array}{ccc} A & \xrightarrow{\alpha} & A' \\ {\scriptstyle f}\uparrow & & \uparrow{\scriptstyle g} \\ B & \xleftarrow[\beta]{} & B' \end{array}$$

commutes in $\mathbf{C}$. Hence $\alpha f \beta = g$ is a factorization of g. Composition is defined by $(\alpha', \beta')(\alpha, \beta) = (\alpha'\alpha, \beta\beta')$.

||

The effective action D on $p:\mathbf{C}\to\mathbf{B}$ in (2.3) gives us the functor

(2.6) $$D:F(\mathbf{C})\to\mathbf{Set},$$

as well denoted by D, which carries the object f of $F(\mathbf{C})$ to the abelian group D_f and which carries $(\alpha,\beta):f\to g$ in $F(\mathbf{C})$ to the induced function

$$D(\alpha,\beta)=\alpha_*\beta^*:D_f\to D_{\alpha f\beta}=D_g.$$

Clearly, $D(\alpha,1)=\alpha_*, D(1,\beta)=\beta^*$.

(2.7) ***Definition.*** We define the **mixed term** Δ of an effective action D on $p:\mathbf{C}\to\mathbf{B}$ by the formula

(a) $$\Delta_{g,f}(\beta,\alpha)=\Phi_{g,f}(\beta,\alpha)-f^*\beta-g_*\alpha,$$

or equivalently by the **distributivity law**:

(b) $$(g+\beta)(f+\alpha)=gf+g_*\alpha+f^*\beta+\Delta_{g,f}(\beta,\alpha).$$

Here $\Delta_{g,f}$ is a function $\Delta_{g,f}:D_{p(g)}\times D_{p(f)}\to D_{p(gf)}$ with $\Delta_{g,f}(0,\alpha)=\Delta_{g,f}(\beta,0)=0$. The action D has a **left** distributivity law if the function $\Delta_{g,f}$ depends only on (g,pf), that is, if $\Delta_{g,f}=\Delta_{g,f_0}$ for all f_0 with $pf_0=pf$. The action D has a **right** distributivity law if $\Delta_{g,f}$ depends only on (pg,f). ||

We mainly consider actions which have nice linear properties:

(2.8) ***Definition.*** An effective action D on $p:\mathbf{C}\to\mathbf{B}$ is **quadratic** if all induced functions ($f\in\mathbf{C}(A,B)$, $g\in\mathbf{C}(B,C)$) $f^*:D_g\to D_{gf}, g_*:D_f\to D_{gf}$ are homomorphisms of abelian groups and if D has a left and right distributivity law such that the mixed term $\Delta:D_g\times D_f\to D_{gf}$ is bilinear. In this case we write the mixed term in the form

$$\Delta(\beta,\alpha)=\beta\odot\alpha.$$

We say that a quadratic action with a trivial mixed term, $\Delta=0$, is a **linear** action. More generally, we call an action D on $p:\mathbf{C}\to\mathbf{B}$ quadratic (resp. linear) if the associated effective action D/I is quadratic (resp. linear), see (2.3). ||

For a quadratic action the functor D in (2.6) gives us a functor $D:F(\mathbf{C})\to\mathbf{Ab}$ into the category of abelian groups, we call such a functor a **natural system of abelian groups on C**. Moreover, we get the following properties of quadratic actions. Let

$$A\underset{f}{\longrightarrow}B\underset{g}{\longrightarrow}C\underset{h}{\longrightarrow}D$$

be morphisms in $\mathbf{C}$. Moreover, let $\alpha\in D_f$, $\beta\in D_g$ and $\gamma\in D_h$.

(2.9) **Proposition:**

(*a*) $(g+\beta)(f+\alpha)=gf+g_*\alpha+f^*\beta+\beta\odot\alpha$,

(*b*) $(f+\alpha)^*\beta=f^*\beta+\beta\odot\alpha$,

(c) $(g+\beta)_*\alpha = g_*\alpha + \beta \odot \alpha$,
(d) $h_*(\beta \odot \alpha) = (h_*\beta) \odot \alpha$,
(e) $f^*(\gamma \odot \beta) = \gamma \odot (f^*\beta)$,
(f) $(g^*\gamma) \odot \alpha = \gamma \odot (g_*\alpha)$,
(g) $\gamma \odot (\beta \odot \alpha) = (\gamma \odot \beta) \odot \alpha$.

Thus the mixed term $\odot$ *is an associative bilinear pairing.*

Proof of (2.9). (a), (b) and (c) correspond to (2.4). We derive from (2.4)(3)

$$\begin{array}{c} h_*(f+\alpha)^*\beta = h_*(f^*\beta + \beta \odot \alpha), \\ \| \\ (f+\alpha)^*h_*\beta = f^*h_*\beta + (h_*\beta) \odot \alpha. \end{array}$$

Thus (d) follows from linearity of h_*. Similarly we obtain (e) by

$$\begin{array}{c} (h+\gamma)_*f^*\beta = h_*f^*\beta + \gamma \odot f^*\beta, \\ \| \\ f^*(h+\gamma)_*\beta = f^*(h_*\beta + \gamma \odot \beta). \end{array}$$

We derive (f) from

$$\begin{array}{l} (g(f+\alpha))^*\gamma = (f+\alpha)^*g^*\gamma = f^*g^*\gamma + (g^*\gamma) \odot \alpha, \\ \quad \| \\ (gf + g_*\alpha)^*\gamma = (gf)^*\gamma + \gamma \odot g_*\gamma. \end{array}$$

Moreover, we obtain (g) by:

$$(h+\gamma)(g+\beta) = hg + h_*\beta + g^*\gamma + \gamma \odot \beta.$$

Therefore

$$\begin{array}{l} (h+\gamma)_*(g+\beta)_*\alpha = (hg)_*\gamma + (h_*\beta + g^*\gamma + \gamma \odot \beta) \odot \alpha, \\ \quad \| \\ (h+\gamma)_*(g_*\alpha + \beta \odot \alpha) = (h+\gamma)_*g_*\alpha + (h+\gamma)_*\beta \odot \alpha \\ \quad = h_*g_*\alpha + \gamma \odot g_*\alpha + h_*(\beta \odot \alpha) + \gamma \odot (\beta \odot \alpha). \end{array}$$

Thus from (d) and (f) we derive (g) since $\odot$ is bilinear. □

Let D be an effective action on $p:\mathbf{C} \to \mathbf{B}$. For an object A in $\mathbf{C}$ and its identity 1_A we write

$$(2.10) \qquad \begin{cases} D_A = D_{1_A} \\ \Delta_A = \Delta_{1_A,1_A}: D_A \times D_A \longrightarrow D_A, \text{ with} \\ (1_A + \beta)(1_A + \alpha) = 1_A + \beta + \alpha + \Delta_A(\beta, \alpha). \end{cases}$$

This yields the exact sequence of groups, see (1.6),

$$0 \longrightarrow (D_A, \circ) \xrightarrow{1^+} E_{\mathbf{C}}(A) \xrightarrow{p} E_{\mathbf{B}}(pA),$$

where the group multiplication $\circ$ on D_A is given by the formula

$$\alpha \circ \beta = \alpha + \beta + \Delta_A(\alpha, \beta), \tag{2}$$

and where the injective homomorphism 1^+ is defined by $1^+(\alpha) = 1_A + \alpha$. The zero element $0 \in D_A$ is also the neutral element of the group $(D_A, \circ)$. If D is a quadratic action we see by (2.9) that D_A is equipped with an associative $\mathbb{Z}$-bilinear multiplication $\odot: D_A \times D_A \to D_A$ with $\Delta_A(\alpha, \beta) = \alpha \odot \beta$.

(2.11) **Proposition.** *Assume $p: \mathbf{C} \to \mathbf{B}$ satisfies the realizability condition and assume D is an action on p for which the induced functions are homomorphisms. Then p is a detecting functor if and only if the mixed term satisfies the condition that for all objects A in $\mathbf{C}$ and for all $\beta \in D_A$ there exists $\alpha \in D_A$ such that*

$$\alpha + \beta + \Delta_A(\alpha, \beta) = 0 = \alpha + \beta + \Delta_A(\beta, \alpha). \tag{$*$}$$

(2.12) **Corollary.** *Assume $p: \mathbf{C} \to \mathbf{B}$ satisfies the realizability condition and assume D is a quadratic action on p such that $\odot$ on D_A is nilpotent for all objects A. Then p is a detecting functor.*

Here we say $\odot$ is **nilpotent** on D_A if for all $\beta \in D_A$ there is an n such that the n-fold product $\beta^{\odot n} = \beta \odot \cdots \odot \beta = 0$ is trivial.

Proof of (2.12). p is a detecting functor iff for $\beta \in D_A$ there is $\alpha \in D_A$ with

$$\alpha + \beta + \alpha \odot \beta = 0 = \alpha + \beta + \beta \odot \alpha$$

We take $\alpha = -\beta + \beta^{\odot 2} - \cdots + (-1)^n \beta^{\odot n} + \cdots$ □

Proof of (2.11). We may assume that p is the identity on objects. If p is a detecting functor we know that $1 + \beta \in \mathbf{C}(A, A)$ is an equivalence for all $\beta \in D_A$. Thus there exists $\alpha \in D_A$ with

$$\left.\begin{aligned} (1+\alpha)(1+\beta) &= 1 + \alpha + \beta + \Delta_A(\alpha, \beta) = 1, \\ (1+\beta(1+\alpha) &= 1 + \beta + \alpha + \Delta_A(\beta, \alpha) = 1. \end{aligned}\right\} \tag{1}$$

Therefore ($*$) is satisfied. Now assume ($*$). Then (1) shows that $1 + \beta$ is an equivalence and thus $(1+\beta)_*$ is an isomorphism. We have to prove that each 'realization' of an equivalence in $\mathbf{B}$ is an equivalence in $\mathbf{C}$. Let $\bar{f} \in \mathbf{B}(A, B)$, $\bar{g} \in \mathbf{B}(B, A)$ be equivalences with

$$\bar{f}\bar{g} = 1, \quad \bar{g}\bar{f} = 1, \tag{2}$$

and let $pf = \bar{f}, pg = \bar{g}$. We know by (2) that there are $\beta \in D_B$ and $\alpha \in D_A$ such that

$$fg = 1 + \beta, \quad gf = 1 + \alpha. \tag{3}$$

We have to show that there exists $\delta \in D_g$ with

$$f(g + \delta) = 1, \quad (g + \delta)f = 1. \tag{4}$$

This is equivalent to

$$f_*\delta = -\beta, \quad f^*\delta = -\alpha. \tag{5}$$

Now (3) shows that $f_*g_* = (1+\beta)_*$ and $g_*f_* = (1+\alpha)_*$ are isomorphisms. Therefore $f_*: D_g \to D_B$ is surjective and $f_*: D_A \to D_f$ is injective. Now let $\delta \in (f_*)^{-1}(-\beta)$. Then δ satisfies (5) since we prove $f^*(\delta) = -\alpha$. In fact, by (3) we get $fgf = f + f^*\beta = f + f_*\alpha$, and thus $f^*\beta = f_*\alpha$. Now (5) holds since f_* is injective and since

$$f_*f^*(\delta) = f^*f_*(\delta) = f^*(-\beta) = f_*(-\alpha).$$

Here we use the assumption that f^*, f_* are homomorphisms. □

(2.13) ***Example.*** Any **ring** R yields a quadratic action on the functor $\mathbf{R} \to *$ where $*$ is the trivial category and where $\mathbf{R}$ is the category with a single object $*$ given by the multiplication $R = \mathbf{R}(*, *)$. The action is given by addition in R.

(2.14) ***Example.*** Let **Gr** be the category of groups and group homomorphisms and let $\mathbf{Ab} \subset \mathbf{Gr}$ be the full subcategory of abelian groups. Then the projection functor

$$\mathrm{pr}: \mathbf{C} = \mathbf{Gr} \times \mathbf{Ab} \to \mathbf{Gr}$$

admits a quadratic action D as follows. For objects $A = (A_1, A_2)$ and $B = (B_1, B_2)$ in $\mathbf{C}$ the abelian group $D(A, B) = \mathrm{Hom}(A_2, B_2)$ acts on the morphism set

$$\mathbf{C}(A, B) = \mathrm{Hom}(A_1, B_1) \times \mathrm{Hom}(A_2, B_2),$$

by the formula $(f, g) + g' = (f, g + g')$.

§ 3 Linear extensions of categories

Let $\mathbf{C}$ be a category. Recall that a **natural system of abelian groups** D on $\mathbf{C}$ is a functor

(3.1) $$D: F(\mathbf{C}) \to \mathbf{Ab}, \quad D(f) = D_f,$$

from the category of factorization (2.5) to the category of abelian groups. For example a bifunctor $D: \mathbf{C}^{\mathrm{op}} \times \mathbf{C} \to \mathbf{Ab}$ yields **a** natural system by $D_f = D(A, B)$ for $f \in \mathbf{C}(A, B)$. Moreover, we write

$$D_f = D(A, \varphi, B) \tag{$*$}$$

provided a functor $\phi: \mathbf{C} \to \mathbf{Coef}$ is given such that $D_f = D_g$ for all $f, g \in \mathbf{C}(A, B)$ with $\phi f = \varphi = \phi g$. A functor $p: \mathbf{C} \to \mathbf{B}$ yields an obvious functor $F(p): F(\mathbf{C}) \to F(\mathbf{B})$. Therefore a natural system D on $\mathbf{B}$ gives us the natural system $p^*D = DF(p)$ on $\mathbf{C}$.

(3.2) ***Definition.*** Let D be a natural system on the category $\mathbf{B}$. We call a category $\mathbf{C}$ a **linear extension of B by** D, and we write

$$D+ \rightarrowtail \mathbf{C} \xrightarrow{p} \twoheadrightarrow \mathbf{B},$$

if the following properties are satisfied. The categories $\mathbf{C}$ and $\mathbf{B}$ have the same classes of objects and p is a full functor which is the identity on objects. For each morphism $f: A \to B$ in $\mathbf{B}$ the group D_f acts transitively and effectively on the subset $p^{-1}(f)$ of $\mathbf{C}(A, B)$; the action is denoted by $f_0 + \alpha$ for $f_0 \in p^{-1}(f)$ and $\alpha \in D_f$. Moreover, the **linear distributivity law**

$$(f_0 + \alpha)(g_0 + \beta) = f_0 g_0 + f_* \beta + g^* \alpha$$

is satisfied, $g_0 \in p^{-1}(g)$, $\beta \in D_g$, where the induced functions f_* and g^* are given by the natural system D.

We call $D+ \rightarrowtail \mathbf{C} \xrightarrow{p} \twoheadrightarrow \mathbf{B}$ a (**weak**) **linear extension** if $p\mathbf{C} \to \mathbf{B}$ is an equivalence of categories and if $D+ \rightarrowtail \mathbf{C} \to p\mathbf{C}$ is a linear extension as above. ||

Clearly, a linear action D on $p: \mathbf{C} \to \mathbf{K}$ as defined in (2.8) yields a linear extension

(3.3) $$(D/I)+ \to \mathbf{C} \to p\mathbf{C},$$

where $p\mathbf{C}$ is the image category of p defined in (1.2) and where D/I is the natural system on $p\mathbf{C}$ given as in (2.3). By (2.12) a linear extension as in (3.2) satisfies the sufficiency condition (1.3)(a).

(3.4) ***Definition.*** Let D be a natural system of abelian groups on $\mathbf{B}$. We say two linear extensions p, q of $\mathbf{B}$ by D are **equivalent** if there is a commutative diagram

$$\begin{array}{ccccc} D+ & \longrightarrow & \mathbf{C} & \xrightarrow{p} & \mathbf{B} \\ \downarrow{\scriptstyle 1} & & \downarrow{\scriptstyle \varepsilon} & & \downarrow{\scriptstyle 1} \\ D+ & \longrightarrow & \mathbf{C}' & \xrightarrow[q]{} & \mathbf{B} \end{array},$$

where ε is a D-equivariant isomorphism of categories. That is, ε induces for all $f \in \mathbf{B}(A, B)$ the bijection $\varepsilon: p^{-1}(f) \approx q^{-1}(f)$ with the property

$$\varepsilon(\bar{f} + \alpha) = \varepsilon(\bar{f}) + \alpha$$

for $p\bar{f} = f$ and $\alpha \in D_f$. ||

A linear extensions p is a *split extension* if there is a functor $s: \mathbf{B} \to \mathbf{C}$ for which $ps = 1$ is the identity functor on $\mathbf{B}$.

(3.5) **Proposition.** *Let* **B** *be a category and let* D *be a natural system of groups on* **B**. *Then there exists a split extension* $D+\longrightarrow \mathbf{B}\times D\xrightarrow{p}\mathbf{B}$ *of* **B** *by* D *(which we call the* **semi direct** *product of* **B** *and* D*) and two split extensions of* **B** *by* D *are equivalent.*

Proof. We define $\mathbf{K}=\mathbf{B}\times D$ as follows: Objects of **K** are the objects of **B**. The morphism sets of **K** are

$$\mathbf{K}(A,B)=\bigcup_{f\in\mathbf{B}(A,B)} f\times D_f,$$

and composition is defined by $(g,\beta)(f,\alpha)=(gf,g_*\alpha+f^*\beta)$.

It is easily seen that **K** is a category and that the projection $pr_1:\mathbf{K}\to\mathbf{B}$ is a split extension of **B** by D. We set $s(f)=(f,0)$. If $p:\mathbf{C}\to\mathbf{B}$ is a further split extension of **B** by D with splitting s we obtain the equivariant isomorphism $\mathbf{B}\times D\xrightarrow{\varepsilon}\mathbf{C}$ by $\varepsilon(A)=s(A)$ on objects and $\varepsilon(f,\alpha)=s(f)+\alpha$ on morphisms. □

Next we show that linear extensions of categories correspond exactly to extensions of groups:

(3.6) $$0\longrightarrow D\xrightarrow{i}E\xrightarrow{p}G\longrightarrow 0,$$

where D is a G-module with $x^g=i^{-1}(g_0^{-1}i(x)g_0)$ for $g_0\in p^{-1}(g)$. We call $h:G\to\mathrm{Aut}(D)$ with $h(g^{-1})(x)=x^g$ the **associated homomorphism** of the extension. Two extensions E,E' as in (3.6) are **equivalent** if there is an isomorphism $\varepsilon:E\cong E'$ of groups with $p\varepsilon=p$, $\varepsilon i=i$.

(3.7) ***Example.*** Each extension of groups (3.6) yields a linear extension of categories

$$\bar{D}+\longrightarrow\mathbf{E}\xrightarrow{p}\mathbf{G}.$$

Here **E** is the category with one object $*$ and with morphism set $\mathbf{E}(*,*)=E$. The functor p is given by p in (3.6). We define the action $\bar{D}+$ by $\bar{D}(g)=D$ for $g\in\mathbf{G}(*,*)=G$ and

$$g_0+x=g_0\cdot i(x)$$

for $g_0\in p^{-1}(g)$ and $x\in D$. Thus we get

$$\begin{aligned}(g_0+x)(f_0+y)&=g_0\cdot i(x)\cdot f_0\cdot i(y)\\&=g_0f_0\cdot(f_0^{-1}i(x)f_0)\cdot i(y)\\&=g_0f_0+x^f+y=g_0f_0+f^*(x)+g_*y.\end{aligned}$$

The induced maps are $f^*(x)=x^f$ and $g_*(y)=y$ for $x,y\in\bar{D}$. ||

On the other hand, a linear extension $D+\to \mathbf{C}\xrightarrow{p}\mathbf{B}$ of the category $\mathbf{B}$ by $\mathbf{D}$ gives us for each object A in $\mathbf{B}$ the extension of groups (see (2.10))

$$(3.8) \qquad 0\longrightarrow D_A \xrightarrow{1^+} E_{\mathbf{C}}(A)\xrightarrow{p} E_{\mathbf{B}}(pA)\longrightarrow 1,$$

where $E_{\mathbf{C}}(A)$ is the group of self-equivalences of A in $\mathbf{C}$. Here $1^+(\alpha)=1_A+\alpha$, $D_A=D_{1_A}$.

(3.9) **Proposition.** *(3.8) is an extension of groups where D_A is a right $E_{\mathbf{B}}(pA)$-module by $x^q=(g^{-1})_* g^*(x)=g^*(g^{-1})_*(x)$.*

Proof. p in (3.8) is surjective since p is a detecting functor. 1^+ is a homomorphism since we have

$$\begin{aligned} 1+(\alpha+\beta) &= 1_A+\alpha+\beta = 1+1^*\alpha+1_*\beta \\ &= (1+\alpha)(1+\beta)=1^+(\alpha)\cdot 1^+(\beta). \qquad (*) \end{aligned}$$

1^+ is injective and $\operatorname{Im} 1^+=\ker p$ since D_A acts effectively and transitively on $p^{-1}(1_A)$. Moreover, we have

$$\begin{aligned} 1^+(x^g) &= g_0^{-1}1^+(x)g_0, \quad g_0\in p^{-1}(g), \\ &= g_0^{-1}(1+x)g_0 \\ &= 1+(g^{-1})_* g^*(x). \end{aligned}$$

□

§4 Linear coverings of categories and exact sequences for functors

The notion of a linear covering of a category arises naturally by the exact sequences in towers of categories which play a central role in this book.

(4.1) ***Definition.*** Let $\mathbf{C}$ be a category and let H be a natural system of abelian groups on $\mathbf{C}$. We call $\mathbf{L}$ and the sequence

$$\mathbf{L}\xrightarrow{j}\mathbf{C}\xrightarrow{\mathfrak{D}} H$$

a **linear covering of C by** H if the following properties are satisfied:

$\mathbf{L}$ is a category and j is a full and faithful functor. (1)

On classes of objects the functor j is surjective, $j\colon Ob(\mathbf{L})\twoheadrightarrow Ob(\mathbf{C})$, and for $\bar{X}\in Ob(\mathbf{C})$ the class $j^{-1}(X)$ of all objects X in $\mathbf{L}$ with $j(X)=\bar{X}$ is a set. Moreover, the group $H(1_{\bar{X}})$ (given by the natural system H) acts transitively and effectively on the set $j^{-1}(\bar{X})$, we denote the action by $X+\xi\in j^{-1}(\bar{X})$ for $X\in j^{-1}(\bar{X})$, $\xi\in H(1_{\bar{X}})$. (2)

For objects X, Y in $\mathbf{L}$ and for a morphism $f:j(X)\to j(Y)$ in $\mathbf{C}$ an **obstruction element** $\mathfrak{O}_{X,Y}(f)\in H(f)$ is defind with the **derivation property** (3)

$$\mathfrak{O}_{X,Z}(gf) = g_*\mathfrak{O}_{X,Y}(f) + f^*\mathfrak{O}_{Y,Z}(g) \tag{$*$}$$

for $gf:j(X)\longrightarrow j(Y)\longrightarrow j(Z)$ in $\mathbf{C}$ and with

$$\mathfrak{O}_{X+\xi,Y+\eta}(f) = \mathfrak{O}_{X,Y}(f) + (f_*\xi - f^*\eta) \tag{$**$}$$

for $\xi\in H(1_{j(X)})$, $\eta\in H(1_{j(Y)})$. ||

A linear covering gives us the inclusion

(4.2) $$j^{-1}:\mathbf{L}^{\mathfrak{O}} = \text{kernel}(\mathfrak{O}) \subset \mathbf{L}.$$

The objects of $\mathbf{L}^{\mathfrak{O}}$ are the same as those of $\mathbf{L}$ and the morphisms $X\to Y$ in $\mathbf{L}^{\mathfrak{O}}$ are all morphisms $f:j(X)\to j(Y)$ with $\mathfrak{O}(f)=0$. We say that the **linear covering is split** if there is a functor $s:\mathbf{C}\to\mathbf{L}^{\mathfrak{O}}$ such that the composite

(4.3) $$\mathbf{C} \xrightarrow{s} \mathbf{L}^{\mathfrak{O}} \subset \mathbf{L} \xrightarrow{j} \mathbf{C}$$

is the identical functor on $\mathbf{C}$.

(4.4) ***Definition.*** We say that linear coverings

$$\mathbf{L} \xrightarrow{j} \mathbf{C} \xrightarrow{\mathfrak{O}} H \quad \text{and} \quad \mathbf{L}' \xrightarrow{j} \mathbf{C} \xrightarrow{\mathfrak{O}'} H$$

of $\mathbf{C}$ by H are **equivalent** if there is a functor $\varphi:\mathbf{L}\to\mathbf{L}'$ with the following properties

$$j'\varphi = j, \tag{1}$$
$$\varphi(X+\xi) = (\varphi X) + \xi, \tag{2}$$
$$\mathfrak{O}_{X,Y}(f) = \mathfrak{O}'_{\varphi X,\varphi Y}(f), \tag{3}$$

for objects X, Y in $\mathbf{L}$ and for a morphism $f:j(X)\to j(Y)$ in $\mathbf{C}$. The right-hand side of (2) and (3) is well defined since $j'(\varphi X) = j(X)$ by (1). Clearly, φ is actually an isomorphism of categories by (1), (2) and (4.1). ||

(4.5) **Proposition.** *Let* $\mathbf{C}$ *be a category and let* H *be a natural system on* $\mathbf{C}$*. Then there exists a split linear covering of* $\mathbf{C}$ *by* H *and two such split linear coverings are equivalent.*

Proof. We define the split linear covering $\mathbf{C}_0 \xrightarrow{j} \mathbf{C} \to H$ as follows: objects of $\mathbf{C}_0$ are pairs (X,ξ) with $\xi\in H(1_X)$, $X\in Ob(\mathbf{C})$. The action is defined by $(X,\xi)+\xi' = (X,\xi+\xi')$. The morphisms $(X,\xi)\to(Y,\eta)$ in $\mathbf{C}_0$ are the morphisms $X\to Y$ in $\mathbf{C}$. We set $j(\mathbf{X},\xi) = X$ and j is the identity on morphisms. Moreover,

$$\mathfrak{O}_{(X,\xi),(Y,\eta)}(f) = f_*\xi - f^*\eta.$$

For a split extension as in (4.3) we obtain φ by $\varphi(X,\xi)=s(X)+\xi$ and $\varphi(f)=s(f)$. □

We now introduce the concept of an exact sequence for functors. Many examples of such exact sequences are discussed in the Chapters VI, VII, VIII and IX below. An exact sequence combines the notions of a linear extension and of a linear covering.

Let $\lambda:\mathbf{A}\to\mathbf{B}$ be a functor. The **image category** $\lambda\mathbf{A}$ is defined as a quotient category of $\mathbf{A}$, see (1.2). The objects of $\lambda\mathbf{A}$ are the same as in $\mathbf{A}$. Moreover, we define the **reduced image category** $\mathbf{A}(\lambda)$ as follows: objects are the equivalence classes in $Ob(\mathbf{A})/\sim$ where we set $\sim\, = \overset{\lambda}{\sim}$ with

(4.6) $$X\overset{\lambda}{\sim}Y\Leftrightarrow\begin{cases}\exists\text{ equivalence } f:X\to Y\text{ in }\mathbf{A}\\ \text{with }\lambda X=\lambda Y\text{ and }\lambda f=1.\end{cases}$$

We denote by $\{X\}^\lambda$ the equivalence class of X in $Ob(\mathbf{A})/\sim$. By (4.6) the object λX in $\mathbf{B}$ depends only on $\{X\}^\lambda$. Morphisms $\{X\}^\lambda\to\{Y\}^\lambda$ in $\mathbf{A}(\lambda)$ are all morphisms $F:\lambda X\to\lambda Y$ in $\mathbf{B}$ which are realizable in $\mathbf{A}$ (that is, for F exists $f:X\to Y$ with $F=\lambda f$). Clearly, one has the canonical functor

(4.7) $$e:\lambda\mathbf{A}\xrightarrow{\sim}\mathbf{A}(\lambda),$$

which is an equivalence of categories.

We also use the **full image** $\mathbf{B}_\lambda$ of λ which is the full subcategory of $\mathbf{B}$ consisting of objects $\lambda A, A\in Ob(\mathbf{A})$. Moreover the **enlarged full image** $\mathbf{B}(\lambda)$ of λ is the following category. Objects are the same as in $\mathbf{A}(\lambda)$, morphisms $\{X\}^\lambda\to\{Y\}^\lambda$ are all morphisms $\lambda X\to\lambda Y$ in $\mathbf{B}$. Now we have the canonical functor

(4.8) $$j:\mathbf{B}(\lambda)\longrightarrow\mathbf{B}_\lambda,$$

which is the identity on morphisms and which satisfies $j\{X\}^\lambda=\lambda X$ on objects. Thus j is full and faithful and j is surjective on classes of objects.

By (4.6) and (4.7) we get the following factorization of the functor λ

(4.9) $$\lambda:\mathbf{A}\xrightarrow{q}\lambda\mathbf{A}\underset{\sim}{\xrightarrow{e}}\mathbf{A}(\lambda)\xrightarrow{k}\mathbf{B}(\lambda)\xrightarrow{j}\mathbf{B}_\lambda\xrightarrow{i}\mathbf{B}.$$

Here q is the quotient functor and i is the full inclusion. The functor k is the identity on objects and is the inclusion on morphism sets.

(4.10) ***Definition.*** Let $\lambda:\mathbf{A}\to\mathbf{B}$ be a functor and let D and H be natural systems (see (3.1)) of abelian groups on $\lambda\mathbf{A}$ and $\mathbf{B}_\lambda$ respectively. We call the sequence

$$D+\longrightarrow \mathbf{A}\xrightarrow{\lambda}\mathbf{B}\xrightarrow{\mathfrak{O}}H$$

an **exact sequence for** λ if the following properties are satisfies.

(a) The sequence $D/I+\to\mathbf{A}\xrightarrow{q}\lambda\mathbf{A}$ is a linear extension of categories as in (3.3), here I denotes the isotropy groups of the *linear action* D on λ, see (2.8).

(b) For all objects X, Y in $\mathbf{A}$ and morphisms $f:\lambda X\to\lambda Y$ in $\mathbf{B}$ an *obstruction element* $\mathfrak{O}_{X,Y}(f)\in H(f)$ is given such that $\mathfrak{O}_{X,Y}(f)=0$ if and only if there is a morphism $F:X\to Y$ in $\mathbf{A}$ with $\lambda F=f$. This is the *obstruction property* of $\mathfrak{O}$.

(c) $\mathfrak{O}$ has the *derivation property*

$$\mathfrak{O}_{X,Z}(gf)=g_*\mathfrak{O}_{X,Y}(f)+f^*\mathfrak{O}_{Y,Z}(g)\quad\text{for } f:\lambda X\to\lambda Y,\ g:\lambda Y\to\lambda Z.$$

(d) For all objects X in $\mathbf{A}$ and for all $\alpha\in H(1_{\lambda X})$ there is an object Y in $\mathbf{A}$ with $\lambda Y=\lambda X$ and $\mathfrak{O}_{X,Y}(1)=\alpha$; we write $X=Y+\alpha$ in this case. This is the *transivity property* of $\mathfrak{O}$. From (c) we derive for $f:\lambda X\to\lambda Y$

$$\mathfrak{O}_{X+\alpha,Y+\beta}(f)=\mathfrak{O}_{X,Y}(f)+(f_*\alpha-f^*\beta),\ \text{where } \alpha\in H(1_{\lambda X}),\ \beta\in H(1_{\lambda Y}).$$

(e) **Lemma.** *The properties* (a), (b), (c), (d) *above imply that*

$$\mathbf{B}(\lambda)\xrightarrow{j}\mathbf{B}_\lambda\xrightarrow{\mathfrak{O}}H$$

is a linear covering with $\mathbf{B}(\lambda)^{\mathfrak{O}}=\mathbf{A}(\lambda)$. *Conversely, if such a linear covering is given then* (b), (c) *and* (d) *are satisfied.* ||

Proof. Assume first that a linear covering is given. Then we define

$$\mathfrak{O}_{X,Y}(f)=\mathfrak{O}_{\{X\}^\lambda,\{Y\}^\lambda}(f)\tag{1}$$

Clearly, (b) holds by $\mathbf{B}(\lambda)^{\mathfrak{O}}=\mathbf{A}(\lambda)$, and (c) holds by (4.1)(3). Next we denote by $X+\alpha$ an object in $\mathbf{A}$ with $\{X+\alpha\}^\lambda=\{X\}^\lambda+\alpha$, see (4.1)(2). Then, clearly, (d) is satisfied by (4.1)(3).

Next assume that (a), (b), (c), and (d) are satisfied. The derivation property (c) and the obstruction property (b) imply that $\mathfrak{O}_{X,Y}(f)$ depends only on $(\{X\}^\lambda,\{Y\}^\lambda,f)$. In fact, for $g:X'\to X$ with $\lambda g=1$ we get $\mathfrak{O}_{X',X}(1)=0$ by the obstruction property. Therefore the derivation property yields

$$\begin{aligned}\mathfrak{O}_{X',Y}(f\circ 1)&=1^*\mathfrak{O}_{X,Y}(f)+f_*\mathfrak{O}_{X',X}(1)\\&=\mathfrak{O}_{X,Y}(f).\end{aligned}\tag{2}$$

Next we show that the equivalence class

$$\{X\}^\lambda+\alpha=\{X+\alpha\}^\lambda\tag{3}$$

is well defined by $\{X\}^\lambda$ and α and does not depend on the choice of $X+\alpha$. Let $f:X \xrightarrow{\sim} Y$ with $\lambda f = 1$. Then $\mathfrak{O}_{X,Y}(1)=0$ by (b) and therefore by (d)

$$\mathfrak{O}_{X+\alpha,Y+\alpha}(1)=\alpha-\alpha=0. \tag{4}$$

Hence there is $\bar{f}:X+\alpha \to Y+\alpha$ by (b) with $\lambda\bar{f}=1$. By (2.12) and (a) we know that $\bar{f}$ is an equivalence whence we get $\{X+\alpha\}^\lambda=\{Y+\alpha\}^\lambda$. Moreover, (3) defines an action of $H(1_{\lambda X})$ on the set $\{\{Y\}^\lambda : \lambda Y = \lambda X\}$ which is transitive and effective (this completes the proof of the lemma). We first show that (3) is an action. We have

$$\mathfrak{O}_{X+0,X}(1)=\mathfrak{O}_{X,X}(1)+0=0. \tag{5}$$

Therefore there is an equivalence $f:X+0 \xrightarrow{\sim} X$ with $\lambda f=1$, hence $\{X\}^\lambda+0=\{X\}^\lambda$. Moreover, we have

$$\mathfrak{O}_{(X+\alpha)+\beta,X+(\alpha+\beta)}(1)=\mathfrak{O}_{X+\alpha,X}(1)+\beta-(\alpha+\beta)=0. \tag{6}$$

This yields an equivalence $f:(X+\alpha)+\beta \xrightarrow{\sim} X+(\alpha+\beta)$ with $\lambda f=1$, hence $(\{X\}^\lambda+\alpha)+\beta=\{X\}^\lambda+(\alpha+\beta)$. The action is effective since $\{X\}^\lambda=\{X\}^\lambda+\alpha=\{X+\alpha\}^\lambda$ yields a map $f:X\to X+\alpha$ with $\lambda f=1$. Hence $0=\mathfrak{O}_{X,X+\alpha}(1)=-\alpha$ by (b) and (d). The action is transitive since for $\lambda Y=\lambda X$ we get $\alpha=\mathfrak{O}_{X,Y}(1)$ and

$$\mathfrak{O}_{X,Y+\alpha}(1)=\mathfrak{O}_{X,Y}(1)-\alpha=0 \tag{7}$$

This yields $f:X \xrightarrow{\sim} Y+\alpha$ with $\lambda f=1$ by (b) and (a), see (2.12). Therefore $\{X\}^\lambda=\{Y\}^\lambda+\alpha$. □

(4.11) **Proposition.** *The functor λ in an exact sequence* (4.10) *satisfies the sufficiency condition, see* (1.3)(*a*). *Moreover groups of automorphisms are embedded in an exact sequence*

$$D(1_{\lambda A}) \xrightarrow{1^+} E_{\mathbf{A}}(A) \xrightarrow{\lambda} E_{\mathbf{B}}(\lambda A) \xrightarrow{\bar{\mathfrak{O}}} H(1_{\lambda A})$$

Here A is an object in **A** *and λ and 1^+, $1^+(\alpha)=1_A+\alpha$, are homomorphisms between groups* (*see* (3.8)). *The group $H(1_{\lambda A})$ is a right $E_{\mathbf{B}}(\lambda A)$-module as in* (3.9) *and $\bar{\mathfrak{O}}$ defined by $\bar{\mathfrak{O}}(f)=(f^{-1})_*\mathfrak{O}(f)$ is a derivation, see* (7.1) *below.*

Proof. We show that $\bar{\mathfrak{O}}$ is actually a derivation as in (7.1).

$$\begin{aligned}\bar{\mathfrak{O}}(fg)&=(g^{-1}f^{-1})_*\mathfrak{O}(fg)\\&=(g^{-1}f^{-1})_*(f_*\mathfrak{O}(g)+g^*\mathfrak{O}(f))\\&=(g^{-1})_*\mathfrak{O}(g)+(g^{-1})_*(f^{-1})_*g^*\mathfrak{O}(f)\\&=\bar{\mathfrak{O}}(g)+g^*(g^{-1})_*\bar{\mathfrak{O}}(f)\\&=\bar{\mathfrak{O}}(f)^g+\bar{\mathfrak{O}}(g)\end{aligned}$$

We have $\bar{\mathfrak{O}}(f)=0$ if and only if $\mathfrak{O}(f)=0$ since $(f^{-1})_*$ is an isomorphism. □

For a functor $\lambda:\mathbf{A}\to\mathbf{B}$ and for an object B in $\mathbf{B}$ we define the **class of realizations of** B

$$\operatorname{Real}(B)=\operatorname{Real}_\lambda(B)=\{(A,b)\,|\,b:\lambda A\cong B\}/\sim.$$

Here we consider all pairs (A,b) where A is an object in $\mathbf{A}$ and where $b:\lambda A\to B$ an isomorphism in the category $\mathbf{B}$. We define an equivalence relation on such pairs by

$$(A,b)\sim(A',b')\Leftrightarrow\exists g:A'\cong A \text{ in } \mathbf{A} \text{ with } \lambda(g)=b^{-1}b'.$$

Let $\{A,b\}$ be the equivalence class of (A,b) in Real (B).

(4.12) **Proposition.** *Assume λ is a functor in an exact sequence* (4.10) *and assume* $\{A,b\}\in\operatorname{Real}_\lambda(B)$. *Then the group $H(1_{\lambda A})$ acts transitively and effectively on* $\operatorname{Real}_\lambda(B)$. *In particular* $\operatorname{Real}_\lambda(B)$ *is a set.*

Proof. Clearly, $A\overset{\lambda}{\sim}A'$ implies $(A,b)\sim(A',b)$. Whence we have the function

$$M=\{\{A'\}^\lambda:\lambda A'=\lambda A\}\to\operatorname{Real}(B) \tag{1}$$

which carries $\{A'\}^\lambda$ to $\{A',b\}$. By (4.10)(*e*) the group $H(1_{\lambda A})$ acts transitively and effectively on the set M. It remains to show that the function (1) is a bijection. Let $\{A'',b''\}\in\operatorname{Real}(B)$. Then we have for $f=b^{-1}b'':\lambda A''\to\lambda A$ the obstruction element

$$\alpha=(f^*)^{-1}\mathfrak{O}_{A'',A}(f)\in H(1_{\lambda A}) \tag{2}$$

where we use the isomorphism $f^*:H(1_{\lambda A})\cong H(f)$. We define $A'=A+\alpha$. Then we have $\lambda A'=\lambda A$ and

$$\mathfrak{O}_{A'',A'}(f)=\mathfrak{O}_{A'',A}(f)-f^*(\alpha)=0 \tag{3}$$

by (4.10)(d). Therefore there is $g:A''\to A'$ in $\mathbf{A}$ with $\lambda g=f$. By (4.11) we see that g is an isomorphism in $\mathbf{A}$. Whence $\{A',b\}=\{A'',b''\}$. This proves that the function (1) is surjective. It is clear by the definitions that the function (1) is injective. Therefore the proof of (4.12) is complete. □

Remark. On each category we have the natural system, 0, which consists only of trivial groups. The sequence $0\longrightarrow\mathbf{A}\overset{\lambda}{\longrightarrow}\mathbf{B}\longrightarrow\mathbf{0}$ is exact if and only if $\mathbf{A}=\lambda\mathbf{A}\xrightarrow{\sim}\mathbf{A}(\lambda)=\mathbf{B}_\lambda$. In particular λ is full and faithful.

A map between exact sequences as in (4.10) preserves all the structure. More precisely consider the diagram

$$
(4.13)\qquad
\begin{array}{ccccccc}
D+ & \longrightarrow & \mathbf{A} & \xrightarrow{\lambda} & \mathbf{B} & \xrightarrow{\mathfrak{D}} & H \\
\downarrow{\scriptstyle \bar{p}} & & \downarrow{\scriptstyle p} & & \downarrow{\scriptstyle q} & & \downarrow{\scriptstyle \bar{q}} \\
E+ & \longrightarrow & \mathbf{C} & \xrightarrow{\lambda'} & \mathbf{K} & \xrightarrow{\mathfrak{D}'} & G
\end{array}
$$

The rows of this diagram are exact sequence for the functors λ and λ' respectively. The diagram is a **map between exact sequences** if the following holds:

p and q are functors and $\lambda' p = q\lambda$ on objects and morphisms. (1)

$\bar{p}: D_f \to E_{pf} (f \in \operatorname{Mor} \mathbf{A})$ is a natural homomorphism between natural systems on $\lambda\mathbf{A}$ such that $p(f + \alpha) = p(f) + \bar{p}(\alpha)$ for $\alpha \in D_f$. We therefore call p a *$\bar{p}$-equivariant* functor. (2)

$\bar{q}: H_f \to G_{qf} (f \in \operatorname{Mor} \mathbf{B}_\lambda)$ is a natural homomorphism between natural systems on $\mathbf{B}_\lambda$ such that $\mathfrak{D}'_{pX,pY}(qf) = \bar{q}\mathfrak{D}_{X,Y}(f)$ for $f: \lambda X \to \lambda Y$. (3)

The following proposition corresponds to the five lemma.

(4.14) **Proposition.** *Consider the map between exact sequences in* (14.13) *and assume* $\bar{q}: H \to G$ and $\bar{p}: D/I \to E/I$ *are natural isomorphism. If* $q: \mathbf{B} \to \mathbf{K}$ *is full and faithful then also* $p: \mathbf{A} \to \mathbf{C}$ *is full and faithful. If* $q: \mathbf{B}_\lambda \to \mathbf{K}_{\lambda'}$ *is an equivalence of categories then also* $p: \mathbf{A} \to \mathbf{C}$ *is an equivalence of categories.*

(4.15) ***Definition.*** A **tower of categories** is a diagram ($i \in \mathbb{Z}$, $M < i \leqq N$)

$$
\begin{array}{ccccc}
 & & \downarrow & & \\
D_i & \longrightarrow & \mathbf{H}_i & \longrightarrow & \Gamma_{i+1} \\
 & & \downarrow{\scriptstyle \lambda} & & \\
D_{i-1} & \longrightarrow & \mathbf{H}_{i-1} & \longrightarrow & \Gamma_i \\
 & & \downarrow & &
\end{array}
$$

where $D_i \to \mathbf{H}_i \to \mathbf{H}_{i-1} \to \Gamma_i$ is an exact sequence for the functor λ. Examples are described in Chapter VI. ||

§ 5 The cohomology of a small category

We introduce the cohomology groups of a small category with coefficients in a natural system, compare also Baues–Wirsching. The cohomology groups in degree 2 and 1 classify linear extensions and linear coverings respectively. We show this in the following sections § 6, § 7.

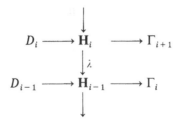

(5.1) ***Definition.*** Let **C** be a small category. We define the **cohomology** $H^n(\mathbf{C}, D)$ of **C with coefficients in the natural system** D by the cohomology of the following cochain complex $\{F^n, \delta\}$. The nth cochain group $F^n = F^n(\mathbf{C}, D)$ is the abelian group of all functions

(a)
$$f: N_n(\mathbf{C}) \to \dot{\bigcup_{\lambda \in \mathrm{Mor}(\mathbf{C})}} D_\lambda, \quad \text{with } f(\lambda_1, \ldots, \lambda_n) \in D_{\lambda_1 \circ \cdots \circ \lambda_n}.$$

Here $N_n(\mathbf{C})$ is the set of sequences $(\lambda_1, \ldots, \lambda_n)$ of n composable morphisms

$$A_0 \xleftarrow[\lambda_1]{} A_1 \longleftarrow \cdots \xleftarrow[\lambda_n]{} A_n$$

in **C** (which are the n-simplices of the **nerve** of **C**). For $n = 0$ let $N_0(\mathbf{C}) = Ob(\mathbf{C})$ be the set of objects in **C** and let $F^0(\mathbf{C}, D)$ be the set of all functions

(a)′
$$f: Ob(\mathbf{C}) \to \dot{\bigcup_{A \in Ob(\mathbf{C})}} D_A$$

with $f(A) \in D_A = D(1_A)$. Addition in F^n is given by adding pointwise in the abelian groups D_f. The coboundary

(b)
$$\delta: F^{n-1} \to F^n$$

is defined by the formula ($n > 1$):

(c)
$$(\delta f)(\lambda_1, \ldots, \lambda_n) = \lambda_{1*} f(\lambda_2, \ldots, \lambda_n) + \sum_{i=1}^{n-1} (-1)^i f(\lambda_1, \ldots, \lambda_i \lambda_{i+1}, \ldots, \lambda_n) + (-1)^n \lambda_n^* f(\lambda_1, \ldots, \lambda_{n-1}).$$

For $n = 1$ the coboundary δ in (b) is given by

(c)′
$$(\delta f)(\lambda) = \lambda_* f(A) - \lambda^* f(B) \text{ for } (\lambda: A \to B) \in N_1(\mathbf{C}).$$

One can check that $\delta f \in F^n$ for $f \in F^{n-1}$ and that $\delta\delta = 0$. ||

We now describe the natural properties of the cohomology. To this end we introduce the *category* **Nat** *of all natural systems*. Objects are pairs $(\mathbf{C}, D)$ where D is a natural system on the small category **C**, see (3.1). Morphisms are pairs

(5.2)
$$(\phi^{op}, \tau): (\mathbf{C}, D) \to (\mathbf{C}', D'),$$

where $\phi: \mathbf{C}' \to \mathbf{C}$ is a functor and where $\tau: \phi^* D \to D'$ is a natural transformation of functors. Here $\phi^* D: F\mathbf{C}' \to Ab$ is given by

(5.3)
$$(\Phi^* D)_f = D_{\phi f} \quad \text{for } f \in \mathrm{Mor}(\mathbf{C}'),$$

and $\alpha_* = \phi(\alpha)_*$, $\beta^* = \phi(\beta)^*$. A natural transformation $t: D \to \tilde{D}$ yields as well the natural transformation

(5.4)
$$\phi^* t: \phi^* D \to \phi^* \tilde{D}.$$

Now morphisms in **Nat** are composed by the formula

$$(5.5)\qquad (\psi^{op},\sigma)(\phi^{op},\tau)=((\phi\psi)^{op},\ \sigma\circ\psi^*\tau).$$

The cohomology introduced above is a **functor**,

$$(5.6)\qquad H^n:\mathbf{Nat}\to\mathbf{Ab}\quad (n\in\mathbb{Z}),$$

which carries the morphism (ϕ^{op},τ) of (5.2) to the induced homomorphism

$$(5.7)\qquad \tau_*\phi^*:H^n(\mathbf{C},D)\to H^n(\mathbf{C}',D'),$$

given on cochains $f\in F^n$ by $(\tau_*\phi^*f)(\lambda'_1,\dots,\lambda'_n)=\tau_{\lambda'}\circ f(\phi\lambda'_1,\dots,\phi\lambda'_n)$ with $\lambda'=\lambda'_1\circ\cdots\circ\lambda'_n$. We have $(\phi^{op},\tau)=(1,\tau)(\phi^{op},1)$ and we write $\phi^*=(\phi^{op},1)_*$ and $(1,\tau)_*=\tau_*$.

(5.8) **Theorem.** *Suppose $\phi:\mathbf{C}'\to\mathbf{C}$ is an equivalence of small categories. Then ϕ induces an isomorphism*

$$\phi^*:H^n(\mathbf{C},D)\cong H^n(\mathbf{C}',\phi^*D)$$

for all natural systems D on $\mathbf{C}$, $n\in\mathbb{Z}$.

For the proof of this result we consider first a natural equivalence

$$t:\phi\cong\psi,\quad \phi,\psi:\mathbf{C}'\to\mathbf{C},$$

which induces an isomorphism of natural systems

$$\begin{cases}\tilde{t}:\phi^*D\cong\psi^*D,\\ \text{with } \tilde{t}=t_*(t^{-1})^*:D_{\phi f}\cong D_{\psi f}.\end{cases}$$

Here we have $\psi f=t(\phi f)t^{-1}$ since t is a natural equivalence.

(5.9) **Lemma.** $\tilde{t}_*\phi^*=\psi^*$ *on* $H^n(\mathbf{C},D)$.

Proof of theorem (5.8). Let $\phi':\mathbf{C}\to\mathbf{C}'$ be a functor and let

$$t:\phi'\phi\cong 1,\quad \tau:\phi\phi'\cong 1$$

be equivalences. Then by (5.9) we have

$$\tilde{t}_*(\phi'\phi)^*=1^*=1\quad\text{and}\quad\tilde{\tau}_*(\phi\phi')^*=1^*=1.$$

Here $\tilde{t}_*$ and $\tilde{\tau}_*$ are isomorphisms and therefore ϕ^* is an isomorphism. □

(5.10) *Proof of lemma* (5.9). We construct a chain homotopy h for the diagram of cochain maps

$$\begin{array}{ccc} & \overset{\phi^*}{\nearrow} & F^*(\mathbf{C}',\phi^*D)\\ F^*(\mathbf{C},D) & & \downarrow{\scriptstyle\tilde{t}_*}\\ & \underset{\psi^*}{\searrow} & F^*(\mathbf{C}',\psi^*D),\end{array}\qquad (1)$$

$$\begin{cases} \tilde{t}_*\phi^* - \psi^* = \delta h + h\delta, \text{ with} \\ h: F^{n+1}(\mathbf{C}, D) \to F^n(\mathbf{C}', \psi^* D). \end{cases} \tag{2}$$

Here h is given by the following formula

$$(hf)(\lambda'_1,\ldots,\lambda'_n) = (t^*)^{-1} \sum_{i=0}^{n} (-1)^i f(\psi\lambda'_1,\ldots,\psi\lambda'_i, t, \phi\lambda'_{i+1},\ldots,\phi\lambda'_n). \tag{3}$$

The terms in the alternating sum correspond to paths in the commutative diagram

$$\begin{array}{ccccccccc} * & \xleftarrow{\phi\lambda'_1} & * & \xleftarrow{\phi\lambda'_2} & \cdots & & * & \xleftarrow{\phi\lambda'_n} & * \\ \downarrow t & & \downarrow t & & \cdots & & \downarrow t & & \downarrow t. \\ * & \xleftarrow[\psi\lambda'_1]{} & * & \xleftarrow[\psi\lambda'_2]{} & \cdots & & * & \xleftarrow[\psi\lambda'_n]{} & * \end{array} \tag{4}$$

A somewhat tedious but straightforward calculation shows that formula (2) is satisfied for h. □

There are various special cases of natural systems which we obtain by the functors:

(5.11) $$F\mathbf{C} \xrightarrow{\pi} \mathbf{C}^{op} \times \mathbf{C} \xrightarrow{p} \mathbf{C} \xrightarrow{q} \pi\mathbf{C} \xrightarrow{o} *$$

Here π and p are the obvious forgetful functors and q is the localization functor for the **fundamental groupoid**

(5.12) $$\pi\mathbf{C} = (\mathrm{Mor}\,\mathbf{C})^{-1}\mathbf{C}, \quad \text{see (II.3.5).}$$

Moreover $*$ in (5.11) is the trivial category consisting of one object and one morphism and o is the trivial functor. Using the functors in (5.11) we get special natural systems on $\mathbf{C}$ by pulling back functors $\mathbf{K} \to \mathbf{Ab}$ where $\mathbf{K}$ is one of the categories in (5.11). Such functors are denoted as follows:

(5.13) ***Definition***

M is a **C-bimodule** if $M: \mathbf{C}^{op} \times \mathbf{C} \to \mathbf{Ab}$.

F is a **C-module** if $F: \mathbf{C} \to \mathbf{Ab}$.

L is a **local system** on $\mathbf{C}$ if $L: \pi\mathbf{C} \to \mathbf{Ab}$.

A is a **trivial system** on $\mathbf{C}$ if A is an abelian group or equivalently if $A: * \to \mathbf{Ab}$.

Clearly we define the **cohomology of C with coefficients in** M, F, L and A respectively by the groups

$$H^n(\mathbf{C}, M) = H^n(\mathbf{C}, \pi^* M), \tag{1}$$

$$H^n(\mathbf{C}, F) = H^n(\mathbf{C}, \pi^* p^* F), \tag{2}$$

$$H^n(\mathbf{C}, L) = H^n(\mathbf{C}, \pi^* p^* q^* L), \tag{3}$$

$$H^n(\mathbf{C}, A) = H^n(\mathbf{C}, \pi^* p^* q^* o^* A). \tag{4}$$

(5.14) ***Remark.*** The cohomology (1) can be identified with the Hochschild–Mitchell cohomology which was found by Mitchell (1972) by imitating the classical ring theory on the level of categories. The cohomology (2) is used by Watts (1965), by Quillen (1973) and by Grothendieck, see, for example, Johnstone (1977) for the definition of topos cohomology. Next the cohomologies (3) and (4) can be identified with the usual singular cohomologies of the classifying space $B(\mathbf{C})$ with local coefficients L, and with coefficients in the abelian group A, respectively, see Quillen (1973).

Our approach generalizes these concepts by taking natural systems as coefficients which are more adapted to categories than the coefficients in (5.13). Indeed, a module (resp. a bimodule) associates on abelian group to an object (resp. to a pair of objects), while a natural system associates an abelian group to each morphism.

(5.15) ***Remark.*** The cohomology (5.1) as well generalizes the *cohomology of a group G*. Let D be a right G-module. Then we have

$$H^n(G, D) = H^n(\mathbf{G}, \bar{D}),$$

where the left-hand side is the usual cohomology of G with coefficients in the G-module D, see for example Cartan–Eilenberg (1956). The right-hand side is the cohomology (5.1) of the category $\mathbf{G}$ with coefficients in the natural system $\bar{D}$ defined by (3.7).

(5.16) ***Remark.*** For a $\mathbf{C}$-module F there is a natural isomorphism

$$H^n(\mathbf{C}, F) = \operatorname{Lim}^n(F).$$

where Lim^n is the derived of the Lim functor, see Roos (1961). In (III.1.11) above we used the derived functor Lim^1 for a very special category $\mathbf{C}$.

(5.17) ***Remark.*** It is clear that each linear extension of a **free category** $\mathbf{F}$ is a split extension. Therefore the result in §6 shows that $H^2(\mathbf{F}, D) = 0$ for all natural systems D. More generally also $H^n(\mathbf{F}, D) = 0$ for $n \geqq 2$ and for any natural system D on $\mathbf{F}$. Moreover, let S *be a subclass of morphisms in* $\mathbf{F}$ and let $S^{-1}\mathbf{F}$ be the localized category, see (II.3.5). Then also

$$H^n(S^{-1}\mathbf{F}, D) = 0 \quad \text{for } n \geqq 2.$$

This is proved in Baues–Wirsching.

Finally, we introduce the cup product for the cohomology groups (5.1).

(5.18) ***Definition.*** Let D, D', D'' be natural systems on the small category **C**. A **pairing**, denoted by

$$\mu:(D, D') \to D'',$$

associates with each 2-chain (f, g) in **C** the homomorphism

$$\mu: D_f \otimes D'_g \to D''_{fg}$$

of abelian groups such that for a 3-chain (f, g, h) and for $x \in D_f$, $y \in D_g$, $y' \in D'_g$, $z \in D'_h$ we have the following formulas where we set $x \cdot y = \mu(x \otimes y)$:

$$(g^* x) \cdot z = x \cdot (g_* z),$$
$$f_*(y \cdot z) = (f_* y) \cdot z,$$
$$h^*(x \cdot y') = x \cdot (h^* y'). \qquad \|$$

(5.19) ***Definition.*** Let $\mu:(D, D') \to D''$ be a pairing of natural systems on **C**. Then we have the **cup product**

$$\cup : H^n(\mathbf{C}, D) \otimes H^m(\mathbf{C}, D') \to H^{n+m}(\mathbf{C}, D''), \tag{1}$$

which is defined on cochains by the formula

$$(f \cup g)(\lambda_1, \ldots, \lambda_{n+m}) = f(\lambda_1, \ldots, \lambda_n) \cdot g(\lambda_{n+1}, \ldots, \lambda_{n+m}), \tag{2}$$

Here the multiplication $x \cdot y = \mu(x \otimes y)$ is defined by the pairing μ as in (5.18) above. We define (1) by

$$\{f\} \cup \{y\} = \{f \cup g\}, \tag{3}$$

where $\{f\}$ denotes the cohomology class represented by the cocycle f. $\|$

(5.20) **Lemma.** $\delta(f \cup g) = (\delta f) \cup g + (-1)^n f \cup (\delta g)$.

The lemma is easily checked by (5.18) and by the formula for δ in (5.1). The lemma implies that the cup product above is a well-defined homomorphism.

(5.21) ***Example.*** Let R be a commutative ring and let $\mathbf{m}_R$ be the (small) category of finitely generated R-modules. Then $\mathrm{Hom}_R : \mathbf{m}_R^{op} \times \mathbf{m}_R \to \mathbf{Ab}$ is a bimodule on $\mathbf{m}_R$ and we have the pairing

$$\mu:(\mathrm{Hom}_R, \mathrm{Hom}_R) \to \mathrm{Hom}_R,$$

which is given by the composition of homomorphisms, that is $\mu(\alpha \otimes \beta) = \alpha \circ \beta$. Therefore we obtain by (5.19) the cup product

$$\cup : H^n(\mathbf{m}_R, \mathrm{Hom}_R) \otimes H^m(\mathbf{m}_R, \mathrm{Hom}_R) \to H^{n+m}(\mathbf{m}_R, \mathrm{Hom}_R)$$

which shows that $H^*(\mathbf{m}_R, \mathrm{Hom}_R)$ is a graded ring.

§6 Classification of linear extensions

The definition of the cohomology groups of a small category with coefficients in a natural system was motivated by the following result, compare (3.2) and (3.4).

(6.1) **Theorem (classification).** *Let D be a natural system on a small category $\mathbf{C}$ and let $M(\mathbf{C}, D)$ be the set of equivalence classes of linear extensions of $\mathbf{C}$ by D. Then there is a canonical bijection*

$$\Psi : M(\mathbf{C}, D) \cong H^2(\mathbf{C}, D),$$

which maps the split extension to the zero element in the cohomology group $H^2(\mathbf{C}, D)$.

(6.2) ***Example.*** Let G be a group and let D be a right G-module. For the natural system $\bar{D}$ on $\mathbf{G}$ in (3.7) the set $M(\mathbf{G}, \bar{D})$ can be identified easily with the set $E(G, D)$ of all equivalence classes of extensions in (3.6). Therefore (6.1) and (5.15) yield the result:

$$E(G, D) = M(\mathbf{G}, \bar{D}) = H^2(\mathbf{G}, \bar{D}) = H^2(G, D).$$

This, in fact, is the well-known classification of group extensions.

Proof of theorem (6.1). Let $p : \mathbf{E} \to \mathbf{C}$ be a linear extension by D. Since p is surjective on morphisms there exists a function

$$s : \mathrm{Mor}(\mathbf{C}) \to \mathrm{Mor}(\mathbf{E}) \tag{1}$$

with $ps = 1$. If we have two such functions s and s' the condition $ps = 1 = ps'$ implies that there is a unique element

$$\left.\begin{array}{l} d \in F^1(\mathbf{C}, D), \\ \text{with } s'(f) = s(f) + d(f), \quad f \in \mathrm{Mor}(\mathbf{C}). \end{array}\right\} \tag{2}$$

Moreover, each $d \in F^1(\mathbf{C}, D)$ gives us by $(s + d)(f) = s(f) + d(f)$ a function $s + d : \mathrm{Mor}(\mathbf{C}) \to \mathrm{Mor}(\mathbf{E})$ with $p(s + d) = ps = 1$. For $(y, x) \in N_2(\mathbf{C})$ the formula

$$s(yx) = s(y)s(x) + \Delta_s(y, x) \tag{3}$$

determines the element

$$\Delta_s \in F^2(\mathbf{C}, D). \tag{4}$$

This element measures the deviation of s from being a functor. If s is a splitting then $\Delta_s = 0$. We now define the function Ψ in (6.1) by

$$\Psi\{\mathbf{E}\} = \{\Delta_s\}. \tag{5}$$

Here $\{\mathbf{E}\} \in M(\mathbf{C}, D)$ is the equivalence class of extension $\mathbf{E}$ and $\{\Delta_s\} \in H^2(\mathbf{C}, \bar{D})$ is the cohomology class represented by the cocycle Δ_s in (4) where s is chosen as in (1). First we have to check the cocycle condition for Δ_s: We compute

$$s((zy)x) = s(z)s(y)s(x) + x^*\Delta_s(z, y) + \Delta_s(zy, x)$$
$$s(z(yx)) = s(z)s(y)s(x) + z_*\Delta_s(y, x) + \Delta_s(z, yx)$$

Therefore associativity of composition implies

$$\begin{aligned} o &= z_*\Delta_s(y, x) - \Delta_s(zy, x) + \Delta_s(z, yx) - x^*\Delta_s(z, y) \\ &= (\delta\Delta_s)(z, y, x), \quad \text{see (5.1)(c).} \end{aligned} \tag{6}$$

Moreover the cohomology class $\{\Delta_s\}$ does not depend on the choice of s: We compute

$$(s + d)(yx) = s(y)s(x) + x^*d(y) + y_*d(x) + \Delta_{s+d}(y, x).$$

Therefore we have by (3)

$$\Delta_s(y, x) - \Delta_{s+d}(y, x) = y_*d(x) - d(yx) + x^*d(y) = (\delta d)(y, x), \quad \text{see (5.1)(c).} \tag{7}$$

In addition, we see that for an equivalence ε we have

$$\Delta_{\varepsilon s} = \Delta_s. \tag{8}$$

By (6), (7) and (8) the function Ψ in (5) is well-defined. The function Ψ is surjective by the following construction: Let $\Delta \in F^2(\mathbf{C}, D)$, $\delta\Delta = 0$. We get an extension

$$\left.\begin{array}{l} p\Delta: \mathbf{E}_\Delta \to \mathbf{C} \\ \text{with } \Psi\{\mathbf{E}_\Delta\} = \{\Delta\}. \end{array}\right\} \tag{9}$$

The morphisms in $\mathbf{E}_\Delta$ are the pairs (f, α) with $f \in \mathrm{Mor}(\mathbf{C})$, $\alpha \in D_f$. The composition in $\mathbf{E}_\Delta$ is defined by

$$(g, \beta)(f, \alpha) = (gf, -\Delta(g, f) + g_*\alpha + f^*\beta). \tag{10}$$

The action of D on $\mathbf{E}_\Delta$ is defined by $(f, \alpha) + \alpha' = (f, \alpha + \alpha')$, $\alpha' \in D_f$.
Since we have an equivalence

$$\left.\begin{array}{l} \mathbf{E}_{\Delta_s} \xrightarrow{\varepsilon} \mathbf{E} \\ \text{with } \varepsilon(f, \alpha) = s(f) + \alpha \end{array}\right\} \tag{11}$$

we see that Ψ is also injective. □

(6.3) ***Remark.*** For a linear extension

$$D + \to \mathbf{E} \to \mathbf{C} \tag{1}$$

the corresponding cohomology class $\Psi\{E\} \in H^2(\mathbf{C}, D)$ has the following *universal property* with respect to the groups of automorphisms in $\mathbf{E}$: For an object A in $\mathbf{E}$ the extension (1) yields the group extension

$$0 \to D_A \to \mathrm{Aut}_{\mathbf{E}}(A) \to \mathrm{Aut}_{\mathbf{C}}(A) \to 0 \tag{2}$$

by restriction. Here $\alpha \in \mathrm{Aut}_C(A)$ acts on $x \in D_A = D(1_A)$ by $x^\alpha = (\alpha^{-1})_* \alpha^*(x)$. The cohomology class corresponding to the extension (2) is given by the image of the class $\Psi\{\mathbf{E}\}$ under the homomorphism

$$H^2(\mathbf{C}, D) \xrightarrow{t_* i^*} H^2(\mathrm{Aut}_C(A), D_A). \tag{3}$$

Here i is the inclusion functor $\mathrm{Aut}_{\mathbf{C}}(A) \hookrightarrow \mathbf{C}$ and $t: i^*D \to \bar{D}_A$ is the isomorphism of natural systems, see (3.7), with

$$t = (\alpha^{-1})_* : D_{i\alpha} \to D(1_A) = \bar{D}_A. \tag{4}$$

We now describe some examples of linear extensions of categories. We first describe an example for the group $H^2(\mathbf{C}, D)$ which can be computed directly by the formula in (5.1).

(6.4) ***Example.*** Consider the category $\mathbf{Q}$ pictured by the commutative square:

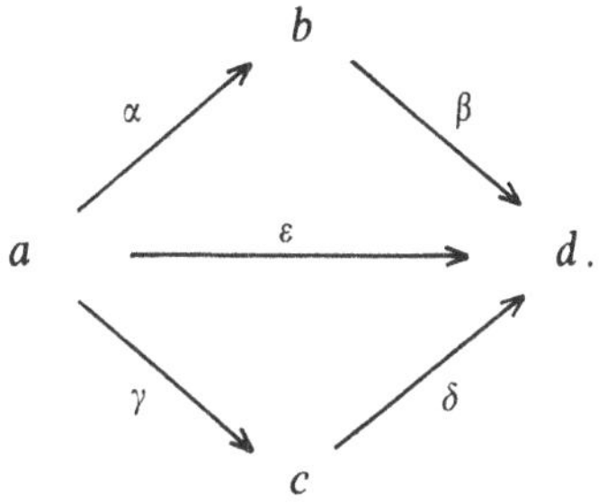

(1)

Then we have for a natural system D on $\mathbf{Q}$ the isomorphism

$$H^2(\mathbf{Q}, D) = D_\varepsilon / (\alpha^* D_\beta + \beta_* D_\alpha + \delta_* D_\gamma + \gamma^* D_\delta). \tag{2}$$

Let $D + \to \mathbf{E} \to \mathbf{C}$ be any linear extension and let ϕ be a commutative square in $\mathbf{C}$; this is a functor $\phi: \mathbf{Q} \to \mathbf{C}$. Then (5.7) yields the element $\phi^*\{\mathbf{C}\}$ in the group $H^2(\mathbf{Q}, \phi^* D)$.

(6.5) ***Example.*** For any prime p there is a canonical linear extension of categories

$$\mathrm{Hom}_{\mathbb{Z}/p} \xrightarrow{+} \mathbf{m}_{\mathbb{Z}/p^2} \xrightarrow{q} \mathbf{m}_{\mathbb{Z}/p}. \tag{1}$$

Here $\mathbf{m}_R$ with $R = \mathbb{Z}/p^2$ or $R = \mathbb{Z}/p$ denotes the (small) category of finitely generated free R-modules. Objects are $R^n = R \oplus \cdots \oplus R$, $n \geqq 1$, and morphisms $R^n \to R^m$ are $(m \times n)$-matrices (α_{ij}) over R, composition is multiplication of matrices. We have the canonical bifunctor in (1)

$$\mathrm{Hom}_R : \mathbf{m}_R^{op} \times \mathbf{m}_R \to \mathbf{Ab}, \tag{2}$$

which carries (R^n, R^m) to the abelian group $\mathrm{Hom}_R(R^n, R^m) = M^{m,n}(R)$ of $(m \times n)$-matrices. The functor q in (1) is reduction mod p. The action $+$ in (1) is given by the formula

$$(\alpha_{ij}) + (\beta_{ij}) = (\alpha_{ij} + p\beta_{ij}) \tag{3}$$

with $(\alpha_{ij}) \in M^{m,n}(\mathbb{Z}/p^2)$ and $(\beta_{ij}) \in M^{m,n}(\mathbb{Z}/p)$. Here $p: \mathbb{Z}/p \to \mathbb{Z}/p^2$ maps $\{1\}$ to $\{p\}$. It is an easy excercise to show that (1) is a well-defined linear extension of categories. This extension is not split. In fact, Hartl computed the cohomology group

$$H^2(\mathbf{m}_{\mathbb{Z}/p^2}, \mathrm{Hom}) \cong \mathbb{Z}/p, \tag{4}$$

and showed that the extension (1) represents a generator of this group via (6.1). We now restrict the extension in (1) to the group of self equivalences as in (6.3). This yields the group extension

$$M^{n,n}(\mathbb{Z}/p) \rightarrowtail GL(n, \mathbb{Z}/p^2) \twoheadrightarrow GL(n, \mathbb{Z}/p). \tag{5}$$

W. Meyer proved that the extension of groups has a splitting if and only if $(n-1)(p-1) \leqq 2$. The extension (5) played a role in recent work of Friedlander–Dwyer on the cohomology of $GL(n, \mathbb{Z}/p)$.

In the next example we describe a more general procedure for the construction of linear extensions of categories.

(6.6) ***Example*** (***categories defined by central extension of groups***). Let $\mathbf{C}$ be a (small) category and let

$$\left.\begin{aligned} A&: \mathbf{C} \to \mathbf{Ab}, \\ G&: \mathbf{C} \to \mathbf{Gr}, \end{aligned}\right\} \tag{1}$$

be functors from $\mathbf{C}$ to the category, $\mathbf{Ab}$, of abelian groups and to the category, $\mathbf{Gr}$, of groups respectively. We choose for each object X in $\mathbf{C}$ a central extension

$$A(X) \rightarrowtail E_X \twoheadrightarrow G(X) \tag{2}$$

of groups. We do not assume that E is a functor on $\mathbf{C}$. By use of the data (1) and (2) we obtain the linear extension of categories

$$\mathrm{Hom}(G-, A-) \xrightarrow{+} \mathbf{E} \xrightarrow{p} p\mathbf{E} \tag{3}$$

as follows. Objects of $\mathbf{E}$ are the same as in $\mathbf{C}$ and p is the identity on objects. Morphisms in $\mathbf{E}$ are pairs $(f, \varphi): X \to Y$ where $f: X \to Y$ is a morphism in $\mathbf{C}$ and where φ is a homomorphism of groups for which the diagram

$$\begin{array}{ccccc} A(X) & \rightarrowtail & E_X & \overset{p}{\rightarrowtail} & G(X) \\ \downarrow{\scriptstyle A(f)} & & \downarrow{\scriptstyle \varphi} & & \downarrow{\scriptstyle G(f)} \\ A(Y) & \overset{i}{\rightarrowtail} & E_Y & \twoheadrightarrow & G(Y) \end{array} \tag{4}$$

commutes. Composition is defined by $(f,\varphi)(f',\varphi')=(ff',\varphi\varphi')$. The functor $p{:}\mathbf{E}\to\mathbf{C}$ in (3) is the forgetful functor which carries (f,φ) to f. The bifunctor

$$\mathrm{Hom}(G-,A-){:}\mathbf{C}^{op}\times\mathbf{C}\to\mathbf{Ab} \tag{5}$$

carries (X,Y) to the abelian group of homomorphisms $\mathrm{Hom}(GX,AY)$. For $\alpha\in\mathrm{Hom}(GX,AY)$ we define the action $+$ in (3) by

$$(f,\varphi)+\alpha=(f,\varphi+i\alpha p). \tag{6}$$

Since the rows of (4) are central extensions we see that (6) is a well defined morphism $X\to Y$ in $\mathbf{E}$. Now it is easy to check that (3) is a welldefined linear extension of categories. ||

(6.7) ***Example.*** Let $\mathbf{C}$ be the (small) category of finitely generated abelian groups and let p be a prime. For each object A in $\mathbf{C}$ we have the canonical homomorphism

$$A*\mathbb{Z}/p=\{a\in A\,|\,pa=0\}\subset A\longrightarrow\!\!\!\!\!\rightarrow \mathbb{Z}/pA=A\otimes\mathbb{Z}/p, \tag{1}$$

which determines the extension of abelian groups

$$A\otimes\mathbb{Z}/p\rightarrowtail E_A\twoheadrightarrow A*\mathbb{Z}/p \tag{2}$$

since $\mathrm{Hom}(A*\mathbb{Z}/p, A\otimes\mathbb{Z}/p)=\mathrm{Ext}(A*\mathbb{Z}/p, A\otimes\mathbb{Z}/p)$. As in (6.6) we thus have the linear extension of categories determined by (2):

$$\mathrm{Hom}(-*\mathbb{Z}/p, -\otimes\mathbb{Z}/p)\overset{+}{\longrightarrow}\mathbf{E}_p\to\mathbf{C}. \tag{3}$$

In fact, the extension of categories in (6.5) is the restriction of (3) to the subcategory $\mathbf{m}_{\mathbb{Z}/p}\subset\mathbf{C}$. By a result of Hartl we have the cohomology group

$$H^2(\mathbf{C},\mathrm{Hom}(-*\mathbb{Z}/p, -\otimes\mathbb{Z}/p))\cong\mathbb{Z}/p, \tag{4}$$

and the linear extension (3) represents a generator in this group via (6.1). For $p=2$ we can identify the extension (3) with the full homotopy category of Moore spaces in degree $\geqq 3$. See (V.3a.8) below.

§7 Classification of linear coverings

A derivation from a group G into a right G-module A is a function $d{:}G\to A$ with the property

$$d(xy)=(dx)^y+dy. \tag{7.1}$$

An inner derivation $i{:}G \to A$ is one for which there exists an element $a \in A$ with $i(x) = a - a^x$. It is a classical result that

(7.2) $$H^1(G, A) = \mathrm{Der}\,(G, A)/\mathrm{Ider}\,(G, A)$$

where Der and Ider denote the abelian groups of derivations and of inner derivations respectively. Compare for example Hilton–Stammbach.

We now consider derivations from a small category $\mathbf{C}$ into a natural system D on $\mathbf{C}$ and show that the cohomology $H^1(\mathbf{C}, D)$ can be described similarly as in (7.2). In the following definition we use the groups $F^n(\mathbf{C}, D)$ defined in (5.1).

(7.3) ***Definition.*** *A derivation* $d{:}\mathbf{C} \to D$ is a function in $F^1(\mathbf{C}, D)$ with $d(xy) = x_*(dy) + y^*(dx)$ An *inner derivation* $i{:}\mathbf{C} \to D$ is one for which there exists an element $a \in F^0(\mathbf{C}, D)$ such that for $x{:}A \to B$ $i(x) = x_* a(A) - x^* a(B)$. ‖

For example, a linear covering $\mathbf{L} \xrightarrow{j} \mathbf{C} \xrightarrow{\mathfrak{D}} D$ of $\mathbf{C}$ by D, see (4.1), yields a derivation $\mathfrak{D}{:}\mathbf{L} \to j^* D$.

(7.4) ***Example.*** Let $\mathbf{G}$ and $\bar{D}$ be defined as in (3.7) then a derivation $\mathbf{G} \to \bar{D}$ is exactly given by a derivation $G \to D$. The same holds for inner derivations.

We denote by Der $(\mathbf{C}, D)$ and Ider $(\mathbf{C}, D)$ the abelian groups of all derivations and of all inner derivations $\mathbf{C} \to D$ respectively. These are actually functors

(7.5) $$\mathrm{Der},\quad \mathrm{Ider}{:}\mathbf{Nat} \to \mathbf{Ab}$$

which are defined on morphisms (ϕ^{op}, τ) exactly as in (5.7).

(7.6) **Proposition.** *There is a natural isomorphism*

$$H^1(\mathbf{C}, D) = \mathrm{Der}(\mathbf{C}, D)/\mathrm{Ider}(\mathbf{C}, D)$$

of functors on **Nat** *which carries the cohomology class* $\{f\}$ *to the class* $\{f\}$.

This is clear by the definition in (5.1) which shows that $f \in F^1(\mathbf{C}, D)$ is a derivation iff $\delta f = 0$. Moreover, f is an inner derivation iff $f = \delta(g)$ with $g \in F^0(\mathbf{C}, D)$.

(7.7) **Theorem (classification).** *Let H be a natural system of abelian groups on the small category* $\mathbf{C}$ *and let* $N(\mathbf{C}, H)$ *be the set of equivalence classes of linear coverings of* $\mathbf{C}$ *by* H, *see* §4. *Then there is a canonical bijection*

$$N(\mathbf{C}, H) \underset{\psi}{=} H^1(\mathbf{C}, H)$$

which maps the split linear covering to the zero element in the abelian group $H^1(\mathbf{C}, H)$, *see* (4.5).

This result is due to Unsöld.

Proof. Let $\mathbf{L} = (L \to \mathbf{C} \to H)$ be a linear covering. We choose a function $s: Ob(\mathbf{C}) \to Ob(L)$ with $j(sX) = X$ and we define the derivation

$$\left.\begin{aligned} &\mathfrak{D}_s: \mathbf{C} \to H \\ &\mathfrak{D}_s(f) = \mathfrak{D}_{sX,sY}(f),\ f: X \to Y \in \mathbf{C}. \end{aligned}\right\} \tag{1}$$

Formula (4.1)(3) shows that the cohomology class $\{\mathfrak{D}_s\}$ does not depend on the choice of s. Moreover $\{\mathfrak{D}_s\} = 0$ if $\mathbf{L}$ is the split linear covering.

If $\mathbf{L}$ and $\mathbf{L}'$ are equivalent linear coverings as in (4.4) we have the function φs with $j'\varphi sX = X$ and clearly by (4.4)(3) we get $\mathfrak{D}_s = \mathfrak{D}_{\varphi s}$. This shows that the function ψ in (7.7) is well defined by

$$\psi\{\mathbf{L}\} = \{\mathfrak{D}_s\}. \tag{2}$$

Here $\{\mathbf{L}\}$ denotes the equivalence class of $\mathbf{L}$. We now show that ψ is a bijection. Let $d: \mathbf{C} \to H$ be any derivation. Then we define the linear covering

$$\mathbf{L}_d \xrightarrow{j} \mathbf{C} \xrightarrow{\mathfrak{D}} H \tag{3}$$

as follows. Objects in $\mathbf{L}_d$ are pairs (X, ξ) with $\xi \in H(1_X)$. Clearly, $\xi' \in H(1_X)$ acts by $(X, \xi) + \xi' = (X, \xi + \xi')$. Morphisms $(X, \xi) \to (Y, \eta)$ in $\mathbf{L}_d$ are the same as morphisms $X \to Y$ in $\mathbf{C}$ and we define $j(X, \xi) = X$. We define the derivation $\mathfrak{D}$ in (3) by

$$\left.\begin{aligned} &\mathfrak{D}: \mathbf{L}_d \to j^*H \\ &\mathfrak{D}_{(X,\xi),(Y,\eta)}(f) = d(f) + (f_*\xi - f^*\eta). \end{aligned}\right\} \tag{4}$$

Clearly, for s with $s(X) = (X, 0)$ we get

$$\mathfrak{D}_s = d. \tag{5}$$

Thus it is enough to show that ψ^{-1} with $\psi^{-1}\{d\} = \{\mathbf{L}_d\}$ is a well-defined function. Let $\lambda \in F^0(\mathbf{C}, H)$, see (7.3). The function λ associates with each object X in $\mathbf{C}$ an element $\lambda X \in H(1_X)$ and yields the inner derivation

$$i(f) = f_*\lambda X - f^*\lambda Y$$

for $f: X \to Y \in \mathbf{C}$. We show that there is an equivalence

$$\varphi: \mathbf{L}_d \to \mathbf{L}_{d+i}. \tag{6}$$

Clearly, φ is defined on objects by $\varphi(X, \xi) = (X, \xi + \lambda X)$ and φ is the identity on morphisms. Therefore (1) and (2) in (4.4) are obviously satisfied. Moreover, for $\mathfrak{D}': \mathbf{L}_{d+i} \to j'^*H$, given by (4), we get

$$\begin{aligned} \mathfrak{D}'_{(X,\xi),(Y,\eta)}(f) &= (d+i)(f) + (f_*\xi - f^*\eta) \\ &= d(f) + f_*(\xi + \lambda X) - f^*(\eta + \lambda Y) \\ &= \mathfrak{D}_{(X,\xi+\lambda X),(Y,\eta+\lambda Y)}(f) \end{aligned}$$

so that also (4.4)(3) is satisfied. This completes the proof of (7.7). □

V

Maps between mapping cones

We study maps between mapping cones, $C_f \to C_g$. In particular, we study the properties of the action of $[\Sigma A, C_g]$ on the set $[C_f, C_g]$, $f: A \to X$. This action leads to natural group actions on a subcategory **PAIR** of $Ho(\mathbf{C})$. The general concept of a natural group action in Chapter IV is mainly motivated by the properties of the category **PAIR**. We introduce subcategories

$$\mathbf{PRIN} \subset \mathbf{TWIST} \subset \mathbf{PAIR} \subset Ho(\mathbf{C}).$$

PRIN contains the principal maps and **TWIST** the twisted maps between mapping cones. We show that these categories can be described as linear extensions of model categories $\mathbf{Prin}/\simeq$ and $\mathbf{Twist}/\simeq$ respectively.

In many applications, for example in topology, it is possible to compute the model categories, but it is much harder to compute the categories **PRIN** and **TWIST**; since there is an extension problem. A result in §5 allows under suitable conditions the computation of the natural equivalence relation $\simeq$ on the model categories **Prin** and **Twist**.

In the sections §7...§10 we describe some results in topology for which the concepts of this chapter are relevant. For example, we show that problems of J.H.C. Whitehead can be solved by use of twisted maps as discussed in §2. Moreover, we prove the general suspension theorem under D and the general loop theorem over D which imply applications of the abstract theory in topology.

§1 Group actions on the category PAIR

Let $\mathbf{C}$ be a cofibration category with an initial object $*$. We have the canonical functor (see (II.1.3))

(1.1) $$Ho(\mathbf{Pair}(\mathbf{C})) \to Ho(\mathbf{C})$$

which carries the pair $(i_X: Y \to X) = (X, Y)$ in **Pair**(**C**) to the object X in **C**.

(1.2) ***Definition of the category*** **PAIR**. Objects (C_f, Y) are principal cofibrations $Y \rightarrowtail C_f$ in **C** with attaching map $f \in [X, Y]$ where $X = \Sigma X'$ is a suspension in **C**, compare (II.8.3). We denote the object (C_f, Y) also by C_f or simply by f. A morphism $C_f \to C_g$ in **PAIR** is a homotopy class in $Ho(\mathbf{C})$ in the image of the functor (1.1). Let **PAIR** (f, g) be the set of all such morphisms; this set is the image of the function

$$[(C_f, Y), (C_g, B)] \to [C_f, C_g]$$

given by (1.1). In general, **PAIR** (f, g) is a proper subset of $[C_f, C_g]$. In this chapter we may assume for all (C_f, Y) in **PAIR** that the objects X, Y, and C_f are fibrant and cofibrant in **C** and that $f: X \to Y$ is a map in **C**. ||

The assumption on X in (1.2), $X = \Sigma X'$, implies that $[\Sigma X, U]$ is an abelian group for all U. The action (II.8.10) restricts to the subset **PAIR** (f, g) of $[C_f, C_g]$ and yields the action

$$\mathbf{PAIR}\,(f, g) \times [\Sigma X, C_g] \xrightarrow{+} \mathbf{PAIR}\,(f, g) \tag{1.3}$$

of the abelian group $[\Sigma X, C_g] = D(f, g)$. We denote this action by D. With the notation in (IV.2.1) and (IV.2.7) we get

(1.4) **Proposition.** *The action D is a group action on the category* **PAIR** *and D has a left distributivity law.*

Proof of (1.4). Let $F_1: C_f \to C_g$ and $F_2: C_g \to C_h$ be maps in **Pair**(**C**). Then we have the class (see (II.8.1)),

$$\xi = \{F_1 \pi_f\} \in \pi_1^X(C_g, B). \tag{1.5}$$

We derive from (II.12.8) that the action (1.3) has the following distributivity law:

$$\begin{aligned} (F_2 + \beta)(F_1 + \alpha) &= F_2 F_1 + (\nabla\xi)^*(\beta, F_2) + (F_2 + \beta)\alpha \\ &= F_2 F_1 + (\nabla\xi)^*(\beta, F_2) + F_2\alpha + (\nabla\alpha)^*(\beta, F_2), \end{aligned} \tag{1.6}$$

where $\nabla\xi$ and $\nabla\alpha$ are defined in (II.§ 12). The induced functions and the mixed term of the action are

$$\begin{aligned} F_1^*(\beta) &= (\nabla\xi)^*(\beta, F_2), \\ F_{2*}(\alpha) &= F_2\alpha, \\ \Delta_{F_2, F_1}(\beta, \alpha) &= (\nabla\alpha)^*(\beta, F_2). \end{aligned} \tag{1.7}$$

Thus Δ_{F_2, F_1} does not depend on F_1. Therefore formula (1.6) is a left distributivity law. □

Remark. The proof makes use of the element ξ in (1.5) which we only have for a pair map F. This is the reason for the definition of **PAIR** (f, g) via the image of the functor (1.1).

§2 Principal and twisted maps between mapping cones

The following construction of a map $C_f \to C_g$ between mapping cones is classical. We assume that C_g is fibrant. Suppose we have a diagram of maps in **C**

(2.1)

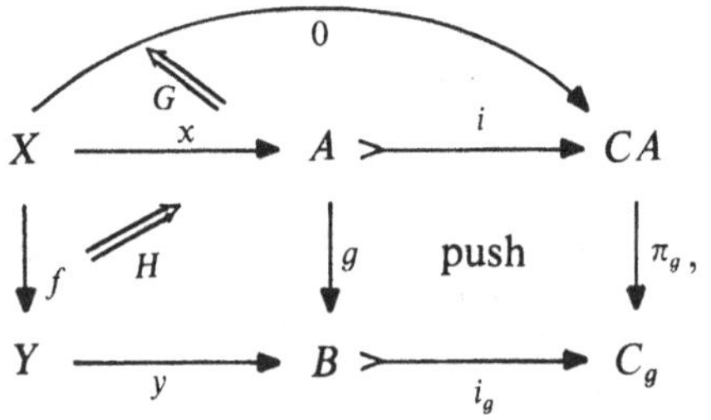

and homotopies $H: yf \simeq gx$ and $G: ix \simeq O$ rel $*$. Then we obtain

(2.2) $$F = C(x, y, H, G): C_f \to C_g$$

by

$$\begin{cases} Fi_f = i_g y, \text{ and} \\ F\pi_f = i_g H + \pi_g G. \end{cases}$$

Here $\pi_f: CX \to C_f$ is the map in (II.8.1) and $F\pi_f$ is defined by addition of homotopies. The map $G: CX \to CA$, given by the homotopy G, is an extension of x. This shows that the map F in (2.2) is well defined up to homotopy rel Y by the track classes of H and G, see (II.8.6). If x is a based map we have a canonical choice for G (up to homotopy rel X) by a lifting in the diagram

$$
\begin{array}{ccccc}
X & \xrightarrow{\;x\;} & A & \longrightarrow & CA \\
\Big\downarrow & & {\scriptstyle G}\nearrow & & {\scriptstyle o}\Big\downarrow \scriptstyle{\sim} \,, \\
CX & \xrightarrow{\quad o \quad} & & * &
\end{array}
$$

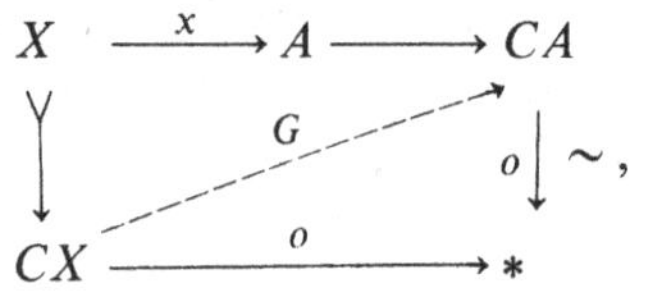

see (II.1.11). We call the map F in (2.2) or its homotopy class rel Y a **principal map** between mapping cones. The map F corresponds to f'' in (II.9.1).

We now introduce a further method of defining a map between mapping cones. This method generalizes the construction in (2.1). Suppose the diagram of maps and homotopies

(2.3)

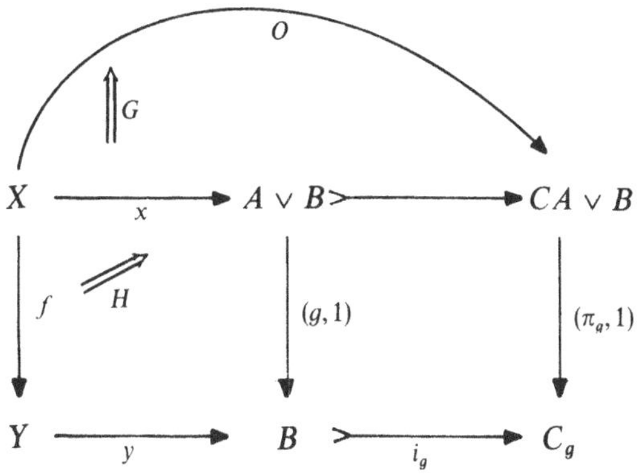

is given. Then we define the map

(2.4) $$F = C(x, y, H, G): C_f \to C_g$$

by

$$\begin{cases} Fi_f = i_g y, \text{ and} \\ F\pi_f = i_g H + (\pi_g, 1)G. \end{cases}$$

We call F in (2.4) or its homotopy class rel Y *a* **twisted map** between mapping cones. Again F is well defined up to homotopy rel Y by the track classes of H and G. Clearly, a principal map is also a twisted map.

We have the following characterization of twisted maps and principal maps respectively. Consider the commutative diagram

(2.5)
$$\begin{array}{ccc} \pi_1^X(CA, A) & \longrightarrow & \pi_1^X(CA \vee B, A \vee B) \\ & {\scriptstyle \pi_{g*}} \searrow \quad \swarrow {\scriptstyle (\pi_g, 1)_*} & \\ & \pi_1^X(C_g, B) & \end{array}$$

(2.6) **Proposition.** *Let* $F:(C_f, Y) \to (C_g, B)$ *be a pair map. By* F *we have the element* $\{F\pi_f\} \in \pi_1^X(C_g, B)$. *Then* F *is twisted* iff $\{F\pi_f\} \in \text{Image}\,(\pi_g, 1)_*$ *and* F *is principal* iff $\{F\pi_f\} \in \text{image}\,(\pi_g)_*$.

Proof of (2.6). Clearly, for G in (2.3) we have

$$(\pi_g, 1)_*\{G\} = \{F\pi_f\}. \tag{1}$$

Now assume a pair map $F:(C_f, Y) \to (C_g, B)$ and G is given such that (1) holds. Then we have a homotopy K for which the following diagram commutes.

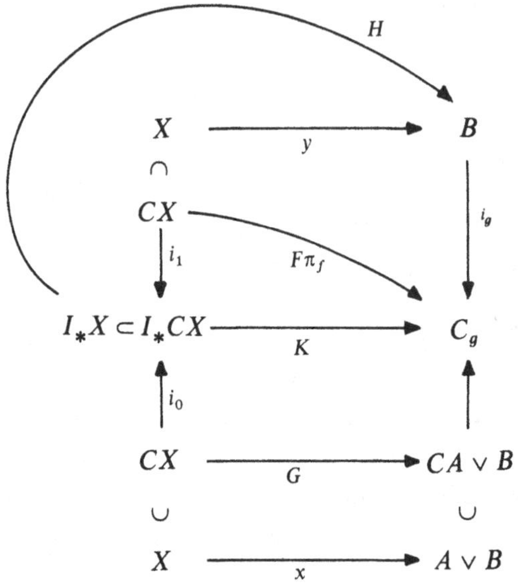

We claim that

$$F_1 = C(x, y, H, G) \simeq F \text{ rel } Y. \tag{2}$$

Indeed this is true since the difference

$$d(F, F_1) = -F\pi_f + F_1\pi_f : \Sigma X \to C_g \tag{3}$$

is nullhomotopic as follows from the existence of K, compare (II.8.13). □

In particular, if π_{g*} or $(\pi_g, 1)_*$ in (2.5) is *surjective* all pair maps $F:(C_f, Y) \to (C_g, B)$ are principal or twisted respectively.

We deduce from (2.6) the following facts: A pair map F is *twisted* iff there is a commutative diagram in $Ho\,\mathbf{Pair}(\mathbf{C})$

(2.7)
$$\begin{array}{ccc} (CX, X) & \xrightarrow{G} & (CA \vee B, A \vee B) \\ \downarrow{\scriptstyle \pi_f} & & \downarrow{\scriptstyle (\pi_g, 1)} \\ (C_f, Y) & \xrightarrow{F} & (C_g, B) \end{array}$$

Moreover, F is *principal* iff there is a commutative diagram

(2.8)
$$\begin{array}{ccc} (CX, X) & \xrightarrow{G} & (CA, A) \\ \downarrow{\scriptstyle \pi_f} & & \downarrow{\scriptstyle \pi_g} \\ (C_f, Y) & \xrightarrow{F} & (C_g, B) \end{array}$$

in $Ho\,\mathbf{Pair}(\mathbf{C})$.

(2.9) **Proposition.** *The composition of twisted maps is twisted, the composition of principal maps is principal.*

Proof. For principal maps this is clear by (2.8). Moreover, for twisted maps we have by (2.7) the commutative diagram in $Ho\,\mathbf{Pair}(\mathbf{C})$

$$\begin{array}{ccc} (CX \vee Y, X \vee Y) & \xrightarrow{\bar{G}} & (CA \vee B, A \vee B) \\ \downarrow{\scriptstyle (\pi_f, 1)} & & \downarrow{\scriptstyle (\pi_g, 1)} \\ (C_f, Y) & \xrightarrow{F} & (C_g, B) \end{array} \tag{2.10}$$

where $\bar{G}|CX = G$ and $\bar{G}|Y = i_B(F|Y)$. Here $i_B: B \rightarrowtail A \vee B$ is the inclusion. □

(2.11) ***Remark.*** Diagram (2.10) shows that twisted maps are compatible with the functional suspension in (II.11.7). This follows by naturality of diagram (II.11.6). Also we see by diagram (2.10) that the twisted map F corresponds to a principal map in the category $\mathbf{B} = \mathbf{C}^B$. In fact, the map $F \vee 1: C_f \vee B \to C_g \vee B$ is a map between mapping cones in $\mathbf{B}$ and is a principal map in $\mathbf{B}$ by (2.10). Compare (II.11.5).

§3 A linear group action

Let **PAIR** be the category in §1. We now define the subcategories

(3.1) $$\mathbf{PRIN} \subset \mathbf{TWIST} \subset \mathbf{PAIR} \subset Ho(\mathbf{C}).$$

Let $\lambda: C_f = (C_f, Y) \to C_g = (C_g, B)$ be a morphism in **PAIR**. We say λ is **principal** or **twisted** if λ can be represented by a principal or twisted map $C_f \to C_g$ respectively. By (2.9) we see that the principal morphisms in **PAIR** form a subcategory which we denote by **PRIN**. Also the twisted morphisms in **PAIR** form a subcategory which we denote by **TWIST**. The objects in **PRIN** and **TWIST** are the same as in **PAIR**, see (1.2). Let

(3.2) $$\mathbf{PRIN}(f,g) \subset \mathbf{TWIST}(f,g) \subset \mathbf{PAIR}(f,g) \subset [C_f, C_g]$$

be the subsets of principal and twisted maps respectively, compare (1.3). These sets are sets of morphisms in the categories (3.1). We define natural group actions on **TWIST** and on **PRIN** by restriction of the action in (1.4) as follows. Let

(3.3) $$\Gamma(f,g) = \mathrm{Image}\,\{i_{g*}: [\Sigma X, B] \to [\Sigma X, C_g]\}.$$

By the action (1.4) we obtain

$$\begin{array}{ccc} \mathbf{TWIST}(f,g) \times \Gamma(f,g) & \xrightarrow{+} & \mathbf{TWIST}(f,g) \\ \cup & & \cup \\ \mathbf{PRIN}(f,g) \times \Gamma(f,g) & \xrightarrow{+} & \mathbf{PRIN}(f,g) \end{array}$$

We denote this action by Γ. We point out that $\Gamma(f,g)$ in (3.3) is not a bifunctor on **TWIST**. With the notation in (IV.2.8) we get the

(3.4) **Proposition.** *Γ is a linear group action on the categories* **PRIN** *and* **TWIST** *respectively.*

Proposition (3.4) is a consequence of (4.5) below. The natural group action Γ gives us the quotient categories, (IV.2.2),

(3.5) $$\mathbf{PRIN}/\Gamma \subset \mathbf{TWIST}/\Gamma.$$

It is possible to describe these quotient categories in a different way. To this end we introduce categories

(3.6) $$\mathbf{Prin} \subset \mathbf{Twist}$$

which we call the **model categories** for **PRIN** and **TWIST** respectively. The objects in these categories are the same as in **PAIR**, see (1.2). A morphism $(\xi,\eta): f \to g$ in **Prin** is a pair $(\xi,\eta) \in [X,A]_0 \times [Y,B]$ for which the diagram

(3.7) $$\begin{array}{ccc} & * & \\ {}^{0}\nearrow & & \nwarrow{}^{0} \\ X & \xrightarrow[x]{} & A \\ \downarrow{\scriptstyle f} & & \downarrow{\scriptstyle g} \\ Y & \xrightarrow{y} & B \end{array}$$

with $x \in \xi$, $y \in \eta$ is homotopy commutative rel $*$, compare (II.9.8). A morphism $(\xi,\eta): f \to g$ in **Twist** is a pair $(\xi,\eta) \in [X, A \vee B]_2 \times [Y,B]$, for which the diagram

(3.8) $$\begin{array}{ccc} & B & \\ {}^{0}\nearrow & & \nwarrow{}^{(0,1)} \\ X & \xrightarrow{x} & A \vee B \\ \downarrow{\scriptstyle f} & & \downarrow{\scriptstyle (g,1)} \\ Y & \xrightarrow{y} & B \end{array}$$

with $x \in \xi$, $y \in \eta$ is homotopy commutative rel $*$. Here x is trivial on B, compare (II.11.4). The functor (3.6) is defined by $(\xi,\eta) \mapsto (i_A\xi,\eta)$ where $i_A: A \to A \vee B$ is the inclusion. With the inclusion $i_B: B \to A \vee B$ we define the composition in

Twist by

(3.9) $$(\xi,\eta)(\xi',\eta')=((\xi,i_B\eta)\xi',\eta\eta').$$

We now describe a commutative diagram of functors

(3.10) $$\begin{array}{ccc}\mathbf{Prin} & \subset & \mathbf{Twist}\\ {\scriptstyle q}\downarrow & & {\scriptstyle p}\downarrow\\ \mathbf{PRIN}/\Gamma & \subset & \mathbf{TWIST}/\Gamma\end{array}$$

where in fact the functors q and p are quotient functors. For (ξ,η) in (3.7) there exist homotopies H and G as in (2.2). We set $q(\xi,\eta)=\{C(x,y,H,G)\}$ compare (2.2). Here $\{\ \}$ denotes the equivalence class in $\mathbf{PRIN}/\Gamma$. Moreover, for (ξ,η) in (3.8) there exist a homotopy G and a homotopy H as in (2.3). We define $p(\xi,\eta)=\{C(x,y,H,G)\}$. Clearly, by definition of principal and twisted maps in § 2 the functors q and p are surjective on morphism sets. Let $\simeq$ be the natural equivalence relation induced by q and p respectively. Then (3.10) gives us the isomorphism of categories

(3.11) **Proposition.** $\mathbf{Prin}/\simeq\ =\mathbf{PRIN}/\Gamma$, *and* $\mathbf{Twist}/\simeq\ =\mathbf{TWIST}/\Gamma$.

Proof. We have to check that q and p in (3.10) are well defined functors. We do this for p. Let $F_1=C(x,y,H,G)$, $F_2=C(x,y,H',G')$ be maps associated to (ξ,η) by (3.8). We have to show that $p(\xi,\eta)=\{F_1\}=\{F_2\}$ is well defined. Equivalently, we have to show $d(F_1,F_2)\in\operatorname{Image}(i_{g*}:[\Sigma X,B]\to[\Sigma X,C_g])$, compare (3.3) and (II.8.15). By addition of tracks we see

$$\begin{aligned}d(F_1,F_2)&=-(\pi_g,1)G-i_gH+i_gH'+(\pi_g,1)G'\\&=-(\pi_g,1)G+i_g(\alpha)+(\pi_g,1)G+(\pi_g,1)(\gamma),\end{aligned}\qquad(*)$$

where $\alpha=-H+H':\Sigma_*X\to B$ and $\gamma=-G+G':\Sigma X\to CA\vee B$. Since X is a suspension (compare the definition of **PAIR**) we have the equivalence $\Sigma_*X\longrightarrow\Sigma X\vee X$ in $Ho(\mathbf{C}^X)$, see (II.10.17). Therefore there is $\beta:\Sigma X\to B$ with

$$\alpha:\Sigma_*X\xrightarrow{\sim}\Sigma X\vee X\xrightarrow{\beta,(g,1)x}B,\qquad(**)$$

We deduce from $(*)$ and $(**)$ $d(F_1,F_2)=i_{g*}(\beta+p_2\gamma)$, where $p_2:CA\vee B\to B$ is the projection. □

We derive from (3.11) and (3.4).

(3.12) **Theorem.** *We have a commutative diagram of linear extensions of categories*

$$\begin{array}{ccccc} \Gamma/I & + \longrightarrow & \mathbf{TWIST} & \xrightarrow{p} & \mathbf{Twist}/\simeq \\ \| & & \uparrow & & \uparrow \\ \Gamma/I & + \longrightarrow & \mathbf{PRIN} & \xrightarrow{q} & \mathbf{Prin}/\simeq \end{array} \tag{1}$$

We say that (ξ, η) is **associated** to the twisted map F if $p\{F\} = \{\xi, \eta\}$. In case $F:(C_f, Y) \to (C_g, B)$ is a pair map we obtain a pair (ξ, η) associated to F by the restriction $\eta = F|Y: Y \to B$ and by an element $\xi \in [X, A \vee B]_2$ with

$$F_*\{\pi_f\} = (\pi_g, 1)_* \partial^{-1}(\xi). \tag{2}$$

Therefore (II.12.5) and (II.12.6) show

$$\nabla_F = \nabla F_*\{\pi_f\} = (1 \vee i_g)_* E\xi. \tag{3}$$

Theorem (3.12) describes precisely the connection between the category **TWIST** and the model category **Twist**. In many applications it is possible to compute the category **Twist** but it is much harder to compute the category **TWIST**. For this we have to solve three difficult problems:

(3.13) ***Problems***

(1) *The homotopy problem*: compute the equivalence relation $\simeq$ on **Twist**!
(2) *The isotropy problem*: compute the isotropy groups of the action Γ!
(3) *The extension problem*: if (1) and (2) are solved, determine the extension class $\{\mathbf{TWIST}\} \in H^2(\mathbf{TWIST}/\simeq, \Gamma/I)$ in (IV.6.1)!

We will exhibit examples for which we can solve these problems (at least for certain small subcategories of **TWIST**). Clearly, we have similar problems for the computation of **PRIN**.

(3.14) ***Notation.*** Let $\mathfrak{X}$ be a class of objects in **PAIR**. We write **PAIR** $(\mathfrak{X})$ for the full subcategory of **PAIR** consisting of objects in $\mathfrak{X}$. In the same way we define **TWIST** $(\mathfrak{X})$, **Twist** $(\mathfrak{X})$, **PRIN** $(\mathfrak{X})$ and **Prin** $(\mathfrak{X})$ respectively.

Now recall the definition of a model functor in (I.1.10).

(3.15) **Proposition.** *A model functor $\alpha: \mathbf{C} \to \mathbf{K}$ with $\alpha * = *$ induces a commutative diagram of functors*

$$\begin{array}{ccc} Ho(\mathbf{C}) & \xrightarrow{Ho(\alpha)} & Ho(\mathbf{K}) \\ \cup & & \cup \\ \mathbf{PAIR_C} & \overset{a}{- - - \rightarrow} & \mathbf{PAIR_K} \\ \cup & & \cup \\ \mathbf{TWIST_C} & - - - \rightarrow & \mathbf{TWIST_K} \\ \cup & & \cup \\ \mathbf{PRIN_C} & - - - \rightarrow & \mathbf{PRIN_K} \end{array}$$

Proof. Since α is compatible with push outs we know that $\alpha Y \rightarrowtail M\alpha C_f \xrightarrow{\sim} \alpha C_f$ is a principal cofibration in **K** with attaching map $\alpha f \in [\alpha X, \alpha Y]$ where $\alpha X \sim \Sigma M\alpha X'$ is equivalent to a suspension in **K**. The functor a carries (C_f, Y) to $(M\alpha C_f, \alpha Y)$. Since diagram (2.5) is compatible with α we see that a carries twisted maps to twisted maps, and carries principal maps to principal maps. We also denote the functor a by α. □

(3.16) **Proposition.** *A model functor* $\alpha: \mathbf{C} \to \mathbf{K}$ *with* $\alpha * = *$ *induces a commutative diagram of linear extensions of categories*:

$$\begin{array}{ccc} \Gamma/I & \xrightarrow{\alpha} & \Gamma/I \\ {\scriptstyle +}\big\downarrow & & \big\downarrow{\scriptstyle +} \\ \mathbf{TWIST_C} & \xrightarrow[\alpha]{} & \mathbf{TWIST_K} \\ \big\downarrow & & \big\downarrow \\ \mathbf{Twist_C}/\simeq & \xrightarrow[\alpha]{} & \mathbf{Twist_K}/\simeq \end{array}$$

The left-hand side is the extension (3.12) in the cofibration category **C**. The functor $\alpha: \mathbf{Twist_C} \to \mathbf{Twist_K}$ is given by $\alpha(\xi, \eta) = (\alpha\xi, \alpha\eta)$. The proposition is a consequence of (II.8.27). A similar result holds for **PRIN**.

§3a Appendix: the homotopy category of Moore spaces in degree *n*

A **Moore-space** $M(A, n)$ in degree $n \geqq 2$ is a simply connected CW-space together with an isomorphism

$$H_i M(A, n) = \begin{cases} A & i = n, \\ 0 & i \neq n. \end{cases} \tag{3a.1}$$

Here H_i is the (reduced) singular homology. The homotopy type of $M(A, n)$ is well defined by the abelian group A and by the degree n (the construction of $M(A, n)$, however, is not natural in A since the linear extension of categories in (3a.2) below is not split).

We choose for each pair (A, n) a pointed Moore space $M(A, n)$. Let $\mathbf{M}^n$ be the full subcategory of $\mathbf{Top}^*/\simeq$ consisting of the Moore spaces $M(A, n)$ where A is an abelian group. Then we get the isomorphic linear extensions of categories

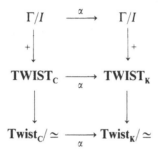

(3a.2)
$$\begin{array}{ccccc} E^n & + \longrightarrow & \mathbf{M}^n & \xrightarrow{H_n} & \mathbf{Ab} \\ \uparrow \cong & & \uparrow \cong & & \uparrow \cong \\ \Gamma/I & + \longrightarrow & \mathbf{PRIN}(\mathfrak{X}) & \xrightarrow{q} & \mathbf{Prin}(\mathfrak{X})/\simeq \end{array} .$$

Here H_n is the homology functor which carries $M(A,n)$ to A, recall that **Ab** denotes the category of abelian groups. The natural system E^n on **Ab** is the bimodule

(3a.3)
$$E^n(A,B) = \begin{cases} \operatorname{Ext}(A, \Gamma B), & n = 2, \\ \operatorname{Ext}(A, B \otimes \mathbb{Z}/2), & n \geqq 3. \end{cases}$$

Here Γ is the *quadratic functor* of J.H.C. Whitehead with $\Gamma(A) = \pi_3 M(A,2)$.

The class $\mathfrak{X}$ of maps in **Top*** in (3a.2) is given as follows. For each A we choose a short exact sequence $\bigoplus_M \mathbb{Z} \overset{d}{\rightarrowtail} \bigoplus_N \mathbb{Z} \twoheadrightarrow A$ and we choose a map

(3a.4)
$$f : \bigvee_M S^n \to \bigvee_N S^n,$$

which represents d. Then we have the mapping cone $M(A,n) = C_f$. It is easy to check that maps between such mapping cones are homotopic to principal maps. This yields the isomorphism of categories $\mathbf{M}^n \cong \mathbf{PRIN}(\mathfrak{X})$ in (3a.2). The bottom row in (3a.2) is given by the linear extension in (3.12). One can check that $\mathbf{Prin}(\mathfrak{X})/\simeq$ can be identified with the full subcategory of $\mathbf{Chain}_{\mathbb{Z}}/\simeq$ consisting of the cellular chain complexes $C_*(M(A,n))$. The homology yields the isomorphism H in (3a.2).

We obtain the top row of (3a.2) by the following **universal coefficient theorem** (compare Hilton (1965)) which we easily derive from the cofiber sequence for $C_f = M(A,n)$.

(3a.5) **Proposition.** *Let U be a pointed space. Then we have the short exact sequence*

$$\operatorname{Ext}(A, \pi_{n+1}U) \rightarrowtail [M(A,n), U] \overset{p}{\twoheadrightarrow} \operatorname{Hom}(A, \pi_n U)$$

For $n \geqq 3$ this is an exact sequence of abelian groups; for $n = 2$ the left-hand group acts freely on the set $[M(A,2), U]$ such that each subset $p^{-1}p(x)$ is the orbit of x.

For $U = M(B,n)$ we have $\pi_n M(B,n) = B$ and $\pi_{n+1} M(B,n) = (\Gamma B$ for $n = 2$ and $B \otimes \mathbb{Z}/2$ for $n \geqq 3)$. Thus (3a.5) gives us the short exact sequence

(3a.6)
$$E^n(A,B) \overset{+}{\rightarrowtail} [M(A,n), M(B,n)] \overset{p}{\twoheadrightarrow} \operatorname{Hom}(A,B).$$

Here p can be identified with the homology functor H_n. It is easy to see that the action $+$ of $E^n(A,B)$ satisfies the linear distributivity law. Hence (3a.6) as well yields the linear extension in (3a.2).

For $n \geqq 3$ there is an algebraic description of the extension (3a.2) as follows. For the homotopy group $\pi_{n+2}(M(A,n))$, $n \geqq 3$, we have the short exact sequence

(3a.7) $$A \otimes \mathbb{Z}/2 \rightarrowtail \pi_{n+2}(M(A,n)) \to A * \mathbb{Z}/2$$

which, in fact, represents the extension of groups E_A in (IV. 6.7) where we set $p = 2$. We obtain (3a.7) by considering the fiber sequence of f in (3a.4). For the linear extension of categories $\mathbf{E}_2$ in (IV.6.7) we now get the isomorphism of extensions ($n \geqq 3$)

(3a.8) $$\begin{array}{ccccc} \mathrm{Hom}(-*\mathbb{Z}/2, -\otimes\mathbb{Z}/2) & \xrightarrow{+} & \mathbf{E}_2 & \longrightarrow & \mathbf{Ab} \\ \cong\uparrow & & \cong\uparrow \pi_{n+2} & & \| \\ \mathrm{Ext}(-, -\otimes\mathbb{Z}/2) & \xrightarrow{+} & \mathbf{M}^n & \longrightarrow & \mathbf{Ab} \end{array}$$

where we use the natural isomorphism

(3a.9) $$\mathrm{Ext}(A, B \otimes \mathbb{Z}/2) = \mathrm{Hom}(A * \mathbb{Z}/2, B \otimes \mathbb{Z}/2)$$

Now (IV.6.7) and (IV.6.5) shows that the extension (3a.2) is *not split*. The isomorphism (3a.8) is our best algebraic description of the category $\mathbf{M}^n$, $n \geqq 3$. Barratt (1954) computed the category $\mathbf{M}^n$ in terms of generators and relations. For $n = 2$ we do not have a nice algebraic description of $\mathbf{M}^2$ as in (3a.8).

Let **Abf** be the small category of finitely generated abelian groups, then the element

(32.10) $$\{\mathbf{M}^n\} \in H^2(\mathbf{Abf}, E^n) \cong \mathbb{Z}/2\mathbb{Z}, \quad n \geqq 2,$$

is a generator of $\mathbb{Z}/2\mathbb{Z}$, see Hartl.

§4 A quadratic group action

We define quadratic group actions on the categories **PRIN** and **TWIST** as follows. Let C_f and C_g be objects in **PAIR** with $f: X \to Y$, $g: A \to B$. We define the subgroup

(4.1) $$E(f,g) = \mathrm{Image}\, E_g \subset [\Sigma X, C_g]$$

by the image of the functional suspension in (II.11.7). Moreover, we define the subgroup

(4.2) $$\bar{E}(f,g) = E_g(i_A \,\mathrm{kernel}\, g_*) \subset [\Sigma X, C_g],$$

where $g_*: [X, A] \to [X, B]$ and where $i_A: A \to A \vee B$ is the inclusion. Clearly, we have by definitions of E_g and Γ the inclusions

(4.3) $$\Gamma(f,g) \subset \bar{E}(f,g) \subset E(f,g).$$

We claim that the action (1.4) gives us actions

(4.4) $$\begin{array}{ccc} \mathbf{TWIST}(f,g) \times E(f,g) & \xrightarrow{+} & \mathbf{TWIST}(f,g) \\ \cup & & \cup \\ \mathbf{PRIN}(f,g) \times \bar{E}(f,g) & \xrightarrow{+} & \mathbf{PRIN}(f,g) \end{array},$$

which we denote by $E+$ and $\bar{E}+$ respectively.

(4.5) **Proposition.** *$E+$ and $\bar{E}+$ are quadratic group actions on* **TWIST** *and* **PRIN** *respectively.*

Compare the definition of a quadratic action in (IV.2.8). The functor π_1 on the homotopy category of 2-dimensional CW-complexes admits a quadratic action which is an example for (4.5). Compare (VI.8.3) where we set $n=2$.

Proof. From (2.6) we derive that the actions in (4.4) are well defined by (1.4) since $(F+\alpha)_*\{\pi_f\} = F_*\{\pi_f\} + j\alpha$ for a pair map $F:(C_f, Y) \to (C_g, B)$ and for $\alpha \in [\Sigma X, C_g]$, ($X$ is a suspension). Now let $\alpha \in E(f,g)$, $\beta \in E(g,h)$ and let $C_f \xrightarrow[F_1]{} C_g \xrightarrow[F_2]{} C_h$ be morphisms in **TWIST** with

$$p(\xi_i, \eta_i) = \{F_i\}, \quad (i=1,2), \tag{1}$$

see (3.10). From (II.12.3) and (II.12.5) we deduce that $\nabla\xi$ ($\xi = F_{1*}\{\pi_f\}$) and $\nabla\alpha$ in (1.6) and (1.7) are given by

$$\nabla\xi = (1 \vee i_g)_* E\xi_1, \tag{2}$$

$$\nabla\alpha = (1 \vee i_g)_* E\hat{\alpha} \quad \text{for } \alpha \in E_g(\hat{\alpha}). \tag{3}$$

Thus we derive from (1.6) the following distributivity law of the action $E+$ on **TWIST**.

$$(F_2+\beta)(F_1+\alpha) = F_2F_1 + (E\xi_1)^*(\beta, i_{h*}\eta_2) + F_{2*}\alpha + (E\hat{\alpha})^*(\beta, i_{h*}\eta_2). \tag{4}$$

For the proof that $E+$ is natural we have to show that the summands in (4) are elements of $E(f,h)$. Since F_2 is twisted we see by naturality of the functional suspension (compare (2.11)) that

$$F_{2*}\alpha \in E_h((\xi_2, i_2\eta_2)\hat{\alpha}) \subset E(f,h). \tag{5}$$

By similar naturality arguments we see for $\beta \in E_h(\hat{\beta})$

$$(E\xi_1)^*(\beta, i_{h*}\eta_2) \in E_h(\xi_1^*(\hat{\beta}, i_2\eta_2)), \tag{6}$$

$$(E\hat{\alpha})^*(\beta, i_{h*}\eta_2) \in E_h(\hat{\alpha}^*(\hat{\beta}, i_2\eta_2)). \tag{7}$$

For (6) and (7) we use the commutative diagram in *Ho* **Pair(C)**:

$$\begin{array}{ccc} (CA \vee B, A \vee B) & \xrightarrow[(\hat{\beta}, i_2\eta_2)]{} & (CV \vee W, V \vee W) \\ \downarrow{\scriptstyle (\pi_0, 1)} & & \downarrow{\scriptstyle (\pi_h, 1)} \\ (\Sigma A \vee B, B) & \xrightarrow[(\beta, i_h\eta_2)]{} & (C_h, W) \end{array}$$

By (5), (6) and (7) we see that $E+$ is in fact a natural group action, see (IV.2.1). Moreover, the induced functions and the mixed term respectively are given as follows:

(4.6) $$\begin{cases} F_{2*}(\alpha) = F_2\alpha, \\ F_1^*(\beta) = (E\xi_1)^*(\beta, i_h\eta_2), \\ \beta \odot \alpha = (E\hat{\alpha})^*(\beta, i_h\eta_2) \quad \text{for } \alpha \in E_g(\hat{\alpha}). \end{cases}$$

It follows from (II.11.17) that $E+$ has a quadratic distributivity law. If F_1 and F_2 are in **PRIN** and $\alpha \in \bar{E}(f, g)$, $\beta \in \bar{E}(g, h)$ we obtain a distributivity law with

(4.7) $$\begin{cases} F_{2*}(\alpha) = F_2\alpha, \\ F_1^*(\beta) = (\Sigma\xi_1)^*\beta, \quad q(\xi_1, \eta_1) = \{F_1\}, \\ \beta \odot \alpha = (\Sigma\hat{\alpha})^*\beta, \quad \alpha \in E_g(i_1\hat{\alpha}). \end{cases}$$ □

From (4.3) and (3.12) we derive the commutative diagram of quadratic actions

(4.8) $$\begin{array}{ccccc} (E/\Gamma)+ & \longrightarrow & \mathbf{Twist}/\simeq & \xrightarrow{\bar{p}} & \mathbf{TWIST}/E \\ \uparrow & & \uparrow & & \uparrow \\ (\bar{E}/\Gamma)+ & \longrightarrow & \mathbf{Prin}/\simeq & \xrightarrow{\bar{q}} & \mathbf{PRIN}/\bar{E} \end{array}.$$

§ 5 The equivalence problem

We use the quadratic action

(5.1) $$(E/\Gamma)+ \longrightarrow \mathbf{Twist}/\simeq \xrightarrow{\bar{p}} \mathbf{TWIST}/E$$

in (4.8). Let

(5.2) $$(\xi, \eta), (\xi_1, \eta_1) \in \mathbf{Twist}(f, g),$$

where $f: X \to Y$, $g: A \to B$. By definition of $\bar{p}$ we have the following lemma:

(5.3) **Lemma.** *We have $\bar{p}(\xi, \eta) = \bar{p}(\xi_1, \eta_1)$ if and only if the following two conditions (a) and (b) hold:*

(*a*) $$i_{g*}\eta = i_{g*}\eta_1.$$

(*b*) *There are twisted maps* $F, F_1: C_f \to C_g$ *associated to* (ξ, η) *and* (ξ_1, η_1) *and there is a homotopy* $H: Fi_f \simeq F_1 i_f$ *such that*

$$d(F, H, F_1) \in E_g(\delta)$$

for some $\delta \in \pi_0^X(A \vee B)_2$.

Compare (II.8.15).

(5.4) ***Definition.*** If $\bar{p}(\xi, \eta) = \bar{p}(\xi_1, \eta_1)$ let $\bar{d}(\xi, \eta, \xi_1, \eta_1) \subset \pi_0^X(A \vee B)_2$ be the set of all δ as in (*b*) of (5.3). ||

Then we have the following *characterization of the equivalence relation* $\simeq$ *on* **Twist**:

(5.5) $$(\xi, \eta) \simeq (\xi_1, \eta_1) \Leftrightarrow \begin{cases} \bar{p}(\xi, \eta) = \bar{p}(\xi_1, \eta_1), \text{ and} \\ 0 \in \bar{d}(\xi, \eta, \xi_1, \eta_1). \end{cases}$$

This follows since $E_g(0) = \Gamma(f, g)$, see (II.11.7).

We now assume that C_f is a double mapping cone as in the following diagram, compare (II.§ 13).

(5.6) $$\begin{array}{ccccc} & & C_f & \xrightarrow[F, F_1]{} & C_g \\ & & \cup & & \cup \\ X & \xrightarrow{f} C_r = & Y & \underset{\eta, \eta_1}{\dashrightarrow} & B \\ & & \cup & & \\ & Q & \xrightarrow{r} T & & \end{array}$$

Here X is a suspension, but Q is a based object which needs not to be a suspension.

Remark. If $T = *$ the map r needs not to be the trivial map, compare (II.8.2). If $T = *$ and if r is the trivial map, $r = 0$, then $C_r = Y = \Sigma Q$ is a suspension. This leads to an interesting special case of the result in (5.7) below.

The following theorem, in which we use the *notation of* (II.13.8), is our main tool for the computation of the equivalence relation $\simeq$ on **Twist**, see (3.13) (1) and (5.5).

(5.7) **Theorem.** *Suppose that* $* = T$ *and that*

$$(\pi_g, 1)_*: \pi_2^Q(CA \vee B, A \vee B) \to \pi_2^Q(C_g, B)$$

is surjective. Then we have

$$\bar{p}(\xi,\eta)=\bar{p}(\xi_1,\eta_1)\Leftrightarrow i_{g*}\eta=i_{g*}\eta_1$$

and $\bar{d}(\xi,\eta,\xi_1,\eta_1)$ *is the set of all elements*

$$\delta=\xi_1-\xi+(\bar{\nabla}f)^*(-\alpha,i_2\eta_1)+\beta,$$

where $\alpha\in\pi_1^Q(A\vee B)_2$ *and* $\beta\in\pi_0^X(A\vee B)_2$ *satisfy*

$$\eta+(g,1)_*\alpha=\eta_1 \text{ by the action (1.4), and}$$
$$\partial^{-1}\beta\in\text{kernel }(\pi_g,1)_*$$

with $(\pi_g,1)_*:\pi_1^X(CA\vee B,A\vee B)\to\pi_1^X(C_g,B)$.

(5.8) ***Remark.*** We can replace the assumption $*=T$ in (5.7) by the following assumption: Suppose for all maps $y,y_1:C_r\to B$ each track class $\{H\}$ of a homotopy $H:i_gy\simeq i_gy_1$ rel $*$ contains a pair map

$$(I_*C_r,I_*T)\to(C_g,B),\quad \text{see (II.8.12).}$$

Proof of (5.7). We prove that with the assumption in (5.7) condition (*a*) in (5.3) implies condition (*b*) in (5.3). Let

$$i_{g*}\eta=i_{g*}\eta_1,\tag{1}$$

and let F,F_1 be maps associated to (ξ,η) and (ξ_1,η_1) respectively. By (1) there is a homotopy

$$H:I_*C_r\to C_g,\quad H:Fi_f\simeq F_1i_f\,\text{rel}\,*.\tag{2}$$

By (5.8) we can assume that H is a pair map:

$$H:(I_*C_r,\dot{I}_*C_r)\to(C_g,B).\tag{3}$$

Here $I_*C_r=C_{w_r}$ is a mapping cone by (II.8.12). Since by assumption $(\pi_g,1)_*$ is surjective onto $\pi_2^Q(C_g,B)$ we know by (2.6) that H is a twisted map. Therefore also

$$G:(\dot{I}_*C_f,\dot{I}_*C_r)\to(C_g,B)\tag{4}$$

is a twisted map with respect to $\dot{I}_*C_f=C_W$, $W:X\vee\Sigma Q\vee X\to\dot{I}_*C_r$. Here we define G by $G|_{I_*C_r}=H$ and $G|_{i_0C_f}=F$, $G|_{i_1C_f}=F_1$. By (II.13.7) we know $w_f\in E_W(\bar{\xi})$ with $\bar{\xi}$ defined in (II.13.8). By naturality of the functional suspension with respect to twisted maps we obtain for

$$d_G=G_*W_f=d(F,H,F_1)\tag{5}$$

the result

$$d_G\in E_g(-\xi+(\bar{\nabla}f)^*(-\alpha,i_2\eta_1)+\xi_1),\tag{6}$$

where $\alpha\in\partial(\pi_g,1)_*^{-1}H^*\{\pi_{W_r}\}$. This shows $d_G\in E_g(\delta)$ for some δ and thus (*b*) in (5.3) is satisfied. For α we have

$$(g,1)_*\alpha = w_r^*(H|I^{\cdot}_*C_r) = d(F|Y,\bar{H},F_1|Y). \tag{7}$$

Here $\bar{H}:I_*T\to B$ is the restriction of H in (3). By (II.8.15) and (7) we get

$$\eta + (g,1)_*\alpha = \eta_1. \tag{8}$$

Thus we conclude from (6) and (5.4) that $\bar{d}(\xi,\eta,\xi_1,\eta_1)$ consists only of elements δ as described in (5.7).

On the other hand, if $\alpha\in\pi_1^Q(A\vee B)_2$ satisfies (8), we know that there exists a homotopy

$$\bar{H}:F|T\simeq F_1|T\,\mathrm{rel}\,*$$

with $d(F|Y,\bar{H},F_1|Y)=(g,a)_*\alpha$. This is equivalent to the homotopy commutativity of

$$\begin{array}{ccc} \Sigma Q & \xrightarrow{\alpha} & A\vee B \\ {\scriptstyle w_r}\downarrow & & \downarrow{\scriptstyle (g,1)} \\ I^{\cdot}_{Y_0}C_r & \xrightarrow[a]{} & B \end{array}$$

where $a=(F|Y,\bar{H},F_1|Y)$. This shows that

$$(\alpha,a)\in\mathbf{Twist}(w_r,g).$$

Therefore there is $H:C_{w_r}\to C_g$ associated to (α,a). This shows that each element δ as described in (5.7) is an element of the set $\bar{d}(\xi,\eta,\xi_1,\eta_1)$. □

Theorem (5.7) gives us the following result on isotropy groups: Let

$$F:C_f\to C_g$$

be a twisted map as in (5.6) which is associated to $(\xi,\eta)\in\mathbf{Twist}(f,g)$. Let $I(F)\subset[\Sigma X,C_g]$ be the *isotropy group in F* of the action

$$[C_f,C_g]\times[\Sigma X,C_g]\xrightarrow{+}[C_f,C_g]$$

in (1.4). Clearly, $I(F)=I_0(\eta)$ depends only on $i_{g*}\eta=i_f^*F\in[Y,C_g]$. By the homomorphism in (II.11.6) we obtain the subgroup

(5.9) $$I_E(\eta)=\partial(\pi_g,1)_*^{-1}jI_0(\eta)\in\pi_0^X(A\vee B)_2.$$

This is the inverse image of $I_0(\eta)$ under the functional suspension E_g. Since by (5.3)

(5.10) $$I_E(\eta)=\bar{d}(\xi,\eta,\xi,\eta)$$

we derive from (5.7):

(5.11) **Corollary.** *Suppose the assumptions in* (5.7) *are satisfied. Then* $I_E(\eta)$ *is the subgroup of all elements*

$$\delta = (\bar{\nabla} f)^*(-\alpha, i_2\eta) + \beta,$$

where $\alpha \in \pi_1^Q(A \vee B)_2$, $\beta \in \pi_0^X(A \vee B)_2$ *satisfy*

$$\left.\begin{array}{l} \eta + (g,1)_*\alpha = \eta, \text{ that is } (g,1)_*\alpha \in I_0(\eta), \\ \partial^{-1}\beta \in \text{kernel } (\pi_g, 1)_*. \end{array}\right\} \qquad (*)$$

§6 Maps between fiber spaces in a fibration category

Let **F** be a fibration category with a final object $*$. We obtain the subcategory

(6.1) $$\mathbf{PAIR} \subset Ho(\mathbf{F})$$

which is dual to the category (1.2). Objects are pairs $(P_f|X)$, or maps $f: X \to Y$, where $Y = \Omega Y'$ is a loop object. (We may assume that X, Y, P_f are fibrant and cofibrant in **F**.) For objects f and $g: A \to B$ in **PAIR** the subset

$$\mathbf{PAIR}(f,g) \subset [P_f, P_g]$$

of morphisms in **PAIR** consists of all homotopy classes $\{F\}$ which can be represented by a map $F:(P_f|X) \to (P_g|A)$ in **Pair(F)**, see (1.2). By the action dual to (II.8.10) we obtain the action $D+$:

(6.2) $$\mathbf{PAIR}(f,g) \times [P_f, \Omega B] \xrightarrow{+} \mathbf{PAIR}(f,g)$$

Here $[P_f, \Omega B]$ is an abelian group since we assume $B = \Omega B'$. We derive from (1.5):

(6.3) **Proposition.** *The action* $D+$ *in* (6.2) *is a group action on the category* **PAIR** *and* $D+$ *has a right distributivity law.*

Compare (IV.§2). Principal maps and twisted maps in $\mathbf{PAIR}(f,g)$ are constructed dually to (2.2) and (2.4) respectively. By (2.6) we have the following result where we use the commutative diagram (compare (II.14.8)):

$$\begin{array}{ccc} \pi_B^1(WY|Y) & \xrightarrow{p_1^*} & \pi_B^1(WY \times X|Y \times X) \\ & {}_{\pi_f^*}\searrow \quad \swarrow_{(\pi_f,1)^*} & \\ & \pi_B^1(P_f|X) & \end{array}$$

(6.4) **Proposition.** *Let* $F:(P_f|X) \to (P_g|A)$ *be a pair map. By* F *we have the element* $\{\pi_g F\} \in \pi_B^1(P_f|X)$ *which satisfies:*

$$F \text{ is twisted} \Leftrightarrow \{\pi_g F\} \in \text{image}(\pi_f, 1)^*,$$
$$F \text{ is principal} \Leftrightarrow \{\pi_g F\} \in \text{image}(\pi_f)^*.$$

Now principal and twisted maps yield subcategories

$$\mathbf{PRIN} \subset \mathbf{TWIST} \subset \mathbf{PAIR} \subset Ho(\mathbf{F}),$$

as in (3.1). We define dually to (3.3) the groups

(6.5) $$\Gamma(f,g) = \text{Image}\{q^*:[X,\Omega B] \to [P_f, \Omega B]\}$$

where $q:P_f \twoheadrightarrow X$ is the pair $(P_f|X)$. By (6.2) this group acts linearly on the subcategories **PRIN** and **TWIST** respectively, see (3.4). Moreover, by (3.12) we obtain the following result:

(6.6) **Theorem.** *We have a commutative diagram of linear extensions of categories*

$$\begin{array}{ccccc} \Gamma/I & + \longrightarrow & \mathbf{TWIST} & \xrightarrow{p} & \mathbf{Twist}/\simeq \\ \| & & \uparrow & & \uparrow \\ \Gamma/I & + \longrightarrow & \mathbf{PRIN} & \xrightarrow[q]{} & \mathbf{Prin}/\simeq \end{array}$$

Here the model categories **Prin** and **Twist** are given as follows: objects are the same as in **PAIR**. Maps $(\xi,\eta):f \to g$ in **Prin** are given by commutative diagrams

(6.7) $$\begin{array}{ccc} X & \xrightarrow[\xi]{} & A \\ {\scriptstyle f}\downarrow & & \downarrow{\scriptstyle g} \\ Y & \xrightarrow{\eta} & B \\ {\scriptstyle 0}\nwarrow & & \nearrow{\scriptstyle 0} \\ & * & \end{array}$$

in $Ho(\mathbf{F})$. Next morphisms $(\xi,\eta):f \to g$ in **Twist** are commutative diagrams

(6.8) $$\begin{array}{ccc} X & \xrightarrow[\xi]{} & A \\ \downarrow{\scriptstyle (f,1)} & & \downarrow{\scriptstyle g} \\ Y \times X & \xrightarrow[\eta]{} & B \\ {\scriptstyle (0,1)}\nwarrow & & \nearrow{\scriptstyle 0} \\ & X & \end{array}$$

in $Ho(\mathbf{F})$, equivalently we have

$$\begin{cases} (\xi,\eta) \in [X,A] \times [Y \times X, B]_2, \\ (f,1)^*\eta = g_*\xi. \end{cases}$$

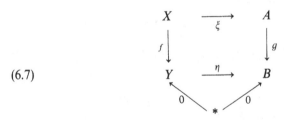

There is an obvious law of composition in **Twist** which is dual to (3.9), namely

(6.9) $$(\alpha, \beta)(\xi, \eta) = (\alpha\xi, \beta(\eta, \xi p_2)),$$

where $p_2: Y \times X \to X$ is the projection.

Now the results dual to the results in §5 are available. They can be used for the solution of the equivalence problem and of the isotropy problem in certain cases, see (3.13). Examples are described in §9 below.

§7 The homotopy type of a mapping cone in topology

We consider maps between mapping cones in the cofibration category of topological spaces. As an application of the general ideas in this chapter we obtain results on the following two fundamental problems.

(7.1) Describe conditions on the maps f and g which imply that the mapping cones C_f and C_g are homotopy equivalent!

(7.2) Compute the set of homotopy classes of maps, $[C_f, C_g]$, between mapping cones! In particular, compute the subset **TWIST**(f,g) of twisted maps and determine the law of composition for elements in such sets! Compute the groups of homotopy equivalences of C_f!

In general these problems are extremely difficult but under certain restrictions on f and g we can apply the abstract theory. In fact, we apply the theory in the category $\mathbf{Top}^D$ of *spaces under* $D, D \neq \phi$. It will be convenient for the reader to assume first that $D = *$ is a point. The results below are, in particular, of importance for this special case.

By (I.5.1) and by (II.1.4) we know that $\mathbf{Top}^D$ *is a cofibration category*. If $D = *$ is a point, this is just the category of basepoint preserving maps; in this case $*$ is also the final object and therefore each well pointed space is based. In general, a based object in $\mathbf{Top}^D$ is given by a cofibration $i: D \rightarrowtail A$ in **Top** and by a retraction $r: A \to D$, $ri = 1_D$. For example, if D has a base point and if A' is well pointed, then $A = (D \rightarrowtail D \vee A' \to D)$ is a based object in $\mathbf{Top}^D$. Now let

(7.3) $$f: X \to Y, \quad g: A \to B$$

be maps in $\mathbf{Top}^D$ and assume X and A are based objects which are suspensions in $\mathbf{Top}^D$ and assume Y and B are cofibrant in $\mathbf{Top}^D$. As in any cofibration category we have the inclusions

(7.4) $$\mathbf{PRIN}(f,g) \subset \mathbf{TWIST}(f,g) \subset \mathbf{PAIR}(f,g) \subset [C_f, C_g]^D$$

Here C_f and C_g are mapping cones in $\mathbf{Top}^D$ and $[C_f, C_g]^D$ is the set of homotopy classes of maps under D in $(\mathbf{Top}^D)_c/\simeq$. Below we describe criteria under which the inclusions of (7.4) are actually bijections. To this end we need a result which is a far reaching generalization of the Freudenthal suspension theorem.

(7.5) ***Definition.*** Let $D \rightarrowtail A$ be a cofibration in **Top**. We say that (A, D) is a-**connected** if the homotopy groups $\pi_i(A, D; d_0)$ vanish for all $i \leqq a$ and $d_0 \in D$. We write

$$\dim(A - D) = \dim(A, D) \leqq n$$

if there is a homotopy equivalence $A \simeq A'$ under D where A' is obtained from D by a well-ordered succession of attaching cells of dimension $\leqq n$, see (I.0.12). ||

(7.6) **Theorem (general suspension theorem under *D*).** *Let X and A be based objects in* $\mathbf{Top}^D$ *and let $g: A \to B$ be a map in* $\mathbf{Top}^D$. *Let X and D be path connected spaces in* **Top**. *Assume that (A, D) is $(a-1)$-connected. Then the map*

$$(\pi_g, 1)_*: \pi_1^X(CA \vee B, A \vee B) \to \pi_1^X(C_g, B)$$

is a bijection if $\dim(X, D) < 2a - 1$ *and is a surjection if* $\dim(X, D) \leqq 2a - 1$. *Here the map $(\pi_g, 1)_*$ is defined in the cofibration category* $\mathbf{Top}^D$ *by* (2.5).

Addendum. *Let X and A be based objects in* $\mathbf{Top}^D$ *and let $D \subset B' \subset B$. Suppose X and D are path connected spaces in* **Top**, *that (A, D) is $(a-1)$-connected, and that (B, B') is $(b-1)$-connected. Then the inclusion $i: B' \subset B$ induces the map*

$$(1 \vee i)_*: \pi_1^X(CA \vee B', A \vee B') \to \pi_1^X(CA \vee B, A \vee B),$$

which is a bijection if $\dim(X, D) < a + b - 2$ *and is surjective if* $\dim(X, D) \leqq a + b - 2$. *The map $(1 \vee i)_*$ is defined in the cofibration category* $\mathbf{Top}^D$ *(in particular, $A \vee B = A \bigcup_D B$).*

For $B' = D$ the addendum is a result on the map

$$(i_1)_*: \pi_1^X(CA, A) \to \pi_1^X(CA \vee B, A \vee B)$$

in (2.5) with $\mathbf{C} = \mathbf{Top}^D$. The results in (7.6) are also of interest for $D = *$.

Remark. If $D = *$, $B = *$ and if X is a sphere than (7.6) is equivalent to the classical Freudenthal suspension theorem. When $g = 0$ theorem (7.6) is a result on the partial suspension. In Baues (1975) $(\pi_g, 1)_*$ (with $D = *$) is actually embedded in an EHP sequence which can be used for the computation of the image and of the kernel of $(\pi_g, 1)_*$ in a metastable range. This is useful if we want to apply (5.7).

Now consider again the inclusions in (7.4). If the inclusion $B \subset C_g$ induces a surjection $[Y, B]^D \twoheadrightarrow [Y, C_g]^D$ of homotopy sets in $(\mathbf{Top}^D)_c/\simeq$ we know

(7.7) $$\mathbf{PAIR}(f, g) = [C_f, C_g]^D.$$

By the cellular approximation theorem equation (7.7) is satisfied if (A, D) is $(a-1)$-connected and if $\dim(Y, D) \leqq a$. (For $D = *$ we use this in (8.8), but equation (7.7) also holds in (8.7) below by a different argument.)

Next we derive from (2.6)

$$\textbf{TWIST}(f, g) = \textbf{PAIR}(f, g) \tag{7.8}$$

if $(\pi_g, 1)_*$ in (7.6) is surjective; this, for example, is satisfied provided that (A, D) is $(a-1)$-connected and $\dim(X, D) \leqq 2a - 1$.

Moreover, we derive from (2.6) that

$$\textbf{PRIN}(f, g) = \textbf{TWIST}(f, g) \tag{7.9}$$

in case $(i_1)_*$ in (7.6) is surjective; this, for example, holds if (A, D) is $(a-1)$-connected, (B, D) is $(b-1)$-connected and if $\dim(X, D) \leqq a + b - 2$.

There are numerous applications of the facts (7.7), (7.8) and (7.9) with respect to the problems in (7.1) and (7.2). We now discuss a simple application on the realizability of abstract homology homomorphisms.

Recall that a map $(\xi, \eta): f \to g$ in **Twist** is given by a homotopy commutative diagram in $(\mathbf{Top}^D)_c$:

$$\begin{array}{ccc} & B & \\ {\scriptstyle 0}\nearrow & & \nwarrow{\scriptstyle (0,1)} \\ X & \xrightarrow{\ \xi\ } & A \vee B = A \bigcup_D B \\ {\scriptstyle f}\downarrow & & \downarrow{\scriptstyle (g,1)} \\ Y & \xrightarrow{\ \eta\ } & B \end{array}, \tag{7.10}$$

where $0: X \xrightarrow{0} D \rightarrowtail B$ is the trivial map. The pair (ξ, η) yields associated twisted maps $F: C_f \to C_g$ in $\mathbf{Top}^D$ which induce the following commutative diagram with exact rows:

$$\begin{array}{ccccccccc} H_i(X, D) & \xrightarrow{f_*} & H_i(Y, D) & \longrightarrow & H_i(C_f, D) & \longrightarrow & H_{i-1}(X, D) & \xrightarrow{f_*} & \\ \downarrow{\scriptstyle \xi_*} & & & & & & \downarrow{\scriptstyle \xi_*} & & \\ H_i(A \vee B, B) & & \downarrow{\scriptstyle \eta_*} & & \downarrow{\scriptstyle \varphi = F_*} & & H_{i-1}(A \vee B, B) & & \\ \wr\| \, i & & & & & & \wr\| \, i & & \\ H_i(A, D) & \xrightarrow[g_*]{} & H_i(B, D) & \longrightarrow & H_i(C_g, D) & \longrightarrow & H_{i-1}(A, D) & \xrightarrow[g_*]{} & \end{array}$$

(7.11)

Here i is the excision isomorphism of integral singular homology.

Now assume

$$\begin{cases} \dim(Y,D) \leqq n, \quad \dim(B,D) \leqq n \text{ and,} \\ (X,D) \text{ and } (A,D) \text{ are } (n-1)\text{-connected.} \end{cases} \tag{$*$}$$

Then φ in (7.11) is actually determined by ξ_* and η_*. In this case we write $\varphi = (\xi, \eta)_*$.

(7.12) **Theorem.** *Assume* $(*)$ *above is satisfied and assume* $\dim(X,D) \leqq 2n-1$. *Then (a) and (b) holds:*

(a) An abstract homomorphism $\varphi: H_*(C_f, D) \to H_*(C_g, D)$ *is realizable by a map* $C_f \to C_g$ *in* $\mathbf{Top}^D$ *if and only if there exists*

$$(\xi, \eta): f \to g \in \mathbf{Twist} \quad \textit{as in } (7.10)$$

with $\varphi = (\xi, \eta)_*$.

(b) Assume in addition that C_f *and* C_g *are simply connected CW-spaces in* **Top**. *Then there is a homotopy equivalence*

$$C_f \simeq C_g \quad \text{under } D$$

if and only if there exists $(\xi, \eta): f \to g \in \mathbf{Twist}$ *such that* $(\xi, \eta)_*: H_*(C_f, D) \to H_*(C_g, D)$ *is an isomorphism.*

Proposition (*b*) essentially is a consequence of (*a*).

For the convenience of the reader we deduce from (7.11) the following special case ($D = *$).

(7.13) **Corollary.** *We write* $X = X_n^k$ *if* X *is a CW-complex with cells only in dimension* d *with* $n \leqq d \leqq k$. *Consider maps in* $\mathbf{Top}^*$:

$$f: X = X_n^{2n-1} \to Y = Y_2^n, \textit{ and}$$
$$g: A = A_n^{2n-1} \to B = B_2^n.$$

Then there is a homotopy equivalence $C_f \simeq C_g$ *in* $\mathbf{Top}^*$ *if and only if there is a homotopy commutative diagram in* $\mathbf{Top}^*$

$$\begin{array}{ccc} & B & \\ {\scriptstyle 0}\nearrow & & \nwarrow{\scriptstyle (0,1)} \\ X & \xrightarrow{\ \xi\ } & A \vee B \\ {\scriptstyle f}\downarrow & & \downarrow{\scriptstyle (g,1)} \\ Y & \xrightarrow{\ \eta\ } & B \end{array}$$

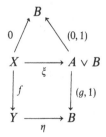

(where $0{:}X \to * \in B$ *is the trivial map) such that* $\varphi = (\xi,\eta)_*$ *is an isomorphism on homology* (*see* (7.11) *where we set* $D = *$).

For example, when all spaces in (7.13) are one-point-unions of spheres the proposition is an efficient criterion for the solution of problem (7.1).

The general suspension theorem (7.6) shows that the assumptions in (5.7) are frequently satisfied. Therefore, under certain restrictions on f and g, we can compute the equivalence relation $\simeq$ on **Twist** (f,g) by (5.5). This solves the equivalence problem in (3.13) for **Twist** (f,g). Also the isotropy problem in (3.13) can be solved similarly by (5.11). This implies results on the group of homotopy equivalences of a mapping cone C_f which are available in a better range than correspondingly results in the literature, see, for example, Oka–Sawashita–Sugawara. As an example we consider maps f which are elements of the following class $\mathfrak{X}$.

(7.14) ***Definition.*** Let $a \geqq 2$ and let $\mathfrak{X}$ be a class of maps in **Top*** with the following properties: each map $f \in \mathfrak{X}$ is a map between suspensions,

$$f{:}\Sigma A \to \Sigma B,$$

where A and B are CW-complexes and where ΣA is $(a-1)$-connected, $\dim(\Sigma A) \leqq 2a-1$, and $\dim(\Sigma B) \leqq 2a-1$. ||

By (7.8) the class $\mathfrak{X}$ in (7.14) satisfies

(7.15) $$\mathbf{TWIST}(\mathfrak{X}) = \mathbf{PAIR}(\mathfrak{X}) \subset \mathbf{Top}^*/\simeq.$$

Here **PAIR**$(\mathfrak{X})$ is the subcategory of all homotopy classes $C_f \to C_g$ $(f, g \in \mathfrak{X})$ which can be represented by pair maps, see (1.2). When $\mathfrak{X}$ satisfies (7.14) we can solve the isotropy problem and the homotopy problem (3.13) for the linear extension of categories

(7.16) $$\Gamma/I \to \mathbf{TWIST}(\mathfrak{X}) \to \mathbf{Twist}(\mathfrak{X})/\simeq.$$

In fact, we derive easily from (7.6) (where we set $D = *$) and from (5.7) and (II.13.10) the following result:

(7.17) **Theorem.** *Let* $\mathfrak{X}$ *be a class as in* (7.14) *and let* $f{:}\Sigma A \to \Sigma B$, $g{:}\Sigma X \to \Sigma Y$ *be elements in* $\mathfrak{X}$. *The set*

$$\mathbf{Twist}(f,g) \subset [\Sigma A, \Sigma X \vee \Sigma Y]_2 \times [\Sigma B, \Sigma Y] \tag{1}$$

consists of all elements (ξ,η) *with* $(f,1)_*\xi = f^*\eta$ *and we have* $(\xi,\eta) \simeq (\xi_1,\eta_1)$ *if and only if there exist* α, β *with the following properties:*

$$\begin{cases} \alpha \in \pi_1^B(\Sigma X \vee \Sigma Y)_2, \quad \beta \in \pi_1^A(\Sigma X \vee \Sigma Y)_2, \\ \eta + (g,1)_*\alpha = \eta_1 \\ 0 = \xi_1 - \xi + (\bar{\nabla} f)^*(-\alpha, i_2\eta_1) + \beta, \\ 0 = (\pi_g, 1)_*\partial^{-1}(\beta) \quad \text{in } \pi_2^A(C_g, \Sigma Y). \end{cases} \tag{2}$$

The last condition implies $\beta = 0$ *if* $\dim(\Sigma A) < 2a - 1$. *Moreover, the natural system* Γ/I on **Twist**$(\mathfrak{X})/\simeq$ *is given by*

$$(\Gamma/I)(\xi,\eta) = [\Sigma^2 A, \Sigma Y]/I(f,\eta,g), \tag{3}$$

where $I(f,\eta,g)$ *is the subgroup of all elements* β *with*

$$i_{g*}(\beta) \in \operatorname{image} \nabla(i_g\eta, f). \tag{4}$$

Here $i_g : \Sigma Y \subset C_g$ *is the inclusion and*

$$\left.\begin{array}{l} \nabla(i_g\eta, f) : [\Sigma^2 B, C_g] \to [\Sigma^2 A, C_g], \\ \nabla(i_g\eta, f)(\gamma) = (\gamma, i_g\eta)E(\nabla f). \end{array}\right\} \tag{5}$$

Recall that for $f \in \mathfrak{X}$ we have the differences

$$\begin{cases} \nabla f = -f^*(i_2) + f^*(i_2 + i_1) : \Sigma A \to \Sigma B \vee \Sigma B, \\ \bar{\nabla} f = f^*(i_2 + i_1) - f^*(i_2) : \Sigma A \to \Sigma B \vee \Sigma B, \\ E\nabla f : \Sigma^2 A \to \Sigma^2 B \vee \Sigma B. \end{cases}$$

(7.18) ***Remark.***

(A) The images of the two homomorphisms

$$(g,1)_* : \pi_2^A(\Sigma X \vee \Sigma Y)_2 \to \pi_2^A(\Sigma Y),$$
$$\nabla(\eta, f) : [\Sigma^2 B, \Sigma Y] \to [\Sigma^2 A, \Sigma Y]$$

with $\nabla(\eta,f)\ (\beta) = (\beta,\eta)\ (E\nabla f)$ are always contained in $I(f,\eta,g)$. If $\dim(\Sigma A) < 2a - 1$ and if

$$i_{g*} : [\Sigma^2 B, \Sigma Y] \to [\Sigma^2 B, C_g]$$

is surjective then $I(f,\eta,g)$ is actually the subgroup generated by the images of $(g,1)_*$ and $\nabla(\eta,f)$ above.

(B) For $\dim(\Sigma A) = 2a - 1$ the condition on β in (7.17) (2) is equivalent to

$$\beta \in \operatorname{image}[i_1, i_1 - i_2 g]_* : \pi_1^A(\Sigma X \wedge X) \to \pi_1^A(\Sigma X \vee \Sigma Y)_2,$$

provided that ΣB is simply connected. Compare Baues (1975).

By (7.7) we derive from (7.16) the following result on the *group of homotopy equivalences* $\operatorname{Aut}(C_f)^*$ in **Top***$/\simeq$.

(7.19) **Corollary.** *Let* $f : \Sigma A \to \Sigma B$ *be an element in* $\mathfrak{X}$ *with* $\dim \Sigma B \leq a$, *see* (7.14). *Then we have the exact sequence of groups*

$$[\Sigma^2 A, \Sigma B]/I(f,1,f) \rightarrowtail \operatorname{Aut}(C_f)^* \twoheadrightarrow T(f)$$

Here $T(f)$ *is the subgroup of units in the monoid* **Twist**$(f,f)/\simeq$.

(7.20) ***Remark.*** The group $T(f)$ in (7.17) consists of all $\{(\xi,\eta)\}$ for which $(\xi,\eta)_*$ in (7.11)($D=*$) is an isomorphism provided that C_f is simply connected.

We compute examples for (7.19) in the next section.

§7a Appendix: proof of the general suspension theorem under D

For the proof of (7.6) we use the following excision theorem of Blakers–Massey (1952), compare also tom Dieck–Kamps-Puppe and Gray.

(7a.1) **Theorem.** *Let $X=X_1\cup X_2$, $Y=X_1\cap X_2$ and assume $Y\subset X_1$, $Y\subset X_2$ are cofibrations in* **Top**. *Suppose (X_1,Y) is $(n-1)$ connected and (X_2,Y) is $(m-1)$ connected, see (7.5). Then the inclusion $(X_1,Y)\subset(X,X_2)$ induces for $y_0\in Y$ the map*

$$\pi_r(X_1,Y,y_0)\to\pi_r(X,X_2,y_0)$$

between homotopy groups. This map is a bijection for $r<m+n-2$ and a surjection for $r\leqq m+n-2$.

We first show

(7a.2) **Lemma.** *Let $d_0\in D$ be a basepoint of D. Then (7.6) is true if $X=(D\rightarrowtail D\vee S^r\to D)$ where S^r is a sphere. Clearly,* $\dim(X,D)=r$ *in this case.*

Proof of (7a.2). For a pair (U,V) in $\mathbf{Top}^D$ and for X in (7a.2) we have

$$\pi_1^X(U,V)=\pi_{r+1}(U,V,d_0). \tag{1}$$

The basepoint d_0 in $V\subset U$ is given by $D\to V$. By (1) the result in (7.6) is a statement on relative homotopy groups in **Top**. Here we can use the Blakers–Massey theorem as follows: for the mapping cone

$$X=C_g=C_Dg=B\bigcup_g C_DA\quad\text{in }\mathbf{Top}^D, \tag{2}$$

considered as a space in **Top**, we define subspaces

$$\left.\begin{aligned}X_1&=\{x\in C_g\,|\,x\in B\quad\text{or }x=(t,b)\in C_DA,\ t\geqq 1/2\},\\ X_2&=\{x\in C_g\,|\,x\in B\quad\text{or }x=(t,b)\in C_DA,\ t\leqq 1/2\}.\end{aligned}\right\} \tag{3}$$

Here C_DA is the cone CA in $\mathbf{Top}^D$ which is given by the following push out in **Top**:

$$\begin{array}{ccccc} & & 1 \times A & \xrightarrow{0} & D \\ & & \downarrow & & \Big| \\ I \times D & \rightarrowtail & I \times A & \text{push} & \Big| \\ \downarrow pr & \text{push} & \downarrow & & \downarrow \\ D & \rightarrowtail & I_D A & \longrightarrow & C_D A (= CA \text{ in } \mathbf{Top}^D) \end{array} \tag{4}$$

(If $D = *$ then $C_D A$ is the reduced cone on A). Now homeomorphisms $[0, 1/2] = I$ and $[1/2, 1] = I$ induce the homeomorphisms in **Top**

$$\left.\begin{array}{l} X_1 = C_D A \bigcup_D B (= CA \vee B \text{ in } \mathbf{Top}^D) \\ X_2 = I_D A \bigcup_g B = Z_g \end{array}\right\} \tag{5}$$

where Z_g is the mapping cylinder in $\mathbf{Top}^D$. We can sketch the situation as

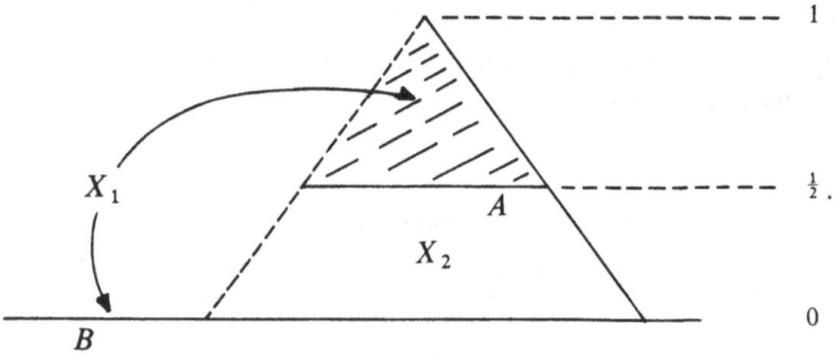

We clearly have $X_1 \cup X_2 = X$ and

$$X_1 \cap X_2 = A \bigcup_D B (= A \vee B \text{ in } \mathbf{Top}^D). \tag{6}$$

Moreover, $X_1 \cap X_2 \subset X_1$ and $X_1 \cap X_2 \subset X_2$ are cofibrations. Therefore we can apply the Blakers–Massey theorem to the commutative diagram

$$\begin{array}{ccc} \pi_{r+1}(CA \vee B, A \vee B) & \xrightarrow{(\pi_g, 1)_*} & \pi_{r+1}(C_g, B) \\ \uparrow \cong & & \downarrow \cong \\ \pi_{r+1}(X_1, X_1 \cap X_2) & \longrightarrow & \pi_{r+1}(X, X_2) \end{array} . \tag{7}$$

We now show that $(X_1, X_1 \cap X_2)$ and $(X_2, X_1 \cap X_2)$ are a-connected (the proposition (7a.2) is therewith a consequence of (7a.1)). Consider the exact sequences

$$0 \to \pi_n(CA \vee B, A \vee B) \xrightarrow{\partial} \pi_{n-1}(A \vee B) \underset{i_{2*}}{\overset{(0,1)_*}{\rightleftarrows}} \pi_{n-1}(B) \to 0. \tag{8}$$

$$0 \to \pi_n(Z_g, A \vee B) \xrightarrow{\partial} \pi_{n-1}(A \vee B) \underset{i_{2*}}{\overset{(g,1)_*}{\rightleftarrows}} \pi_{n-1}(B) \to 0. \tag{9}$$

These are portions of the long exact sequences of the indicated pairs in **Top** (where, however, Z_g and $A \vee B = A \bigcup_D B$ are constructed in $\mathbf{Top}^D$, see (5) and (6)). Here we use the homotopy equivalences $Z_g \to B$ and $CA \vee B \to B$. Therefore the inclusion i_2 yields a splitting in (8) and (9). Since (A, D) is $(a-1)$-connected we see that i_{2*} in (8) and (9) is surjective for $n-1 \leqq a-1$. Therefore ∂ in (8) and (9) is trivial for $n \leqq a$ and this shows that $(X_1, X_1 \cap X_2)$ and $(X_2, X_1 \cap X_2)$ are a-connected. Also the addendum of (7.6) is an easy consequence of the Blakers–Massey theorem by (1): consider $X_1 = CA \vee B'$, $X_2 = A \vee B$. □

Theorem (7.6) is a consequence of (7a.2) and of the following lemma:

(7a.3) **Lemma.** *Let X be a based object in* $\mathbf{Top}^D$ *and let* $p:(U, V) \to (U', V')$ *be a map in* $\mathbf{Pair}(\mathbf{Top}^D)$. *We suppose that X and D path connected spaces in* **Top**. *Consider the induced map*

$$p_*: \pi_1^X(U, V) \to \pi_1^X(U', V'),$$

which is defined in the cofibration category $\mathbf{Top}^D$. *Then the proposition*

$$\begin{cases} p_* \text{ is surjective for } \dim(X, D) \leqq n \\ \text{and } p_* \text{ is bijective for } \dim(X, D) < n \end{cases} \tag{*}$$

is satisfied provided that $(*)$ *is true for all* $X = (D \rightarrowtail D \vee S^r \to D)$ *when* S^r *is a sphere.*

Proof. Let $* \in D$ be a basepoint. Since X and D are path connected we may assume that (X, D) is a relative CW-complex in which all attaching maps

$$f: S^{r-1} \to X^{r-1}, \quad f(*) = *,$$

are basepoint preserving ($r \geqq 1$) and for which $X^0 = D$. Here (X^r, D) is the relative r-skeleton. For simplicity let

$$X^r = X^{r-1} \bigcup_f CS^{r-1} \tag{1}$$

be the mapping cone in $\mathbf{Top}^*$ of f. (The argument is similar if there are many r-cells in $X^r - X^{r-1}$.) We observe that X^r is a based object by

$$D \rightarrowtail X^r \subset X \xrightarrow{0} D. \tag{2}$$

The projection $I_* X^r \to C_D X^r$ in (4) of (7a.2) yields the push out diagram of pairs

$$\begin{array}{ccc} (I_*CS^{r-1}, o \times CS^{r-1}) & \longrightarrow & (C_DX^r, X^r) \\ \cup_i & & \cup \\ (C^+S^{r-1}, o \times S^{r-1}) & \xrightarrow[g]{} & (C_DX^{r-1}, X^{r-1}) \end{array} \tag{3}$$

Here we set $I_*X = I \times X/I \times *$ and $C^+S^{r-1} = I_*S^{r-1} \cup 1 \times CS^{r-1}$. We observe that the left hand inclusion i is equivalent to a cone in **Pair(Top*)**, see (II.7.5) (1), and that we have an equivalence

$$h: CS^{r-1} \sim C^+S^{r-1} \text{ under } S^{r-1}.$$

Therefore (C_DX^r, X^r) is the mapping cone in **Pair(Top^D)** with the attaching map

$$(CS^{r-1} \vee D, S^{r-1} \vee D) \rightarrow (C_DX^{r-1}, X^{r-1})$$

defined by gh. Now the exactness of the cofibration sequence in **Pair(Top^D)** and the five lemma inductively yield the result. □

§8 Example: homotopy theory of the 2-stem and of the 3-stem

Let $\mathbf{A}_n^k$ be the full subcategory of $\mathbf{Top}^*/\simeq$ consisting of $(n-1)$-connected $(n+k)$-dimensional CW-complexes. The suspension Σ gives us the sequence of functors

(8.1) $$\mathbf{A}_2^k \xrightarrow{\Sigma} \mathbf{A}_3^k \xrightarrow{\Sigma} \cdots \xrightarrow{\Sigma} \mathbf{A}_n^k \xrightarrow{\Sigma} \mathbf{A}_{n+1}^k \rightarrow \cdots$$

which we call the **k-stem of homotopy categories**. The Freudenthal suspension theorem shows that for $k+1<n$ the functor $\Sigma: \mathbf{A}_n^k \rightarrow \mathbf{A}_{n+1}^k$ is an equivalence of categories. Moreover, for $k+1=n$ this functor is full and a $1-1$ correspondence of homotopy types, and each object in $\mathbf{A}_{k+1}^k$ is homotopy equivalent to a suspension. We say that the homotopy types of $\mathbf{A}_n^k$ are **stable** if $k+1 \leqq n$ and we say that the morphisms of $\mathbf{A}_n^k$ are **stable** if $k+1<n$.

The spheres S^n and S^{k+n} are objects in $\mathbf{A}_n^k$. Therefore we can restrict the functors (8.1) to the morphism sets

(8.2) $$\pi_{k+n}(S^n) = [S^{k+n}, S^n]$$

which are the **homotopy groups of spheres**. They form the k-stem

(8.3) $$\pi_{k+2}(S^2) \xrightarrow{\Sigma} \pi_{k+3}(S^3) \xrightarrow{\Sigma} \cdots \pi_{k+n}(S^n) \xrightarrow{\Sigma} \pi_{k+n+1}(S^{n+1})$$

The limit of this sequence of abelian groups is the **stable homotopy group** π_k^S which is isomorphic to $\pi_{k+n}(S^n)$ for $k+1<n$.

The computation of the k-stem is a classical and principal problem of homotopy theory. The k-stem of homotopy groups of spheres now is known for fairly large k, for example one can find a complete list for $k \leqq 19$ in Toda's book. The k-stem of homotopy types, however, is still mysterious even for very small k. We derive from (IX.2.23) below the

(8.4) **Lemma.** *A CW-complex in* $\mathbf{A}_n^k$ *is homotopy equivalent to a CW-complex* X *with cells only in dimension* n, $n+1,\ldots,n+k$.

Hence, for $k=0$ an object in $\mathbf{A}_n^0$ is just a one point union of n-spheres or equivalently a Moore-space $M(F,n)$ of a free abelian group. This shows that the homotopy theory of the 0-stem is equivalent to the linear algebra of free abelian groups.

For $k=1$ a homotopy type in $\mathbf{A}_n^1$ is given by a one point union of Moore spaces $M(A,n) \vee M(F,n+1)$ where F is a free abelian groups and where A is an arbitrary abelian group. In particular the category $\mathbf{M}^n$ of Moore spaces in degree n is a full subcategory of $\mathbf{A}_n^1$. Similarly, as in § 3a one can describe the category $\mathbf{A}_n^1$ by a linear extension.

For $k=2$ the classification of homotopy types in the 2-stem was achieved by J.H.C. Whitehead in 1949; in particular, he classified 1-connected 4-dimensional CW-complexes in $\mathbf{A}_2^2$. We describe new proofs of this result in this book. Steenrod, in his review of Whitehead's paper, pointed out the problem to compute also the maps in $\mathbf{A}_2^2$ in terms of the classifying invariants. We now show that this problem can be solved by use of theorem (7.17) above.

Let $\mathfrak{X}A_n^2$, $n \geqq 2$, be the class of all maps

$$f: \bigvee_A S^{n+1} \vee \bigvee_B S^n \to \bigvee_C S^{n+1} \vee \bigvee_D S^n$$

in $\mathbf{Top}^*/\simeq$ (where A, B, C, D are index sets) such that f induces an injective map on the nth homology group H_n.

(8.5) **Theorem.** *The inclusions*

$$\mathbf{TWIST}(\mathfrak{X}A_2^2) \xrightarrow{\sim} \mathbf{A}_2^2, \quad \textit{and}$$

$$\mathbf{PRIN}(\mathfrak{X}A_n^2) \xrightarrow{\sim} \mathbf{A}_n^2, \quad n \geqslant 3,$$

are equivalences of categories.

The class $\mathfrak{X}A_n^2$ satisfies the conditions in (7.14) and for $n \geqq 3$ we have $\mathbf{TWIST}(\mathfrak{X}A_n^2) = \mathbf{PRIN}(\mathfrak{X}A_n^2)$. Therefore we obtain by theorem (7.17) a complete solution of the isotropy problem and of the homotopy problem, see (3.13). This yields by (8.5) an algebraic description of the categories $\mathbf{A}_n^2$

in terms of linear extensions of categories. The extension problem (3.13), however, is not solved. We leave it as an exercise to describe the categories **Twist** $(\mathfrak{X}A_2^2)$ and **Prin** $(\mathfrak{X}A_n^2)$ in algebraic terms and to compute the homotopy relation on these categories.

Proof of (8.5): We only prove the result for $\mathbf{A}_2^2$. For X in $\mathbf{A}_2^2$ we find a map $f \in \mathfrak{X}A_2^2$ and a homotopy equivalence $C_f \simeq X$ as follows. By (8.4) we assume that $X^1 = *$. Let $C_*X = (C_*, d)$ be the cellular chain complex of X. Let $C_3'' = \text{kernel}(d_3)$. Since d_3C_3 is free abelian we find a splitting of $d_3: C_3 \twoheadrightarrow d_3C_3$. Let C_3' be the image of this splitting so that $C_3 = C_3'' \oplus C_3'$. The restriction $d_3': C_3' \to C_2$ of d_3 is injective. The 3-skeleton of X is the mapping cone of

$$g = i_2 d_3': M(C_3', 2) \to Y = M(C_3'', 3) \vee M(C_2, 2). \tag{1}$$

Below we show that the inclusion $i_g: Y \subset C_g \simeq X^3$ induces a surjective map

$$i_{g*}: \pi_3 Y \twoheadrightarrow \pi_3 C_g. \tag{2}$$

Therefore the attaching map of 4-cells in X is homotopic to a map

$$M(C_4, 3) \xrightarrow{h} Y \subset C_g \simeq X^3. \tag{3}$$

Hence the map

$$f = (h, g): M(C_4, 3) \vee M(C_3', 2) \to Y \tag{4}$$

is a map in the class $\mathfrak{X}A_2^2$ with $C_f \simeq X$. For the surjectivity of i_{g*} consider the commutative diagram

$$\begin{array}{ccccccc} \pi_3 Y & \xrightarrow{i_{g*}} & \pi_3 C_g & \xrightarrow{j} & \pi_3(C_g, Y) & \xrightarrow{\partial} & \pi_2(Y) = C_2 \\ & & & & \cong \uparrow & \nearrow_{(g,1)_*} & \uparrow d_3' \\ & & & & \pi_2(M \vee Y)_2 & = \pi_2(M) = & C_3', \end{array} \tag{5}$$

where $M = M(C_3', 2)$ and where the isomorphism is given by (7.6). Since d_3' is injective also ∂ is injective, hence by exactness i_{g*} is surjective. By surjectivity of i_{g*} we see that each map $F: C_k \to C_f$ $(k, f \in \mathfrak{X}A_2^2)$ is homotopic to a pair map. Whence the theorem follows from (7.8). □

(8.6) ***Remark.*** Theorem (8.5) has many consequences for the homotopy theory of the 2-stem. For example, one can enumerate all homotopy classes of maps between objects in $\mathbf{A}_n^2$, or one can compute the group of homotopy equivalences of an object in $\mathbf{A}_n^2$. Such explicit applications are described in Baues (1984). We point out that a result as in (8.5) as well holds in the category $\mathbf{DA}_*$(flat) of chain algebras and that one can study the functor

$SC_*\Omega$ (I.7.29) on $\mathbf{A}_2^2$ by the result in (3.16). We leave this as an exercise. We also can use (3.16) for the computation of the suspension functor Σ in the 2-stem (8.1). In addition to (8.5) we can use the tower of categories in (VI.6.2) below for the computation of $\mathbf{A}_n^2$.

As an example, for (7.19) we show the

(8.7) **Theorem.** *Let M be a simply connected closed 4-dimensional manifold. Then the group of homotopy equivalences* $\mathrm{Aut}(M)$ *in* $\mathbf{Top}/\simeq$ *in embedded in the short exact sequence of groups*

$$(\mathbb{Z}/2)^{n+\delta} \rightarrowtail \mathrm{Aut}(M) \overset{H}{\twoheadrightarrow} \mathrm{Aut}(H^*(M), \cup)^{op}$$

Here the right-hand group is the automorphism group of the cohomology ring of M and H is the cohomology functor. The left-hand group is the $\mathbb{Z}/2$-vector space of dimension $n+\delta$ where n is the rank of the free abelian group H_2M. Moreover, $\delta = -1$ if the intersection form of M is odd, and $\delta = 0$ if this form is even.

This result was independently obtained by Quinn.

Proof of (8.7). Since M is simply connected we have $\mathrm{Aut}(M) = \mathrm{Aut}(M)^*$. Recall from the proof of (II.16.8) that $M = C_f$ is a mapping cone, $f \in \mathfrak{X}A_2^2$. It is easy to see by (7.17) that

$$\mathbf{Twist}(f,f) = \mathbf{Twist}(f,f)/\simeq,$$

and that the group of units in $\mathrm{Twist}(f,f)$ is $\mathrm{Aut}(H^*(M), \cup)^{op}$. We can use (7.18) (with $a = 3$) for the computation of the kernel

$$\pi_4(\bigvee_B S^2)/I(f,1,f) = (\mathbb{Z}/2)^{n+\delta}.$$

For the computation of ∇f see the proof of (II.16.8). In (2) we also use the Hilton–Milnor theorem for the description of $\pi_4(\bigvee_B S^2)$ where B is a basis of H_2. □

Remark. Kahn considers the group $\mathrm{Aut}(M)$ of an $(n-1)$-corrected $2n$-manifold. His result can be easily derived from (7.19) as well, since $M \simeq C_f$ where $f: S^{2n-1} \to \bigvee S^n$.

Next we consider the homotopy theory of the 3-stem. Various authors worked on the classification of homotopy types in the stable 3-stem, so Shiraiwa, Chang, and Chow, compare also Baues (1984). Using again (7.17) we actually obtain a description of the homotopy categories in the 3-stem as follows. Let $\mathfrak{X}A_n^3$ be the class of all maps

$$f: M(A, n+2) \vee M(B, n+1) \to M(C, n+1) \vee M(D, n)$$

in **Top*** where A and C are free abelian groups and where B and D are arbitrary abelian groups.

(8.8) **Theorem.** *The inclusions*

$$\mathbf{TWIST}(\mathfrak{X}A_2^3) \xrightarrow{\sim} \mathbf{A}_2^3, \quad \textit{and}$$

$$\mathbf{PRIN}(\mathfrak{X}A_n^3) \xrightarrow{\sim} \mathbf{A}_n^3, \quad n \geqslant 3,$$

are equivalences of categories.

Proof. The result follows from (VII.3.1) below. □

The class $\mathfrak{X}A_n^3$ satisfies the conditions in (7.14) and for $n \geqq 3$ we have $\mathbf{TWIST}(\mathfrak{X}A_n^3) = \mathbf{PRIN}(\mathfrak{X}A_n^3)$. Therefore theorem (7.17) yields a solution of the isotropy problem and of the homotopy problem in (3.13) for the categories in (8.8). Hence we get a description of these categories as linear extensions. The extension problem is not solved. We can compute $\mathbf{Twist}(\mathfrak{X}A_2^3)$ and $\mathbf{Prin}(\mathfrak{X}A_n^3)$, $n \geqq 3$, in purely algebraic terms. The computation is fairly intricate and elaborate and will appear elsewhere. This as well yields the classification of homotopy types in the 3-stem, compare also Baues (1984). We point out that (8.8) also gives us the classification of maps in $\mathbf{A}_n^3$, $n \geqslant 2$. One can use (7.18) for the computation of the isotropy groups of the action Γ. In particular, one can compute the groups $\mathrm{Aut}(X)$ of homotopy equivalences by (7.19) for each CW-complex in the 3-stem.

§9 Example: the group of homotopy equivalences of the connected sum $(S^1 \times S^3) \# (S^2 \times S^2)$.

The connected sum $M = (S^1 \times S^3) \# (S^2 \times S^2)$ has the homotopy type of the mapping cone C_f where

$$(9.1) \qquad \begin{cases} f: S^3 \to S^1 \vee S^3 \vee S^2 \vee S^2 = B, \\ f = [i_3, i_1] + [i_2, j_2]. \end{cases}$$

Here i_1, i_3, i_2, j_2 are the inclusions and $[\ ,\ ]$ denotes the Whitehead product. With $a = 3$ we see that f satisfies the condition in (7.14) and that we can apply (7.19). This yields the result below on the group of homotopy equivalences of M. The following remark shows that the determination of groups of homotopy equivalences is of importance for the classification of manifolds.

Remark. The manifold $M = (S^1 \times S^3) \# (S^2 \times S^2)$ plays a role in proposition

3.2 of Cappel–Shaneson who consider the group $S(M)$ of smoothings of M and who found a subgroup $\mathbb{Z}/2 \subset S(M)$. The group of homotopy equivalences, $\mathrm{Aut}(M)$, acts on $S(M)$ such that the orbits are the s-cobordism classes of all manifolds homotopy equivalent to M. M. Kreck suggested to me to compute $\mathrm{Aut}(M)$ since this might be helpful for deciding whether the non trivial element in $\mathbb{Z}/2 \subset S(M)$ yields a non trivial s-cobordism class or not.

(9.2) **Theorem.** *Let $* \in M$ be a base point. Then there is a short exact sequence of groups*

$$(\mathbb{Z}/2)^3 \rightarrowtail \mathrm{Aut}(M)^* \xrightarrow{\lambda} T(M).$$

Here $\mathrm{Aut}(M)^*$ *is the group of homotopy equivalences of M in* **Top***$/\simeq$ *and $T(M)$ is the group defined algebraicly below in* (9.7).

Remark. The generators of the kernel of λ above are the elements

$$(1 + ii_3\Sigma\eta), \quad (1 + ii_2\eta\Sigma\eta), \quad (1 + ij_2\eta\Sigma\eta),$$

where 1 is the identity of $C_f = M$, where $i: B \subset C_f$ is the inclusion, and where $i_3: S^3 \subset B$, $i_2, j_2: S^2 \subset B$ are the inclusions, see (9.1). Moreover $\eta: S^3 \to S^2$ is the Hopf map.

In the definition of the group $T(M)$ we use the following notation: Let $\pi = \pi_1 M = \mathbb{Z}$ be the fundamental group and let $R = \mathbb{Z}[\pi]$ be the groupring of $\pi = \mathbb{Z}$. Thus R is the free abelian group generated by the elements $[n]$, $n \in \mathbb{Z}$. The element $[0] = 1$ is the unit of R. Let $\varepsilon: R \to \mathbb{Z}$ be the augmentation with $\varepsilon([n]) = 1$ for $n \in \mathbb{Z}$. Now let A be an R-module, we write the action of $\xi \in R$ on $a \in A$ by a^ξ. By Whitehead's quadratic functor Γ (with $\Gamma(A) = \pi_3 M(A, 2) = H_4 K(A, 2)$) we obtain the R-module $\Gamma(A)$ with the action of π determined by the functor Γ.

Let $\mathrm{Aut}_\tau(A)$ be the set of τ-equivariant automorphisms of A where $\tau: R \to R$ is a ring isomorphism. This means $\alpha \in \mathrm{Aut}_\tau(A)$, $\alpha: A \to A$, satisfies $\alpha(x^\xi) = (\alpha x)^{\tau\xi}$. We will use the special ring isomorphisms

(9.3) $$\tau = \tau_{\eta_1} = \begin{cases} \text{identity of } R & \text{if } \eta_1 = 1, \\ (-1)_\#: R \to R & \text{if } \eta_1 = -1, \end{cases}$$

where $(-1)_\#[n] = [-n]$, $n \in \mathbb{Z}$.

Moreover, we use the structure elements

(9.4) $$\begin{cases} v = -[0] + [1] = -1 + [1] \in R, \\ w = [e_1, e_2] \in \Gamma(R \oplus R), \end{cases}$$

where $\{e_1 = (1, 0),\ e_2 = (0, 1)\}$ is the canonical basis in $R \oplus R$ and where $w = [e_1, e_2]$ is the Whitehead product.

(9.5) ***Definition.*** We define algebraicly the *set* $t(M)$ to be set of all pairs (ξ,η) with the following properties (1)···(4):

$$\xi \in R, \tag{1}$$

$$\eta = (\eta_1, \eta_2, \eta_3, \eta_\Gamma) \quad \text{with,} \tag{2}$$

$$\begin{cases} \eta_1 \in \{1, -1\}, \\ \eta_2 \in \mathrm{Aut}_\tau(R \oplus R) \quad \text{where } \tau = \tau_{\eta_1}, \text{ see (9.3)}, \\ \eta_3 \in R, \ \varepsilon(\eta_3) \in \{1, -1\}, \\ \eta_\Gamma \in \Gamma(R \oplus R). \end{cases}$$

$$v \cdot \xi = \tau(v) \cdot \eta_3 \quad \text{in } R. \tag{3}$$

$$w^\xi = \Gamma(\eta_2)(w) + (\eta_\Gamma)^{\tau(v)} \quad \text{in } \Gamma(R \oplus R). \tag{4}$$

In (3) and (4) we use the structure elements in (9.4). ||

(9.6) ***Definition.*** We define a **homotopy relation** $\simeq$ on the set $t(M)$ as follows: let $(\xi,\eta) \simeq (\xi',\eta')$ if and only if there exists $\alpha \in R$ with

$$\begin{cases} \xi' = \xi + \tau(v)\cdot\alpha, \ \tau = \tau_{\eta_1} \\ \eta_1' = \eta_1 \\ \eta_2' = \eta_2 \\ \eta_3' = \eta_3 + v \cdot \alpha \\ \eta_\Gamma' = \eta_\Gamma + \omega^\alpha \end{cases}$$

This corresponds to an action of the abelian group R on $t(M)$. ||

(9.7) ***Definition.*** The quotient set

$$T(M) = t(M)/\simeq \tag{1}$$

is a group with the following multiplication; we denote by $\{(\xi,\eta)\}$ the homotopy class of $(\xi,\eta) \in t(M)$ in $T(M)$:

$$\{(\xi,\eta)\}\cdot\{(\xi',\eta')\} = \{(\xi\cdot\xi', \eta\cdot\eta')\}. \tag{2}$$

Here $\eta'' = \eta\cdot\eta'$ has the coordinates

$$\left.\begin{array}{ll} \eta_1'' = \eta_1 \cdot \eta_1' & \text{(in the group } \{1,-1\}), \\ \eta_2'' = \eta_2 \cdot \eta_2' & \text{(composition of automorphisms)}, \\ \eta_3'' = \eta_3 \cdot \eta_3' & \text{(in the ring } R), \\ \eta_\Gamma'' = \Gamma(\eta_2)(\eta_\Gamma') + (\eta_\Gamma)^{\eta_3'}. & \end{array}\right\} \tag{3}$$

Proof of (9.2). We apply (7.19) and (7.18), $a = 3$. First consider the universal covering $\hat{M}$ of M where e^4 is the 4-cell of M,

$$\hat{M} = \mathbb{R} \bigcup_{\mathbb{Z}} (\mathbb{Z} \times (S^3 \vee S^2 \vee S^2)) \cup (\mathbb{Z} \times e^4). \tag{1}$$

Since $\mathbb{R}$ is contractible we get

$$\hat{M} \simeq \hat{M}/\mathbb{R} = C_{\tilde{f}}, \quad \text{where} \tag{2}$$

$$\tilde{f}: \bigvee_{\mathbb{Z}} S^3 \to \bigvee_{\mathbb{Z}} S^3 \vee S^2 \vee S^2 = \hat{B}, \tag{3}$$

$$\tilde{f}^n = -i_3^n + i_3^{n+1} + [i_2^n, j_2^n], \quad n \in \mathbb{Z}. \tag{4}$$

Here $\tilde{f}$ is derived from the formula for f in (9.1). We have $[i_3, i_1] = -i_3 + i_3^{i_1}$; this corresponds to the term $-i_3^n + i_3^{n+1}$ in (4). The cellular chain complex of $\hat{M}/\mathbb{R}$ is

$$\begin{array}{ccccc} \hat{C}_2 & \longleftarrow & \hat{C}_3 & \longleftarrow & \hat{C}_4 \\ \| & & \| & & \| \\ R \oplus R & \xleftarrow{0} & R & \xleftarrow{d} & R \end{array} \tag{5}$$

where $d[n] = -[n] + [n+1]$ as follows from (4). Therefore we get

$$H_n(\hat{M}) = \left.\begin{array}{ll} R \oplus R & n = 2, \\ \mathbb{Z} & n = 3, \\ 0 & n = 4. \end{array}\right\} \tag{6}$$

In fact $H_4 \hat{M} = 0$ since $\hat{M}$ is an open manifold; one can also check that d in (5) is injective.

Moreover, the sequence

$$R \xrightarrow{d} R \xrightarrow{\varepsilon} \mathbb{Z} \to 0 \tag{7}$$

is exact. Let

$$B = S^1 \vee S^3 \vee S^2 \vee S^2. \tag{8}$$

Then the injectivity of d implies that

$$\hat{C}_4 = \pi_4(C_f, B) \to \pi_3 B \to \pi_3(B, B^2) = \hat{C}_3$$

is injective and therefore

$$\pi_4 B \twoheadrightarrow \pi_4 C_f, \tag{9}$$

is surjective. This shows by (7.18) that

$$I(f, 1, f) = \text{image}(f, 1)_* + \text{image}\, \nabla(1, f). \tag{10}$$

Here we use $\nabla f = \bar{\nabla} f \in \pi_3(B' \vee B'')_2$, $B' = B'' = B$. We have by (9.1)

$$\begin{aligned} \bar{\nabla} f &= f^*(i'' + i') - f^*(i'') \\ &= [i_3'' + i_3', i_1'' + i_1'] + [i_2'' + i_2', j_2'' + j_2'] - [i_3'', i_1''] - [i_2'', j_2''] \\ &= [i_3', i_1'' + i_1'] - [i_1', i_3''^{i_1''}] + [i_2', j_2''] - [j_2', i_2''] + [i_2', j_2']. \end{aligned} \tag{11}$$

Thus we get

$$E\bar{\nabla} f = E\nabla f = [\Sigma i_3', i_1''] - [\Sigma i_1', i_3''^{i_1''}] + [\Sigma i_2', j_2''] - [\Sigma j_2', i_2'']. \tag{12}$$

Compare (II.15a.10). We now can compute (10). The image of $\nabla(1,f)$ in $\pi_4(B)\cong\pi_4(\hat{B})$ is generated by all elements (13)$\cdots$(16):

$$-\alpha_4+\alpha_4^{[1]}, \qquad \alpha_4\in\pi_4\hat{B}, \tag{13}$$

$$-[\alpha_2,i_3^1], \qquad \alpha_2\in\pi_2\hat{B}, \tag{14}$$

$$+[\alpha_3,j_2^0], \qquad \alpha_3\in\pi_3\hat{B}, \tag{15}$$

$$-[\beta_3,i_2^0], \qquad \beta_3\in\pi_3\hat{B}. \tag{16}$$

This follows from (12). We use inclusions as in (4). Consider the following congruences, $\equiv$, module image $(\nabla(1,f))$ deduced from (13)$\cdots$(16) ($n\in\mathbb{Z}$)

$$\begin{aligned}\alpha_4&\equiv\alpha_4^{[1]}\equiv\alpha_4^{[2]}\equiv\cdots\equiv\alpha_4^{[n]},\\ 0&\equiv[\alpha_2,i_3^1]\equiv[\alpha_2,i_3^1]^{[n]}=[\alpha_2^{[n]},i_3^{n+1}],\\ 0&\equiv[\alpha_3,j_2^0]\equiv[\alpha_3,j_2^0]^{[n]}=[\alpha_3^{[n]},j_2^n],\\ 0&\equiv[\beta_3,i_2^0]\equiv[\beta_3,i_2^0]^{[n]}=[\beta_3^{[n]},i_2^n].\end{aligned}$$

Since $\alpha\longmapsto\alpha^{[n]}$ is an isomorphism we see that for all $n\in\mathbb{Z}$ we have

$$\left.\begin{aligned}-\alpha_4+\alpha_4^{[n]}&\equiv 0 &\quad \alpha_4\in\pi_4\hat{B},\\ [\alpha_2,i_3^n]&\equiv 0 & \alpha_2\in\pi_2\hat{B},\\ [\alpha_3,j_2^n]&\equiv 0 & \alpha_3\in\pi_3\hat{B},\\ [\beta_3,i_2^n]&\equiv 0 & \beta_3\in\pi_3\hat{B}.\end{aligned}\right\} \tag{17}$$

This implies that all Whitehead products in $\pi_4\hat{B}$ are congruent to 0; these generate the kernel of the suspension homomorphism

$$\Sigma:\pi_4\hat{B}\to\pi_5\Sigma\hat{B}=(R\oplus R\oplus R)\otimes\mathbb{Z}/2. \tag{18}$$

By (17) we thus get

$$\pi_4\hat{B}/\text{image }\nabla(1,f)=(\mathbb{Z}/2)^3. \tag{19}$$

Moreover image $(f,1)_*\subset$ image $\nabla(1,f)$ as follows from (4). Therefore we have by (10):

$$\pi_4(B)/I(f,1,f)=(\mathbb{Z}/2)^3. \tag{20}$$

This yields by (7.19) the *kernel* of λ *in the exact sequence* (9.2) and it proves the remark following (9.2).

Next we show that

$$T(M)=T(f)=\text{group of units in }\mathbf{Twist}(f,f)/\simeq. \tag{21}$$

We show that

$$t(M)\subset\mathbf{Twist}(f,f) \tag{22}$$

is the subset of all (ξ,η) which induce an isomorphism on the fundamental group and on the homology group (6). Thus the homotopy classes $\{(\xi,\eta)\}$

with $(\xi,\eta)\in t(M)$ are exactly the units in $\mathbf{Twist}(f,f)/\simeq$. By definition, $\mathbf{Twist}(f,f)$ is the subset of all (ξ,η)

$$(\xi,\eta)\in[S^3,S^3\vee B]_2\times[B,B] \tag{23}$$

with $(f,1)_*\xi=f^*\eta$. Here we have

$$[S^3,S^3\vee B]_2=R, \tag{24}$$

$$[B,B]=\pi_1 B\times\pi_3 B\times\pi_2 B\times\pi_2 B, \tag{25}$$

where $\pi_1 B=\mathbb{Z}$, $\pi_2 B=R\oplus R$ and where

$$\pi_3 B=R\oplus\Gamma(R\oplus R). \tag{26}$$

From $\eta=(\eta_1,\eta^3,\eta^2,\bar\eta^2)$ we get $\eta^3=(\eta_3,\eta_\Gamma)$ and we get

$$\eta_2:R\oplus R\to R\oplus R,\quad \eta_2(e_1)=\eta^{2'},\quad \eta_2(e_2)=\bar\eta^2. \tag{27}$$

Thus η yields the element η described in (9.5). Here η_2 is the induced homomorphism on $H_2\hat M$ and $\varepsilon(\eta_3)$ is the induced homomorphism on $H_3\hat M=\mathbb{Z}$ as follows from (7). Now $(f,1)_*\xi=f^*\eta$ is equivalent to (9.5)(3), (4), in fact

$$(f,1)_*\xi=(v\cdot\xi,w^\xi),$$
$$f^*\eta=(\tau(v)\cdot\eta_3,\Gamma(\eta_2)(w)+\eta_\Gamma^{\tau(v)}),$$

in $R\oplus\Gamma(R\oplus R)$. Similarly, we see that the homotopy relation in (7.17) corresponds exactly to the homotopy relation in (9.6), here we use (11). q.e.d. □

§ 10 The homotopy type of a fiber space in topology

We now describe the results on fiber spaces which are dual to the corresponding results on mapping cones in §7. We consider maps between fiber spaces in topology. Using duality we apply the results on the category **Twist** in the same way as in §7. This yields some new results on the following problems which are dual to the problems in (7.1) and (7.2) respectively.

(10.1) Describe conditions on the maps f and g which imply that the fiber spaces P_f and P_g are homotopy equivalent!

(10.2) Compute the set of homotopy classes of maps $[P_f,P_g]$, and compute the group of homotopy equivalences of P_f!

Again we use the results in §6; that is, we transform these results into the dual language of a fibration category and we apply them in the category $\mathbf{Top}_D$ of spaces over D. By (I.5.2) and (II.1.4) we know that $\mathbf{Top}_D$ is a fibration category. If $D=*$ is a point we have $\mathbf{Top}_*=\mathbf{Top}$ since $*$ is the final object of **Top**. The spaces with basepoint are the based objects in $\mathbf{Top}_* = \mathbf{Top}$. In general a based object in $\mathbf{Top}_D$ is given by a fibration $p: A \twoheadrightarrow D$ in **Top** and by a section $0: D \to A$, $P0 = 1_D$. For example, if F has a basepoint, then $(D \xrightarrow{(0,1)} F \times D \twoheadrightarrow D)$ is a based object in $\mathbf{Top}_D$ where $0: D \to * \in F$. Now let

$$(10.3) \qquad f: X \to Y, \quad g: A \to B$$

be maps in $\mathbf{Top}_D$ and assume Y and B are based objects, which are loop objects in $\mathbf{Top}_D$. We suppose that all spaces are CW-spaces in **Top**. As in any fibration category we have the inclusions

$$(10.4) \qquad \mathbf{PRIN}(f,g) \subset \mathbf{TWIST}(f,g) \subset \mathbf{PAIR}(f,g) \subset [P_f, P_g]_D.$$

Here P_f and P_g are *fiber spaces in* $\mathbf{Top}_D$ and $[P_f, P_g]_D$ is the set of homotopy classes of maps in $(\mathbf{Top}_D)_f/\simeq$, see (II.§8, §14).

We want to describe criteria under which the inclusions of (10.4) are actually bijections. To this end we need a result which in a sense is the Eckmann–Hilton dual of the general suspension theorem under D in (7.6), (this result, however, is not obtained just by the use of the opposite category).

(10.5) ***Definition.*** Let $p: B \twoheadrightarrow D$ be a fibration in **Top**. We say $p = (B|D)$ is b-**connected** if for all $d_0 \in D$ the fiber $p^{-1}(d_0)$ is a $(b-1)$-connected space in **Top**. This is the case iff (Z_p, B) is b-connected in the sense of (7.5); here Z_p denotes the mapping cylinder of p. We define the **homotopy dimension** by $\operatorname{hodim}(B|D) \leq N$ if $p^{-1}(d_0)$ is path connected and if $\pi_n(p^{-1}(d_0)) = 0$ for $n > N$. We write $\operatorname{hodim}(B|*) = \operatorname{hodim}(B)$. For example, $\operatorname{hodim}(K(\pi, n)) = n$. ||

Thus $\operatorname{hodim}(B|D)$ is the top dimension of a non trivial homotopy group of the (pathconnected) fiber of $B \twoheadrightarrow D$. Moreover $\operatorname{hodim}(B)$ is the top dimension of a non trivial homotopy group of the (path connected) space B.

(10.6) **Theorem (general loop theorem over D).** *All spaces in this theorem are CW-spaces. Let Y and B be based objects in* $\mathbf{Top}_D$ *and let $f: X \to Y$ be a map in* $\mathbf{Top}_D$. *Assume that D and B are path connected and that $B \twoheadrightarrow D$ induces an isomorphism on fundamental groups. If $(Y|D)$ is y-connected, then the map*

$$(\pi_f, 1)^*: \pi_B^1(WY \times X | Y \times X) \to \pi_B^1(P_f|X)$$

is a bijection for $\operatorname{hodim}(B|D) < 2y - 1$ *and is an injection for* $\operatorname{hodim}(B|D) \leq 2y - 1$. *The map $(\pi_f, 1)^*$ is defined in the fibration category* $\mathbf{Top}_D$ *as in* (6.4).

Addendum. *Let Y and B be based objects in* $\mathbf{Top}_D$ *and let $X \twoheadrightarrow X' \twoheadrightarrow D$ be fibrations in* **Top**. *Assume that X, X', Y and D are CW-spaces and assume that D and B are path connected and that $B \twoheadrightarrow D$ induces an isomorphism on fundamental groups. Moreover, suppose that $(Y|D)$ is y-connected and that*

$(X|X')$ is x-connected. Then the map

$$(1 \times p)^*: \pi_B^1(WY \times X' | Y \times X') \to \pi_B^1(WY \times X | Y \times X)$$

is bijective for hodim $(B|D) < x + y$ and injective for hodim $(B|D) \leqslant x + y$. The map $(1 \times p)^*$ is defined in the category $\mathbf{Top}_D$ by the fibration $p: X \twoheadrightarrow X'$.

For $X' = D$ this is a result on the map

$$p_1^*: \pi_B^1(WY | Y) \to \pi_B^1(WY \times X | Y \times X)$$

in the diagram near (6.4).

(10.6)′ ***Notation.*** We say that $(B|D)$ *is* **good** if B and D are path connected CW-spaces and if $B \twoheadrightarrow D$ induces an isomorphism on fundamental groups.

Remark. If $D = *$ and $X = *$ we have $P_f = \Omega Y$ and $\pi_B^1(P_f|*) = [\Omega Y, \Omega B]$. In this case the map $(\pi_f, 1)^*$ is given by the *loop functor*

$$\Omega: [Y, B] \overset{\partial}{\cong} \pi_B^1(WY|Y) \to [\Omega Y, \Omega B],$$

which carries $\xi: Y \to B$ to $\Omega\xi: \Omega Y \to \Omega B$.

Next we describe applications of the general loop theorem; we proceed in the same way as in § 7. Consider again the inclusions in (10.4). We have

$$\text{(10.7)} \qquad \mathbf{PAIR}(f, g) = [P_f, P_g]_D,$$

provided that the projection $p: P_f \twoheadrightarrow X$ induces a surjection $p^*: [X, A]_D \twoheadrightarrow [P_f, A]_D$ of homotopy sets in $(\mathbf{Top}_D)_f/\simeq$. The surjectivity of p^* corresponds to the existence of liftings F in the commutative diagram

$$\begin{array}{ccccc} P_f & \longrightarrow & & & A \\ \downarrow & & \overset{F}{\nearrow} & & \downarrow \\ Z_p & \longrightarrow & X & \twoheadrightarrow & D \end{array}$$

Thus obstruction theory shows that (10.7) is satisfied if $(P_f|X)$ is $(y-1)$-connected and if hodim $(A|D) < y - 1$. Here $(P_f|X)$ is $(y-1)$-connected if $(Y|D)$ is y-connected.

Next we derive from (6.4) that

$$\text{(10.8)} \qquad \mathbf{TWIST}(f, g) = \mathbf{PAIR}(f, g)$$

if $(\pi_f, 1)^*$ in (10.6) is surjective. This for example is satisfied if $(Y|D)$ in y-connected and if hodim $(B|D) < 2y - 1$ where $(B|D)$ is good.

Moreover, we derive from (6.4) that

$$\text{(10.9)} \qquad \mathbf{PRIN}(f, g) = \mathbf{TWIST}(f, g),$$

provided that p_1^* in (10.6) is surjective. This, for example, holds if $(Y|D)$ is

y-connected, $(X|D)$ is x-connected and if hodim $(B|D) < x + y$ where $(B|D)$ is good.

An immediate consequence of (10.7) and (10.8) is a result which is dual to (7.11). This result yields partial solutions of the problems in (10.1) and (10.2). Recall that a map $(\xi, \eta): f \to g$ in **Twist** is given by a diagram

(10.10)
$$\begin{array}{ccc} X & \xrightarrow[\xi]{} & A \\ \downarrow{\scriptstyle (f,1)} & & \downarrow{\scriptstyle g} \\ Y \times X & \xrightarrow[\eta]{} & B \\ {\scriptstyle (0,1)}\nwarrow & & \nearrow{\scriptstyle 0} \\ & X & \end{array}$$

which is homotopy commutative in $\mathbf{Top}_D$. Here $o: X \twoheadrightarrow D \xrightarrow{o} B$ is trivial map.

For a pair $(X|D) = (p: X \twoheadrightarrow D)$ in **Top** we define **the homotopy groups**

(10.11)
$$\pi_n(X|D) = \pi_n(Z_p, X) = \pi_{n-1}(F),$$

where F is the fiber of the fibration p. Here we assume that basepoints in D and F are chosen. If $(X|D)$ is good as in (10.6)′ then the groups do not depend on the choice of basepoints since in this case the fiber F is simply connected.

The pair of maps (ξ, η) in (10.10) yields associated twisted maps $F: P_f \to P_g$ (see §2) which induce the commutative diagram below of homotopy groups. We assume that $(A|D)$, $(B|D)$, $(X|D)$ and $(Y|D)$ are good.

(10.12)
$$\begin{array}{ccccccc} \xrightarrow{f_*} \pi_{i+1}(Y|D) & \xrightarrow{\partial} & \pi_i(P_f|D) & \xrightarrow{p_*} & \pi_i(X|D) & \xrightarrow{f_*} & \pi_i(Y|D) \\ \wr\| & & & & & & \wr\| \\ \pi_{i+1}(Y \times X|X) & & \downarrow{\scriptstyle F_* = \varphi} & & \downarrow{\scriptstyle \xi_*} & & \pi_i(Y \times X|X) \\ \downarrow & & & & & & \downarrow{\scriptstyle \eta_*} \\ \xrightarrow[g_*]{} \pi_{i+1}(B|D) & \xrightarrow[\partial]{} & \pi_i(P_g|D) & \xrightarrow[p_*]{} & \pi_i(A|D) & \xrightarrow[g_*]{} & \pi_i(B|D) \end{array}$$

The rows of the diagram are exact sequences induced by the fibrations $P_f \to X$ and $P_g \to A$ respectively. The isomorphisms i are canonically given since the corresponding fibers coincide. If $(Y|D)$ and $(B|D)$ are n-connected and if hodim $(X|D)$, hodim $(A|D) < n - 1$ we see that φ is actually determined by ξ_* and η_* in (10.12). In this case we write $\varphi = (\xi, \eta)_*$, in fact we have

$$\varphi = \begin{cases} \partial \eta_* i \partial^{-1} & i \geqq n, \\ p_*^{-1} \xi_* p_* & i < n. \end{cases}$$

(10.13) **Theorem.** *Consider the maps* f, g *in* (10.3) *and assume that* $(A|D)$, $(B|D)$, $(X|D)$ *and* $(Y|D)$ *are good in the sense of* (10.6)′. *Moreover, assume that* $(Y|D)$ *and* $(B|D)$ *are n-connected, that hodim* $(X|D)$, *hodim* $(A|D) < n-1$ *and hodim* $(B|D) < 2n-1$. *Then we get:*

(*a*) *An abstract homomorphism* $\varphi: \pi_*(P_f|D) \to \pi_*(P_g|D)$ *is realisable by a map* $P_f \to P_g$ *in* $\mathbf{Top}_D$ *if and only if there exists*

$$(\xi, \eta): f \to g \in \mathbf{Twist}$$

with $\varphi = (\xi, \eta)_*$.

(*b*) *There is a homotopy equivalence* $P_f \simeq P_g$ *in* $\mathbf{Top}_D$ *if and only if there exists* $(\xi, \eta): f \to g \in \mathbf{Twist}$ *such that* $\varphi = (\xi, \eta)_*$ *is an isomorphism.*

Using the Whitehead theorem proposition (*b*) is a consequence of (*a*). As in (7.13) we have the following special case of (10.13) which is a good criterion for a solution of problem (10.1).

(10.14) **Corollary.** *We write* $X = X(k, n)$ *if* X *is a CW-space with non-trivial homotopy groups* π_d *only in dimension* d, $k \leq d \leq n$. *Consider maps in* $\mathbf{Top}^*$ $(n \geq 4)$

$$f: X = X(2, n-2) \to Y = Y(n, 2n-2),$$
$$g: A = A(2, n-2) \to B = B(n, 2n-2).$$

Then there is a homotopy equivalence $P_f \simeq P_g$ *if and only if there is a homotopy commutative diagram in* $\mathbf{Top}$

$$\begin{array}{ccc} X & \xrightarrow[\xi]{} & A \\ \downarrow{\scriptstyle (f,1)} & & \downarrow{\scriptstyle g} \\ Y \times X & \xrightarrow[\eta]{} & B \\ {\scriptstyle (0,1)}\nwarrow & & \nearrow{\scriptstyle 0} \\ & X & \end{array}$$

for which ξ *and* $\eta | Y: Y \to B$ *are homotopy equivalences in* $\mathbf{Top}$.

(10.15) ***Example.*** Consider maps

$$g, f: K(\mathbb{Z}, 2) = \mathbb{C}P_\infty \to K(\mathbb{Z}, 4) \times K(\mathbb{Z}, 6). \tag{1}$$

The homotopy class of f is given by

$$f = (f_4 u^2, f_6 u^3) \quad (f_4, f_6 \in \mathbb{Z}) \tag{2}$$

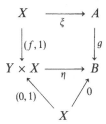

where $u \in H^2(K(\mathbb{Z},2),\mathbb{Z})$ is the fundamental class. We get by (10.14) with $n=4$:

$$P_f \simeq P_g \Leftrightarrow \left\{ \begin{array}{l} \exists \varepsilon, \tau \in \{+1,-1\}, N \in \mathbb{Z} \text{ such that} \\ g_4 = \varepsilon f_4, \\ g_6 = \tau f_6 + N f_4. \end{array} \right\} \tag{3}$$

Proof. Since $[K(\mathbb{Z},4), K(\mathbb{Z},6)] = 0$ we see that the map η in (10.14) is given by maps

$$\begin{array}{ll} \eta_4 \cdot 1 : K(\mathbb{Z},4) \to K(\mathbb{Z},4), & \eta_4 \in \mathbb{Z}, \\ \eta_6 \cdot 1 : K(\mathbb{Z},6) \to K(\mathbb{Z},6), & \eta_6 \in \mathbb{Z}, \\ N \cdot k : K(\mathbb{Z},4) \wedge K(\mathbb{Z},2) \to K(\mathbb{Z},6), & N \in \mathbb{Z}. \end{array}$$

Since $\eta | Y$ is a homotopy equivalence we have $\eta_4 = \pm 1$, $\eta_6 = \pm 1$. Moreover, $\xi = \xi_2 \cdot 1$ with $\xi_2 = \pm 1$. Now the equation $\eta(f,1) = g\xi$ yields the result in (3), compare (II.15a.7). □

Next we dualize the result in (7.17).

(10.16) ***Definition.*** Let $y \geq 2$ and let $\mathfrak{X}$ be a class of maps in **Top** with the following properties: each map $f \in \mathfrak{X}$ is a map between simply connected loop spaces

$$f : \Omega X \to \Omega Y,$$

where X and Y are CW-spaces and where ΩY is $(y-1)$-connected and where hodim $(\Omega X) < 2y - 1$, hodim $(\Omega Y) < 2y - 1$. ||

By (10.8) the class $\mathfrak{X}$ satisfies

(10.17) $$\mathbf{TWIST}(\mathfrak{X}) = \mathbf{PAIR}(\mathfrak{X}) \subset \mathbf{Top}/\simeq.$$

Here **PAIR** $(\mathfrak{X})$ is the subcategory of all homotopy classes $F : P_f \to P_g$ $(f, g \in \mathfrak{X})$ in **Top**$/\simeq$ for which there exists a homotopy commutative diagram

$$\begin{array}{ccc} P_f & \xrightarrow{F} & P_g \\ {\scriptstyle p_f}\downarrow & & \downarrow{\scriptstyle p_g} \\ \Omega X & \dashrightarrow & \Omega A \end{array}$$

with $g : \Omega A \to \Omega B$, $f : \Omega X \to \Omega Y$.

When $\mathfrak{X}$ satisfies (10.16) we can solve the isotropy problem and the homotopy problem for the linear extension of categories (see (6.6)):

(10.18) $$\Gamma/I \to \mathbf{TWIST}(\mathfrak{X}) \to \mathbf{Twist}(\mathfrak{X})/\simeq.$$

(10.19) **Theorem.** *The set*

$$\mathbf{Twist}(f,g) \subset [\Omega X, \Omega A] \times [\Omega Y \times \Omega X, \Omega B]_2 \tag{1}$$

consists of all elements (ξ, η) *with* $g\xi = \eta(f, 1)$. *We have* $(\xi, \eta) \simeq (\xi_1, \eta_1)$ *if and only if there exists* α *with the following properties*:

$$\left.\begin{aligned} &\alpha \in [\Omega Y \times \Omega X, \Omega A]_2, \\ &\xi + \alpha(f, 1) = \xi_1, \\ &0 = \eta_1 - \eta + (\nabla g)^*(-\alpha, \xi_1 p_2). \end{aligned}\right\} \tag{2}$$

Moreover, the natural system Γ/I *on* **Twist**$(\mathfrak{X})/\simeq$ *is given by*

$$(\Gamma/I)(\xi, \eta) = [\Omega X, \Omega^2 B]/I(f, \xi, g). \tag{3}$$

Here $I(f, \xi, g)$ *is the subgroup of all elements* β *with*

$$p_f^*(\beta) \in \text{image}\, \nabla(\xi p_f, g), \tag{4}$$

where $p_f: P_f \twoheadrightarrow \Omega X$ *is the projection and where*

$$\left.\begin{aligned} &\nabla(\xi p_f, g): [P_f, \Omega^2 A] \to [P_f, \Omega^2 B], \\ &\nabla(\xi p_f, g)(\gamma) = L(\nabla g)(\gamma, \xi p_f). \end{aligned}\right\} \tag{5}$$

Recall that we have the difference

$$\nabla g = -gp_2 + g(p_2 + p_1): \Omega A \times \Omega A \to \Omega B,$$

where $p_2 + p_1: \Omega A \times \Omega A \to \Omega A$ maps (σ, τ) to $\tau + \sigma$. Moreover the partial loop operation L yields

$$L(\nabla g): \Omega^2 A \times \Omega A \to \Omega^2 B.$$

The theorem above is an easy consequence of (10.6), (5.7) and (II.13.10). We point out that $\nabla f = \bar{\nabla} f$ for $f \in \mathfrak{X}$ since Y is an H-space by the assumption on ΩY in (10.16).

(10.20) **Corollary.** *Let* $f: \Omega X \to \Omega Y$ *be a map between simply connected loop spaces, where* X *and* Y *are CW-spaces. Moreover, suppose* ΩY *is* $(y-1)$ *connected, hodim* $(\Omega X) < y - 1$ *and hodim* $(\Omega Y) < 2y - 1$. *Then we have the short exact sequence of groups*

$$0 \to [\Omega X, \Omega^2 Y]/I(f, 1, f) \to \text{Aut}(P_f) \to T(f) \to 0.$$

Here $T(f)$ *is the subgroup of units in* **Twist**(f, f). *We have* $(\xi, \eta) \in T(f)$ *if and only if* ξ *and* $\eta | \Omega Y: \Omega Y \to \Omega Y$ *are homotopy equivalences.*

Proof. We can apply (10.7) since hodim $(\Omega X) < y - 1$. Moreover this implies that α in (10.19)(2) is trivial. □

Remark. The exact sequence in the corollary was obtained by Nomura (1966) provided that ΩX is $(x-1)$-connected and hodim $(\Omega Y) < y + x$. In this case $\eta | \Omega Y$ determines η and **Twist**$(f, f) =$ **Prin**(f, f). Compare also Rutter (1970).

(10.21) ***Example.*** Let $f_4, f_6 \in \mathbb{Z}$ and let

$$f = (f_4 u^2, f_6 u^3): K(\mathbb{Z},2) \to K(\mathbb{Z},4) \times K(\mathbb{Z},6) \tag{1}$$

be a map as in (10.15). Then we have by (10.20)

$$\mathrm{Aut}(P_f) = T(f), \tag{2}$$

where $T(f)$ as a set consists of all tuple $(\xi_2, \eta_4, \eta_6, N)$ with

$$\left.\begin{array}{l} \xi_2, \eta_4, \eta_6 \in \{1, -1\}, \\ N \in \mathbb{Z}, \\ f_4 = \eta_4 f_4, \\ f_6 \xi_2 = f_6 \eta_6 + N f_4. \end{array}\right\} \tag{3}$$

Multiplication in the group $T(f)$ is given by

$$(\xi_2, \eta_4, \eta_6, N)(\xi_2', \eta_4', \eta_6', N') = (\xi_2\xi_2', \eta_4\eta_4', \eta_6\eta_6', \eta_6 N' + N\xi_2'). \tag{4}$$

§ 10a Appendix: proof of the general loop theorem over D

We will use the following theorem which plays a role dual to the Blakers–Massey theorem.

(10a.1) **Lemma.** *Let $A \subset X$ be a cofibration and let $p: E \twoheadrightarrow X$ be a fibration in* **Top** *with fiber F. Let $E_A \twoheadrightarrow A$ be the restricted fibration and let G be a local system of coefficients on X. If (X, A) is n-connected and if F is r-connected, then*

$$p^*: \hat{H}^k(X, A, G) \to \hat{H}^k(E, E_A, p^*G)$$

is an isomorphism for $k < n + r + 2$ and is a monomorphism for $k \leqq n + r + 2$.

Proof. Without loss of generality all spaces are CW-complexes. We may assume that $X - A$ has only cells in dimension $\geqq n+1$ and that $F - *$ has only cells in dimension $\geqq r+1$. Therefore $E - E_A$ has cells of the form $e \times *$ (where e is a cell in $X - A$) and cells of dimension $\geqq n + r + 2$. This yields the result. Compare also (6.4.8) in Baues (1977). □

For the proof of (10.6) we first show that the following special case is satisfied. Let $\pi = \pi_1 D$ and let A be π-module. Then we obtain a based object in $\mathbf{Top}_D$ by the pull back

(10a.2)
$$\begin{array}{ccc} L(A,n) & \longrightarrow & L(A,n) \\ \downarrow \uparrow o & \text{pull} & \downarrow \uparrow s \\ D & \xrightarrow[d]{} & K(\pi,1) \end{array} .$$

Here d is a map which induces the identity on π_1 and $L(A, n)$ is the space considered in (III.§6).

(10a.3) **Lemma.** *Theorem* (10.6) *is true if B is a based object of the form $L(A, n)$ in* (10a.2). *Clearly hodim* $(B|D) = n$ *in this case.*

Proof of (10a.3). Consider the pull back diagram (in $\mathbf{Top}_D$):

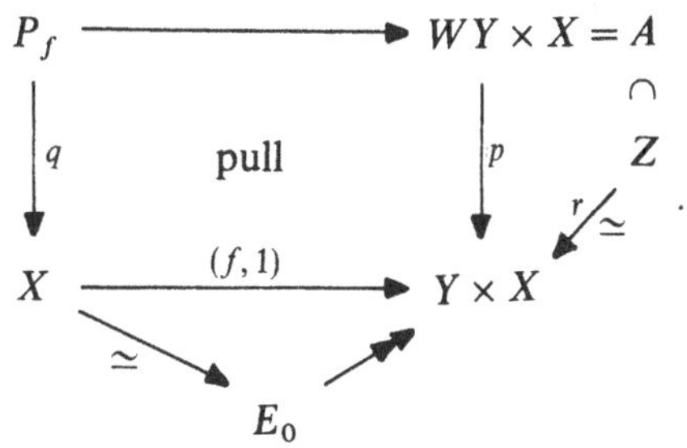

(1)

Here we turn $(f, 1)$ into a fibration in **Top** and we turn p into a cofibration in **Top**. Let $E \twoheadrightarrow Z$ be given by r^*E_0 and apply (10a.1) to

$$p_E:(E, E_A) \to (Z, A). \tag{2}$$

The fiber F of $E \to Z$ is the fiber of $(f, 1)$. Since $(f, 1)$ is a section of the fibration $Y \times_D X \to X$ we see that F is $(y-2)$-connected since $(Y|D)$ is y-connected. For the same reason the pair (Z, A) is $(y-1)$-connected; here the fiber of $A \subset Z$ is the fiber of $(0, 1): X \to Y \times X$. Now (10a.1) shows that (2) induces a homomorphism

$$p_E^*: \hat{H}^n(Z, A, G) \to \hat{H}^n(E, E_A, p_E^*G), \tag{3}$$

which is an isomorphism for

$$n < (y-1) + (y-2) + 2 = 2y - 1, \tag{4}$$

and which is a monomorphism for $n \leq 2y - 1$. Next we observe that there is a homotopy equivalence

$$(Z_q, P_f) \simeq (E, E_A) \text{ in } \mathbf{Pair\ (Top)}. \tag{5}$$

Moreover for $B = L(A, n)$ we have by (III.§6)

$$\left.\begin{aligned} \hat{H}^n(Z_q, P_f) &= \pi_B^1(P_f|X), \\ \hat{H}^n(Z, A) &= \pi_B^1(WY \times X | Y \times X), \end{aligned}\right\} \tag{6}$$

and by (5) the map $(p_E)^*$ in (3) is equivalent to $(\pi_f, 1)^*$ in (10.6). This proves the proposition in (10a.3). □

The following proof is essentially dual to the argument in the proof of (7a.3).

Proof of (10.6). Consider the based object

$$D \xrightarrow{o} B \xrightarrow{p} D \tag{1}$$

in (10.6). By the assumptions the fiber F of p is simply connected. We now use the Postnikov decomposition of p, see (III.§ 7). Thus we may assume that p has the factorization

$$p{:}B = B_N \twoheadrightarrow \cdots B_n \twoheadrightarrow B_{n-1} \twoheadrightarrow \cdots B_1 = D, \tag{2}$$

where each $B_n \twoheadrightarrow B_{n-1}$ is a principal fibration in $\mathbf{Top}_D$ with classifying map

$$f_n{:}B_{n-1} \to L_n = L(\pi_n, n+1), \quad n \geqslant 2. \tag{3}$$

Here $\pi_n = \pi_n F$ is a $\pi_1 D$-module by (1) and L_n is the based object in $\mathbf{Top}_D$ in (10a.2) above. By the composition

$$o{:}D \xrightarrow{o} B \twoheadrightarrow B_n \tag{4}$$

each B_n is a based object in $\mathbf{Top}_D$. Now consider the diagram

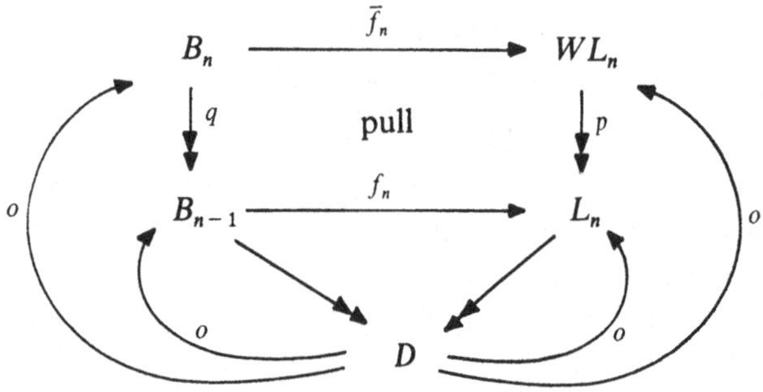

(5)

Here f_n is based up to homotopy in $\mathbf{Top}_D$ since

$$f_n 0 = p\bar{f}_n 0 \simeq p0 = 0 \text{ over } D. \tag{6}$$

In fact, we have a homotopy $\bar{f}_n 0 \simeq 0$ since $WL_n \twoheadrightarrow D$ is a homotopy equivalence over D. Diagram (5) yields based objects

$$\bar{L}_n = (D \xrightarrow{f_n 0} L_n \twoheadrightarrow D)$$

$$\overline{WL_n} = (D \xrightarrow{\bar{f}_n 0} WL_n \twoheadrightarrow D)$$

for which we have the pull back diagram in **Pair** $(\mathbf{Top}_D)$:

$$\begin{array}{ccc} (WB_n | B_n) & \xrightarrow{W\bar{f}_n} & (W\overline{WL_n} | WL_n) \\ \downarrow p & & \downarrow \bar{q} \\ (WB_{n-1} | B_{n-1}) & \xrightarrow[Wf_n]{} & (W\bar{L}_n | L_n) \simeq (WL_n | L_n) \end{array} \tag{7}$$

Using the homotopy in (6) we see that $W\bar{L}_n$ is homotopy equivalent over L_n to WL_n. Moreover $\bar{q}$ is dual to a cone, see in (II.7.5)(1). This shows that p in (7) is actually a principal fibration in **Pair** (**Top**$_D$) with classifying map Wf_n. Now the exactness of the fibration sequence in **Pair** (**Top**$_D$) and the five lemma inductively yield the proposition in (10.6) by use of (10a.3). In a similary way we prove the addendum of (10.6). □

VI

Homotopy theory of CW-complexes

In the chapter we continue the work of J.H.C. Whitehead on the combinatorial homotopy theory of CW-complexes. We show that there is a '*CW-tower of categories*' which approximates the homotopy category (under D) of relative CW-complexes (X, D). The tower is a new useful tool for the homotopy classification problems; it is a special case of the tower $\mathbf{TWIST}^c_*/\simeq$ in the next chapter.

Most of the results in the classical paper 'Combinatorial homotopy II' of Whitehead are immediate and special consequences of the CW-tower; we also deduce the final theorem in Whitehead's paper 'Simple homotopy types'. We not only give conceptually new and easy proofs of these results but also obtain generalizations to the relative case.

For example, we show that (for $\pi_2(D) = 0$) the homotopy types under D of 3-dimensional relative CW-complexes (X, D) are classified by the purely algebraic 3-dimensional crossed chain complexes under the group $\pi_1(D)$. This result is due to Whitehead for $D = *$.

Moreover, we describe finiteness obstructions for relative CW-complexes (X, D) which for $D = *$ coincide with those of Wall (1966).

We also derive new results from the CW-tower concerning the homotopy classification problems. In particular, applications on the classification of homotopy types, on the group of homotopy equivalences, and on the realizability of chain maps are described.

§ 1 Crossed chain complexes

Crossed chain complexes (under the trivial group) are the '*homotopy systems*' introduced and studied by Whitehead is his paper 'Combinatorial homotopy

II'. Moreover, they are special *crossed complexes* as defined by Brown–Higgins. We introduce the category of crossed chain complexes (under a group) since this is the algebraic bottom category for a tower of categories which approximates the homotopy category of CW-complexes.

Let D be a well-pointed path connected CW-space and let $\mathbf{Top}^D$ be the cofibration category of topological spaces under D. We consider the subcategory

(1.1) $$\mathbf{CW}_0^D \subset \mathbf{Top}^D.$$

The objects of $\mathbf{CW}_0^D$ are relative CW-complexes $X = (X, D)$ with *skeletal filtration*

$$D = X^0 \subset X^1 \subset \cdots \subset \varinjlim X^n = X. \tag{1}$$

Here X^n is obtained from X^{n-1} $(n \geq 1)$ by attaching n-cells. We derive from $D = X^0$ that X is a path connected. Moreover, we have attaching maps in $\mathbf{Top}^*$

$$f_n : \Sigma^{n-1} Z_n^+ = \bigvee_{Z_n} S^{n-1} \to X^{n-1} \tag{2}$$

and homotopy equivalences $C_{f_n} \simeq X^n$ under X^{n-1}. Here Z_n denotes the discrete set of n-cells of $X - D$. Since $D = X^0$ is path connected we can assume that $f_1 = 0$ is the trivial map so that

$$X^1 \simeq (\Sigma Z_1^+) \vee D \quad \text{under } D. \tag{3}$$

This shows that X is a complex in the cofibration category $\mathbf{Top}^*$, compare (III.3.1) and (III.5.3).

Remark. We can consider (X, D) as a complex in $\mathbf{Top}^D$ with attaching maps $(f_n, i): \Sigma^{n-1} Z_n^+ \vee D \to X^{n-1}$ where $i: D \subset X^{n-1}$ is the inclusions and where $(D \subset \Sigma^{n-1} Z_n^+ \vee D \xrightarrow{0,1} D)$ is a based object in $\mathbf{Top}^D$. (4)

The morphisms $f: X \to U$ in $\mathbf{CW}_0^D$ are **cellular maps** or, equivalently, filtration preserving maps,

$$f(X^n) \subset U^n, \tag{5}$$

under D. The k-homotopies, defined in (III.1.5) yield a natural equivalence relation, $\overset{k}{\simeq}$, on $\mathbf{CW}_0^D$. For $k = 1$ we know by the cellular approximation theorem that

$$\mathbf{CW}_0^D / \overset{1}{\simeq} = \mathbf{CW}_0^D / \simeq \; \subset \mathbf{Top}^D / \simeq \tag{6}$$

is a *full* subcategory of the homotopy category $\mathbf{Top}^D/\simeq$ of maps under D.

In this section we study the diagram of functors (where $G = \pi_1(D)$)

$$(1.2)\qquad \begin{array}{ccc} \mathbf{CW}_0^D/\simeq & \xrightarrow{\rho = C_*^{S^0}} & \\ \hat{C}_* \downarrow & & \mathbf{H}(G)/\simeq' \\ \mathbf{Chain}_{\hat{\mathbb{Z}}}/\simeq & \xleftarrow{C} & \end{array}$$

and a natural isomorphism $\hat{C}_* = C\rho$. Here $\hat{C}_*$ carries the relative CW-complex (X, D) to the *(relative) cellular chain complex*

$$\hat{C}_*(X, D) = C_*(\hat{X}, \hat{D}) \tag{1}$$

of the universal covering $\hat{X}$ of X; this functor is defined in (III.5.6). The functor ρ carries (X, D) to the *(fundamental) crossed chain complex* $\rho(X, D) = C_*^{S^0}\{X^n\}$ given by

$$\cdots \xrightarrow{d_4} \pi_3(X^3, X^2) \xrightarrow{d_3} \pi_2(X^2, X^1) \xrightarrow{d_2} \pi_1(X^1) \tag{2}$$

with $\rho_n = \pi_n(X^n, X^{n-1})$, see (III.10.6). Here we have

$$\pi = \pi_1(X) = \text{cokernel}(d_2). \tag{3}$$

To this end we define the algebraic category of crossed chain complexes as follows.

(1.3) ***Definition.*** A **crossed chain complex** $\rho = \{\rho_n, d_n\}$ is a sequence of homomorphisms between groups

$$\xrightarrow{d_4} \rho_3 \xrightarrow{d_3} \rho_2 \xrightarrow{d_2} \rho_1 \twoheadrightarrow \pi \tag{1}$$

such that $d_{n-1}d_n = 0$ for $n \geqq 3$ and such that the following properties are satisfied. The homomorphism d_2 is a crossed module the cokernel of which is $\pi = \rho_1/d_2\rho_2$, see (II.7.14). Hence kernel (d_2) is a right π-module. Moreover, $d_n (n \geqq 4)$ and $d_3: \rho_3 \to \text{kernel}(d_2)$ are homomorphisms between right π-modules. A **crossed chain map** $f: \rho \to \rho'$ between crossed chain complexes is a family of homomorphisms between groups $(n \geqq 1)$

$$f_n: \rho_n \to \rho'_n \quad \text{with } f_{n-1}d_n = d_n f_n, \tag{2}$$

such that f_2 is f_1-equivariant and such that f_n is $\bar{f}_1$-equivariant, where $\bar{f}_1: \pi \to \pi'$ is induced by f_1, $n \geqq 3$. Hence (f_2, f_1) is a map between crossed modules, see (II.7.14). For two chain maps $f, g: \rho \to \rho'$ a **homotopy** $\alpha: f \simeq g$ is given by a sequence of functions

$$\alpha_n: \rho_n \to \rho'_{n+1} \quad (n \geqq 1) \tag{3}$$

with the property ($n \geqq 2$)

$$-f_1+g_1=d_2\alpha_1 \quad \text{and} \quad -f_n+g_n=d\alpha+\alpha d, \tag{4}$$

where α_n is homomorphism between groups which is f_1-equivariant for $n=2$ and which is $\bar{f}_1$-equivariant for $n>2$. Moreover, α_1 is an f_1-crossed homomorphism, that is,

$$\alpha_1(x+y)=(\alpha_1 x)^{f_1 y}+\alpha_1 y (x, y \in \rho_1). \tag{5}$$

||

More generally, we use the following notation on crossed homomorphisms. Let π be a group and let M be a group on which π acts from the right via automorphisms. We denote the action of $y \in \pi$ on $x \in M$ by x^y. Moreover, let $\eta: N \to \pi$ be a homomorphism between groups. An η-**crossed homomorphism** $A: N \to M$ is a function which satisfies

(1.4) $$A(x+y)=(Ax)^{(\eta y)}+(Ay).$$

As usual we write the group structure additively, $x, y \in N$. We call A a **crossed homomorphism** if η is the identity; this is a **derivation** if M is abelian, see (IV.7.1).

The fundamental crossed chain complex $\rho(X, D)$ in (1.2) has various '*freeness properties*' which we describe as follows in (1.5), (1.7) and (1.12) respectively. For $n \geqq 1$ we have the function

(1.5) $$hp_*^{-1}: \rho_n=\pi_n(X^n, X^{n-1}) \overset{\cong}{\longleftarrow} \pi_n(\hat{X}^n, \hat{X}^{n-1}) \underset{h}{\longrightarrow} H_n(\hat{X}^n, \hat{X}^{n-1}),$$

where the Hurewicz map h is surjective for $n=2$ and bijective for $n>2$. For $n \geqq 2$ the function hp_*^{-1} is a homomorphism of groups which is $(\lambda: \pi_1 X^1 \to \pi_1 X)$-equivariant for $n=2$ and which is $\pi_1(X)$-equivariant for $n \geqq 3$. This shows by (III.5.7) that ρ_n is a free $\pi_1(X)$-module for $n \geqq 3$. A basis Z_n is given by the inclusion

$$Z_n \subset \rho_n=\pi_n(X^n, X^{n-1}), \quad e \longmapsto c_e, \tag{1}$$

which carries the cell e to the *characteristic map* c_e; here we use the homotopy equivalence $C_{f_n} \simeq X^n$ in (1.1)(2). The inclusion (1) is given for $n \geqq 1$. This yields via hp_*^{-1} in (1.5) the inclusion

$$Z_n \subset \hat{C}_n(X, D), \quad e \longmapsto \hat{e}=hp_*^{-1}(c_e), \tag{2}$$

which gives us a basis Z_n of the free $\pi_1(X)$-module $\hat{C}_n(X, D)$ for $n \geqq 1$.

Next we consider $\rho_1=\pi_1(X^1)$. Let $\langle Z_1 \rangle$ be the **free group** generated by the set Z_1 and let $A * B$ be the **free product** of the groups A and B (this is the sum in the category of groups). By the Van Kampen theorem we know that for well pointed spaces A, B

(1.6) $$\pi_1(A \vee B)=\pi_1(A) * \pi_1(B).$$

This gives us by (1.1)(3) the isomorphism of groups ($G=\pi_1(D)$)

$$\rho_1=\pi_1X^1\cong\pi_1(\Sigma Z_1^+\vee D)=\langle Z_1\rangle * G. \tag{1.7}$$

We call $\langle Z_1\rangle * G$ a **free group under** G **with basis** Z_1. The attaching map f_2 of 2-cells (1.1)(2) gives us the induced map on fundamental groups

$$f=\pi_1(f_2):\langle Z_2\rangle\to\pi_1X^1\cong\langle Z_1\rangle * G. \tag{1.8}$$

The homotopy class of f_2 in **Top*** can be identified with this homomorphism f. Moreover, we have the commutative diagram in the category of groups

$$\begin{array}{ccc} \pi_2(X^2,X^1) & \xrightarrow{d_2} & \pi_1X^1 \\ {\scriptstyle i}\uparrow & & \| \\ \langle Z_2\rangle & \xrightarrow[f]{} & \langle Z_1\rangle * G \end{array}, \tag{1.9}$$

where i is given by the inclusion $Z_2\subset\pi_2(X^2,X^1)$ in (1.5). This is clear since $d_2\bar{f}_e$ is the attaching map of e. The **normal subgroup** NfZ_2 generated by fZ_2 in $\langle Z_1\rangle * G$ is the image of d_2 so that

$$\pi_1X=\operatorname{cokernel}(d_2)=(\langle Z_1\rangle * G)/NfZ_2. \tag{1.10}$$

Therefore we call f a **presentation under** G of the group $\pi_1(X)$. If $D=*$ ($G=0$) this is the usual presentation of the fundamental group given by the attaching map of 2-cells. For the freeness property of (1.9) we need the following definition (see (II.7.14)).

(1.11) ***Definition.*** A **free crossed module** (d,i) **associated to** $f:\langle Z\rangle\to\rho_1$ is defined by the following universal property. For each crossed module d' and for each commutative diagram of unbroken arrows

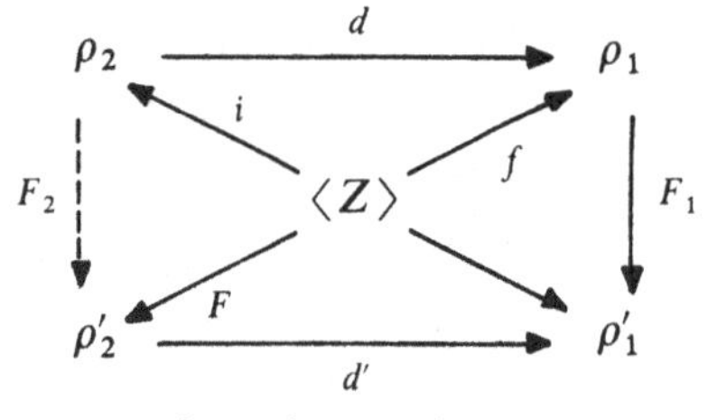

in the category of groups there is a unique map $(F_2,F_1):d\to d'$ between crossed modules such that $F_2i=F$. We call $i:Z\dot{\subset}\rho_2$ a **basis** of the free crossed module d. ||

By a result of J.H.C. Whitehead (1949) we know:

(1.12) **Theorem.** *(d_2, i) in (1.9) is the free crossed module associated to the presentation f of the fundamental group. Moreover, the kernel of hp_*^{-1} for $n = 2$, see (1.5), is the commutator subgroup of $\pi_2(X^2, X^1)$.*

We study further properties of free crossed modules below. With the freeness properties in (1.5), (1.7), and (1.12) we define the category **H**(G) in (1.2) as follows.

(1.13) ***Definition of the category* H(*G*).** Let G be a group. Objects of **H**(G) are **free crossed chain complexes under** G given by a crossed chain complex $\rho = (\rho_n, d_n)$ with the freeness properties (a), (b), (c),

(a) ρ_1 is a free group under G with a basis $Z_1 \subset \rho_1$, $\rho_1 = \langle Z_1 \rangle * G$,
(b) $d_2: \rho_2 \to \rho_1$ is a free crossed module associated to $f: \langle Z_2 \rangle \to \rho_1$ with basis $Z_2 \subset \rho_2$, and
(c) for $n \geqq 3$ the π-module ρ_n with $\pi = \rho_1 / d_2 \rho_2$ is a free π-module with basis $Z_n \subset \rho_n$.

A **morphism** $f: \rho \to \rho'$ in **H**(G) is a crossed chain map for which f_1 is the identity on G, see (1.3)(2). Moreover, a **homotopy** $\alpha: f \simeq g$ is a homotopy as in (1.3)(3) for which α_1 is trivial on G, $\alpha_1 | G = 0$. An object ρ in **H**(G) has **dimension** $\leqq n$ if $\rho_k = 0$ for $k > n$. Objects of dimension 2 are **free crossed modules under** G. ||

By use of the properties (1.5), (1.7) and (1.12) we see that $\rho(X, D)$ is an object in **H**(G). Hence the functor $\rho: CW_0^D \to$ **H**(G) is well defined by (1.2)(2). We now define the functor C in (1.2).

(1.14) ***Definition of the chain functor C on* H(*G*).** Let $\rho = (\rho_n, d_n)$ be on object in **H**(G) with basis $Z_n \subset \rho_n$. We define $C(\rho) = (\pi, C_*)$ in **Chain**$_{\mathbb{Z}}^{\wedge}$ by the group $\pi = \rho_1 / d_2 \rho_2$ and by the free π-module

$$C_n = \bigoplus_{Z_n} \mathbb{Z}[\pi], \tag{1}$$

generated by Z_n. Let $\lambda: \rho_1 \twoheadrightarrow \pi$ be the quotient map. We get maps

$$h_n: \rho_n \to C_n \quad (n \geqq 1), \tag{2}$$

as follows. For $n \geqq 2$ the map h_n is the unique λ-equivariant homomorphism of groups with $h_n(e) = e$ for $e \in Z_n$. For $n = 1$ the map h_1 is the unique λ-crossed homomorphism with $h_1(g) = 0$ for $g \in G$ and $h_1(e) = e$ for $e \in Z_1$.

We define the **boundaries** $\partial_n: C_n \to C_{n-1}$ of the chain complex C_* by those π-equivariant maps which make the following diagram commute

$$\begin{array}{ccccccccc} \longrightarrow & \rho_4 & \xrightarrow{d_4} & \rho_3 & \xrightarrow{d_3} & \rho_2 & \xrightarrow{d_2} & \rho_1 \\ & \cong\downarrow h_4 & & \cong\downarrow h_3 & & \downarrow h_2 & & \downarrow h_1 \\ \longrightarrow & C_4 & \xrightarrow[\partial_4]{} & C_3 & \xrightarrow[\partial_3]{} & C_2 & \xrightarrow[\partial_2]{} & C_1 \end{array} \tag{3}$$

There is, in fact, a unique ∂_2 with $h_1 d_2 = \partial_2 h_2$. For a map $f:\rho \to \rho'$ in $\mathbf{H}(G)$, which induces $\bar{f}_1:\pi \to \pi'$, let

$$C(f):C(\rho) \to C(\rho') \tag{4}$$

be the unique $\bar{f}_1$-equivariant homomorphism for which $h'_n f_n = C_n(f) h_n$, $n \geqq 1$. This completes the definition of the functor $C:\mathbf{H}(G) \to \mathbf{Chain}_{\mathbb{Z}}^{\wedge}$. ||

(1.15) **Proposition.** *The chain functor C on $\mathbf{H}(G)$ is well defined and there is a canonical natural isomorphism $C\rho(X,D) = \hat{C}_*(X,D)$ of chain complexes in $\mathbf{Chain}_{\mathbb{Z}}^{\wedge}$. Moreover, the functors ρ and C induce functors on homotopy categories as in* (1.2).

Proof. The natural isomorphism of $\pi_1(X)$-modules $C_* = C\rho(X,D) = \hat{C}_*(X,D)$ is given by $e \mapsto \hat{e}$, $e \in Z_n$, $n \geqq 1$. Equivalently, the following diagrams are commutative ($n \geqq 2$)

$$\begin{array}{ccc} \pi_1 X^1 & \longrightarrow & \pi_1(X^1, D) \\ \downarrow h_1 & & \downarrow hp_*^{-1} \\ C_1 & = & \hat{C}_1(X,D) \end{array} \qquad \begin{array}{ccc} \rho_n & = & \pi_n(X^n, X^{n-1}) \\ \downarrow h_n & & \downarrow hp_*^{-1} \\ C_n & = & \hat{C}_n(X,D) \end{array} .$$

In (1.16) below we show that the differential ∂_2 of $C\rho(X)$ coincides with the differential ∂_2 of $\hat{C}_*(X,D)$. Similarly, one shows that the induced maps on C_1 coincide with the induced maps on $\hat{C}_1(X,D)$. This shows that $C\rho = \hat{C}_*$ is a natural isomorphism. One can check that ρ and C are compatible with homotopies. For this consider the cylinder $I_D X$ in $\mathbf{CW}_0^D$, which is a relative CW-complex under D. We obtain a cylinder I in $\mathbf{H}(G)$ by computing $I(\rho(X)) = \rho(I_D X)$. This cylinder I induces the homotopy relation $\simeq$ on $\mathbf{H}(G)$ defined in (1.3)(3) and (1.13). We leave the details to the reader, compare also §3. □

(1.16) **Lemma.** *Let f be the attaching map of* 2*-cells in* (1.9) *and let ∂_2 be the differential of $\hat{C}_*(X,D)$. Then we have $\partial_2 \hat{e} = h_1 f(e)$ for $e \in Z_2$.*

Proof. For $f(e) \in \langle Z_1 \rangle * G$ we have the description

$$f(e) = y = \alpha_0 + y_1 + \alpha_1 + y_2 + \cdots + \alpha_{n-1} + y_n + \alpha_n, \tag{1}$$

where $\alpha_i \in G$, $y_i \in Z_1$ or $-y_i \in Z_1$. Let $\lambda: \rho_1 = \langle Z_1 \rangle * G \to \pi_1(X)$ be the quotient map. We consider (1) as a path in X, let $\hat{y}$ be the covering path of y which ends in $* \in \hat{X}$. This path can be computed by the rule

$$(x+y)\hat{} = \hat{x} \cdot \lambda(y) + \hat{y}. \tag{2}$$

In $\hat{C}_1(X, D)$ all elements $\hat{\alpha}_i \cdot \lambda(\beta)$ vanish since these paths lie in $\hat{D}$. Formula (2) corresponds to the formula for a crossed homomorphism. For $x \in Z_1$ we have the path $\hat{x}$ which corresponds to the generator $\hat{x} \in \hat{C}_1(X, D)$. Thus (2) shows

$$\partial_2(\hat{e}) = h_1 f(e), \tag{3}$$

where $h_1: \langle Z_1 \rangle * G \to \hat{C}_1(X, D)$ is the λ-crossed homomorphism with $h_1(x) = \hat{x}$ for $x \in Z_1$ and $h_1(\alpha) = 0$ for $\alpha \in G$. This is exactly the map h_1 which we used in (1.14)(2). More explicitly, we derive from (1) and (3) the formula

(1.17)
$$\begin{cases} \partial_2(\hat{e}) = \sum_{k=1}^{n} \hat{y}_k \cdot \lambda(\beta_k) \in \hat{C}_1 X \\ \text{with} \quad \beta_k = \tilde{y}_k + \alpha_k + y_{k+1} + \alpha_{k+1} + \cdots + y_n + \alpha_n, \\ \text{and} \quad \tilde{y} = 0 \quad \text{if } y \in Z_1 \quad \text{and} \quad \tilde{y} = y \quad \text{if } -y \in Z_1. \end{cases}$$ □

This formula is the well known *Reidemeister–Fox derivative* of $f(e)$, compare also Brown–Huebschmann. We now describe the connection of this formula with the difference construction ∇f and with the partial suspension which are concepts available in any cofibration category, see (II.§12). In particular, these concepts do not depend on the existence of a universal covering.

We first introduce some notation on groups. For groups A and B let A^{*B} be the **normal subgroup in $A * B$ generated by** A, this is the kernel of the projection $p_2 = 0 * 1: A * B \to B$. Recall that we write the group structure additively. The group multiplication is $+$, the inverse is $-$, the neutral element is 0. The trivial group is 0 and $0: A \to 0 \to G$ is the trivial homomorphism. For $a, b \in G$ the **commutator** is $(a, b) = -a - b + a + b = -a + a^b$ with $a^b = -b + a + b$.

(1.18) **Lemma.** *For $x \in A^{*B}$ there exist $a_i \in A$ and $b_i \in B$ such that $x = a_1^{b_1} + \cdots + a_n^{b_n}$.*

Proof. For $x \in A * B$ there exist $a_i \in A$, $\beta_i \in B$ with $x = \beta_0 + a_1 + \beta_1 + \cdots + a_n + \beta_n$. Since x is trivial on B we know $\beta_0 + \beta_1 + \cdots + \beta_n = 0$. Now we set $b_i = \beta_i + \cdots + \beta_n$ $(i = 1, \ldots, n)$, then the proposition is satisfied. □

We derive from (1.6) and (1.18) the isomorphisms of groups (B a well pointed space and Z a set)

(1.19)
$$\pi_1(\Sigma Z^+ \vee B)_2 = \langle Z \rangle^{*\pi_1 B} \underset{\chi}{=} \langle Z \times \pi_1 B \rangle.$$

Here $\pi_1(A \vee B)_2 = \text{kernel}(0, 1)_*$ is given by the projection $p_2 = (0, 1)$: $A \vee B \to B$ which induces $p_2 = 0 * 1$ on $\pi_1 A * \pi_1 B$. The isomorphism χ carries the generator $(x, \alpha) \in Z \times \pi_1 B$ of the free group $\langle Z \times \pi_1 B \rangle$ to the element x^α.

Now let $f_2 = f: \Sigma Z_2^+ \to X^1$ be the attaching map of 2-cells in (1.1)(2). We consider $X^1 = \Sigma Z_1^+ \vee D$ as the mapping cone of the trivial map f_1, see (1.1) (3). Then the **difference**

$$\text{(1.20)} \qquad \nabla f = -i_2 f + (i_2 + i_1) f \in [\Sigma Z_2^+, \Sigma Z_1^+ \vee X^1]_2$$

is defined as in (II.12.2). The map ∇f is trivial on X^1 and hence ∇f induces on fundamental groups the homomorphism

$$\nabla f: \langle Z_2 \rangle \to \langle Z_1 \rangle^{*\rho_1} = \pi_1(\Sigma Z_1^+ \vee X^1)_2. \tag{1}$$

We denote this homomorphism as well by ∇f since it is an algebraic equivalent of the homotopy class in (1.20). The isomorphism in (1) with $\rho_1 = \pi_1(X^1)$ is given by (1.19). We now show that ∇f in (1), restricted to generators, is the composite map

$$\nabla f: Z_2 \xrightarrow[f]{} \langle Z_1 \rangle * G = \rho_1 \xrightarrow[\nabla]{} \langle Z_1 \rangle^{*\rho_1}, \tag{2}$$

where ∇ is the unique crossed homomorphism which satisfies $\nabla(x) = x$ for $x \in Z_1$ and $\nabla(\alpha) = 0$ for $\alpha \in G$.

Proof of (2). Let $Z_1 = Z_1' = Z_1''$. Then i_2 and $(i_2 + i_1)$ in (1.20) correspond to homomorphisms between groups ($G = \pi_1 D$)

$$i_2, i_2 + i_1: \langle Z_1 \rangle * G \to \langle Z_1' \rangle * \langle Z_1'' \rangle * G, \tag{3}$$

which extend the identity on G and which satisfy $i_2(y) = y''$ and $(i_2 + i_1)(y) = y'' + y'$ for $y \in Z_1$. Here y', y'' are the elements in $\langle Z_1' \rangle$ and $\langle Z_1'' \rangle$, respectively, given by $y \in Z_1$. For $-y \in Z_1$ we get

$$\begin{aligned}(i_2 + i_1)(y) &= -(i_2 + i_1)(-y)\\ &= -((-y)'' + (-y)') = y' + y''.\end{aligned} \tag{4}$$

Now let y be a word in $\langle Z_1 \rangle * G$ as in (1.16) (1). We write y'' and y' for the corresponding words in $\langle Z_1'' \rangle * G$ and $\langle Z_1' \rangle * G$, respectively, and we set

$$x^y = -y'' + x' + y'' \in \langle Z_1' \rangle * \langle Z_2'' \rangle * G$$

for $x, y \in \langle Z_1 \rangle * G$. Suppose we have $f(e) = y$, $e \in Z_2$, then the definition of ∇f in (1.20) yields

$$\begin{aligned}(\nabla f)(e) = &-(\alpha_0 + y_1'' + \alpha_1 + \cdots + y_n'' + \alpha_n)\\ &+ \alpha_0 + (i_2 + i_1) y_1 + \alpha_1 + \cdots + (i_2 + i_1) y_n + \alpha_n.\end{aligned} \tag{5}$$

Using the elements β_k defined in (1.17) we see by (4) and (5)

$$(\nabla f)(e) = y_1^{\beta_1} + y_2^{\beta_2} + \cdots + y_n^{\beta_n}. \tag{6}$$

On the other hand, the definition of ∇ in (2) shows $\nabla(-x) = -x$ for $-x \in Z_1$, hence $0 = \nabla(-x+x) = (\nabla(-x))^x + \nabla(x)$, and therefore $\nabla(x) = -(-x)^x = x^x$ for $-x \in Z_1$. Now one can check by (6) that $(\nabla f)(e) = \nabla(f(e))$ and the proof of (2) is complete. □

We derive from (2) that the following diagram is commutative (see (1.9))

$$\begin{array}{ccccc}
\pi_2(X^2, X^1) & & \xrightarrow{\quad d_2 \quad} & & \pi_1 X^1 = \rho^1 \\
 & \cup & & \swarrow \nabla & \\
h_2 \downarrow & Z_2 & \xrightarrow[\nabla f]{} \langle Z_1 \rangle^{*\rho_1} & & \downarrow h_1 \\
 & \cap & & \searrow \bar{E}_f & \\
\hat{C}_2(X, D) & & \xrightarrow[\partial_2]{\qquad} & & \hat{C}_1(X, D)
\end{array}$$

Here $\bar{E}_f$ is the homomorphism of groups with $\bar{E}_f(x^\alpha) = \hat{x} \cdot \lambda(\alpha)$ for $x \in Z_1, \alpha \in \rho_1$, $\lambda: \rho_1 \twoheadrightarrow \pi_1(x)$, see (1.5)(2). By definition of h_1 and ∇, respectively, we immediately get $\bar{E}_f \nabla = h_1$. The operator $\bar{E}_f$ can be described by using the partial suspension E since the diagram

$$\begin{array}{ccc}
\pi_1(\Sigma Z_1^+ \vee X^1)_2 & = & \langle Z_1 \rangle^{*\rho_1} \\
(1 \vee i)_* E \downarrow & & \downarrow \bar{E}_f \\
\pi_2(\Sigma^2 Z_1 \vee X^2)_2 & \underset{\tau}{=} & \hat{C}_1(X, D)
\end{array} \tag{8}$$

commutes. Here τ is the isomorphism of $\pi_1(x)$-modules as in the proof of (III.5.9). Moreover, E is the *partial suspension* in (II.11.8) and $i: X^1 \subset X^2$ is the inclusion. One readily checks, by using (II.11.13), that (8) commutes.

Remark. From (7) and (8) we deduce that $(1 \vee i)_*(E \nabla f)$ represents ∂_2, this proves the commutativity of diagram (III.5.9)(1) for $n = 1$. In similar way one can show that (III.5.9)(2) commutes. (9)

§2 The functional suspension and the Peiffer group for 2-dimensional CW-complexes

We show that the classical Peiffer group of a presentation is connected with the functional suspension defined in (II.11.6). Let (X, D) be a CW-complex in $\mathbf{CW}_0^D$ as in (1.1). The attaching map

(2.1) $$f = f_2 : \Sigma Z_2^+ \to X^1 = \Sigma Z_1^+ \vee D$$

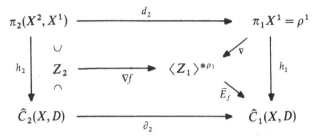

of 2-cells with $X^2 = C_f$ gives us the following commutative diagram which is a special case of diagram (II.11.6) for the definition of the functional suspension.

$$
\begin{array}{ccccccc}
 & & & & \pi_2(C\Sigma Z_2^+ \vee X^1, \Sigma Z_2^+ \vee X^1) & \xrightarrow[\partial]{\cong} & \pi_1(\Sigma Z_2^+ \vee X^1)_2 \\
 & & & & \downarrow{\scriptstyle (\pi_f, 1)_*} & & \downarrow{\scriptstyle (f,1)_*} \\
\pi_2 X^1 & \xrightarrow[i]{} & \pi_2 X^2 & \xrightarrow[j]{} & \pi_2(X^2, X^1) & \xrightarrow[d_2 = \partial]{} & \pi_1(X^1). \\
 & & & & \downarrow{\scriptstyle hp_*^{-1}} & & \\
 & & & & H_2(\hat{X}^2, \hat{X}^1) = \hat{C}_2(X, D) & &
\end{array}
\tag{2.2}
$$

We now describe various properties of this diagram. We know, that $(\pi_f, 1)_*$ is surjective by the 'general suspension theorem' (V.7.6). Moreover, we know that hp_*^{-1} is surjective and that

(2.3) $$\operatorname{kernel}(hp_*^{-1}) = [\rho_2, \rho_2] \subset \rho_2 = \pi_2(X^2, X^1)$$

is the commutator subgroup of ρ_2 denoted by $[\rho_2, \rho_2]$, see (1.12) above.

We derive from (1.19) the isomorphisms

(2.5) $$\pi_1(\Sigma Z_2^+ \vee X^1)_2 = \langle Z_2 \rangle^{*\rho_1} \underset{\chi}{=} \langle Z_2 \times \rho_1 \rangle,$$

where $\rho_1 = \pi_1 X^1$. This shows that $(f, 1)_*$ in (2.2) satisfies

(2.6) $$\operatorname{image}(f, 1)_* = Nf(Z_2) \subset \pi_1 X^1,$$

where $Nf(Z_2)$ is the normal subgroup generated by the subset $f(Z_2)$, see (1.8). We derive from Whitehead's result (1.12) the

(2.7) **Proposition.** *The kernel of* $(\pi_f, 1)_* \partial^{-1}$ *in* $\pi_1(\Sigma Z_2^+ \vee X^1)_2 = \langle Z_2 \rangle^{*\rho_1}$ *is the normal subgroup generated by the Peiffer elements*

$$\langle x^\alpha, y^\beta \rangle_f = -x^\alpha - y^\beta + x^\alpha + y^{\beta - \alpha + f(x) + \alpha},$$

where $x, y \in Z_2$, α, $\beta \in \rho_1$.

Proof. Let P_f be the *Peiffer group*; this is the normal subgroup of $\langle Z_2 \rangle^{*\rho_1}$ generated by the Peiffer elements. Then the homomorphism

$$\rho_2^f = \langle Z_2 \rangle^{*\rho_1} / P_f \to \rho_1, \tag{$*$}$$

induced by $(f, 1)_* = f * 1$ is the *free crossed module associated to* $f : \langle Z_2 \rangle \to \rho_1$. This follows, since the Peiffer elements correspond exactly to the defining equations of a crossed module and thus ($*$) has the universal property in (1.11). On the other hand, $\pi_2(X^2, X^1) \to \pi_1 X^1$ is the free crossed module by (1.12), hence we obtain the result in (2.7). Compare also Brown–Huebschmann.

□

Next we consider the composition $\bar{E}_f = hp_*^{-1}(\pi_f, 1)_* \partial^{-1}$ in (2.2) which is part of the diagram

(2.8)
$$\begin{array}{ccc} \pi_1(\Sigma Z_2^+ \vee X^1)_2 & = & \langle Z_2 \rangle^{*\rho_1} \\ \Big\downarrow (1 \vee i)_* E & & \Big\downarrow \bar{E}_f \\ \pi_2(\Sigma^2 Z_2^+ \vee X^1)_2 & \underset{\tau}{=} & \hat{C}_2(X, D) \end{array}$$

Here τ is the isomorphism of $\pi_1(X)$-modules in the proof of (III.5.9). Moreover, E is the partial suspension and $i: X^1 \subset X^2$ is the inclusion, compare (1.20)(8). One readily checks that diagram (2.8) commutes (by using (II.11.13)) and that the homomorphism $\bar{E}_f$ of groups satisfies $\bar{E}_f(x^\alpha) = \hat{x} \cdot \lambda(\alpha)$ where $x \in Z_2$, $\alpha \in \rho_1$, $\lambda: \rho_1 \twoheadrightarrow \pi_1(x)$. Here $\hat{x}$ is an element in the basis of $\hat{C}_2(X, D)$, see (1.5)(2), and $\hat{x} \cdot \lambda(\alpha)$ is given by the action of $\pi_1(x)$.

(2.9) ***Definition.*** For the relative CW-complex (X, D) in $\mathbf{CW}_0^D$ we define the group

$$\begin{aligned} \tilde{\Gamma}_2(X, D) &= j^{-1}(\ker(d_2) \cap \ker(hp_*^{-1})) \\ &= j^{-1}(\ker(d_2) \cap [\rho_2, \rho_2]) \\ &= E_f(\ker(f, 1)_* \cap \ker(1 \vee i)_* E), \end{aligned}$$

which is a subgroup of $\pi_2(X^2)$. Here we use the operators in (2.2) and (2.8). Moreover, $E_f = j^{-1}(\pi_f, 1)_* \partial^{-1}$ is the functional suspension, compare (II.11.7). The second equation is a consequence of (2.3) and the third equation follows from (2.8) by the surjectivity of $(\pi_f, 1)_*$. ||

We point out that the third equation is well defined in any cofibration category.

(2.10) **Proposition.** *For $D = *$ the group $\tilde{\Gamma}_2(X, *) = 0$ is trivial. Moreover, for $D = *$ the Peiffer group P_f in* (2.7) *satisfies $P_f = \ker(f, 1)_* \cap \ker(1 \vee i)_* E$.*

Proof. This follows since for $X^1 = \Sigma Z_1^+$ we have $\pi_2 X^1 = 0$ and since hp_*^{-1} induces an isomorphism $\pi_2 X^2 \cong H_2 \hat{X}^2$ where $\pi_2 X^2 \cong \ker(d_2)$. □

In general, the group $\tilde{\Gamma}_2(X, D)$ is not trivial. We have the short exact sequence of $\pi_1(X)$-modules

(2.11)
$$0 \to \Gamma_2(X, D) \to \tilde{\Gamma}_2(X, D) \xrightarrow{j} j\tilde{\Gamma}_2(X, D) \to 0,$$

where $\Gamma_2(X, D) = \text{image}(\pi_2 X^1 \to \pi_2 X^2)$, see (III.11.6) and where $j\tilde{\Gamma}_2(X, D) = \ker(d_2) \cap [\rho_2, \rho_2]$ by definition in (2.9). The sequence (2.11) is natural for maps in the homotopy category $\mathbf{CW}_0^D/\simeq$.

(2.12) **Proposition.**

(*A*) *Let* $Nf(Z_2) = d_2(\rho_2) \subset \pi_1 X^1$ *be the normal subgroup generated by the subset* $f(Z_2)$, *see* (2.6). *The second homology of this group with integral coefficients is*

$$j\tilde{\Gamma}_2(X, D) \cong H_2(NfZ_2, \mathbb{Z}).$$

(*B*) *Let* $i: G = \pi_1(D) \to \pi = \pi_1(X)$ *be induced by the inclusion* $D \subset X$. *Then*

$$\begin{aligned}\Gamma_2(X, D) &\cong \hat{H}_2(D; i^*\mathbb{Z}[\pi]) \\ &= \pi_2(D) \underset{\mathbb{Z}[G]}{\otimes} i^*\mathbb{Z}[\pi].\end{aligned}$$

(2.13) **Addendum.** *If* $\pi_2(D) = 0$ *then* $\Gamma_2(X, D) = 0$. *Moreover for* $D = *$ *we have*

$$\Gamma_2(X, *) = j\tilde{\Gamma}_2(X, *) = \tilde{\Gamma}_2(X, *) = 0.$$

Proof of (2.12). Formula (A) is based on the result of Ratcliffe (1980) that ∂_* in (2) below is trivial. Consider the short exact sequence

$$0 \to R \to \rho_2 \xrightarrow{\partial} N \to 0 \tag{1}$$

of groups where $N = d_2\rho_2 = NfZ_2$. This sequence induces the exact sequence

$$H_2\rho_2 \xrightarrow{\partial_*} H_2 N \to R \to \mathrm{Ab}(\rho_2) \to \mathrm{Ab}(N), \tag{2}$$

where Ab = abelianization. A different proof is due to Ellis. For the addendum we point out that for $D = *$ the group N is free and hence $H_2 N = 0$, compare also (2.10).

We now proof (B). For this let $\rho_1 = \pi_1 X^1$ and let $j: \pi_1 D \to \rho_1$ be induced by $D \subset X^1$. Then we have the following commutative diagram, compare (III.§5).

$$\begin{array}{ccc} \pi_2(X^1) & \longrightarrow & \pi_2(X^2) \\ \| & & \| \\ \hat{H}_2(X^1, \mathbb{Z}[\rho_1]) & \longrightarrow & \hat{H}_2(X^2, \mathbb{Z}[\pi]) \\ \uparrow j_* & & \uparrow i_* \\ \hat{H}_2(D, j^*\mathbb{Z}[\rho_1]) & \underset{(\lambda_\#)_*}{\twoheadrightarrow} & \hat{H}_2(D, i^*\mathbb{Z}[\pi]) \end{array} \tag{3}$$

Here $\lambda: \rho_1 \twoheadrightarrow \pi$ is the quotient map. We have

$$\hat{H}_2(D, \mathbb{Z}[G]) \underset{\mathbb{Z}[G]}{\otimes} \mathbb{Z}[\pi] = \hat{H}_2(D, i^*\mathbb{Z}[\pi]), \tag{4}$$

by a Künneth-formula. Here $\hat{H}_2(D, \mathbb{Z}[G]) = \pi_2 D$ by the Hurewicz-isomorphism which is also used in (3). Moreover, (4) shows that $(\lambda_\#)_*$ in (3) is surjective. By the exact sequences of pairs we see that j_* in (3) is surjective and that i_* in (3) is injective. □

§3 The homotopy category of 2-dimensional CW-complexes under D

We consider the full subcategory $(\mathbf{CW}_0^D)^2$ of $\mathbf{CW}_0^D$ consisting of 2-dimensional relative CW-complexes (X, D) with $D = X^0 \subset X^1 \subset X^2 = X$, see (1.1). We have the functor

(3.1) $$\rho:(\mathbf{CW}_0^D)^2/\simeq \;\to \mathbf{H}(G)^2/\simeq,$$

where $\mathbf{H}(G)^2$ is the full subcategory of $\mathbf{H}(G)$ consisting of free crossed modules, $G = \pi_1(D)$. We show that ρ in (3.1) is a detecting functor, see (IV.1.5), and that ρ is actually a nice example for the linear extension of categories $\mathbf{TWIST}(\mathfrak{X})$ described in (V.3.12). Each object (X^2, D) in $(\mathbf{CW}_0^D)^2$ is the mapping cone in the cofibration category $\mathbf{Top}^D$ of an attaching map

(3.2) $$(f, i_2):\Sigma Z_2^+ \vee D \to \Sigma Z_1^+ \vee D$$

in $\mathbf{Top}^D/\simeq$, see (1.1)(4). Let $\mathfrak{X}$ be the class of all such attaching maps (f, i_2) where Z_1 and Z_2 are discrete sets. By (V.3.12) we have the linear extension of categories for $\mathbf{TWIST}(\mathfrak{X})$ in the top row of the diagram

(3.3) $$\begin{array}{ccccc} \Gamma/I & \xrightarrow{+} & \mathbf{TWIST}(\mathfrak{X}) & \xrightarrow{\lambda} & \mathbf{Twist}(\mathfrak{X})/\simeq \\ \cong\big\downarrow \bar{j} & & \sim\big\downarrow j & & \sim\big\downarrow \bar{\rho} \\ H^2\Gamma_2/I & \xrightarrow{+} & (\mathbf{CW}_0^D)^2/\simeq & \xrightarrow[\rho]{} & \mathbf{H}(G)^2/\simeq \end{array} .$$

Here j is the inclusion functor which is an equivalence of categories since $(\pi_f, 1)_*$ in (2.2) is surjective, in fact, this shows that each map in $\mathbf{CW}_0^D$ is actually a twisted map by (V.7.8).

(3.4) **Theorem.** *There is a commutative diagram as in (3.3) where $\bar{\rho}$ is an equivalence of categories and where $\bar{j}$ is an isomorphism of natural systems. The natural system $H^2\Gamma_2$ is given by*

$$H^2\Gamma_2(f) = \hat{H}^2(X, D, \varphi^*\Gamma_2(Y, D)).$$

Here $f:X \to Y$ is a map in $(\mathbf{CW}_0^D)^2/\simeq$ which induces $\varphi = f_*:\pi_1 X \to \pi_1 Y$. The isotropy group I in f is the kernel of the map

$$\hat{H}^2(X, D; \varphi^*\Gamma_2(Y, D)) \to \hat{H}^2(X, D; \varphi^*\pi_2 Y),$$

induced by the inclusion $\Gamma_2(Y, D) \subset \pi_2 Y$, see (2.11).

In (6.2) below we describe the generalization of this theorem for the category $\mathbf{CW}_0^D/\simeq$.

(3.5) **Corollary.** *Let $\pi_2(D) = 0$. Then*

$$\rho:(\mathbf{CW}_0^D)^2/\simeq \;\xrightarrow{\sim} \mathbf{H}(G)^2/\simeq$$

is an equivalence of categories.

This is a consequence of (3.4) since $\pi_2(D)=0$ implies $\Gamma_2(Y,D)\pm 0$ and therefore $\Gamma/I=0$ in (3.3). For $D=*$ the corollary is due to J.H.C. Whitehead (1949) who used different methods. The corollary shows that for $\pi_2(D)=0$ the homotopy theory of 2-dimensional CW-complexes under D, including the homotopy classification of mappings, is equivalent to the purely algebraic homotopy theory of free crossed modules under the group $G=\pi_1 D$.

For the proof of (3.4) we first give an algebraic characterization of the category **Twist**($\mathfrak{X}$). By definition in (V.3.8) a morphism $(\xi,\eta):(f,i_2)\to(g,i_2)$ in **Twist**($\mathfrak{X}$) can be identified with the following commutative diagram in **Top***$/\simeq$.

(3.6)

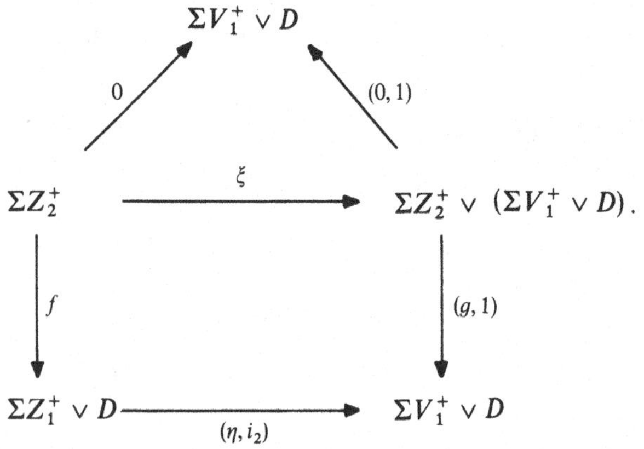

Here Z_n and V_n $(n=1,2)$ are the discrete sets of cells. This diagram can be identified with the corresponding diagram in the category of groups obtained by applying the functor π_1:

(3.7)

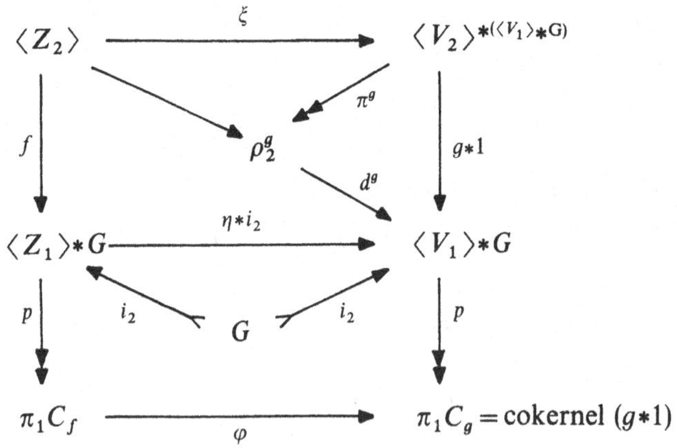

Here $d^g:\rho_2^g\to\langle V_1\rangle * G$ is the free crossed module associated to g, see (2.7)(*).

(3.8) ***Remark.*** Clearly each pair (ξ,η) with $(g*1)\xi=(\eta*i_2)f$ induces a homomorphism φ on fundamental groups such that diagram (3.7) commutes. If $F:C_f\to C_g$ is a map under D associated to (ξ,η) then $\varphi=\pi_1(F)$. On the other hand, diagram (3.7) shows that for each homomorphism φ under G there is η and ξ such that diagram (3.7) commutes. This is clear by exactness of the right column in (3.7); first choose η as a lift of $\varphi\circ p$, then choose ξ as a lift of $(\eta*i_2)f$. Therefore each homomorphism φ under G is realisable by a map $F:C_f\to C_g$ under D.

We have the functor

$$\bar{\rho}:\mathbf{Twist}(\mathfrak{X})\to\mathbf{H}(G)^2, \tag{3.9}$$

which carries the object g to the free crossed module $d^g:\rho_2^g\to\rho_1$ in diagram (3.7) and which carries the morphism (ξ,η) in (3.6) to the map (F_2,F_1) where $F_1=\eta*i_2$ and where F_2 is given by $F=\pi^g\xi$ in (3.7) as in (1.11). It is easy to check that $\bar{\rho}$ is a well-defined functor.

(3.10) **Proposition.** *We have a natural equivalence relation* $\sim$ *on* $\mathbf{Twist}(\mathfrak{X})$ *by setting* $(\xi,\eta)\sim(\xi',\eta')\Leftrightarrow(\pi^g\xi=\pi^g\xi'$ and $\eta=\eta')$. *Moreover* $\bar{\rho}$ *above induces the isomorphism of categories*

$$\bar{\rho}:\mathbf{Twist}(\mathfrak{X})/\sim\;=\mathbf{H}(G)^2.$$

(3.11) *Proof of* (3.4). It is clear that for $\bar{\rho}$ in (3.9) we have a natural isomorphism $\bar{\rho}\lambda=\rho j$ in (3.3). We now show that $\bar{\rho}$ in (3.3) is an equivalence of homotopy categories. For this we use the general result in (V.5.7) which implies that $(\xi,\eta)\simeq(\xi',\eta')$ in $\mathbf{Twist}(\mathfrak{X})$ if and only if there is

$$\alpha:\langle Z_1\rangle\to\langle V_2\rangle^{*(\langle V_1\rangle*G)},$$

such that (1) and (2) are satisfied

$$\eta+(g,1)_*\alpha=\eta', \tag{1}$$

$$0=\pi^g(\xi'-\xi+(\bar{\nabla}f)^*(-\alpha,i_2\eta')). \tag{2}$$

Let $(F_2,F_1)=\bar{\rho}(\xi,\eta)$ and $(F_2',F_1')=\bar{\rho}(\xi',\eta')$ and let

$$A:\langle Z_1\rangle*G\to\rho_2^g \tag{3}$$

be the F_1-crossed homomorphism with $A(z)=\pi^g\alpha(z)$ for $z\in Z_1$ and $A(G)=0$. Then (1) is equivalent to

$$-F_1+F_1'=d^gA. \tag{4}$$

Moreover (2) is equivalent to

$$-F_2+F_2'=Ad^f. \tag{5}$$

We see this as follows. First we observe that (5) is equivalent to

$$-F'_2+F_2=Bd^f, \tag{6}$$

where $B:\langle Z_1\rangle * G\to\rho_2^g$ is the F'_1-crossed homomorphism with $B(z)=-A(z)$ for $z\in Z_1$, and with $B(G)=0$. On the other hand, by definition of $\bar{\nabla}f$ in (II.13.7) we see that (2) is equivalent to

$$-i_2F'_1f+(\xi-\xi')+i_2F'_1f\equiv(\nabla f)^*(-\alpha,i_2\eta') \tag{7}$$

module $P_g=\text{kernel}\,\pi^g$. Here the left-hand side corresponds to

$$\begin{aligned}(F_2-F'_2)^{F'_1d^f}&=(F_2-F'_2)^{d^gF'_2}\\&=-F'_2+(F_2-F'_2)+F'_2=-F'_2+F_2,\quad\text{see (II.7.14).}\end{aligned}$$

The right-hand side of (7) corresponds to Bd^f. Therefore (7) is equivalent to (6). □

The proof shows that the technical result (V.5.7) is consistent with the definition of homotopies in (1.3). We leave it to the reader to give a more direct proof of (3.4). The proposition on the isotropy group in (3.4) follows from (II.13.10) and (1.20)(9), see also (5.16) below.

§4 The tower of categories for the homotopy category of crossed chain complexes

In this section we study the chain functor

(4.1) $$C:\mathbf{H}(G)/\simeq\;\to\mathbf{Chain}_{\mathbb{Z}}^{\wedge}/\simeq,$$

defined in (1.14). We show that C can be nicely described by a tower of categories. All results in this section are purely algebraic.

The definition of C shows that a morphism $f:\rho\to\rho'$ in $\mathbf{H}(G)$ is determined in degree $\geqq 3$ by the induced chain map $Cf:C\rho\to C\rho'$ in $\mathbf{Chain}_{\mathbb{Z}}^{\wedge}$. If $\xi=Cf$ is fixed we have only a restricted choice of defining $f_1:\rho_1\to\rho'_1$ and $f_2:\rho_2\to\rho'_2$ such that $Cf=\xi$. We now study the possible choices of f_1 and f_2 respectively. This leads to the definition of the following categories $\mathbf{H}_2(G)$ and $\mathbf{H}_1(G)$. We call the objects of $\mathbf{H}_2(G)$ and $\mathbf{H}_1(G)$ **homotopy systems of order** 2 and **homotopy systems of order** 1 respectively.

(4.2) ***Definition.*** *Objects in* $\mathbf{H}_2(G)$ are pairs (C,ρ) where $\rho=(d_2:\rho_2\to\rho_1)$ is a free crossed module under G in $\mathbf{H}(G)^2$ and where C is a chain complex in $\mathbf{Chain}_{\mathbb{Z}}^{\wedge}$ of free $\bar{\rho}_1$-modules, $\bar{\rho}_1=\pi(\rho)=\rho_1/d_2\rho_2$, which coincides with $C(\rho)$ in degree $\leqq 2$. Moreover, the object (C,ρ) satisfies the following *cocycle condition*. For (C,ρ) there exists a homomorphism $d_3:C_3\to\text{kernel}(d_2)$ of $\bar{\rho}_1$-modules such that the diagram

$$\begin{array}{ccccc} C_3 & \xrightarrow{d_3} & \rho_2 & \xrightarrow{d_2} & \rho_1 \\ & \searrow^{\partial_3} & \downarrow{\scriptstyle h_2} & & \\ & & C_2 & & \end{array}$$

commutes.

A *morphism*

$$(\xi, \eta):(C, \rho) \to (C', \rho')$$

is a pair with

$$\begin{cases} \eta: \rho_1 = \langle B \rangle * G \to \rho'_1 = \langle B' \rangle * G, \\ \xi: C \to C'. \end{cases}$$

Here η is homomorphism of groups under G which induces $\varphi: \bar{\rho}_1 \to \bar{\rho}'_1$ and ξ is a φ-equivariant chain map such that the following conditions (1) and (2) are satisfied:

The diagram

$$\begin{array}{ccc} \rho_1 & \xrightarrow{\eta} & \rho'_1 \\ {\scriptstyle h_1}\downarrow & & \downarrow{\scriptstyle h_1} \\ C_1 & \xrightarrow{\xi_1} & C'_1 \end{array} \tag{1}$$

commutes.

There is an η-equivariant homomorphism $\bar{\xi}_2$ such that the diagram

$$\begin{array}{ccc} \rho_1 & \xrightarrow{\eta} & \rho'_1 \\ {\scriptstyle d_2}\uparrow & & \uparrow{\scriptstyle d'_2} \\ \rho_2 & \overset{\bar{\xi}_2}{\dashrightarrow} & \rho'_2 \\ {\scriptstyle h_2}\downarrow & & \downarrow{\scriptstyle h_2} \\ C_2 & \xrightarrow{\xi_2} & C'_2 \end{array} \tag{2}$$

commutes.

We have a *homotopy relation* $\simeq$ on $\mathbf{H}_2(G)$ as follows: maps (ξ, η), $(\xi', \eta'):(C, \rho) \to (C', \rho')$ in $\mathbf{H}_2(G)$ are homotopic if there exist maps

$$\begin{cases} \bar{\alpha} : \rho_1 = \langle B \rangle * G \to \rho'_2, \\ \alpha_n: C_n \to C'_{n+1}, \quad n \geqq 2, \end{cases}$$

such that (3, 4, 5, 6) are satisfied:

$$\left.\begin{array}{l}\bar{\alpha}(x)=0 \quad \text{for } x\in G, \quad \text{and} \\ \bar{\alpha}(x+y)=(\bar{\alpha}x)^{\eta y}+(\bar{\alpha}y) \quad \text{for } x,\, y\in\rho_1\end{array}\right\} \tag{3}$$

$$-\eta+\eta'=d_2\bar{\alpha} \tag{4}$$

Condition (4) implies that η and η' induce the same homomorphism $\varphi:\bar{\rho}_1\to\bar{\rho}'_1$.

$$\alpha_n \text{ is a } \varphi\text{-equivariant homomorphism} \tag{5}$$

$$-\xi_n+\xi'_n=d\alpha_n+\alpha_{n-1}d \quad \text{for } n\geqq 2 \tag{6}$$

In (6) the φ-equivariant map α_1 is determined by $\bar{\alpha}$ via

$$h_2\bar{\alpha}=\alpha_1 h_1. \tag{7}$$

We call $(\alpha,\bar{\alpha}):(\xi,\eta)\simeq(\xi',\eta')$ a homotopy in $\mathbf{H}_2(G)$. In (VII.1.7) below we will see that $\mathbf{H}_2(G)$ corresponds to the category $\mathbf{TWIST}_2^c(\mathfrak{X})$ which is defined in any cofibration category. ||

(4.3) ***Notation.*** Let M be a set on which π acts from the right. Then the **π-modul** $\mathbb{Z}[M]$ is the free abelian group generated by element $[m]$, $m\in M$. The action of π is given by $[m]\cdot\alpha=[m\alpha]$ for $\alpha\in\pi$. This generalizes the notation on group rings in (III.§ 5). A φ-equivariant map $f:M\to M'$ with $\varphi:\pi\to\pi'$ induces the φ-equivariant homomorphism $f_\#:\mathbb{Z}[M]\to\mathbb{Z}[M']$ with $f_\#[m]=[fm]$.

(4.4) ***Definition.*** Let $\mathbf{H}_1(G)$ be the following category. *Objects* are tuple

$$X=(\pi,C,\partial,i)=(\pi_X,C_X,\partial_X,i_X) \tag{1}$$

with the following properties (a), (b) and (c).

(a) π is a group and $i:G\to\pi$ is a homomorphism.

(b) (π,C) is a chain complex in $\mathbf{Chain}_{\mathbb{Z}}^{\wedge}$ of free π-modules with $C_n=0$ for $n\leqq 0$.

(c) $$\partial:H_1C\to\mathbb{Z}[iG\backslash\pi]=H_0(G,i^*\mathbb{Z}[\pi])$$

is a homomorphism of right π-modules; here $iG\backslash\pi$ denotes the set of left cosets $(iG)+x$, $x\in\pi$.

A *morphism* $f:X\to Y$ in $\mathbf{H}_1(G)$ is a φ-equivariant chain map in $\mathbf{Chain}_{\mathbb{Z}}^{\wedge}$ which preserves the structure, that is:

(d) $\varphi:\pi_X\to\pi_Y$ is a homomorphism under G; $\varphi i_X=i_Y$. We write $\varphi\in\mathrm{Hom}(\pi_X,\pi_Y)^G$

(e) The following diagram commutes.

$$\begin{array}{ccc} H_1 C_X & \xrightarrow{f_*} & H_1 C_Y \\ \partial_X \downarrow & & \downarrow \partial_Y \\ \mathbb{Z}[iG\backslash\pi_X] & \xrightarrow[\varphi_\#]{} & \mathbb{Z}[iG\backslash\pi_Y] \end{array}$$

Here $\varphi_\#$ is defined as in (4.3). By (d) we see that φ induces a map of left cosets. We say that maps in $\mathbf{H}_1(G)$ are homotopic if they are homotopic in $\mathbf{Chain}_{\mathbb{Z}}^{\wedge}$, see (III.§ 5). ∥

There are canonical functors as in the commutative diagram

(4.5) $$\begin{array}{ccccccc} C{:}\mathbf{H}(G) & \xrightarrow{\lambda} & \mathbf{H}_2(G) & \xrightarrow{\lambda} & \mathbf{H}_1(G) & \overset{i}{\rightarrowtail} & \mathbf{Chain}_{\mathbb{Z}}^{\wedge} \\ \downarrow & & \downarrow & & \downarrow & & \downarrow \\ \mathbf{H}(G)/\simeq & \xrightarrow{\lambda} & \mathbf{H}_2(G)/\simeq & \xrightarrow{\lambda} & \mathbf{H}_1(G)/\simeq & \overset{i}{\rightarrowtail} & \mathbf{Chain}_{\mathbb{Z}}^{\wedge}/\simeq \end{array}.$$

Here i is the inclusion functor and the vertical arrows are the quotient functors. The **functor** λ on $\mathbf{H}(G)$ carries the homotopy system ρ in $\mathbf{H}(G)$ to the object

$$\lambda(\rho) = (C\rho, \rho^2) \quad \text{in } \mathbf{H}_2(G), \tag{1}$$

where ρ^2 is the 2-skeleton of ρ obtained by forgetting ρ^n, $n \geqq 3$. We point out that $\lambda(\rho)$ does not determine ρ completely since there might be a freedom of choice for $d_3{:}\rho_3 \to \rho_2$. For the trivial group $G = 0$, however, $\lambda(\rho)$ determines ρ.

Next we define the **functor** $\boldsymbol{\lambda}$ on $\mathbf{H}_2(G)$ which carries the object (C, ρ) in $\mathbf{H}_2(G)$ to the object

$$\lambda(C, \rho) = (\pi(\rho), C, \partial, i) \quad \text{in } \mathbf{H}_1(G). \tag{2}$$

Here $\pi(\rho) = \rho_1/d_2\rho_2$ is determined by the free crossed module ρ under G with

$$\rho = \{d_2{:}\rho_2 \to \rho_1 = \langle Z_1 \rangle * G\}.$$

The map i is the composition

$$i{:}G \to \langle Z_1 \rangle * G = \rho_1 \overset{q}{\twoheadrightarrow} \pi(\rho) = \pi$$

where q is the quotient map.

The boundary ∂ is defined by

$$\left.\begin{aligned}\partial: C_1 = C(\rho)_1 = \bigoplus_{Z_1} \mathbb{Z}[\pi] \to \mathbb{Z}[iG\backslash\pi] \\ \partial(z) = 1 - [iG + q(z)]\end{aligned}\right\} \tag{3}$$

where $(iG + q(z)) \in iG\backslash\pi$ is the coset of $q(z)$ for $z \in Z_1$. Clearly, $1 = [iG + 0]$.

In (IV. §4) we defined exact sequences for a functor λ which form towers of categories, see (IV.4.15).

(4.6) **Theorem.** *There is a tower of categories*

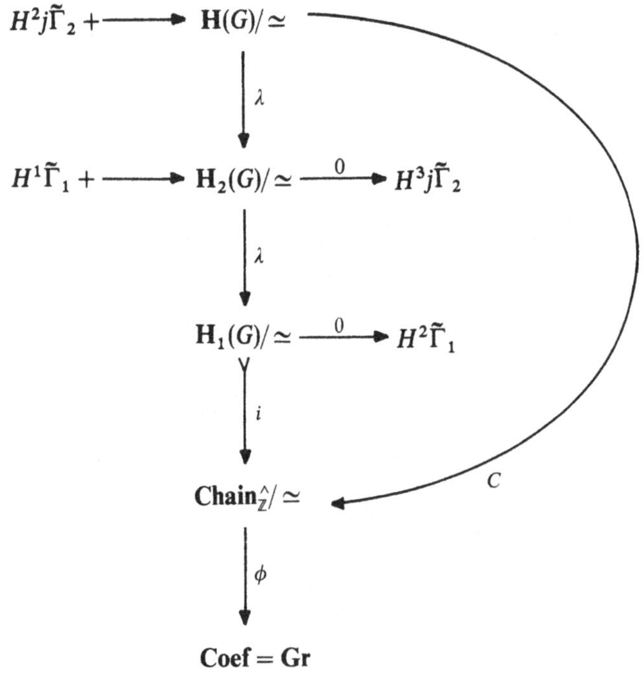

The functor ϕ in (4.6) is the forgetful functor which carries (π, C) to π; **Gr** denotes the category of groups.

We now define the natural systems of groups used in (4.6), compare the notation in (IV.3.1). For a homomorphism $i: G \to \pi$ between groups we have the right $\mathbb{Z}[\pi]$-module

$$\tilde{\Gamma}_1(i) = \hat{H}_1(G, i^*\mathbb{Z}[\pi]). \tag{1}$$

This is the homology of the group G with coefficients in the left G-module $i^*\mathbb{Z}[\pi]$ determined by i. Clearly, $\tilde{\Gamma}_1$ is covariant functor on homomorphisms

$\pi \to \pi'$ under G. We define the natural system $H^n\tilde{\Gamma}_1$ on $\mathbf{H}_1(G)/\simeq$ by

$$H^n\tilde{\Gamma}_1(X, \varphi, Y) = \hat{H}^n(X, \varphi^*\tilde{\Gamma}_1(Y') \\ = \hat{H}^n(C_X, \varphi^*\tilde{\Gamma}_1(i_Y)). \tag{2}$$

Here X and Y are objects in $\mathbf{H}_1(G)$ and $\varphi:\pi_X \to \pi_Y$ is a homomorphism of groups. There are obvious induced maps for (2).

Next we define the natural system $H^n j\tilde{\Gamma}_2$ on $\mathbf{H}_2(G)/\simeq$. First we have the functor

$$\left.\begin{aligned} j\tilde{\Gamma}_2 &: \mathbf{H}_2(G)/\simeq \to \mathbf{Mod}_{\mathbb{Z}}^{\wedge} \text{(see (III.5.1))} \\ j\tilde{\Gamma}_2(C, \xi) &= H_2(d_2\xi_2) \\ &= \text{kernel}\,(h_2) \cap \text{kernel}\,(d_2) \\ &= [\rho_2, \rho_2] \cap \text{kernel}\,(d_2), \end{aligned}\right\} \tag{3}$$

where ρ is a free crossed module under G with $d_2:\rho_2 \to \rho_1$. This group corresponds to $j\tilde{\Gamma}_2(X, D)$ in (2.11). We obtain the induced map for $f:\rho \to \rho'$ obviously by the restriction $f:d_2\rho_2 \to d_2\rho_2'$. One can check there $j\tilde{\Gamma}_2$ is a well-defined functor on the homotopy category in (3). We now define the natural system $H^n j\tilde{\Gamma}_2$ by

$$(H^n j\tilde{\Gamma}_2)(\bar{X}, \varphi, \bar{Y}) = \hat{H}^n(\bar{X}, \varphi^* j\tilde{\Gamma}_2(\bar{Y})) \\ = \hat{H}^n(C, \varphi^* j\tilde{\Gamma}_2(\bar{Y})). \tag{4}$$

Here $\bar{X} = (C, \rho)$ and $\bar{Y}$ are objects in $\mathbf{H}_2(\mathbf{G})$. By (3) there are obvious induced maps for the natural system (4).

In addition to (4.6) we prove:

(4.7) **Addendum.** *There are exact sequences*

$$Z^2 j\tilde{\Gamma}_2 \xrightarrow{+} \mathbf{H}(G) \xrightarrow{\lambda} \mathbf{H}_2(G) \xrightarrow{\mathfrak{O}} H^3 j\tilde{\Gamma}_2$$

$$\mathbf{H}_2(G) \to \mathbf{H}_1(G) \xrightarrow{\mathfrak{O}} H^2 j\tilde{\Gamma}_1$$

where the obstructions are the same as in (4.6) *and where* $Z^2 j\tilde{\Gamma}_2$ *denotes the cocycles for* (4) *above.*

One readily sees that the inclusion functor i in (4.6) satisfies the sufficiency condition, compare (IV.1.3)(a). Therefore we deduce from (4.6) and (IV.4.11)

(4.8) **Corollary.** *The functor* $C:\mathbf{H}(G)/\simeq \to \mathbf{Chain}_{\mathbb{Z}}^{\wedge}/\simeq$ *satisfies the sufficiency condition.*

Moreover, we derive from (4.6) the following special result for the trivial group $G = 0$.

(4.9) **Corollary.** *The functor* $\lambda:\mathbf{H}(0)/\simeq\;\cong\mathbf{H}_2(0)/\simeq$ *is an isomorphism of categories. Moreover,* $\lambda\lambda:\mathbf{H}(0)/\simeq\;\to\mathbf{H}_1(0)/\simeq$ *is full and faithful and* $\lambda\lambda:\mathbf{H}(0)\to\mathbf{H}_1(0)$ *is full.*

Proof. This is clear since for the trivial group $G=0$ we have $\tilde{\Gamma}_1=0$ and $\tilde{\Gamma}_2=0$, see (2.13), and therefore the natural system in (4.6) are trivial. □

Remark. Corollary (4.9) is proved by J.H.C. Whitehead (1949)′. Theorem (4.6) is the generalization of Whitehead's result for non trivial G. A different generalization recently was obtained by Brown–Higgins (1984).

In the rest of this section we prove (4.6).

(4.10) *Proof of* (4.6). We first define the action $H^1\tilde{\Gamma}_1+$ and we prove exactness of the sequence

(A) $$H^1\tilde{\Gamma}_1+\to\mathbf{H}_2(G)/\simeq\;\to\mathbf{H}_1(G)/\simeq.$$

Let $(\xi,\eta):X=(C,\rho)\to Y=(C',\rho')$ be a map in $\mathbf{H}_2(G)$ and let

$$\lambda(\xi,\eta)=(\varphi,\xi)=C\to C' \tag{1}$$

be the corresponding map in $\mathbf{H}_1(G)$. For the bases Z'_2 in ρ'_2 and for the basis Z'_1 in ρ'_1 we get as in (1.9) the function

$$g:Z'_2\subset\rho'_2\to\rho'_1=\langle Z'_1\rangle *G. \tag{2}$$

Let $D=K(G,1)$ be the Eilenberg–Mac Lane space of G and let

$$g:\Sigma^1(Z'_2)^+\to\Sigma(Z'_1)^+\vee D=Y^1 \tag{3}$$

be a map corresponding to (2). Then we have a complex in CW_0^D by

$$Y=C_g,\; D\subset Y^1\subset Y^2=Y. \tag{4}$$

For the universal covering $\hat{Y}$ of Y and for $\hat{D}=p^{-1}D\subset\hat{Y}$ we get

$$H_1(\hat{D})=\tilde{\Gamma}_1(i_Y)\quad\text{see (4.6)(1).} \tag{5}$$

This follows from $\hat{D}=\tilde{D}\times_G\pi_1(Y)$ where $\tilde{D}$ is the universal covering of D; consider the cell structure of $\hat{D}$.

We derive from (5)

(4.11) **Lemma.** *For* $i:G\to\pi=\pi_1(Y)$ *we have* $\tilde{\Gamma}_1(i)=\mathbb{Z}[iG\backslash\pi]\otimes Ab(\ker(i))$. *Here Ab is the abelianization. In particular we have* $\tilde{\Gamma}_1(i)=0$ *if i is injective.*

Indeed, the path components of $\hat{D}$ correspond to the cosets $iG\backslash\pi$ and all path components are homotopy equivalent. By computing $\pi_1\hat{D}=\ker(i)$ we get (4.11).

For $C'=C(\rho')$ there is the commutative diagram

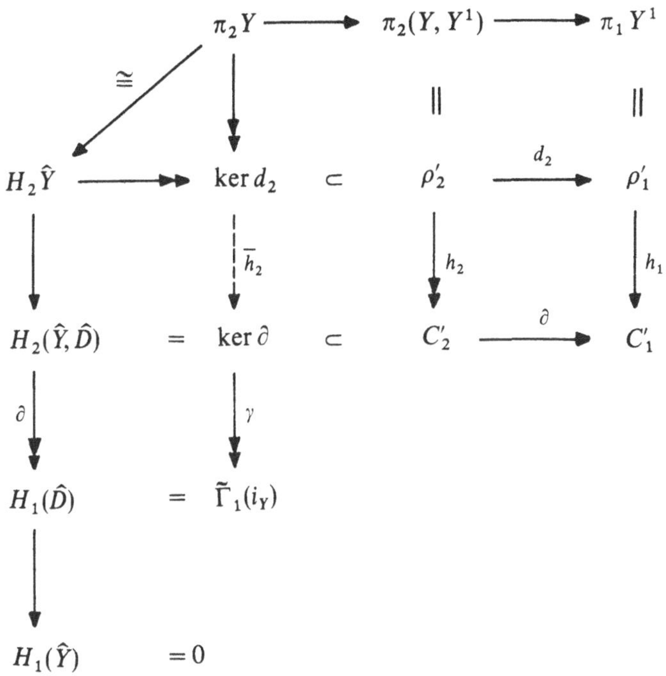

(5)′

Here the left-hand column and the top row are exact sequences. We know that h_2 is surjective, the map $\bar{h}_2$, however, in general is not surjective since the cokernel of $\bar{h}_2$ is $H_1(\hat{D}) = \tilde{\Gamma}_1(i_Y)$.

Now assume that an element

$$\left.\begin{array}{l}\{\beta\} \in \hat{H}^1(C, \varphi^* \tilde{\Gamma}_1(\rho')), \text{ with} \\ \beta : C_1 \to \tilde{\Gamma}_1(\rho')\end{array}\right\} \tag{6}$$

is given. We choose $\bar{\beta}$, $\bar{\bar{\beta}}$ such that the following diagram commutes (here $\bar{\beta}$ satisfies (4.2)(3) and therefore it is enough to choose $\bar{\beta}$ on generators):

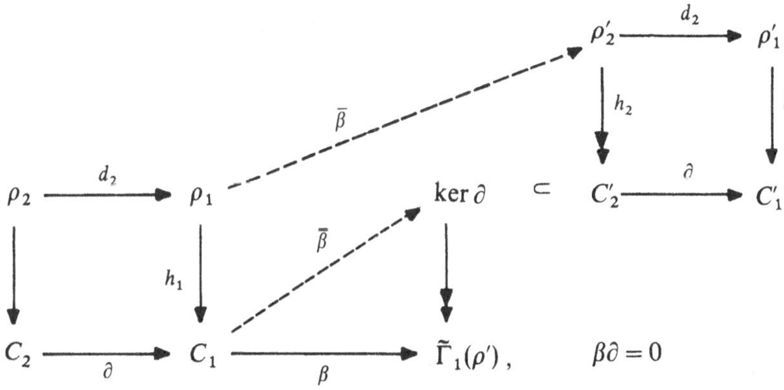

(7)

We now define the action $(H^1\tilde{\Gamma}_1)+$ by

$$\{(\xi,\eta)\}+\{\beta\}=\{(\xi+\bar{\bar{\beta}}\partial,\eta+d_2\bar{\beta})\}. \tag{8}$$

Here $\{(\xi,\eta)\}$ denotes the homotopy class of (ξ,η) in (1) in $\mathbf{H}_2(G)/\simeq$.

We define the chain map $\xi+\bar{\bar{\beta}}\partial$ in (8) by

$$(\xi+\bar{\bar{\beta}}\partial)_n=\begin{cases}\xi_n & n\neq 2\\ \xi_2+\bar{\bar{\beta}}\partial & \text{for } n=2.\end{cases} \tag{9}$$

It is clear by (7) that $\xi+\bar{\bar{\beta}}\partial$ is a chain map since $\partial\bar{\bar{\beta}}=0$. Next we check that the action (8) is well defined. The pair $(\xi+\bar{\bar{\beta}}\partial,\ \eta+d_2\bar{\beta})$ is a morphism in $\mathbf{H}_2(G)$ since we have

$$\begin{aligned}h_1(\eta+d_2\bar{\beta})&=h_1\eta+h_1d_2\bar{\beta}\\&=h_1\eta+\partial h_2\bar{\beta}=h_1\eta.\end{aligned} \tag{10}$$

Moreover, let $\bar{\xi}_2$ be a choice for (ξ,η) as in (4.2)(2). Then $\bar{\xi}_2+\bar{\beta}d_2$ is a choice for $(\xi+\bar{\bar{\beta}}\partial,\ \eta+d_2\bar{\beta})$ since we have

$$\begin{aligned}h_2(\bar{\xi}_2+\bar{\beta}d_2)&=h_2\bar{\xi}_2+h_2\bar{\beta}d_2\\&=\xi_2h_2+\bar{\bar{\beta}}\partial h_2\\&=(\xi_2+\bar{\bar{\beta}}\partial)h_2,\end{aligned} \tag{11}$$

and since we have

$$d_2(\bar{\xi}_2+\bar{\beta}d_2)=d_2\bar{\xi}_2+d_2\bar{\beta}d_2=\eta d_2+d_2\bar{\beta}d_2=(\eta+d_2\bar{\beta})d_2. \tag{12}$$

Next we see that a homotopy $(\alpha,\bar{\alpha})$: $(\xi,\eta)\simeq(\xi',\eta')$ in $\mathbf{H}_2(G)$ yields the homotopy

$$(\alpha,\bar{\alpha}_0):(\xi+\bar{\bar{\beta}},\eta+d_2\bar{\beta})\simeq(\xi'+\bar{\bar{\beta}},\eta'+d_2\bar{\beta}). \tag{13}$$

Here we set $\bar{\alpha}_0=-\bar{\beta}+\bar{\alpha}+\bar{\beta}$, clearly $\bar{\alpha}_0$ induces α_1 since $h_2\bar{\alpha}=h_2\bar{\alpha}_0$.

Finally, let $\bar{\beta}_0$, $\bar{\bar{\beta}}_0$ be a different choice in (7). Then we get the homotopy

$$(\alpha,\bar{\alpha}):(\xi+\bar{\bar{\beta}},\eta+d_2\bar{\beta})\simeq(\xi+\bar{\bar{\beta}}_0,\ \eta+d_2\bar{\beta}_0), \tag{14}$$

where we set $\alpha_n=0$ for $n\geqq 2$ and we set $\bar{\alpha}=\bar{\beta}-\bar{\beta}_0$. Then $\bar{\alpha}$ yields α_1 by $h_2\bar{\alpha}=\alpha_1h_1$.

The results in (10)$\cdots$(14) show that the action (8) is welldefined. Moreover, the action in (8) has linear distributivity law since $\partial\bar{\bar{\beta}}=0$.

We now prove that the sequence in (A) is actually exact. Assume we have a homotopy in $\mathbf{Chain}_{\mathbb{Z}}^{\wedge}$

$$\left.\begin{aligned}&\alpha:\lambda(\xi,\eta)\simeq\lambda(\xi',\eta'),\\&\alpha_n:C_n\to C'_{n+1},\\&-\xi'_n+\xi_n=d\alpha_n+\alpha_{n-1}d\end{aligned}\right\} \tag{15}$$

This also implies that η and η' induce the same homomorphism $\varphi:\pi\to\pi'$.

Therefore we can choose b

$$\left.\begin{aligned} &b:\rho_1 \to \rho_2', \text{ with}\\ &-\eta + \eta' = d_2 b.\end{aligned}\right\} \tag{16}$$

Here b satisfies (4.2)(3) and therefore it is enough to choose b on a basis of ρ_1. For b there is a unique b_1

$$\left.\begin{aligned} &b_1:C_1 \to C_2', \text{ with}\\ &h_2 b = b_1 h_1,\ -\xi_1 + \xi_1' = \partial b_1.\end{aligned}\right\} \tag{17}$$

On the other hand, by (15) we have $-\xi_1 + \xi_1' = \partial\alpha_1$ and thus we get

$$\left.\begin{aligned} &\partial\alpha_1 = \partial b_1, \text{ and}\\ &\bar{\bar{\beta}} = b_1 - \alpha_1,\ \partial\bar{\bar{\beta}} = 0.\end{aligned}\right\} \tag{18}$$

For $\bar{\bar{\beta}}$ we choose $\bar{\beta}$ as in (7) with

$$h_2\bar{\beta} = \bar{\bar{\beta}} h_1. \tag{19}$$

Now we obtain a homotopy in $\mathbf{H}_2(G)$

$$(a, \bar{\alpha}):(\xi + \bar{\bar{\beta}}\partial, \eta + d_2\bar{\beta}) \simeq (\xi', \eta'). \tag{20}$$

Here we set $\bar{\alpha} = -\bar{\beta} + b$ and

$$a_n = \begin{cases} \alpha_n & \text{for } n \neq 1,\\ \alpha_1 - b_1 + \alpha_1 & \text{for } n = 1.\end{cases}$$

We have to check that (20) is actually a homotopy, see (4.2). We obtain (4.2)(4) by

$$\begin{aligned} -(\eta + d_2\bar{\beta}) + \eta' &= -d_2\bar{\beta} - \eta + \eta'\\ &= -d_2\bar{\beta} + d_2 b, \quad \text{see (16)}\\ &= d_2(-\bar{\beta} + b)\\ &= d_2\bar{\alpha}. \end{aligned} \tag{21}$$

Here $\bar{\alpha}$ induces α_1 since

$$\begin{aligned} h_2\bar{\alpha} &= h_2(-\bar{\beta} + b)\\ &= -\bar{\bar{\beta}} h_1 + b_1 h_1, \quad \text{see (19), (17).}\\ &= (-\bar{\bar{\beta}} + b_1)h_1\\ &= \alpha_1 h_1, \quad \text{see (18).} \end{aligned} \tag{22}$$

Moreover, we have

$$\begin{aligned} -\xi_1 + \xi_1' &= d\alpha_1\\ &= d(\alpha_1 - b_1 + \alpha_1), \quad \text{see (18)}\\ &= da_1. \end{aligned} \tag{23}$$

Next we have in degree 2

$$\begin{aligned} -(\xi_2+\bar{\bar{\beta}}\partial)+\xi_2' &= -\bar{\bar{\beta}}\partial+(-\xi_2+\xi_2') \\ &= -(b_1-\alpha_1)\partial+d\alpha_2+\alpha_1 d \quad (\partial=d) \\ &= (\alpha_1-b_1+\alpha_1)d+d\alpha_2 \\ &= a_1 d+da_2. \end{aligned} \tag{24}$$

This completes the proof that $(a,\bar{\alpha})$ in (20) is a well defined homotopy in $\mathbf{H}_2(G)$; therefore (A) is proved. □

Next we define the obstruction and we prove exactness of

(B) $$\mathbf{H}_2(G) \xrightarrow{\lambda} \mathbf{H}_1(G) \xrightarrow{\mathfrak{D}} H^2\tilde{\Gamma}_1.$$

Let $X=(C,\rho)$ be an object in $\mathbf{H}_2(G)$. Then there exists a lifting d_3 in the following diagram since X is in the full image of λ:

$$\begin{array}{ccccccc} & & \rho_2 & \xrightarrow{d_2} & \rho_1 & \xrightarrow{q} & \pi \\ & \overset{d_3}{\nearrow} & \downarrow h_2 & & \downarrow h_1 & & \downarrow h_0. \\ \longrightarrow C_3 & \xrightarrow[\partial]{} & C_2 & \xrightarrow[\partial]{} & C_1 & \xrightarrow[\partial]{} & \mathbb{Z}[iG\backslash\pi] \end{array} \tag{25}$$

Here ∂ denotes the boundary of the chain complex C and q is the quotient map. We define h_0 by

$$h_0(x)=1-[(iG)+x]. \tag{26}$$

This shows that $h_0 q=\partial h_1$, see (4.5)(3). In fact, $(h_0q)(z)=(\partial h_1)(z)$ for $z\in Z_1$ and

$$\begin{aligned} (h_0q)(x+y) &= 1-[iG+qx+qy] \\ &= (1-[iG+qx])[qy]+(1-[iG+qy]) \\ &= (h_0q)(x)\cdot[qy]+(h_0q)(y). \end{aligned}$$

Thus h_0q satisfies the same formula as ∂h_1, see (1.14)(2). We point out that we have the commutative diagram

$$\begin{array}{ccccc} C_2 & \xrightarrow{\partial} & C_1 & \xrightarrow{\partial} & \mathbb{Z}[iG\backslash\pi] \\ & & \downarrow p & \overset{\bar{\partial}}{\nearrow} & \| \\ & & C_1/\partial C_2 & & \| \\ & & \| & & \| \\ H_1(\hat{X}) & \longrightarrow & H_1(\hat{X},\hat{D}) & \rightarrowtail & H_0(\hat{D}) \end{array}, \tag{27}$$

where the bottom row is exact. Here X is a 2-realization of ρ. Since $\hat{X}$ is simply connected we see $H_1(\hat{X}) = 0$ and therefore $\bar{\partial}$ is injective.

Now let $Y = (C', \rho')$ be a further object in $\mathbf{H}_2(G)$ and let

$$(\rho, \xi): \lambda X \to \lambda Y \tag{28}$$

be a map in $\mathbf{H}_1(G)$. Then there is a homomorphism f such that the following diagram of unbroken arrows commutes:

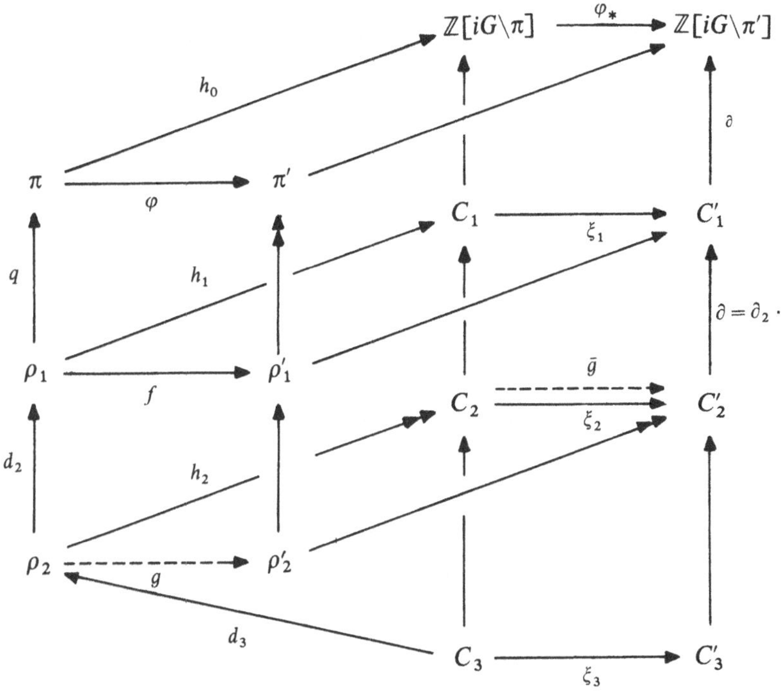

(29)

We find f as follows. We first choose $f_0: \rho_1 \to \rho'_1$ with $q f_0 = \varphi q$. Then f_0 induces a unique map $\bar{f}_0: C_1 \to C'_1$ with $h_1 f_0 = \bar{f}_0 h_1$ and $\partial \bar{f}_0 = \varphi_\# \partial$. Since by assumption on ξ also $\partial \xi_1 = \varphi_\# \partial$ we see

$$\partial(-\bar{f}_0 + \xi_1) = 0. \tag{30}$$

Since $\bar{\partial}$ in (27) is injective there exists $a: C_1 \to C'_2$ with $\partial a = -\bar{f}_0 + \xi_1$. Since h_2 is surjective there exists

$$\alpha: \rho_1 \to \rho'_2, \quad \alpha | G = 0, \tag{31}$$

with $h_2 \alpha = a h_1$. Here we only choose α on generators Z_1 of $\rho_1 = \langle Z_1 \rangle * G$. Now we set

$$f = f_0 + d_2 \alpha. \tag{32}$$

Then we have $qf = qf_0 = \varphi q$ and

$$\begin{aligned} h_1 f &= h_1(f_0 + d_2\alpha) \\ &= h_1 f_0 + h_1 d_2 \alpha \\ &= \bar{f}_0 h_1 + \partial h_2 \alpha \\ &= (\bar{f}_0 + \partial a)h_1 = \xi_1 h_1. \end{aligned}$$

This shows that for f in (32) diagram (29) commutes.

Next we can choose $g:\rho_2 \to \rho_2'$ such that $(g,f):\rho \to \rho'$ is a morphism between free crossed modules. The map g induces $\bar{g}:C_2 \to C_2'$ with

$$h_2 g = \bar{g} h_2, \quad \partial \bar{g} = \xi_1 \partial = \partial \xi_2. \tag{33}$$

Therefore $\partial(-\xi_2 + \bar{g}) = 0$. We now define the **obstruction** for the map (φ, ξ) in (28) by the composition

$$\begin{array}{cccc} \bar{\mathfrak{O}}(\varphi,\xi): C_2 & \longrightarrow & \ker \partial & \xrightarrow[\gamma]{} \tilde{\Gamma}_1(i_Y) \\ & {\scriptstyle -\xi_2+\bar{g}} \searrow & \cap & \\ & & C_2' & \end{array} \tag{34}$$

Here γ is the projection in diagram (5)' above. Now $\bar{\mathfrak{O}}(\varphi, \xi)$ is a cocycle. In fact, we have

$$\begin{aligned} \partial \xi_2 \partial &= \xi_1 \partial \partial = 0, \quad \text{and} \\ \gamma \xi_2 \partial &= \gamma \partial \xi_3 = \gamma h_2 d_3 \xi_3 \\ &= \gamma \bar{h}_2 d_3 \xi_3 = 0, \quad \text{see (5)'}. \end{aligned} \tag{35}$$

On the other hand, we have

$$\begin{aligned} \partial \bar{g} \partial &= \partial \bar{g} h_2 d_3 = \partial h_2 g d_3 \\ &= h_1 d_2 g d_3 = h_1 f d_2 d_3 = 0, \\ \gamma \bar{g} \partial &= \gamma \bar{g} h_2 d_3 = \gamma h_2 g d_3 \\ &= \gamma \bar{h}_2 g d_3 = 0, \quad \text{see (5)'}. \end{aligned} \tag{36}$$

By (35) and (36) we see that $\bar{\mathfrak{O}}(\varphi, \xi)\partial = -\gamma\xi_2\partial + \gamma\bar{g}\partial = 0$.

Now let

$$\mathfrak{O}(\varphi, \xi) = \{\bar{\mathfrak{O}}(\varphi, \xi)\} \in H^2(C, \tilde{\Gamma}_1(i_Y)) \tag{37}$$

be the cohomology class represented by the cocycle (34). We check that this class does not depend on the choice of (f, g) used for the definition in (34). In fact, let f' be a different choice such that (29) commutes and let g' be a lift of f'. Then we have $\alpha:\rho_1 \to \rho_2'$ with $-f + f' = d_2\alpha$ and $h_1 d_2 \alpha = 0$. Moreover,

$$d_2(-\alpha d_2 - g + g') = -d_2\alpha d_2 - f d_2 + f' d_2 = (-d_2\alpha - f + f')d_2 = 0. \tag{38}$$

Thus we get

$$\beta = -\alpha d_2 - g + g' : \rho_2 \to \ker d_2 \subset \rho_2'.$$

Moreover, we have

$$\begin{aligned} -(-\xi_2 + \bar{g}) + (-\xi_2 + \overline{g'}) &= -\bar{g} + \overline{g'} \\ &= -\bar{g} + \overline{(g + \alpha d_2 + \beta)} = a\partial_2 + \bar{\beta}, \end{aligned}$$

where $\gamma\bar{\beta} = 0$. Therefore the cohomology class (37) is well defined by (φ, ξ). Moreover, we check that (37) depends only on the homotopy class of (φ, ξ) in $\mathbf{Chain}_{\mathbb{Z}}^{\wedge}/\simeq$. Let $a_n : \xi \simeq \xi'$ be a homotopy, $a_n : C_n \to C'_{n+1}$.

$$\begin{cases} -\xi_1 + \xi_1' = \partial a_1, \\ -\xi_2 + \xi_2' = \partial a_2 + a_1 \partial. \end{cases}$$

For $a = a_1$ we choose $\alpha : \rho_1 \to \rho_2'$ as in (31). Then we obtain

$$f' = f + d_2\alpha, \quad g' = g + \alpha d_2$$

as a choice for (φ, ξ'). Clearly, $-(-\xi_2 + \bar{g}) + (-\xi_2' + \overline{g'}) = -(-\xi_2 + \xi_2') + a_1\partial_2 = \partial a_2 = \bar{h}_2 d_3 a_2$, where $\gamma\bar{h}_2 = 0$, see (5)′. Therefore $\mathfrak{O}(\varphi, \xi) = \mathfrak{O}(\varphi, \xi')$.

It remains to prove exactness of the sequence in (B). Let (φ, ξ) be given as (28) and suppose $\mathfrak{O}(\varphi, \xi) = 0$. We construct a map

$$(\xi, \eta) : X \to Y \quad \text{in } \mathbf{H}_2(G), \tag{39}$$

where $\eta : \rho_1 \to \rho_1'$, see (4.2). We choose (f, g) as in (34). Then $\mathfrak{O}(\varphi, \xi) = 0$ implies that there is a map β such that

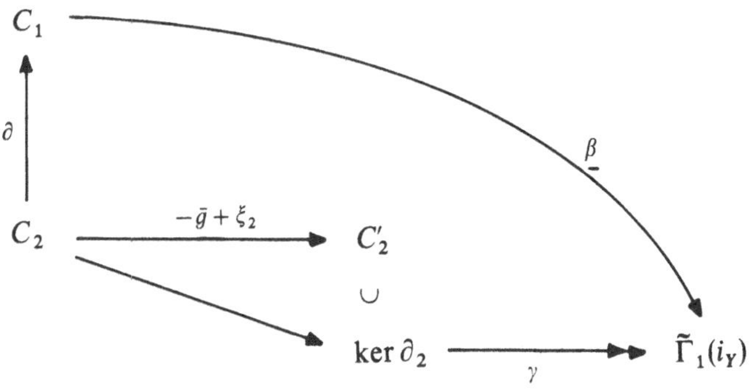

commutes. For β we choose $\bar{\bar{\beta}}$ and $\bar{\beta}$ as in (7) and we set

$$\eta = f + d_2\bar{\beta}. \tag{40}$$

We check (4.2)(1) by

$$\begin{aligned} h_1\eta &= h_1 f + h_1 d_2 \bar{\beta} \\ &= \xi_1 h_1 + \partial_2 h_2 \bar{\beta} \\ &= \xi_1 h_1 + \partial\bar{\bar{\beta}} h_1 = \xi_1 h_1, \quad \text{see (7).} \end{aligned}$$

Moreover we find $\bar{\xi}_2$ in (4.2)(2) as follows: By (39) and (7) we have

$$\gamma(-\bar{\bar{\beta}}\partial - \bar{g} + \xi_2) = 0.$$

Therefore there is a lift $\delta:\rho_2 \to \ker(d_2) \subset \rho'_2$ with $\bar{h}_2\delta = (-\bar{\bar{\beta}}\partial - \bar{g} + \xi_2)h_2$. We set

$$\bar{\xi}_2 = g + \bar{\beta}d_2 + \delta.$$

Then we clearly get $d_2\bar{\xi}_2 = d_2(g + \bar{\beta}d_2) = \eta d_2$ and

$$\begin{aligned} h_2\bar{\xi}_2 &= \bar{g}h_2 + \bar{\bar{\beta}}h_1 d_2 + (-\bar{\bar{\beta}}\partial - \bar{g} + \xi_2)h_2 \\ &= \xi_2 h_2. \end{aligned}$$

This completes the proof of the exactness in (B). □

Next we obtain the exact sequence

(C) $$Z^2 j\tilde{\Gamma}_2 \xrightarrow{+} \mathbf{H}(G) \xrightarrow{\lambda} \mathbf{H}_2(G) \xrightarrow{\mathfrak{O}} H^3 j\tilde{\Gamma}_2$$

as follows: Here $Z^2 j\tilde{\Gamma}_2$ denotes the group of all cocycles

$$\alpha: C_2 \to j\tilde{\Gamma}_2, \quad \alpha\partial = 0.$$

For $f:\rho \to \rho'$ in $\mathbf{H}(G)$ we obtain $(f + \alpha):\rho \to \rho'$ by

$$(f + \alpha)_n = \begin{cases} f_2 + \alpha h_2, & \text{for } n = 2, \\ f_n, & \text{for } n \neq 2. \end{cases} \tag{41}$$

For $(\xi, \eta):\lambda\rho \to \lambda\rho'$ in $\mathbf{H}_2(G)$ let

$$\bar{\mathfrak{O}}(\xi, \eta) = -d_3\xi_3 + \bar{\xi}_2 d_3 : C_3 \to \rho'_2. \tag{42}$$

Here $\bar{\xi}_2$ is chosen as in (4.2)(2). We have $\bar{\mathfrak{O}}(\xi, \eta)\partial_4 = 0$ and $\bar{\mathfrak{O}}(\xi, \eta)$ factors over $j\tilde{\Gamma}_2 \subset \rho'_2$. Therefore the cohomology class

$$\mathfrak{O}(\xi, \eta) = \{\bar{\mathfrak{O}}(\xi, \eta)\} \in H^3(C, \varphi^* j\tilde{\Gamma}_2) \tag{43}$$

is defined. This class does not depend on the choice of $\bar{\xi}_2$ in (42). Now it is easy to see that the sequence (C) is exact. Moreover, the exact sequence in (C) induces the corresponding exact sequence for homotopy categories in (4.6). □

§5 Homotopy systems of order n, $n \geq 3$

In this section we describe the category $\mathbf{H}_n^c$ of homotopy systems of order n and we show that $\mathbf{H}_n^c$ and $\mathbf{H}_{n+1}^c$ are connected by an exact sequence. This gives us the tower of categories which approximates the homotopy category of CW-complexes. In the next section we discuss in detail the properties of this tower.

A complex (X, D) in $\mathbf{CW}_0^D$ has attaching maps, (1.1)(2),

(5.1) $$f_{n+1}: \Sigma^n Z_{n+1}^+ \to X^n \quad \text{with} \quad X^{n+1} = C_{f_{n+1}}.$$

The homotopy class of f_{n+1}, $n \geqq 2$, is given by the homomorphism

(5.2) $$f = f_{n+1}: \hat{C}_{n+1}(X, D) = \bigoplus_{Z_{n+1}} \mathbb{Z}[\pi] \longrightarrow \pi_n(X^n)$$

of $Z[\pi]$-modules, $\pi = \pi_1 X$, where $f(\hat{e}) \in \pi_n(X^n)$ is the attaching map of the cell $e \in Z_{n+1}$. The homomorphism in (5.2) can also be described by the following commutative diagram

(5.3) $$\begin{array}{ccc} \hat{C}_{n+1}(X, D) = & & H_{n+1}(\hat{X}^{n+1}, \hat{X}^n) \\ \Big\downarrow f_{n+1} & & \cong \Big\uparrow h \\ & & \pi_{n+1}(\hat{X}^{n+1}, \hat{X}^n) \\ & & \Big\downarrow p_* \\ \pi_n(X^n) & \xleftarrow{\partial} & \pi_{n+1}(X^{n+1}, X^n) \end{array}$$

(5.4) ***Remark.*** The homomorphism $f_{n+1} \in \mathrm{Hom}_{\mathbb{Z}[\pi]}(\hat{C}_{n+1}(X, D), \pi_n X^n)$ is a cochain in $\hat{C}_*(X, D)$. In fact, f_{n+1} is a cocycle, that is, $f_{n+1} \circ d_{n+2} = 0$ as follows from $\partial\partial = 0$, compare (III. 5.6). Hence the **cocycle condition** for f_{n+1} is satisfied if f_{n+1} is the attaching map of a CW-complex (X, D).

Now let (Y, D) be a further complex in $\mathbf{CW}_0^D$ with attaching maps g_{n+1}. By naturality of the Hurewicz map h, of p_* and of ∂ in (5.3) we see that a cellular map $F: X \to Y$ induces the commutative diagram

(5.5) $$\begin{array}{ccc} \hat{C}_{n+1}(X, D) & \xrightarrow{\xi_{n+1}} & \hat{C}_{n+1}(Y, D) \\ \Big\downarrow f_{n+1} & & \Big\downarrow g_{n+1} \\ \pi_n X^n & \xrightarrow[\eta_*]{} & \pi_n Y^n \end{array} .$$

Here $\xi_{n+1} = F_*$ and $\eta: X^n \to Y^n$ is the restriction of F. For $\varphi = \pi_1 F: \pi_1 X \to \pi_1 Y$ the maps ξ and η_* are φ-equivariant. We say that (ξ_{n+1}, η) is **associated** to the restriction $F: (X^{n+1}, X^n) \to (Y^{n+1}, Y^n)$ of F.

We deduce from the attaching map f_{n+1} the boundary d_{n+1} in $\hat{C}_*(X, D)$ by the composition

(5.6) $$\begin{array}{ccc} \hat{C}_{n+1}(X, D) & \xrightarrow[f_{n+1}]{} & \pi_n X^n \underset{p_*}{\cong} \pi_n \hat{X}^n \\ \Big\downarrow d_{n+1} = \partial(f_{n+1}) & & \Big\downarrow h \\ & & H_n \hat{X}^n \\ & & \cap \\ \hat{C}_n(X, D) & = & H_n(\hat{X}^n, \hat{X}^{n-1}) \end{array}$$

This is true for $n \geqq 2$. For $n = 1$ we have the more complicated formula in (1.20)(7). We now introduce the category of homotopy systems of order $n+1$.

(5.7) ***Definition.*** Let $n \geqq 2$. A **homotopy system of order** $(n+1)$ is a triple (C, f_{n+1}, X^n) where $X^n = (X^n, D)$ is an n-dimensional complex in $\mathbf{CW}_0^D$ and where $(\pi_1 X^n, C)$ is a free chain complex in $\mathbf{Chain}_{\mathbb{Z}}^{\wedge}$ which coincides with $\hat{C}_*(X^n, D)$ in degree $\leqq n$. Moreover, $f_{n+1}: C_{n+1} \to \pi_n(X^n)$ is a homomorphism of $\pi_1(X^n)$-modules with

$$d_{n+1} = \partial(f_{n+1}). \tag{1}$$

Here d_{n+1} is the boundary in C and $\partial(f_{n+1})$ is the composition as in (5.6). **A morphism between homotopy systems** of order $(n+1)$ is a pair (ξ, η) which we write

$$(\xi, \eta): (C, f_{n+1}, X^n) \longrightarrow (C', g_{n+1}, Y^n).$$

Here $\eta: X^n \to Y^n$ is a map in $\mathbf{CW}_0^D / \overset{0}{\simeq}$, see (1.1), and $\xi: C \to C'$ is a $\pi_1(\eta)$-equivariant homomorphism in $\mathbf{Chain}_{\mathbb{Z}}^{\wedge}$ which coincides with $\hat{C}_*\eta$ in degree $\leqq n$ and for which the following diagram commutes:

$$\begin{array}{ccc} C_{n+1} & \xrightarrow{\xi_{n+1}} & C'_{n+1} \\ \downarrow{\scriptstyle f_{n+1}} & & \downarrow{\scriptstyle g_{n+1}} \\ \pi_n X^n & \xrightarrow[\eta_*]{} & \pi_n Y^n \end{array} \tag{2}$$

Let $\mathbf{H}_{n+1}$ be the category of homotopy systems of order $(n+1)$ and of such homomorphisms. Composition is defined by $(\xi, \eta)(\bar{\xi}, \bar{\eta}) = (\xi\bar{\xi}, \eta\bar{\eta})$. Moreover, let $\mathbf{H}_{n+1}^c$ be the full subcategory of $\mathbf{H}_{n+1}$ consisting of objects (C, f_{n+1}, X^n) which satisfy the **cocycle condition** $f_{n+1} d_{n+2} = 0$. ||

We have obvious functors $(n \geqq 3)$

(5.8)
$$\begin{array}{ccccccc} & & & & \mathbf{H}(G) & & \\ & & & {\scriptstyle\rho}\nearrow & & \nwarrow{\scriptstyle\lambda} & \\ \mathbf{CW}_0^D / \overset{0}{\simeq} & \xrightarrow[r_{n+1}]{} & \mathbf{H}_{n+1} & \xrightarrow[\lambda]{} & \mathbf{H}_n & \xrightarrow[\lambda]{} & \mathbf{H}_2 = \mathbf{H}_2(G) \end{array}$$

where $\mathbf{H}(G)$ is the category of crossed chain complexes under $G = \pi_1(D)$ in § 1. Here we set

$$r_{n+1}(X) = (\hat{C}_*(X, D), f_{n+1}, X^n) \tag{1}$$

with f_{n+1} as in (5.3). The functor r_{n+1} carries the cellular map F to the homomorphism $(\hat{C}_* F, \eta)$ where η: $X^n \to Y^n$ is the restriction of F. By (5.5) and (5.6) we see that the functor r_{n+1} is well defined. We define λ by

$$\lambda(C, f_{n+1}, X^n) = (C, f_n, X^{n-1}), \tag{2}$$

where f_n is the attaching map of n-cells in X^n.

Lemma. *The full image of λ in (5.8) is the category $\mathbf{H}_n^c$ consisting of objects which satisfy the cocycle condition.* (3)

Proof. Consider the commutative diagram

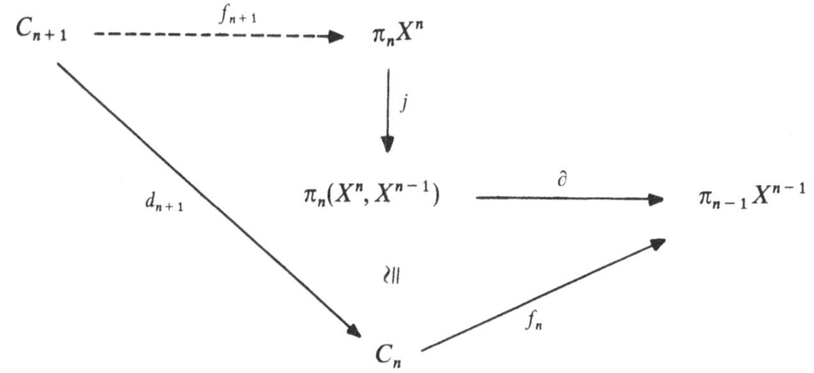

Since kernel $(\partial) =$ image (j) a lifting f_{n+1} of d_{n+1} exists iff $f_n d_{n+1} = 0$. □

Moreover we define ρ in (5.8) by

$$\rho(C, f_n, X^{n-1}) = \rho, \tag{4}$$

where ρ coincides with $\rho(X^n, D)$ in degree $\leqq n$, see § 1, (here X^n is a mapping cone of f_n) and where ρ coincides with C in degree $\geqq n$. Clearly we have $\rho r_n = \rho$.

Next we consider the action $E+$ which we need for the definition of the homotopy relation on the category $\mathbf{H}_n$. An n-dimensional complex (X^n, D) in $\mathbf{CW}_0^D$ is the mapping cone $X^n = C_f$ where $f = f_n$ is the attaching map (5.1). The cooperation $\mu: C_f \to C_f \vee \Sigma^n Z_n^+$ in (II.8.7) gives us the group action

(5.9) $$[X^n, Y]^D \times E(X^n, Y) \xrightarrow{+} [X^n, Y]^D.$$

Here $[X^n, Y]^D$ denotes the set of homotopy classes in $\mathbf{Top}^D/\simeq$. In (5.9) we use the group

$$\begin{aligned} E(X^n, Y) &= [\Sigma^n Z_n^+, Y] \\ &= \mathrm{Hom}_\varphi(\hat{C}_n(X, D), \pi_n Y) \\ &= \mathrm{Hom}_{\mathbb{Z}[\pi]}(\hat{C}_n(X, D), \varphi^* \pi_n Y), \end{aligned} \tag{1}$$

where $\varphi: \pi = \pi_1 X^n \to \pi_1 Y$ is a homomorphism. Let $[X^n, Y]_\varphi^D$ be the subset of all elements is $[X^n, Y]^D$ which induce φ on fundamental groups.

If this subset is not empty the coboundaries in $E(X^n, Y)$ given by $\mathrm{Hom}_{Z[\pi]}(\hat{C}_*(X,D), \varphi^*\pi_n Y)$ are elements in the isotropy group of the action (5.9). Therefore we obtain the action

$$[X^n, Y]^D_\varphi \times \hat{H}^n(X^n, D; \varphi^*\pi_n Y) \to [X^n, Y]^D_\varphi \tag{2}$$

However, only for $n = 2$ this is a transitive and effective action. For $n > 2$ the **isotropy group** in $u_n \in [X^n, Y]^D_\varphi$, given by

$$I(u_n) = \{\xi \in \hat{H}^n(X^n, D, \varphi^*\pi_n Y): u_n + \xi = u_n\}, \tag{3}$$

can be computed by the **spectral sequence** (III.4.18). This shows that $I(u_n)$ is the image of all differentials $d_2, \dots, d_{n-1}$ pictured by the following diagram of the E_2-term

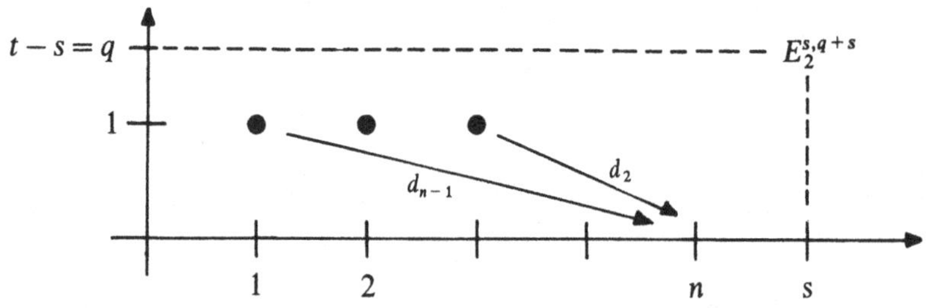

$$E_2^{s,t} = \hat{H}^s(X^n, D, \varphi^*\pi_t Y). \tag{4}$$

Here we use (III.5.9) and (III.4.12). We use the action (5.9) for the following definition of the **homotopy relation,** $\simeq$, on the category $\mathbf{H}_{n+1}$.

(5.10) ***Definition.*** Let $n \geq 2$ and let $(\xi, \eta), (\xi', \eta'):(C, f_{n+1}, X^n) \to (C', g_{n+1}, Y^n)$ be maps in $\mathbf{H}_{n+1}$. We set $(\xi, \eta) \simeq (\xi', \eta')$ if $\pi_1\eta = \pi_1\eta' = \varphi$ and if there exist φ-equivariant homomorphisms $\alpha_{j+1}: C_j \to C'_{j+1}$ $(j \geq n)$, such that

(a) $\{\eta\} + g_{n+1}\alpha_{n+1} = \{\eta'\}$ in $[X^n, Y^n]^D$, and
(b) $\xi'_k - \xi_k = \alpha_k d_k + d_{k+1}\alpha_{k+1}, \quad k \geq n+1$.

The action + in (a) is defined in (5.9). $\{\eta\}$ denotes the homotopy class of η in $[X^n, Y^n]^D$. We write $\alpha:(\xi, \eta) \simeq (\xi', \eta')$. ||

One can check that this homotopy relation is a natural equivalence relation on the category $\mathbf{H}_{n+1}$ so that the homotopy category $\mathbf{H}_{n+1}/\simeq$ is defined. With the notation on exact sequences for functors in (IV.§4) we describe the following crucial result.

(5.11) **Theorem.** *For $n \geqq 2$ there is a commutative diagram of exact sequences*

$$\begin{array}{ccccccc} Z^n\tilde{\Gamma}_n & \xrightarrow{+} & \mathbf{H}^c_{n+1} & \xrightarrow{\lambda} & \mathbf{H}^c_n & \xrightarrow{\mathfrak{D}} & H^{n+1}\tilde{\Gamma}_n \\ \downarrow q & & \downarrow p & & \downarrow p & & \downarrow = \\ H^n\tilde{\Gamma}_n & \xrightarrow{+} & \mathbf{H}^c_{n+1}/\simeq & \xrightarrow{\lambda} & \mathbf{H}^c_n/\simeq & \xrightarrow{\mathfrak{D}} & H^{n+1}\tilde{\Gamma}_n \end{array},$$

where we set $\tilde{\Gamma}_n = \Gamma_n$ for $n \geqq 3$. For $n = 2$ the category $\mathbf{H}^c_2 = \mathbf{H}_2(G)$ with $G = \pi_1(D)$ is the category of homotopy systems of order 2 defined in (4.2). The functor λ is defined in (5.8) and p is the quotient functor.

There is also an exact sequence for $\lambda: \mathbf{H}_{n+1} \to \mathbf{H}_n$, see (VII.1.20). We now define the natural systems of abelian groups used in this theorem. We have the forgetful functor.

$$\phi: \mathbf{H}_n/\simeq \;\to \mathbf{Coef} = \mathbf{Gr}, \tag{1}$$

which carries the object $X = (C, f_n, X^{n-1})$ to the fundamental group $\pi = \pi_1 X$ given by the chain complex (π, C) in $\mathbf{Chain}^{\wedge}_{\mathbb{Z}}$.

Using the notation in (IV.3.1) (*) we define for objects X, Y in $\mathbf{H}_n$ and for a homomorphism $\varphi: \pi_1 X \to \pi_1 Y$ the natural system $H^p\Gamma_n$ on $\mathbf{H}_n/\simeq$ by

$$\begin{aligned} H^p\Gamma_n(X, \varphi, Y) &= \hat{H}^p(X, D; \varphi^*\Gamma_n(Y, D)) \\ &= \hat{H}^p(C, \varphi^*\Gamma_n(Y, D)). \end{aligned} \tag{2}$$

Here we use for $n \geqq 2$

$$\Gamma_n(Y, D) = \Gamma_n(g_n) = \text{image } (\pi_n Y^{n-1} \to \pi_n Y^n), \tag{3}$$

where $Y = (C', g_n, Y^{n-1})$ and $Y^n = C_{g_n}$. We observe that (3) depends only on the homotopy class, g_n, of the attaching map of n-cells in Y^n. There are obvious induced maps for the groups in (2). For $n = 2$ we can replace Γ_2 in (2) by $\tilde{\Gamma}_2$ where we set

$$\tilde{\Gamma}_2(Y, D) = \tilde{\Gamma}_2(Y^2, D), \quad \text{see (2.9)}. \tag{4}$$

Here $Y^2 = C_f$ is a 2-realization of the object $Y = (C, \rho)$ in $\mathbf{H}_2(G)$ given by the attaching map $f: Z_2 \subset \rho_2 \to \rho_1 = \langle Z_1 \rangle * \pi_1(D)$ of 2-cells. Again we have obvious induced maps for the natural system $H^p\tilde{\Gamma}_2$ on $\mathbf{H}_2/\simeq$. Finally, we obtain the natural system $Z^p\Gamma_n$ and $Z^p\tilde{\Gamma}_2$ by taking the **groups of cocycles** for the groups $H^p\Gamma_n$ and $H^p\tilde{\Gamma}_2$ respectively.

Proof of (5.11). Theorem (5.11) is a special case of a result which is available in each cofibration category. This general result is proved in the next Chapter VII, see (VII.2.16). The categories $\mathbf{H}^c_n$ in theorem (5.11) are special cases of the categories $\mathbf{TWIST}^c_n(\mathfrak{X})$ defined in the next chapter. (5)

It is an easy and illustrating exercise to specialize all arguments in the proof of (VII.2.8) for the case of complexes in $\mathbf{CW}_0^D$. This, indeed, yields a direct proof of (5.11). □

With respect to the categories in §4 there is the following additional result: result:

(5.12) **Theorem.** *We have a commutative diagram with exact rows*

$$\begin{array}{ccccccc}
H^2\Gamma_2 & \xrightarrow{+} & \mathbf{H}_3^c/\simeq & \xrightarrow{\rho} & \mathbf{H}(G)/\simeq & \xrightarrow{\mathfrak{D}} & H^3\Gamma_2 \\
\downarrow{\scriptstyle i_*} & & \downarrow{\scriptstyle 1} & & \downarrow{\scriptstyle \lambda} & & \downarrow{\scriptstyle i_x} \\
H^2\tilde{\Gamma}_2 & \xrightarrow{+} & \mathbf{H}_3^c/\simeq & \xrightarrow{\lambda} & \mathbf{H}_2(G)/\simeq & \longrightarrow & H^3\tilde{\Gamma}_2. \\
\downarrow{\scriptstyle j^*} & & \downarrow{\scriptstyle \bar{\rho}} & & \downarrow{\scriptstyle 1} & & \downarrow{\scriptstyle j_*} \\
H^2 j\tilde{\Gamma}_2 & \xrightarrow{+} & \mathbf{H}(G)/\simeq & \xrightarrow{\lambda} & \mathbf{H}_2(G)/\simeq & \longrightarrow & H^3 j\tilde{\Gamma}_2
\end{array}$$

Here the bottom row is the exact sequence in (4.6) and the row in the middle is obtained by (5.11) with $n=2$. The operators i_* and j_* are induced by the exact sequence (2.11).

(5.13) **Corollary.** *Let D be given such that $\pi_2(D)=0$. Then we have for $G=\pi_1(D)$ the equivalence of categories $\rho:\mathbf{H}_3^c/\simeq\ \xrightarrow{\sim} \mathbf{H}(G)/\simeq$*

Proof. Since $\pi_2(D)=0$ we know $\Gamma_2(Y,D)=0$, see (2.13), and hence $H^p\Gamma_2=0$. Therefore ρ is full and faithful, see (IV.4.12). Moreover, each object ρ in $\mathbf{H}(G)$ yields an object (C,f_3,X^2) in $\mathbf{H}_3$ by choosing a 2-realization X^2 of ρ as in (5.11)(4). Then f_3 is determined by d_3 since we have the commutative diagram

$$\begin{array}{ccccccc}
\Gamma_2(X,D) & \longrightarrow & \pi_2 X^2 & \longrightarrow & \pi_2(X^2,X^1) & \longrightarrow & \rho_1 \\
\| & & \uparrow{\scriptstyle f_3} & & \uparrow{\scriptstyle d_3} & \nearrow{\scriptstyle 0} & \\
0 & & C_3 & \cong & \rho_3 & &
\end{array}$$

where the top row is exact. □

The next definition is a modification of the classical *primary obstruction for the extension of mappings*, see for example 4.2.9 in Baues (1977).

(5.14) ***Definition of the obstruction.*** Let $X=(C,f_{n+1},X^n)$ and $Y=(C',g_{n+1},Y^n)$ be objects in $\mathbf{H}_{n+1}^c$, $n\geqq 2$, and let

$$(\xi,\eta):\lambda X\to\lambda Y,\quad (\pi_1\eta=\varphi),$$

be a map in $\mathbf{H}_n^c$. Then there is a map $F: X^n \to Y^n$ associated to (ξ_n, η) as in (5.5). This map induces F_* in the diagram

$$\begin{array}{ccc} C_{n+1} & \xrightarrow[\xi_{n+1}]{} & C'_{n+1} \\ {\scriptstyle f_{n+1}}\downarrow & & \downarrow{\scriptstyle g_{n+1}} \\ \pi_n X^n & \xrightarrow[F_*]{} & \pi_n Y^n \end{array}.$$

This diagram needs not to be commutative. We define

$$\mathfrak{O}(F) = -g_{n+1}\xi_{n+1} + (F_*)f_{n+1} \in \operatorname{Hom}_\varphi(C_{n+1}, \pi_n Y^n). \tag{1}$$

The homomorphism $\mathfrak{O}(F)$ actually factors over

$$\Gamma_n(\lambda Y) = \Gamma_n(g_n) \subset \pi_n(Y^n),$$

and is a cocycle in $\operatorname{Hom}_\varphi(C_{n+1}, \Gamma_n(\lambda Y))$. Therefore $\mathfrak{O}(F)$ represents the cohomology class

$$\mathfrak{O}(\xi, \eta) = \{\mathfrak{O}(F)\} \in H^{n+1}\Gamma_n(\xi, \eta) = \hat{H}^{n+1}(C, \varphi^*\Gamma_n(\lambda Y)). \tag{2}$$

This is the obstruction operator in (5.11) for $n \geqq 3$. Next we consider the case $n = 2$. Let X and Y be given as above and let

$$(\xi, \eta): \lambda X = (C, \rho X^2) \to \lambda Y = (C', \rho Y^2)$$

be a map in $\mathbf{H}_2(G)$ where $\eta: \pi_1 X^1 \to \pi_1 Y^1$. We can choose a map $F: X^2 \to Y^2$ which is a 2-realization of (ξ, η) and we obtain $\mathfrak{O}(F)$ as in (1) where we set $n = 2$. Now however $\mathfrak{O}(F)$ factors over

$$\tilde{\Gamma}_2(\lambda Y) = \tilde{\Gamma}_2(Y^2, D) \subset \pi_2(Y^2),$$

and is a cocycle in $\operatorname{Hom}_\varphi(C_3, \tilde{\Gamma}_2(\lambda Y))$. Therefore $\mathfrak{O}(F)$ represents

$$\mathfrak{O}(\xi, \eta) = \{\mathfrak{O}(F)\} \in H^3\tilde{\Gamma}_2(\xi, \eta) = \hat{H}^3(C, \varphi^*\tilde{\Gamma}_2(\lambda Y)). \tag{3}$$

Finally, let for $n = 2$

$$f: \bar{\rho}(X) \to \bar{\rho}(Y)$$

be a map in $\mathbf{H}(G)$. Again we choose a 2-realization $F: X^2 \to Y^2$ of f and we define $\mathfrak{O}(F)$ as in (1). Then $\mathfrak{O}(F)$ factors over $\Gamma_2(\lambda Y) \subset \pi_2 Y^2$ and is a cocycle in $\operatorname{Hom}_\varphi(C_3, \Gamma_2(\lambda Y))$. This yields the cohomology class

$$\mathfrak{O}(f) = \{\mathfrak{O}(F)\} \in H^3\Gamma_2(f) = \hat{H}^3(C, \varphi^*\Gamma_2(\lambda Y)). \tag{4}$$

We check that (3) and (4) are well defined by the following diagram

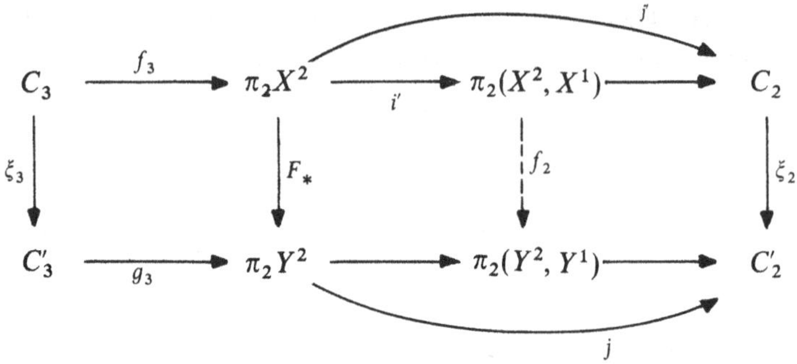

Here we have

$$\Gamma_2(\lambda Y) = \text{kernel } (i) \tag{5}$$

$$\tilde{\Gamma}_2(\lambda Y) = \text{kernel } (j) \tag{6}$$

If F is a 2-realization of (ξ, η) as in (3) then we know that $jF_* = \xi_2 j'$ and therefore $j\mathfrak{O}(F) = 0$ since ξ is a chain map. Thus (6) shows that $\mathfrak{O}(F)$ factors over $\tilde{\Gamma}_2(\lambda Y)$. On the other hand, if F is a 2-realization of f with $Cf = \xi$ as in (4) then we know $iF_* = f_2 i'$ and therefore $i\mathfrak{O}(F) = 0$ since f is a map in $\mathbf{H}(G)$. Thus (5) shows that (4) is well defined. In the same way we see that (2) is well defined.

We point out that the obstruction classes (2), (3), and (4) are defined by cocycles $\mathfrak{O}(F)$ as in (1). This implies the following *relativization property* of the obstruction. Let $A = (C^A, f^A_{n+1}, A^n) \subset X$ be a "subcomplex" of X and assume

$$(\xi^A, \tilde{\eta}^A): A \to Y \text{ in } \mathbf{H}^c_{n+1}$$

is given with $\lambda\{\xi^A, \tilde{\eta}^A) = (\xi, \eta) | \lambda A$. Then F in (1) can be chosen such that $F | A^n = \tilde{\eta}^A$ so that $\mathfrak{O}(F) | C^A_{n+1} = 0$. Whence $\mathfrak{O}(F)$ represents the **relative obstruction**

(7) $$\mathfrak{O}(\xi, \eta; \tilde{\eta}^A) = \{\mathfrak{O}(F)\} \in \hat{H}^{n+1}(X, A; \varphi^* \Gamma_n(\lambda Y)).$$

We have $\mathfrak{O}(\xi, \eta; \tilde{\eta}^A) = 0$ if and only if there exists $(\xi, \tilde{\eta}): X \to Y$ in $\mathbf{H}^c_{n+1}$ with

$$(\xi, \tilde{\eta}) | A = (\xi^A, \tilde{\eta}^A) \quad \text{and} \quad \lambda(\xi, \tilde{\eta}) = (\xi, \eta).$$

This follows as in the proof of (VII.1.17) below. ||

(5.15) ***Definition of the action.*** Let

$$(\xi, \eta): X = (C, f_{n+1}, X^n) \to Y = (C', g_{n+1}, Y^n) \tag{1}$$

be a morphism in $\mathbf{H}^c_{n+1}$, $n \geqq 2$, with $\pi_1\eta = \varphi$. Here $\eta: X^n \to Y^n$ is a cellular map. For a cohomology class

$$\{\alpha\} \in (H^n\Gamma_n)\lambda(\xi,\eta) = \hat{H}^n(C, \varphi^*\Gamma_n(\lambda Y))$$

the cocycle α yields the composition

$$i\alpha: C_n \to \Gamma_n(\lambda Y) \subset \pi_n(Y^n).$$

We now define

$$\{(\xi,\eta)\} + \{\alpha\} = \{(\xi, \eta + i\alpha)\}, \tag{2}$$

where we use the action in (5.9); clearly, $\{(\xi,\eta)\}$ denotes the homotopy class of (ξ,η) in $\mathbf{H}^c_{n+1}/\simeq$. For $n = 2$ we can replace Γ_2 by $\tilde{\Gamma}_2$ above. Then we get the action $H^2\tilde{\Gamma}_2 +$ by (2) as well. ‖

(5.16) ***Computation of the isotropy groups.*** For (ξ,η) in $\mathbf{H}^c_{n+1}$ as in (5.15) we have the isotropy group

$$I(\xi,\eta) \subset \hat{H}^n(C, \varphi^*\Gamma_n(\lambda Y)), \tag{1}$$

which consists of all $\{\alpha\}$ with $\{(\xi,\eta)\} + \{\alpha\} = \{(\xi,\eta)\}$. For $n = 2$ we obtain equally the isotropy group

$$\tilde{I}(\xi,\eta) \subset \hat{H}^2(C, \varphi^*\tilde{\Gamma}_2(\lambda Y)). \tag{2}$$

These groups are related to the isotropy group $I(\eta)$ which is defined for $\eta \in [X^n, Y^n]$ by (5.9)(3) and which can be computed by the spectral sequence (5.9)(4). The inclusion

$$i: \Gamma_n = \Gamma_n(\lambda Y) \subset \pi_n(Y^n)$$

induces the homomorphism

$$\begin{array}{ccc} \hat{H}^n(C, \varphi^*\Gamma_n) & \subset & \hat{H}^n(X^n, D; \varphi^*\Gamma_n) \\ & \searrow^{j} & \downarrow i_* \\ & & \hat{H}^n(X^n, D; \varphi^*\pi_n Y^n) \end{array} \tag{3}$$

For $n = 2$ we obtain a homomorphism $\tilde{j}$ in the same way by replacing Γ_n by $\tilde{\Gamma}_n = \tilde{\Gamma}_n(\lambda Y)$.

Proposition. *There is a subgroup*

$$A_n(X, Y)_\varphi \subset \hat{H}^n(X_n, D, \varphi^*\pi_n Y^n)$$

which does not depend on the choice of $(\xi,\eta): X \to Y$ *with* $\pi_1(\eta) = \varphi$ *such that for* $n \geqq 3$

$$I(\xi, \eta) = j^{-1}(I(\eta) + A_n(X, Y)_\varphi).$$

For $n = 2$ *we get*

$$\begin{aligned} I(\xi, \eta) &= j^{-1}(A_2(X, Y)_\varphi) \text{ and} \\ \tilde{I}(\xi, \eta) &= \tilde{j}^{-1}(A_2(X, Y)_\varphi). \end{aligned} \tag{4}$$

The group $A_n(X, Y)_\varphi$ is defined in the following proof of (4). We can use the formula in (4) for the effective computation of the isotropy groups in certain examples. Recall that $I(\eta) = 0$ for $n = 2$, this yields the isotropy groups for the action in (5.12).

Proof of (4). We only consider $I(\xi, \eta)$. Let $\alpha \in \mathrm{Hom}_\varphi(C_n, \Gamma_n)$ be a cocycle and let $i\alpha: \Sigma^n Z_n^+ \to Y_n$ be a map represented by $i\alpha$ in (5.15)(2). Then we have $\{\alpha\} \in I(\xi, \eta)$ if and only if there is a homotopy

$$\left.\begin{aligned} &\alpha_{j+1}: (\xi, \eta) \simeq (\xi, \eta + i\alpha) \\ &\alpha_{j+1}: C_j \to C'_{j+1}, \quad j \geqq n \end{aligned}\right\} \tag{5}$$

in $\mathbf{H}^c_{n+1}$, see (5.10). This is equivalent to (6) and (7):

$$\{\eta\} + g_{n+1}\alpha_{n+1} = \{\eta'\} \quad \text{in} \quad [X^n, Y^n]^D, \tag{6}$$

and

$$0 = \alpha_k d_k + d_{k+1}\alpha_{k+1}, \quad k \geqq n+1. \tag{7}$$

Now we obtain the **subgroup** $A_n(X, Y)_\varphi$ which consists of all cohomology classes $\{g_{n+1}\alpha_{n+1}\}$ such that there exist α_{j+1}, $j \geqq n$, satisfying (7). □

§6 The tower of categories for the homotopy category of CW-complexes under *D*

In this section we describe our main result on the combinatorial homotopy theory of relative CW-complexes (X, D). When $D = *$ is a point we deduce classical theorems of J.H.C. Whitehead. The result shows that the homotopy category $\mathbf{CW}_0^D/\simeq$, see (1.1)(6), can be approximated by a tower of categories. The properties of this tower yield applications for the following

(6.1) ***Homotopy classification problems***

(1) Classification of finite dimensional homotopy types in $\mathbf{CW}_0^D/\simeq$,
(2) homotopy classification of maps in $\mathbf{CW}_0^D/\simeq$, and
(3) computation of the group of homotopy equivalences in $\mathbf{CW}_0^D/\simeq$.

The following tower of categories combines the results in §§4 and 5 above. Recall that a tower of categories consists of exact sequences for functors with the properties in (IV.4.10), see (IV.4.15).

(6.2) **Theorem.** *The category* $\mathbf{CW}_0^D/\simeq$ *is approximated by the following tower of categories where* $G=\pi_1(D)$, $n \geqq 3$.

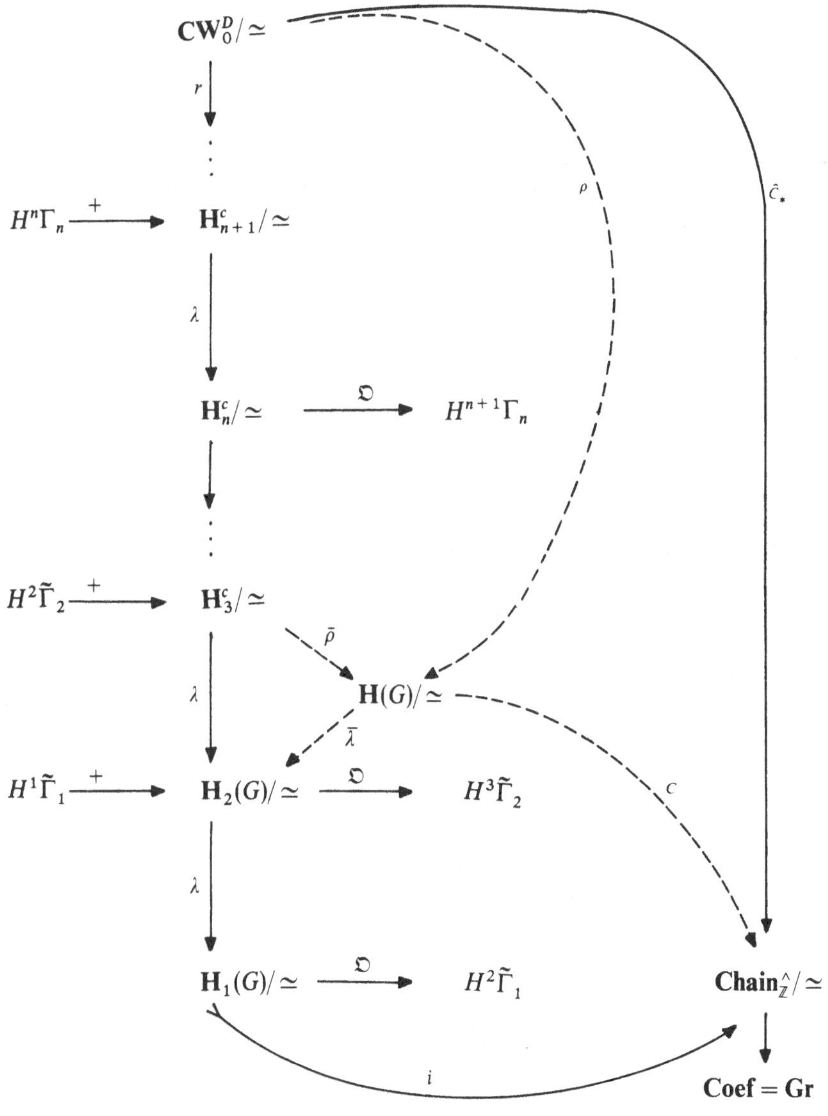

This is a commutative diagram of functors and of exact sequences. The functor i is an inclusion of categories. Moreover, the functors $\bar{\rho}$ and $\bar{\lambda}$ with $\bar{\lambda}\bar{\rho} = \lambda$ are part of the tower of categories

$$\begin{array}{ccccc} H^2\Gamma_2 & \xrightarrow{+} & \mathbf{H}_3^c/\simeq & & \\ & & \downarrow \bar{\rho}=\rho & & \\ H^2 j\tilde{\Gamma}_2 & \xrightarrow{+} & \mathbf{H}(G)/\simeq & \xrightarrow{\mathfrak{O}} & H^3\Gamma_2 \\ & & \downarrow \bar{\lambda}=\lambda & & \\ & & \mathbf{H}_2(G)/\simeq & \xrightarrow{\mathfrak{O}} & H^3 j\tilde{\Gamma}_2 \end{array} \quad . \tag{1}$$

We now describe the natural systems in (6.2) and (1) above. For X, Y in $\mathbf{CW}_0^D$ and for a morphism $F: rX \to rY$ in the category $\mathbf{H}_n^c/\simeq$ the natural system $H^p\Gamma_n$ $(p = n,\ n+1)$ is given by the cohomology groups

$$(H^p\Gamma_n)(F) = \hat{H}^p(X, D; \varphi^*\Gamma_n(Y, D)), \tag{2}$$

where $\varphi = F_*: \pi_1(X) \to \pi_1(Y)$. For $n = 2, 1$ we obtain the corresponding natural systems $H^p\tilde{\Gamma}_n$, $H^p j\tilde{\Gamma}_2$ by replacing Γ_n in (2) by $\tilde{\Gamma}_2$, $\tilde{\Gamma}_1$, and $j\tilde{\Gamma}_2$, respectively, see (4.6).

Since the tower of categories (6.2) is of major importance we describe explicitly some properties of this tower. First we have the following result which is a variant of the Whitehead theorem

(6.3) **Proposition.** *All functors in the diagrams of* (6.2) *satisfy the sufficiency condition, see* (IV.1.3). *In particular, this shows: A map $f: X \to Y$ in $\mathbf{CW}_0^D$ is a homotopy equivalence under D if and only if* (*a*) *and* (*b*) hold:

(*a*) $\varphi = f_*: \pi_1 X \to \pi_n Y$ is isomorphism,
(*b*) $(\varphi, \hat{C}_* f): \hat{C}_*(X, D) \to \hat{C}_*(Y, D)$ *is a homotopy equivalence in* $\mathbf{Chain}_{\mathbb{Z}}^{\wedge}$ *or equivalently $\hat{C}_* f$ induces an isomorphism in homology. Recall that $H_n\hat{C}_*(X, D) = \hat{H}_n(X, D, \mathbb{Z}[\pi])$ with $\pi = \pi_1(X)$.*

This follows immediately from (IV.4.11).

Next we consider the *homotopy classification problem for maps* in $\mathbf{CW}_0^D$. Let X and Y be complexes in $\mathbf{CW}_0^D$ and let φ be a homomorphism for which

$$\text{(6.4)} \qquad \begin{array}{ccc} & \pi_1 D = G & \\ {}^{i_*}\swarrow & & \searrow^{i_*} \\ \pi_1 X & \xrightarrow[\varphi]{} & \pi_1 Y \end{array}$$

commutes. We consider the set of all homotopy classes of maps $X \to Y$ in

$\mathbf{CW}_0^D/\simeq$ which induce φ. This set is denoted by

$$[X, Y]_\varphi^D \subset \operatorname{Mor}(\mathbf{CW}_0^D/\simeq). \tag{1}$$

Similarly we denote by

$$[X, Y]_\varphi^n \subset \operatorname{Mor}(\mathbf{H}_{n/}^c/\simeq), \quad n \geqq 1, \tag{2}$$

the set of all homotopy classes $rX \to rY$ in $\mathbf{H}_{n/}^c/\simeq$ which induce φ. For $n = 2$ and $n = 1$ we set $\mathbf{H}_n^c = \mathbf{H}_n(G)$.

Moreover, we have the subsets

$$[\rho X, \rho Y]_\varphi^G \subset \operatorname{Mor}(\mathbf{H}(G)/\simeq) \tag{3}$$

$$[\hat{C}_*(X, D), \hat{C}_*(Y, D)]_\varphi \subset \operatorname{Mor}(\mathbf{Chain}_{\mathbb{Z}}^{\hat{}}/\simeq) \tag{4}$$

which contain all homotopy classes which induce φ. The sets above might be empty; this depends on the realizability of φ. Now the tower of categories in (6.2) yields the following tower of sets which consists of maps between sets and of group actions denoted by $+$.

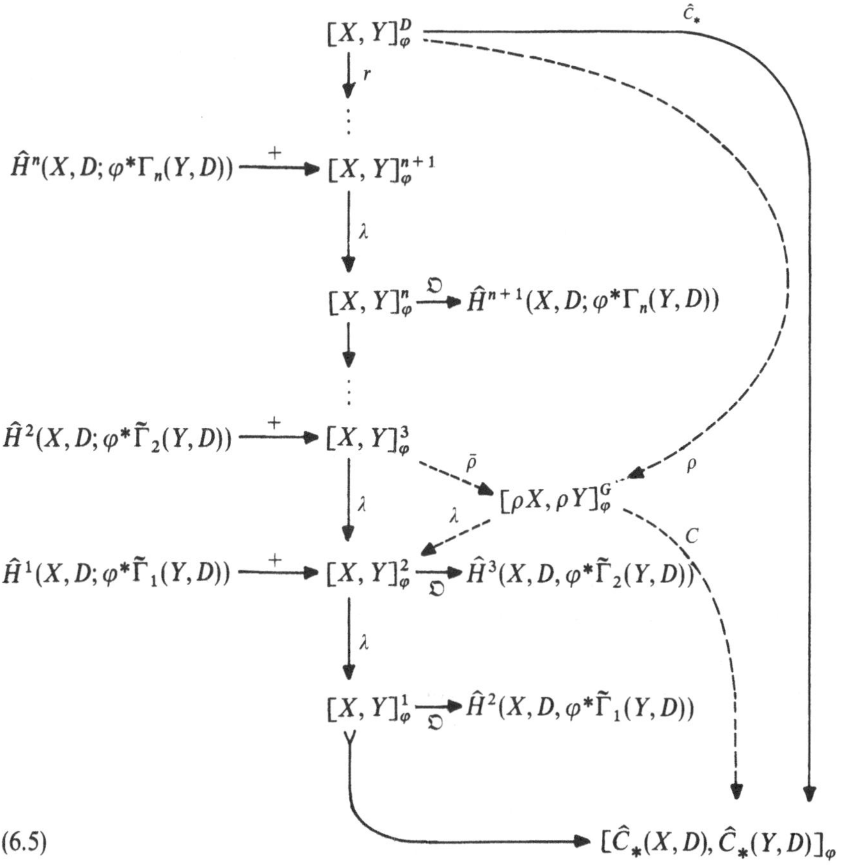

(6.5)

Moreover, we have

$$\begin{array}{ccccc} \hat{H}^2(X,D;\varphi^*\Gamma_2(Y,D)) & \xrightarrow{+} & [X,Y]^3_\varphi & & \\ & & \downarrow \bar{\rho} & & \\ \hat{H}^2(X,D,\varphi^* j\tilde{\Gamma}_2(Y,D) & \xrightarrow{+} & [\rho X,\rho Y]^G_\varphi & \xrightarrow{\mathfrak{O}} & \hat{H}^3(X,D,\varphi^*\Gamma_2(Y,D)) \\ & & \downarrow \bar{\lambda} & & \\ & & [X,Y]^2_\varphi & \xrightarrow{\mathfrak{O}} & \hat{H}^3(X,D;\varphi^* j\tilde{\Gamma}_2(Y,D)) \end{array}$$

(6.5)(a)

These diagrams have the following properties:

$\mathfrak{O}$ is a **derivation**, that is,

$$\mathfrak{O}(fg) = f_*\mathfrak{O}(g) + g^*\mathfrak{O}(f) \tag{1}$$

where fg denotes the composition

$\mathfrak{O}$ has the **obstruction property**, that is, kernel $(\mathfrak{O})$ = image(λ) in (6.5) and kernel $(\mathfrak{O})$ = image $(\bar{\rho})$, kernel $(\mathfrak{O})$ = image $(\bar{\lambda})$ in (6.5)(a). Moreover the obstruction $\mathfrak{O}$ has the relativization property in (5.14)(7). (2)

The group actions denoted by + satisfy the **linear distributivity law**, that is

$$(f+\alpha)(g+\beta) = fg + f_*\beta + g^*\alpha. \tag{3}$$

The group actions have the **exactness property**, that is,

$$\lambda f = \lambda g \Leftrightarrow \exists \alpha \quad \text{with} \quad g = f + \alpha$$

The same holds for $\bar{\rho}$ and $\bar{\lambda}$. (4)

If $X - D$ has only cells in dimension $\leq N$ then the map

$$r:[X,Y]^D_\varphi \to [X,Y]^n_\varphi$$

is bijective for $n = N+1$ and surjective for $n = N$. (5)

The **isotropy groups** of the action + can be computed by (5.16) and by the spectral sequence (5.9)(4). (6)

The obstruction operators in (6.5) and (6.5)(a) yield the following **higher order obstruction**. Let X and Y be complexes in $\mathbf{CW}^D_0$ and let $\varphi:\pi_1 X \to \pi_1 Y$ be a homomorphism. For a morphism $F \in [\rho X, \rho Y]^G_\varphi$ in the category $\mathbf{H}(G)/\simeq$ we get the subsets

(6.6) $$\mathfrak{O}_{n+1}(F) = \mathfrak{O}((\rho_n)^{-1}(F)) \subset \hat{H}^{n+1}(X,D,\varphi^*\Gamma_n(Y,D)),$$

where $\rho_n:[X, Y]^n_\varphi \to [\rho X, \rho Y]^G_\varphi$ is the composition of maps in (6.5) and (6.5)(a), with $\rho_3 = \bar{\rho}$, $n \geqq 2$. Clearly, $\mathfrak{O}_3(F)$ consists of a single element.

Remark. Similarly, we can define higher order obstructions for elements $F_n \in [X, Y]^n_\varphi$, $n = 1, 2$. Here also F_1, F_2, are purely algebraic data.

The higher order obstruction associates with the algebraic morphism F a subset of the cohomology (6.6) which we also assume to be known. The subset $\mathfrak{O}_{n+1}(F)$, however, is defined geometrically and it is a difficult problem to compute this obstruction algebraicly only in terms of appropriate invariants of X and Y.

The obstruction groups in the tower of categories (6.2) have the following property which we derive immediately from (IV.4.12).

(6.7) ***Counting realizations.*** Let X be a object in $\mathbf{H}^c_n/\simeq$ and let $\mathrm{Real}_\lambda(X)$ be the class of realization of X in $\mathbf{H}^c_{n+1}/\simeq$, compare the definition in (IV.4.12). If $\mathrm{Real}_\lambda(X)$ is non empty then the group $\hat{H}^{n+1}(X, D; \Gamma_n(X, D))$ acts transitively and effectively on the set $\mathrm{Real}_\lambda(X)$.

Whence the tower of categories (6.2) inductively yields the enumeration of the set of realizations (in $\mathbf{CW}^D_0/\simeq$) of a crossed chain complex in $\mathbf{H}(G)$ or of a chain complex in $\mathbf{H}_1(G)$. This is a crucial result on the classification of homotopy types. For this we also need the next result (6.8).

By definition in (6.6) we know that F is realizable by a map $f \in [X, Y]^D_\varphi$ with $\rho(f) = F$ if and only if $0 \in \mathfrak{O}_n(F)$ for $n \geqq 3$. Here we assume that $X - D$ is finite dimensional. Therefore we obtain the

(6.8) **Proposition.** *Let X and Y be complexes in $\mathbf{CW}^D_0$ and let $X - D$ be finite dimensional. Then there is a homotopy equivalence $X \simeq Y$ under D if and only if there is an isomorphism φ and a homotopy equivalence $F \in [\rho X, \rho Y]^G_\varphi$ with* $0 \in \mathfrak{O}_n(F)$ for $3 \leqq n \leqq \dim(X - D)$.

We will use this result for the classification of simply connected 4-dimensional polyhedra, see IX.§ 4; we also classified the simply connected 5-dimensional polyhedra by use of (6.8), (details will appear elsewhere).

Remark: Assume D has a trivial second homotopy group, $\pi_2 D = 0$. Then we know $\Gamma_2 = 0$. This clearly implies $\mathfrak{O}_3(F) = 0$ for all F, and $\mathfrak{O}_4(F)$ is non-empty and consists of a single element for all F.

This remark implies by (6.8) the following result which seems to be new.

(6.9) **Theorem.** *Let $\pi_2(D) = 0$. Then the homotopy types of 3-dimensional complexes in $\mathbf{CW}^D_0/\simeq$ are 1–1 corresponded to the algebraic homotopy types of 3-dimensional crossed chain complexes in $\mathbf{H}(G)/\simeq$ where $G = \pi_1(D)$.*

This generalizes the corresponding result for $D = *$ in section 7 of J.H.C. Whitehead (1949) who states that an equivalence class of a 3-dimensional crossed chain complexes in $\mathbf{H}(0)/\simeq$ is an algebraic equivalent of the homotopy type of a 3-dimensional complex in $\mathbf{CW}_0^*/\simeq$. For $\pi_2(D) \neq 0$ the result in (6.9) is not true since the obstruction $\mathfrak{O}_3(F)$ might be non-trivial.

Next we derive from the tower of categories in (6.2) a **structure theorem for the group of homotopy equivalences**. For a complex X in $\mathbf{CW}_0^D$ let

(6.10) $$\mathrm{Aut}(X)^D \subset [X, X]^D \subset \mathrm{Mor}(\mathbf{CW}_0^D/\simeq)$$

be the group of homotopy equivalences under D. For $\pi_1(D) = G$ let A,

(6.11) $$A = \mathrm{Aut}(\pi_1 X)^G \subset \mathrm{Aut}(\pi_1(X)),$$

be the group of all automorphism φ of $\pi_1(X)$ for which (6.4) commutes (where we set $X = Y$).

Now the tower in (6.2) yields the following tower of groups, $n \geqq 3$, where the arrows $\bar{\mathfrak{O}}$ denote derivations and where all the other arrows are homomorphisms between groups.

(6.12)

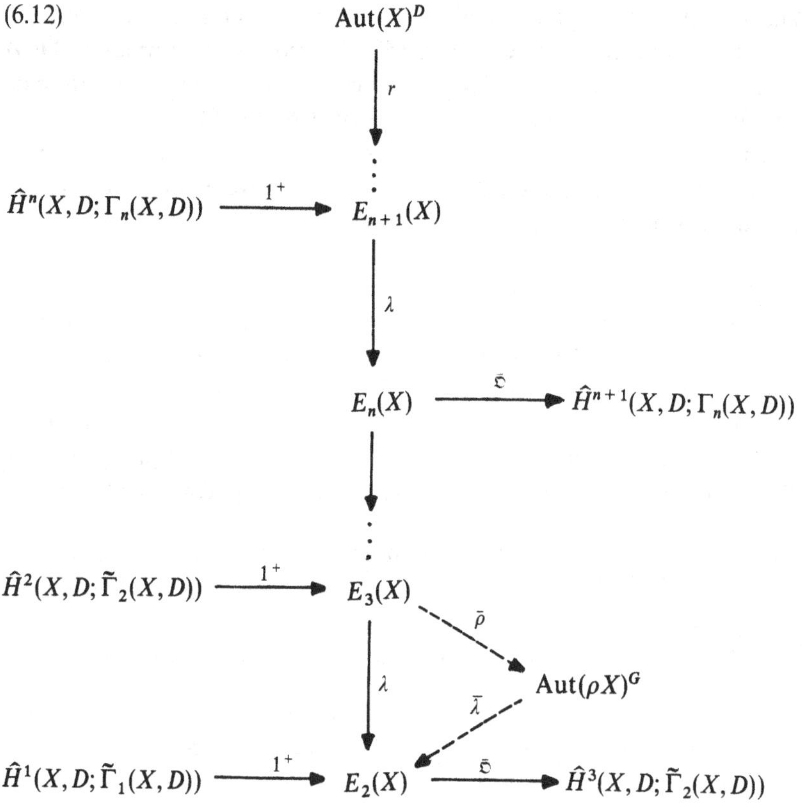

$$\begin{array}{ccc}
\Big\downarrow{\scriptstyle\lambda} & & \\
E_1(X) & \xrightarrow{\;\mho\;} & \hat{H}^2(X, D; \tilde{\Gamma}_1(X, D)) \\
\Big\downarrow & & \\
\operatorname{Aut}(\hat{C}_*(X, D)) & & \\
\Big\downarrow & & \\
\operatorname{Aut}(\pi_1 X) & &
\end{array}$$

Moreover, $\lambda = \bar{\lambda}\bar{\rho}$ is part of the tower

$$\begin{array}{ccccc}
\hat{H}^2(X, D; \Gamma_2(X, D)) & \xrightarrow{1^+} & E_3(X) & & \\
 & & \Big\downarrow{\scriptstyle\bar{\rho}} & & \\
\hat{H}^2(X, D; j\tilde{\Gamma}_2(X, D)) & \xrightarrow{1^+} & \operatorname{Aut}(\rho X)^G & \xrightarrow{\;\mho\;} & \hat{H}^3(X, D; \Gamma_2(X, D)) \\
 & & \Big\downarrow{\scriptstyle\bar{\lambda}} & & \\
 & & E_2(X) & \xrightarrow{\;\mho\;} & \hat{H}^3(X, D; j\tilde{\Gamma}_2(X, D))
\end{array}$$

We denote by

$$E_n(X) \subset \bigcup_{\varphi \in A} [X, X]^n_\varphi \tag{1}$$

the group of equivalences of $r(X)$ in the category $\mathbf{H}^c_n/\simeq$, where $\mathbf{H}^c_n = \mathbf{H}_n(G)$ for $n = 2, 1$. Moreover, the groups

$$\operatorname{Aut}(\rho X)^G \to \operatorname{Aut} \hat{C}_*(X, D) \tag{2}$$

denote the groups of homotopy equivalences in $\mathbf{H}(G)/\simeq$ and $\mathbf{Chain}^{\wedge}_{\mathbb{Z}}/\simeq$ respectively.

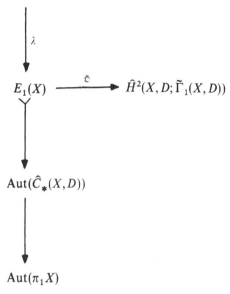

The obstruction operator $\mho$ above is defined as in (IV.4.11) by the corresponding obstruction operator in (6.5) and (6.5)(a) respectively, where we set $X = Y$. Since $\bar{\mathfrak{D}}$ in (6.5) is a derivation we see that kernel $(\bar{\mathfrak{D}})$,

$$\operatorname{kernel}(\bar{\mathfrak{D}}) \subset \bigcup_{\varphi \in A} [X, X]^n_\varphi,$$

is a submonoid. By sufficiency of λ and by (6.5)(2) we get

$$\lambda E_{n+1}(X) = E_n(X) \cap \operatorname{kernel}(\bar{\mathfrak{D}}). \tag{3}$$

Let 1 be the identity in the group $E_{n+1}X$. Then the linear distributivity law in (6.5)(3) shows that $\alpha \longmapsto 1 + \alpha = 1^+(\alpha)$ is a homomorphism of groups. Moreover, exactness in (6.5)(4) shows for $\lambda^{-1}(1) \subset [X, X]^{n+1}_1$,

$$\operatorname{image}(1^+) = \lambda^{-1}(1) \cap E_{n+1}(X). \tag{4}$$

Similar results as in (3) and (4) are true for (6.12)(a). By (3) and (4) we obtain the short exact sequence of groups

$$\frac{\hat{H}^n(X, D; \Gamma_n(X, D))}{\operatorname{kernel}(1^+)} \overset{1^+}{\rightarrowtail} E_{n+1}(X) \overset{\lambda}{\twoheadrightarrow} E_n(X) \cap \operatorname{kernel}(\bar{\mathfrak{D}}). \tag{5}$$

For $n = 1, 2$ we replace Γ_n by $\tilde{\Gamma}_n$ in (5). The associated homomorphism of the extension (5) is induced by

$$\left.\begin{aligned} &h: E_n(X) \to \operatorname{Aut} \hat{H}^n(X, D; \Gamma_n(X, D)), \\ &h(u)(\alpha) = u_*(u^{-1})^*(\alpha) = (u^{-1})_* u^*(\alpha). \end{aligned}\right\} \tag{6}$$

Compare (IV.3.9). This also follows from the linear distributivity law. Moreover, by (6.5)(5) we know

If $\dim(X - D) \leqq N$ then the homomorphism

$$r: \operatorname{Aut}(X)^D \to E_n(X) \tag{7}$$

is **bijective** for $n = N + 1$ and is **surjective** for $n = N$.

The group kernel (1^+) is the isotropy group of the action in $1 \in [X, X]^{n+1}_1$, (8) $\varphi = 1$, which can be computed by (5.16), see (6.5)(6). This is clear by definition of 1^+ in (3).

The properties in (1)$\cdots$(8) show that the tower of groups (6.12) is useful for the computation of $\operatorname{Aut}(X)^D$. *The extension problem* for the groups in (5), however, is not solved and, indeed, there is no general technique known which could be useful for the solution of this extension problem. Recall that there are even more sophisticated *extension problems for categories* which are given by the exact sequences for functors in (6.2) and (6.2)(1), compare (IV.6.1) and (IV.7.7). We show below that most of these extensions are not split.

We derive from (6.12) immediately the

(6.13) **Theorem.** *Let* $\dim(X - D) < \infty$. *Then the kernel of* $\hat{C}_*: \operatorname{Aut}(X)^D \to \operatorname{Aut}(\hat{C}_*(X, D))$ *is a solvable group. Moreover, if* $N \subset \operatorname{Aut}(X)^D$ *is a subgroup* *which acts nilpotently via* (6) *above*, $n \geqq 1$, *and if* $\hat{C}_*(N)$ *is a nilpotent subgroup of* $\operatorname{Aut}(\hat{C}_*(X, D))$ *then the group* N *is nilpotent.*

Remark. For the special case $D = *$ and X simply connected the result in (6.13) follows from theorem 3.3 in Dror–Zabrodsky (1979).

(6.14) ***Example.*** We describe two examples which show that the extension (6.12)(5), in general, does not have a splitting.

(A) Let $X = M(V, m)$ be a Moore space, $m \geqq 3$, for which $V = (\mathbb{Z}/2)^v$ is a $\mathbb{Z}/2$-vector space of dimension v. Then (6.12)(5) yields the commutative diagram with exact columns ($D = *$):

$$\begin{array}{ccc} H^{m+1}(X, \Gamma_{m+1}X) & \cong & \operatorname{Ext}(V, V) \\ \downarrow & & \downarrow \\ \operatorname{Aut}(X)^* = E_{m+1}(X) & \cong & GL_v(\mathbb{Z}/4) \\ \downarrow H_m \quad \downarrow \lambda & & \downarrow p \\ \operatorname{Aut}(V) = E_m(X) & \cong & GL_v(\mathbb{Z}/2) \end{array}$$

Here the homomorphism p between general linear groups is given by reduction mod 2 and the homomorphism H_m is the homology functor. For $v \geqq 4$ the extension λ has no splitting since p has no splitting in this case. Compare (V.§ 3a).

(B) Let $X = S^n \times S^m$ be a product of two spheres $n, m \geqq 2$. Then (6.12)(5) yields the commutative diagram with exact columns

$$\begin{array}{ccc} H^{n+m}(X, \Gamma_{n+m}X) & \twoheadrightarrow & H_{m,n} \\ \downarrow & & \downarrow \\ E_{n+m}(X) & \cong & \operatorname{Aut}(S^n \times S^m) \\ \downarrow \lambda & & \downarrow \\ E_{n+m-1}(X) & \cong & \operatorname{Aut}(S^n \vee S^m) \end{array}$$

Here $H_{m,n}$ is a quotient of $\pi_{n+m}(S^n) \oplus \pi_{n+m}(S^n)$, see (II.16.6). The group extension of image (λ) by $H_{m,n}$ is studied in Sawashita (1974). He shows that this extension is not split for $n = 3$, $m = 5$.

The properties in (6.5) and (6.12), respectively, show that the enumeration of the set $[X, Y]^D$ and the computation of the group $\operatorname{Aut}(X)^D$ can be achieved by the inductive computation of the obstruction operators, $\mathfrak{O}$, and the actions, $+$,. For example, there is the following special case for which the diagrams in (6.5) or (6.12) *collaps.*

(6.15) **Theorem.** *Let X and Y be complexes in* $\mathbf{CW}_0^D$ *and let* $\varphi:\pi_1 X \to \pi_1 Y$ *be given. Suppose that the groups*

$$O = \hat{H}^p(X, D; \varphi^*\Gamma_n(Y, D)) \ (p = n, n+1) \tag{1}$$

are trivial for $2 \leq n \leq \dim(X - D) < \infty$. *Then the functor* ρ *in* (1.2) *yields the bijection*

$$\rho:[X, Y]_\varphi^D = [\rho X, \rho Y]_\varphi^G, \quad \textit{see } (6.5). \tag{2}$$

Moreover, suppose $X = Y$ *and suppose* (1) *is satisfied for* $\varphi = 1$. *Then* ρ *yields the isomorphism of groups*

$$\rho:\mathrm{Aut}(X)^D = \mathrm{Aut}(\rho X)^G, \quad \textit{see } (6.12). \tag{3}$$

This theorem is an immediate consequence of the properties of the diagrams in (6.5) and (6.12) respectively.

(6.16) ***Remark.*** Theorem (6.15) was obtained by J.H.C. Whitehead (1949) for the special case $D = *$ and $\Gamma_n(Y, *) = 0$ for $n = 2, 3, \ldots$. Moreover, the tower above immediately yields for $D = *$ all results of Whitehead (1949)′ on J_m-complexes Y (which are defined by the condition that $\Gamma_n(Y, *) = 0$ for $n = 2, 3, \ldots, m$). J_m-complexes which are simply connected were classified by Adams (1956), see also p. 101 in Hilton (1953).

Theorem (6.15) is just a simple application of the tower of categories in (6.2) where we use the vanishing of actions and obstructions. It is, in fact, fruitful to consider explicit examples for which actions and obstructions do not vanish, we will describe such computations elsewhere.

§7 Small models and obstructions to finiteness for CW-complexes under *D*

In (IV.§1) we defined the sufficiency condition for a functor and we have seen that each functor in the tower of categories for CW-complexes satisfies the sufficiency condition. In this section we show that various functors in this tower actually satisfy a 'strong sufficiency condition'. This is the main step for the construction of small models of complexes under *D* and for the definition of obstructions to finiteness.

(7.1) ***Definition.*** Let $(\mathbf{A}, \simeq)$ and $(\mathbf{B}, \simeq)$ be categories together with natural equivalence relations, $\simeq$, which we call homotopy. Let $\pi:\mathbf{A} \to \mathbf{B}$ be a functor which induces the functor $\bar{\pi}:\mathbf{A}/\simeq \to \mathbf{B}/\simeq$ between homotopy categories. We say that $\bar{\pi}$ satisfies the **strong sufficiency condition** if (a), (b) and (c) hold:

(a) $\bar{\pi}$ satisfies the **sufficiency condition**, that is, a map in $\mathbf{A}$ is a homotopy equivalence if and only if the induced map in $\mathbf{B}$ is a homotopy equivalence.

(b) **Existence of models**: For an object A in $\mathbf{A}$ let $\beta: B \xrightarrow{\simeq} \pi A$ be a homotopy equivalence in $\mathbf{B}$. Then there is an object $M_B A$ in $\mathbf{A}$ together with an isomorphism $i: B \cong \pi(M_B A)$ in $\mathbf{B}$ and together with a map $\alpha: M_B A \to A$ in $\mathbf{A}$ such that the diagram

$$\begin{array}{ccc} \pi(M_B A) & \xrightarrow{\pi(\alpha)} & \pi A \\ & {}_{\cong}\nwarrow^{i} \qquad {}^{\beta}\nearrow_{\simeq} & \\ & B & \end{array}$$

commutes in $\mathbf{B}$. By (a) we know that α is a homotopy equivalence in $\mathbf{A}$. We call $(M_B A, \alpha, i)$ a B-**model** of A.

(c) For $\xi_0: A \to A'$ in $\mathbf{A}$ and $\eta_0 = \pi\xi_0: \pi A \to \pi A'$ we have: if $\eta_0 \simeq \eta_1$ then there is $\xi_0 \simeq \xi_1$ with $\eta_1 = \pi\xi_1$. ||

(7.2) ***Remark.*** We say an object B in $\mathbf{B}$ is $\boldsymbol{\pi}$**-realizable** if there is an object A in $\mathbf{A}$ with $\pi A \cong B$ in $\mathbf{B}$. By condition (b) above we know:

If B is π-realizable then each object B' in the homotopy type of B is π-realizable. (*)

We say a map $f: \pi A \to \pi A'$ is $\boldsymbol{\pi}$**-realizable** if there is $g: A \to A'$ with $\pi g = f$. By condition (c) we know:

If $f: \pi A \to \pi A'$ is π-realizable then each element in the homotopy class of f is π-realizable. (**)

The following lemma is immediate:

(7.3) **Lemma.** *If* $\mathbf{A}/\simeq \xrightarrow{\bar{\pi}} \mathbf{B}/\simeq \xrightarrow{\bar{\tau}} \mathbf{C}/\simeq$ *are both functors which satisfy the strong sufficiency condition then also the composition* $\bar{\tau}\bar{\pi}$ *satisfies the strong sufficiency condition.*

(7.4) **Theorem.** *Consider the tower of categories for the homotopy category* CW_0^D *in* (6.2) *with* $\pi_1(D) = G$. *The functors* $(n \geqq 4)$

$$\lambda: \mathbf{H}_{n+1}^c/\simeq \to \mathbf{H}_n^c/\simeq, \textit{ and}$$
$$\bar{\rho}: \mathbf{H}_3^c/\simeq \to \mathbf{H}(G)/\simeq$$

satisfy the strong sufficiency condition.

Essentially, by (7.3) we thus get

(7.5) **Theorem.** *The functor*

$$\rho : \mathbf{CW}_0^D/\simeq \to \mathbf{H}(G)/\simeq$$

satisfies the strong sufficiency condition.

(7.6) ***Remark.*** For $D = *$ the result in (7.5) corresponds to theorem 17 in the classical paper on simple homotopy types of J.H.C. Whitehead (1950). On the other hand, Wall (1965) proved the same result for $D = *$ by using chain complexes instead of crossed chain complexes. This is possible since for $D = *$ we have the full and faithful functor $\lambda\lambda$ in (4.9). The disadvantage of this approach is the fact that one has to be concerned with the **2-realizability** of chain complexes. This shows that crossed chain complexes are the more natural objects to use in this context. By (7.5) we have a new generalization of these results in the category of spaces under D, $D \neq *$. Moreover, we give a new proof which relies only on simple properties of the obstruction operator in the tower of categories (6.2). This proof can easily be transformed to obtain a proof of the corresponding result for chain algebras (details will appear elsewhere).

(7.7) *Proof of* (7.4) *and* (7.5). The sufficiency condition is already proved in (6.3). Moreover, we obtain (7.1)(c) by the exact sequences in the top row of (5.11) which yield a tower of categories for $\mathbf{CW}_0^D/\overset{0}{\simeq}$. It remains to check the existence of models in (7.1)(b).

Let $Y = (C', g_{n+1}, Y^n)$ be an object in $\mathbf{H}_{n+1}^c$ and let

$$\bar{G} = (\xi, \eta) : X = (C, f_n, X^{n-1}) \to \lambda Y = (C', g_n, Y^{n-1}) \tag{1}$$

be a map in $\mathbf{H}_n^c$ which is an equivalence in $\mathbf{H}_n^c/\simeq$. We have to construct an object

$$M_X Y = (C, f_{n+1}, X^n) \quad \text{in } \mathbf{H}_{n+1}^c, \tag{2}$$

and a map $F : M_X Y \to Y$ in $\mathbf{H}_{n+1}$ with $\lambda(M_X Y) = X$ and $\lambda F = G$. For a basis V_n of C_n we choose a map

$$f_n : \bigvee_{V_n} S^{n-1} \to X^{n-1}, \tag{3}$$

which represents f_n in (1). Let X^n in (2) be the mapping cone of this map. We can choose a map

$$G_1 : X^n \to Y^n, \tag{4}$$

which is associated to (ξ_n, η) in (1). We thus obtain the following commutative diagram of unbroken arrows:

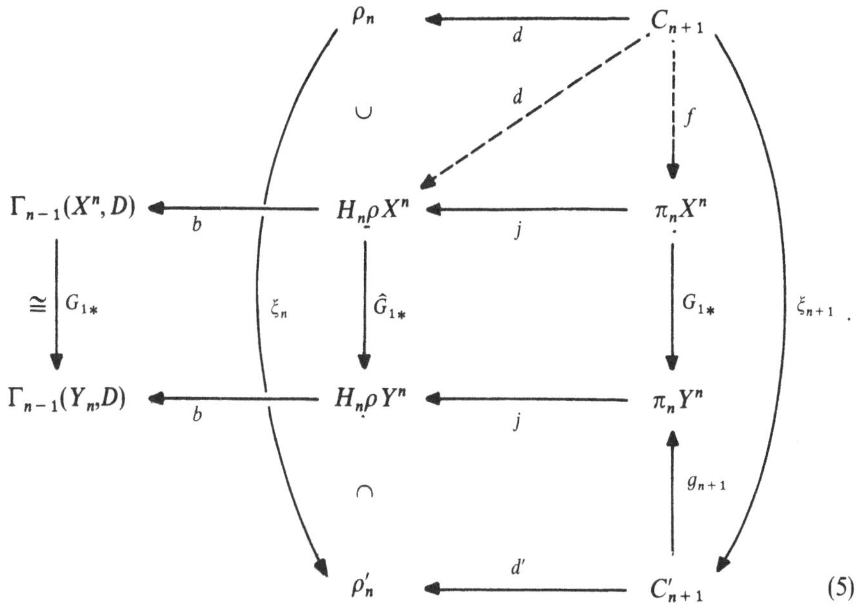

(5)

The rows of this diagram are part of the exact sequence of J.H.C. Whitehead, see (III.11.7). Since $\bar{G}$ in (1) is a homotopy equivalence in $\mathbf{H}_n/\simeq$ we know that the induced map

$$\begin{array}{ccc} \Gamma_k(X^n, D) & \xrightarrow{G_{1*}} & \Gamma_k(Y^n, D) \\ \| & & \| \\ \Gamma_k X & \underset{\cong}{\xrightarrow{\bar{G}_*}} & \Gamma_k \lambda Y \end{array} \tag{6}$$

is an isomorphism for $k \leq n$. Diagram (5) and (6) show that there is a map f with

$$jf = d. \tag{7}$$

In fact, since C_{n+1} is a free $\mathbb{Z}[\pi_1]$-module we obtain f by showing: $bd = 0$. This is true since

$$G_{1*}bd = bjg_{n+1}\xi_{n+1} = 0,$$

where $bj = 0$. By use of f we have the object

$$X_{(0)} = (C, f, X^n) \quad \text{in } \mathbf{H}_{n+1} (\text{not in } \mathbf{H}^c_{n+1}) \tag{8}$$

with $\lambda(X_{(0)}) = X$ as follows from (7). Therefore, the obstruction (VII.1.20)

$$\mathfrak{O}_{X_{(0)}, Y}(\bar{G}) \in \hat{H}^{n+1}(X^{n+1}, D; \varphi^*\Gamma_n(\lambda Y)) \tag{9}$$

is defined where $\varphi = \pi_1(\bar{G})$ is the induced map on fundamental groups. The isomorphism $\bar{G}_*$ in (6) for the coefficients in (9) gives us the element

$$\{\alpha\} = G_*^{-1}\mathfrak{O}_{X_{(0)},Y}(\bar{G}) \in \hat{H}^{n+1}(X^{n+1}, D; \Gamma_n X). \tag{10}$$

Here α is a cocycle which represents the cohomology class $\{\alpha\}$. By α we obtain the composition

$$a: C_{n+1} \xrightarrow{\alpha} \Gamma_n X = \Gamma_n(X^n, D) \xrightarrow[i]{} \pi_n X^n. \tag{11}$$

We define

$$\left.\begin{array}{l} f_{n+1} = f - a: C_{n+1} \to \pi_n X^n, \quad \text{and} \\ M_X Y = (C, f_{n+1}, X^n) \quad \text{in } \mathbf{H}_{n+1}. \end{array}\right\} \tag{12}$$

We again have $\lambda M_X Y = \lambda X_{(0)} = X$. Moreover, we show

$$\mathfrak{O}_{M,Y}(\bar{G}) = 0 \quad \text{with } M = M_X Y. \tag{13}$$

Thus $\bar{G}$ is realizable by a map F in $\mathbf{H}_{n+1}$ with $\lambda F = G$, see (VII.1.20). We have to check (13). We know by (5.14) that the obstruction class (13) is represented by the cocycle

$$G_{1*} f_{n+1} - g_{n+1}\xi_{n+1}: C_{n+1} \to \pi_n Y^n. \tag{14}$$

By definition of f_{n+1} we thus have by (10)

$$\begin{aligned} \mathfrak{O}_{M,Y}(\bar{G}) &= \{G_{1*}(f - a) - g_{n+1}\xi_{n+1}\} \\ &= \{G_{1*} f - g_{n+1}\xi_{n+1}\} - G_{1*}\{\alpha\} \\ &= \mathfrak{O}_{X_{(0)},Y}(\bar{G}) - G_{1*}\{\alpha\} = 0. \end{aligned} \tag{15}$$

By definition in (12) we know that

$$j(f_{n+1})d_{n+2} = j(f - a)d_{n+2} = 0. \tag{16}$$

Therefore, $f_{n+1}d_{n+2}: C_{n+2} \to \Gamma_n X$. Since $\bar{G}_*: \Gamma_n X \longrightarrow \Gamma_n Y$ is an isomorphism the realization F of $\bar{G}$ shows that $f_{n+1}d_{n+2} = 0$. Therefore $M_X Y$ satisfies the cocycle condition and hence $M_X Y$ is an object in $\mathbf{H}_{n+1}^c$. This completes the proof of the strong sufficiency for λ in (7.4). A similar proof holds for $\bar{\rho}$ in (6.2)(1). Moreover, we get (7.5) by an inductive construction as above and by a limit argument. □

The following is an immediate consequence of (7.5):

(7.8) **Corollary.** *Let X be a complex in $\mathbf{CW}_0^D$ with $G = \pi_1(D)$ and let*

$$f: A_* \to \rho(X)$$

be a homotopy equivalence in $\mathbf{H}(G)/\simeq$. Suppose A_* satisfies any combination of *the conditions*

(*i*) A_i *is finitely generated for* $i \leqq m_1$,
(*ii*) A_i *is countably generated for* $i \leqq m_2$,
(*iii*) $A_i = 0$ *for* $i > m_3$.

Then X *is homotopy equivalent in* $\mathbf{CW}_0^D$ *to a CW-complex* Y *satisfying the corresponding conditions*

(*i*) $Y^{m_1} - D$ *has finitely many cells,*
(*ii*) $Y^{m_2} - D$ *has countably many cells,*
(*iii*) $\dim(Y - D) \leqq m_3$.

Also the following result, which for $D = *$ was obtained by Wall (1966), is an easy consequence of (7.5).

We say that a complex X in $\mathbf{CW}_0^D$ is **finite** if $X - D$ has only finitely many cells, (X, D) has **relative dimension** $\leqq n$, if $\dim(X - D) \leqq n$.

(7.9) **Corollary.** *Let* $n \geqq 3$ *and let* $X = (X, D)$ *be a complex in* $\mathbf{CW}_0^D$ *with* $\pi = \pi_1(X)$. *Suppose that:*

(*i*) *the n-skeleton* $X^n - D$ *is finite in* $\mathbf{CW}_0^D$,
(*ii*) $H_i(\hat{X}, \hat{D}) = H_i(\hat{C}_*(X, D)) = 0$ *for* $i > n$, *and*
(*iii*) $d\hat{C}_{n+1}(X, D)$ *is a direct summand of the* π*-module* $\hat{C}_n(X, D)$.

Let B_n *be a complement of this summand in* $\hat{C}_n(X, D)$ *and let*

$$\sigma(X, D) = (-1)^n \{B_n\} \in \tilde{K}_0(\mathbb{Z}[\pi])$$

be the element in the reduced projective class group of π *(see below) which is given by the finitely generated projective* $\mathbb{Z}[\pi]$*-module* B_n. *Then* $\sigma(X, D)$ *is an obstruction, depending only on the homotopy type of* X *under* D *in* $\mathbf{CW}_0^D/\simeq$, *which vanishes if* $X - D$ *is finite in* $\mathbf{CW}_0^D$ *and whose vanishing is sufficient for* X *to be homotopy equivalent in* $\mathbf{CW}_0^D$ *to a finite complex of relative dimension* $\leqq n$.

(7.10) **Addendum.** *If* X *is homotopy equivalent in* $\mathbf{CW}_0^D/\simeq$ *to a complex of relative dimension* $\leqq n$ *then the conditions (ii) and (iii) in (7.9) are satisfied.*

Recall that for a ring Λ the group $\tilde{K}_0(\Lambda)$ is the Grothendieck group of finitely generated projective Λ-modules, modulo free modules; it is known as the **reduced projective class group**. Each finitely generated projective Λ-module P represents a class $\{P\} \in \tilde{K}_0(\Lambda)$ and we have $\{P\} = \{Q\}$ iff there exist free modules F and G with $P \oplus F \cong Q \oplus G$.

For the convenience of the reader we give a proof of (7.9). Using the lemmata below the result is an easy consequence of (7.5). These lemmata are available for any ring Λ.

Recall the notation in (I.§6) on chain complexes. In (I.6.11) we proved:

(7.11) **Lemma.** *Let C and C' be positive projective chain complexes over Λ. Then $f:C\to C'$ is a homotopy equivalence if and only if $f_*:HC\to HC'$ is an isomorphism.*

The following lemma proves the addendum (7.10), compare theorem 6 in Wall (1966).

(7.12) **Lemma.** *A projective positive chain complex A_* is homotopy equivalent to an n-dimensional projective positive chain complex if and only if $H_i(A_*)=0$ for $i>n$ and the image of $d:A_{n+1}\to A_n$ is a direct summand.*

Proof. If the conditions hold, and B_n is a complement to dA_{n+1}, then A_* is equivalent to $0\to B_n\to A_{n-1}\to A_{n-2}\to\cdots$ Conversely, if A_* is equivalent to an n-dimensional complex, it is clear that $H_i(A_*)=0$ for $i>n$. Also

$$H^{n+1}(\mathrm{Hom}_\Lambda(A_*, dA_{n+1}))=0,$$

and $d:A_{n+1}\to dA_{n+1}$ gives an $(n+1)$-cocycle ($d^2=0$), which is thus a coboundary (so factors through A_n) giving a retraction of A_n on dA_{n+1}. □

(7.13) **Lemma.** *Let A_*, B_* be homotopy equivalent finitely generated projective positive chain complexes over Λ. Then there is an isomorphism of Λ-modules*

$$\oplus A_{2i}\oplus B_{2i+1}=\oplus B_{2i}\oplus A_{2i+1}$$

and therefore

$$\sigma(B_*)=\sum_i(-1)^i\{B_i\}\in\tilde{K}_0(\Lambda)$$

is an invariant of the homotopy type of B_*.

Compare Wall (1966), p. 138, and Dyer (1976) theorem (1.2).

Proof of (7.9). Let $\Lambda=\mathbb{Z}[\pi]$. Since $d\hat{C}_{n+1}X$ is a direct summand of $\hat{C}_nX$ we have with B_n in (7.9) the homotopy equivalence

$$\hat{C}_*X\simeq(0\to B_n\to\hat{C}_{n-1}X\to\cdots)=B_*,$$

where B_* is a subcomplex of $\hat{C}_*X$. Since X^n is finite, B_* is a projective finitely generated positive chain complex. By (7.13) the obstruction $\sigma(X)=(-1)^n\{B_n\}=\sigma(B_*)\in\tilde{K}_0(\Lambda)$ depends only on the homotopy type of B_* and thus on the homotopy type of $\hat{C}_*X$ or of X respectively. Moreover, if X is finite the homotopy equivalence $\hat{C}_*X\simeq B_*$ yields, by (7.13), $0=\sigma(\hat{C}_*X)=\sigma(B_*)=\sigma(X,D)$ since $\hat{C}_*X$ is free.

Now assume $\sigma(X) = (-1)^n\{B_n\} = 0$. Then there is a free Λ-module F such that $B_n \oplus F$ is free. We obtain the homotopy equivalence β in the category $\mathbf{H}(G)$, $\rho = \rho(X)$,

$$\begin{array}{ccccccccccc}
B = \{ & \longrightarrow & 0 & \longrightarrow & B_n \oplus F & \xrightarrow{d \oplus 1} & \rho_{n-1} \oplus F & \xrightarrow{d \oplus 0} & \rho_{n-2} & \longrightarrow & \} \\
\beta \big\downarrow \simeq & & & & \big\downarrow 1 \oplus 0 & & & & & & \\
& & & & B_n & & \big\downarrow 1 \oplus 0 & & \big\downarrow 1 & & . \\
& & & & \big\downarrow & & & & & & \\
\rho X = \{ & \longrightarrow & \rho_{n+1} & \longrightarrow & \rho_n & \longrightarrow & \rho_{n-1} & \longrightarrow & \rho_{n-2} & \longrightarrow & \}
\end{array}$$

Here β is a homotopy equivalence since $C(\beta)$ induces an isomorphism in homology and since C on $\mathbf{H}(G)/\simeq$ satisfies the sufficiency condition. Now we obtain by the strong sufficiency a model $M_B(\rho X) = Y$ of X in $\mathbf{CW}_0^D$ with $\rho Y = B$, see (7.5). □

Finally we consider some applications of (7.5) for spaces with trivial fundamental group.

(7.14) **Corollary.** *Let X be a complex in $\mathbf{CW}_0^D$ and let $\pi_1 X = 0 = \pi_1 D$. Suppose that for the integral relative homology groups $H_k(X, D)$ a presentation*

$$0 \to \mathbb{Z}^{r_k} \to \mathbb{Z}^{b_k} \to H_r(X, D) \to 0$$

(b_k generators, r_k relations), $b_1 = r_1 = 0$, is given. Then there is a homotopy equivalence $K \simeq X$ in $\mathbf{CW}_0^D/\simeq$ where $K - D$ has $b_k + r_{k-1}$ k-cells.

For $D = *$ this result is due to Milnor.

Proof. We can assume that $X^1 = D$. Therefore $\pi_2(X^2, X^1)$ is abelian and we can identify

$$\rho X = \hat{C}_*(X, D) = C_*(X, D), \tag{1}$$

where $C_*(X, D)$ is the cellular chain complex of the pair (X, D). There is a $\mathbb{Z}$-free chain complex C and a homotopy equivalence

$$C \simeq C_*(X, D), \tag{2}$$

where C has $b_k + r_{k-1}$ generators in degree k. Since C is a crossed chain complex in $\mathbf{H(0)}$ we obtain the proposition by choosing a model $M_C(X)$.

□

The Eckmann–Hilton decomposition (or **homology decomposition**) of a simply connected space (or of a map between simply connected spaces) is given by (7.14) as well, see Hilton (1965), Eckmann–Hilton (1959). In particular we get:

(7.15) **Corollary.** *Let (X, D) be a complex in $\mathbf{CW}_0^D$ and let $\pi_1 X = 0 = \pi_1 D$. If* *the relative homology $H_*(X, D)$ is a free $\mathbb{Z}$-module we can find a complex (K, D) and a homotopy equivalence $K \simeq X$ in $\mathbf{CW}_0^D/\simeq$ where the cells of $K - D$ are $1-1$ corresponded with the elements of a basis in $H_*(X, D)$. Moreover, $C_*(K, D) = H_*(X, D)$ has trivial differential.*

In this case we call K a **minimal model** of X in $\mathbf{CW}_0^D/\simeq$, compare also (IV.2.18).

§8 *n*-dimensional CW-complexes and $(n-1)$-types

The Postnikov decomposition of an n-dimensional CW-complex X gives us the functor $X \longmapsto P_{n-1}X$ where $P_{n-1}X$ is the $(n-1)$-type of X. There is a canonical quadratic action $E+$ on the functor P_{n-1} which leads to two linear extensions of categories; the one is given by the action $\Gamma+$ which appears also in the tower of categories in § 7 $(D = *)$, the other one yields the action $\bar{\bar{E}}+$ which is related to k-invariants.

Let $\mathbf{CW}_0^*$ be the category of CW-complexes X with $\{*\} = X^0$, see (1.1) where we set $D = *$. We introduce the **n-th Postnikov functor**

(8.1) $$P_n: \mathbf{CW}_0^*/\simeq \;\rightarrow n\text{-}\mathbf{types}$$

Here **n-types** denotes the full subcategory of $\mathbf{Top}^*/\simeq$ consisting of CW-spaces Y with $\pi_i Y = 0$ for $i > n$. For X in $\mathbf{CW}_0^*$ we obtain P_nX by Postnikov decomposition of X as in (III.7.2). We can construct P_nX by *killing homotopy groups*, then P_nX is a CW-complex with $(n+1)$-skeleton

$$X^{n+1} = (P_nX)^{n+1}. \tag{1}$$

For a cellular map $F: X \rightarrow Y$ in $\mathbf{CW}_0^*$ we choose a map $PF^{n+1}: P_nX \rightarrow P_nY$ which extends the restriction $F^{n+1}: X^{n+1} \rightarrow Y^{n+1}$ of F. This is possible since $\pi_i P_n Y = 0$ for $i > n$. The functor P_n in (8.1) carries X to P_nX and carries the homotopy class of F to the homotopy class of PF^{n+1}. Different choices for P_nX yield canonically isomorphic functors P_n. The space

$$P_1(X) = K(\pi_1 X, 1) \tag{2}$$

is an Eilenberg–Mac Lane space and P_1 as a functor is equivalent to the functor π_1. A map

$$p_n: X \rightarrow P_nX, \tag{3}$$

which extends the inclusion $X^{n+1} \subset P_nX$ in (1) is called the **n-th Postnikov section** of X. Clearly, the fiber of $p_1: X \rightarrow K(\pi_1 X, 1)$ is the universal covering of X, more generally the homotopy fiber of $p_n: X \rightarrow P_nX$ is called the **'n-connected covering'** of X. The **Postnikov tower** is given by maps (see (III.7.2))

$$q_n: P_nX \rightarrow P_{n-1}X \quad \text{with } q_n p_n \simeq p_{n-1}. \tag{4}$$

Let $(\mathbf{CW}_0^*)^n$ be the full subcategory of $\mathbf{CW}_0^*$ consisting of n-dimensional CW-complexes. For objects X^n, Y^n in $(\mathbf{CW}_0^*)^n$ the set $[X^n, Y^n]$ is the set of homotopy classes in $\mathbf{Top}^*/\simeq$. As in (5.9)(2), with $D = *$, we have the action $(n \geqq 2)$

(8.2) $$[X^n, Y^n]_\varphi \times \hat{H}^n(X^n, \varphi^*\pi_n Y^n) \xrightarrow{+} [X^n, Y^n]_\varphi$$

where $\varphi: \pi_1 X^n \to \pi_1 Y^n$ is a homomorphism. For $n = 2$ this is a transitive and effective action. For $n > 2$ the isotropy groups of this action can be computed by the spectral sequence in (5.9)(4). The orbit of $u_n \in [X^n, Y^n]_\varphi$ is the subset $j_n^{-1}(j_n(u_n))$ where

$$j_n: [X^n, Y^n] \to [X^{n-1}, Y^n] \tag{1}$$

is induced by the inclusion $X^{n-1} \subset X^n$. For morphisms $F, G: X^n \to Y^n$ in $(\mathbf{CW}_0^*)^n$ we define the **natural equivalence relation**, $\sim$, by

$$F \sim G \Leftrightarrow j_n\{F\} = j_n\{G\}. \tag{2}$$

(8.3) **Proposition.** *The action* (8.2) *is a quadratic action E on the Postnikov functor*

$$P_{n-1}: (\mathbf{CW}_0^*)^n/\simeq \;\to (n-1)\text{-}\mathbf{types}$$

This functor induces the equivalence of categories

$$P_{n-1}: (\mathbf{CW}_0^*)^n/\sim \;\xrightarrow{\sim} (n-1)\text{-}\mathbf{types}$$

This result is originally due to J.H.C. Whitehead who called an equivalence class of objects in $(\mathbf{CW}_0^*)^n/\sim$ an 'n-type', which is now called an $(n-1)$-type.

Proof of (8.3). It is easy to see that we have

$$F \sim G \Leftrightarrow P_{n-1}\{F\} = P_{n-1}\{G\}. \tag{1}$$

This shows that $(\mathbf{CW}_0^*)^n/\sim$ is the image category of P_{n-1}. Hence by (8.2) (1) we see that (8.2) is an action on the functor P_{n-1}, compare (IV.§2), since

$$(\mathbf{CW}_0^*)^n/\sim \;= ((\mathbf{CW}_0^*)^n/\simeq)/E. \tag{2}$$

We derive from (V.4.5) that the action is quadratic. In fact, let $X^n = C_f$, $Y^n = C_g$ where f and g are the attaching maps of n-cells. Then we know

$$[X^n, Y^n] = \mathbf{TWIST}(f, g) \tag{3}$$

by (III.5.11) (a) and hence the action E in (8.2) is an example for the action E in (V.4.5). □

(8.4) ***Remark.*** The functor P_n on $(\mathbf{CW}_0^*)^n/\simeq$ is full and faithful, the functor P_{n-1} on $(\mathbf{CW}_0^*)^n/\simeq$ in (8.3) is full but not faithful.

The quadratic distributivity law of the action (8.2) is given as follows. Let

$$F\in[X^n, Y^n]_{\varphi},\quad G\in[Z^n, X^n]_{\psi},\quad \alpha\in\hat{H}^n(X^n, \varphi^*\pi_n Y^n),\quad \beta\in\hat{H}^n(Z^n, \psi^*\pi_n X^n)$$

be given. Then composition in $(\mathbf{CW}_0^*)^n/\simeq$ satisfies the distributivity law

(8.5) $$(F+\alpha)(G+\beta) = FG + F_*\alpha + G^*\beta + \alpha*\beta$$

in $[Z^n, Y^n]$. Here F_* and G^* are the induced homomorphisms on cohomology groups. The mixed term

(8.6) $$\alpha*\beta\in\hat{H}^n(Z^n, \psi^*\varphi^*\pi_n Y^n)$$

is given by **composition of cocycles**: For $\alpha=\{a\}$ and $\beta=\{b\}$ we have with $X = X^n$

$$\alpha*\beta = \{\hat{C}_n Z^n \xrightarrow[b]{} \pi_n X \underset{p_*}{\cong} \pi_n\hat{X} \xrightarrow[h]{} H_n\hat{X} \subset \hat{C}_n X \xrightarrow[a]{} \pi_n Y^n\}.$$

Here $p:\hat{X}\to X$ is the projection of the universal covering and h is the Hurewicz homomorphism. We derive (8.6) from (V.4.6).

The quadratic distributivity law (8.6) determines two linear groups actions on the category $(\mathbf{CW}_0^*)^n/\simeq$ as follows:

On the one side we can consider elements $\beta=\{b\}$ with $hp_*b=0$. Then clearly $\alpha*\beta=0$ for all α. On the other hand side we can consider elements $\alpha=\{a\}$ with $ah=0$. Then we have $\alpha*\beta=0$ for all β. These two possibilities lead to linear group actions on $(\mathbf{CW}_0^*)^n/\simeq$ which we call $\Gamma+$ and $\bar{\bar{E}}+$ respectively.

We first study $\Gamma+$. For an n-dimensional complex $X=X^n$ we have the group of Whitehead $\Gamma_n(X^n,*)=\Gamma_n X^n$ with

(8.7) $$\Gamma_n X^n = \operatorname{kernel}(\pi_n X = \pi_n\hat{X} \xrightarrow[h]{} H_n(\hat{X}, \mathbb{Z}))$$

Compare (III.11.7). The inclusion $j:\Gamma_n X^n\subset\pi_n X^n$ of $\mathbb{Z}[\pi_1 X]$-modules induces for $\psi:\pi_1 Z\to\pi_1 X$ the homomorphism of cohomology groups $j:\hat{H}^n(Z,\psi^*\Gamma_n X)\to\hat{H}^n(Z,\psi^*\pi_n X)$. Using the action in (8.2) we obtain the action

(8.8) $$[Z^n, X^n]_{\psi}\times\hat{H}^n(Z^n, \psi^*\Gamma_n X^n)\xrightarrow{+}[Z^n, X^n]_{\psi}$$

with $F+\gamma = F+j(\gamma)$. This is again a natural group action on $(\mathbf{CW}_0^*)^n/\simeq$ which we denote by $H^n\Gamma_n$. From (8.8) we deduce that $H^n\Gamma_n$ is in fact a linear action which can be identified with the action $H^n\Gamma_n+$ in (5.15) on the full subcategory

$$r:(\mathbf{CW}_0^*)^n/\simeq\ \subset \mathbf{H}_{n+1}/\simeq.$$

Therefore we have the exact sequence (compare (6.2))

$$H^n\Gamma_n \xrightarrow{+} (\mathbf{CW}_0^*)^n/\simeq \xrightarrow{r} \mathbf{H}_n^n/\simeq \to 0 \tag{8.9}$$

where $\mathbf{H}_n^n$ is the full subcategory of n-dimensional objects in $\mathbf{H}_n$. Clearly, this sequence is a subsequence of the corresponding sequence in (6.2) where $H^{n+1}\Gamma_n$ is trivial on $\mathbf{H}_n^n$. Since the action (8.8) is given by restriction of the action $E+$ in (8.2) we obtain the following commutative diagram ($n \geqq 2$)

$$\begin{array}{ccccc} & & E & \longrightarrow & E/H^n\Gamma_n \\ & {\scriptstyle j}\nearrow & \downarrow{\scriptstyle +} & & \downarrow{\scriptstyle +} \\ H^n\Gamma_n & \xrightarrow{+} & (\mathbf{CW}_0^*)^n/\simeq & \xrightarrow{r} & \mathbf{H}_n^n/\simeq \to 0 \\ & & \downarrow{\scriptstyle P_{n-1}} & & \downarrow{\scriptstyle Q_{n-1}} \\ & & (n-1)\text{-}\mathbf{types} & \xrightarrow{1} & (n-1)\text{-}\mathbf{types} \end{array} \tag{8.10}$$

The row of this diagram for the functor r is given by the exact sequence (8.9), the map j is defined as in (8.8). By (8.3) the functor P_{n-1} induces the functor Q_{n-1}, moreover the cokernel of j, denoted by $E/H^n\Gamma_n$ defines a quadratic action on the functor Q_{n-1}. Hence the two columns of (8.10) describe quadratic extensions of categories. The functor v is a detecting functor, the functor P_{n-1} and Q_{n-1}, however, are not detecting functors.

On the other hand, we obtain by $E+$ an action $\bar{\bar{E}}+$ which fits into the commutative diagram

$$\begin{array}{ccccc} & & E+ & \xrightarrow{\delta} & \bar{E}+ \\ & {\scriptstyle i^*}\nearrow & \downarrow & & \downarrow \\ \bar{\bar{E}}+ & \longrightarrow & (\mathbf{CW}_0^*)^n/\simeq & \xrightarrow[T_n]{} & \mathbf{t}_n^n \subset \mathbf{t}_n \\ & & \downarrow{\scriptstyle P_{n-1}} & & \downarrow{\scriptstyle pr_1} \\ & & (n-1)\text{-}\mathbf{types} & \xrightarrow[1]{} & (n-1)\text{-}\mathbf{types} \end{array} \tag{8.11}$$

The two columns are quadratic extensions and the row is a linear extension of categories, in particular, T_n is a detecting functor.

(8.12) ***Definition of the category*** $\mathbf{t}_n$. An object is given by a triple (Q, π_n, k) where $Q \in (n-1)$-**types**, where π_n is a $\pi_1 Q$-module, and where k is an element $k \in \hat{H}^{n+1}(Q, \pi_n)$. A morphism $(F, \alpha): (Q, \pi_n, k) \to (Q', \pi_n', k')$ is a map $F: Q \to Q'$ in $(n-1)$-**types** together with a $\pi_1(F)$-equivariant homomorphism $\alpha: \pi_n \to \pi_n'$ such that $F^*(k') = \alpha_*(k) \in \hat{H}^{n+1}(Q, \varphi^*\pi_n')$ with $\varphi = \pi_1(F)$.

We have the **enriched Postnikov functor**

(8.13) $$T_n\colon \mathbf{CW}_0^*/\simeq \;\to \mathbf{t}_n$$

which carries a complex X to $T_n(X) = (P_{n-1}(X), \pi_n(X), k_n(X))$ where k_nX is the nth **k-invariant** of X, see (III.7.4). Let $\mathbf{t}_n^n$ be the full subcategory of $\mathbf{t}_n$ consisting of objects which are realizable by an n-dimensional complex X. Naturality of the k-invariants shows that T_n in (8.13) is a well defined functor.

We can describe the k-invariant $k_n(X)$ as follows: We can choose $P_{n-1}X$ with $X^n = (P_{n-1}X)^n$ in such a way that in the following commutative diagram the map f_{n+1}^0 is surjective.

$$\begin{array}{ccccc} & & & & \hat{C}_{n+1}(P_{n-1}X) \\ & & & \swarrow^{f^0_{n+1}} & \downarrow d_{n+1} \\ \pi_nX & \underset{i_*}{\twoheadleftarrow} & \pi_nX^n \underset{p_*}{\cong} \pi_n\hat{X}^n & \underset{h}{\longrightarrow} & H_n\hat{X}^n \subset \hat{C}_nX^n \end{array} \tag{1}$$

The map f_{n+1}^0 is given by the attaching maps of $(n+1)$-cells in $P_{n-1}X$, see (5.3). By (5.4) we know that f_{n+1}^0 is a cocycle. This cocycle represents the cohomology class

$$k_nX = \{i_* f_{n+1}^0\} \in \hat{H}^{n+1}(P_{n-1}X, \pi_nX) \tag{2}$$

which is the **nth k-invariant of the complex** X.

We define the action $\bar{E}+$ on $\mathbf{t}_n^n$ as follows. For complexes X, Y in $(\mathbf{CW}_0^*)^n$ and for a map $(F,\alpha)\colon T_nX \to T_nY$ in $\mathbf{t}_n^n$ let

(8.14) $$\bar{E}_\varphi = \{\beta \in \mathrm{Hom}_\varphi(\pi_nX, \pi_nY) \mid 0 = \beta_* k_nX\}.$$

Clearly, $\bar{E}_\varphi$ acts transitively and effectively on the set of all morphisms $(F,\alpha)\colon T_nX \to T_nY$ in $\mathbf{t}_n^n$ by $(F,\alpha) + \beta = (F, \alpha + \beta)$. This shows that the projection functor pr_1 in (8.11) is a quadratic extension by $\bar{E}+$. We define

(8.15) $$\delta\colon E_\varphi = \hat{H}^n(X, \varphi^*\pi_nY) \twoheadrightarrow \bar{E}_\varphi$$

by $\delta\{c\} = chp_*^{-1}$. Here $c\colon \hat{C}_nX \to \pi_nY$ is a φ-equivalent cocycle and $\delta\{c\}$ is given by composition of c with the Hurewicz map h in (8.13)(1). Since f_{n+1} in (8.13)(1) is surjective we see that δ is surjective.

Moreover, for

(8.16) $$\bar{\bar{E}}_\varphi = \hat{H}(P_{n-1}X, \varphi^*\pi_nY)$$

we have the exact sequence

(8.17) $$0 \to \bar{\bar{E}}_\varphi \underset{i^*}{\longrightarrow} E_\varphi \underset{\delta}{\longrightarrow} \bar{E}_\varphi \to 0$$

Here i^* is induced by the inclusion $X \subset P_{n-1}X$. Since X is the n-skeleton of

$P_{n-1}X$ we see that i^* is injective. By use of (8.13)(1) we see that kernel $\delta = \text{image } i^*$. Now (8.3) and (8.6) show

(8.18) **Proposition:** *Diagram* (8.11) *is a commutative diagram of extensions.* T_n *is a linear extension by* $\bar{\bar{E}}$ *and the columns are quadratic extensions. Since* $\bar{E}$ *acts effectively we see that the isotropy group* $I_F, I_F \subset E_\varphi$, *actually is a subgroup of* $\bar{\bar{E}}_F$.

Let $\text{Aut}(X)^*$ be the group of homotopy equivalences of X in $\mathbf{Top}^*/\simeq$. The linear extension T_n in (8.18) gives us the short exact sequence of groups ($X = X^n$)

(8.19) $$0 \to \hat{H}^n(P_{n-1}X, \pi_n X)/I \xrightarrow{1^+} \text{Aut}(X)^* \longrightarrow \text{Aut}(T_n X) \to 0$$

where $\text{Aut}(T_n X)$ is the group of equivalences of $T_n X$ in the category $\mathbf{t}_n$. The isotropy group I of the action $E+$ in $1 \in [X, X]$ can be computed by the spectral sequence (5.9)(4). This shows that $I = 0$ if $\hat{X}$ is $(n-1)$-connected; in this case X is called a $(\pi_1 X, n)$-complex, see Dyer.

(8.20) ***Definition.*** A (π, n)**-complex** is a path connected n-dimensional CW-complex $X = X^n$ with fundamental group $\pi_1 X \cong \pi$ and with $(n-1)$-connected universal covering $\hat{X}$, hence $\pi_2 X = \cdots = \pi_{n-1} X = 0$. Each 2-dimensional CW-complex is a $(\pi, 2)$-complex. The universal covering $\hat{X}$ of a (π, n)-complex is homotopy equivalent to a one point union of n-spheres. ||

For a (π, n)-complex X we know that $P_{n-1}X = K(\pi, 1)$. Therefore $T_n(X)$ is given by the **algebraic n-type**

(8.21) $$T_n(X) = (\pi_1, \pi_n, k \in \hat{H}^{n+1}(\pi_1, \pi_n)).$$

We leave it to the reader to restrict diagram (8.11) to the subcategory of (π, n)-complexes. We obtain this way all results described in Dyer, Section 6. In particular we obtain by (8.19) for a (π, n)-complex X the exact sequence

(8.22) $$0 \to \hat{H}^n(\pi_1 X, \pi_n X) \xrightarrow{1^+} \text{Aut}(X)^* \xrightarrow[T_n]{} \text{Aut}(T_n X) \to 0$$

where $\text{Aut}(T_n X) \subset \text{Aut}(\pi_1 X) \times \text{Aut}(\pi_n X)$ is the subgroup of all pairs (φ, α) with $\varphi^* k = \alpha_* k$. If $\pi_1 X$ is finite then $\hat{H}^n(\pi_1 X, \pi_n X) = 0$, see Dyer. Compare also Schellenberg.

(8.23) ***Remark on trees of homotopy types.*** Theorem 14 of J.H.C. Whitehead (1950) shows that the homotopy types of finite n-dimensional path connected CW-complexes which have the same $(n-1)$-type Q form a **connected tree** $HT(Q,n)$. The vertices of this tree are the homotopy types $[X^n]$ of n-complexes X^n with $P_{n-1}X^n \simeq Q$. The vertex $[X^n]$ is connected by an edge to vertex $[Y^n]$ if Y^n has the homotopy type of $X^n \vee S^n$. Connectedness of the tree shows that for two finite n-complexes X^n, Y^n with $P_{n-1}X^n \simeq P_{n-1}Y^n$ there exist finite one point unions of spheres $\bigvee_A S^n$, $\bigvee_B S^n$ such that the one point unions

$$X^n \vee \bigvee_A S^n \simeq Y^n \vee \bigvee_B S^n$$

are homotopy equivalent.

VII

Homotopy theory of complexes in a cofibration category

In Chapter VI we described the tower of categories which approximates the homotopy category $\mathbf{CW}_0^D/\simeq$. In this chapter we introduce classes of complexes in a cofibration category which have similar properties as the class of CW-complexes in topology. In particular, the homotopy category of complexes in such a class is approximated by a tower of categories. This is our main result, deduced from the axioms of a cofibration category, which leads to many new theorems on the homotopy classification problems in topology and in various algebraic homotopy theories. It will be very helpful for the reader to compare the abstract theory in this chapter with the applications on *CW*-complexes in Chapter VI. In fact, the main result of Chapter VI on *CW*-complexes is proved here in the context of an abstract cofibration category.

§1 Twisted maps between complexes

In this section we obtain an exact sequence for the functor $\lambda:\mathbf{H}_{n+1}\to\mathbf{H}_n$ in (VI.5.8). We prove this result more generally in any cofibration category $\mathbf{C}$ with an initial object $*$. For this we replace the category $\mathbf{H}_{n+1}$ in (VI.5.7) of homotopy systems by the category $\mathbf{TWIST}_{n+1}(\mathfrak{X})$ of twisted homotopy systems which is defined in any cofibration category. Here $\mathfrak{X}$ is a 'good class of complexes' in $\mathbf{C}$. If $\mathbf{C}=\mathbf{Top}^D$ is the cofibration category of topological spaces under D and if $\mathfrak{X}$ is the class of *CW*-complexes in $\mathbf{CW}_0^D$ then $\mathfrak{X}$ is actually a good class in $\mathbf{Top}^D$ and then we have an equivalence of categories $\mathbf{H}_{n+1}=\mathbf{TWIST}_{n+1}(\mathfrak{X})$ for $n\geqq 1$. Indeed, the category $\mathbf{TWIST}_{n+1}(\mathfrak{X})$ is the precise analogue of the category $\mathbf{H}_{n+1}$ in a cofibration category. We show that there is an exact sequence for the functor

$$\lambda:\mathbf{TWIST}_{n+1}(\mathfrak{X})\to\mathbf{TWIST}_n(\mathfrak{X})$$

This yields a tower of categories which approximates the category **Complex** $(\mathfrak{X})/\overset{0}{\simeq}$. The proof is worked out in a cofibration category. This is rewarding since there are many different good classes of complexes in various cofibration categories for which we can apply our result. Examples of such good classes are described in §3 below. In the following chapter we apply the results of this section in the fibration category of topological spaces.

Recall the definition of a complex in **C** in (III.3.1). For a class of complexes, $\mathfrak{X}$, we have the category **Complex** $(\mathfrak{X})$ which is a full subcategory of **Fil(C)** in (III.1.1). We may assume that all complexes in $\mathfrak{X}$ are actually objects in $\mathbf{Fil(C)}_{cf}$.

In the following definition we use the notation in (II.§11).

(1.1) ***Definition.*** Let $\mathfrak{X}$ be an admissible class of complexes as in (III.5a.5) with $X_0=*$ for $X\in\mathfrak{X}$. We say that $\mathfrak{X}$ is a **good class** if for all X, $Y\in\mathfrak{X}$ with attaching maps $A_i\to X_{i-1}$ and $B_i\to Y_{i-1}$ respectively the following properties $(1)\cdots(4)$ are satisfied.

A_1 is a based object, $A_2=\Sigma A_2'$ is a suspension and $A_n=\Sigma A_n'=\Sigma^2A_n''$ is a double suspension for $n\geqq 3$. (1)

For $m\geqq 3$ each element in $[A_m,B_m\vee Y_1]_2$ and in $[A_m,\Sigma B_{m-1}\vee Y_1]_2$ is a partial suspension. Moreover, the partial suspension is injective on $[A_m,B_m\vee Y_2]_2$ and on $[A_m,\Sigma B_{m-1}\vee Y_2]_2$. (2)

For $n\geqq 2$ and $\varepsilon\in\{1,0\}$ the map

$$(\pi_g,j)_*:\pi_{1-\varepsilon}^{A_{n+\varepsilon}}(CB_n\vee Y_1,B_n\vee Y_1)\to\pi_{1-\varepsilon}^{A_{n+\varepsilon}}(Y_n,Y_{n-1})\qquad(3)$$

is surjective. Here $g:B_n\to Y_{n-1}$ is the attaching map of $C_g=Y_n$ and $j:Y_1\subset Y_{n-1}$ is the inclusion.

For $n\geqq 2$ all maps $f_{n+1}:A_{n+1}\to X_n$ are attaching maps of complexes in $\mathfrak{X}$. (4)

||

(1.2) ***Example.*** Let $(\mathfrak{X}_0^1)^D$ be the class of all CW-complexes in $\mathbf{CW}_0^D$. Then this class is a good class of complexes in the cofibration category $\mathbf{Top}^D$ of topological spaces under D, compare (VI.1.1)(4). This follows readily from the properties of CW-complexes described in (III.5.11). We clearly have the isomorphism of categories $\mathbf{CW}_0^D=\mathbf{Complex}\ (\mathfrak{X}_0^1)^D$.

It will be convenient for the reader to have this example in mind for all proofs and definitions of this chapter.

(1.3) ***Remark.*** Let $\mathfrak{X}$ be a good class as above and let $X, Y \in \mathfrak{X}$. We know that $X_n = C_f$ and $Y_n = C_g$ are mapping cones with attaching maps $f: A_n \to X_{n-1}$ and $g: B_n \to Y_{n-1}$ respectively. The restriction of a map $F: X \to Y$ in **Complex** $(\mathfrak{X})$ gives us the map

$$F^n: (C_f, X_{n-1}) \to (C_g, Y_{n-1})$$

in the category **PAIR**, see (V.1.2). By the assumption (1.1)(3) with $\varepsilon = 0$ we actually know that this map is a **twisted map between mapping cones** in **TWIST** for $n \geqq 2$, compare (V.2.6). We do not assume that F^1 is a twisted map. This is motivated by CW-complexes; in fact, a cellular map between CW-complexes in $\mathbf{CW}_0^D$ is twisted only in degree $\geqq 2$ and not in degree 1.

For an admissible class $\mathfrak{X}$ we have the twisted chain functor

$$\check{K}: \mathbf{Complex}\ (\mathfrak{X})/\overset{0}{\simeq} \to \mathbf{Chain}^\vee(\mathfrak{X}) \tag{1.4}$$

where we set $\check{K}(X) = \check{K}(X, 2) = E\check{k}(X, 2)$ by (III.5a.7). The category $\mathbf{Chain}^\vee(\mathfrak{X})$ denotes the full subcategory of $\mathbf{Chain}^\vee$ consisting of twisted chain complexes EA where

$$A = ((X_2, X_1), A_i, d_i) \tag{1.5}$$

is **compatible with** $\mathfrak{X}$. This means that for A there exists $X, X \in \mathfrak{X}$, such that (X_2, X_1) in A is the 2-skeleton of X and such that the objects A_i in A are given by the attaching maps $A_i \to X_{i-1}$ of X. (We do not assume that also d_i is given by X.)

(1.6) ***Definition.*** Let $\mathfrak{X}$ be a good class of complexes and let $n \geqq 1$. A **twisted homotopy system of order** $(n+1)$ (or equivalently an $(n+1)$**-system** for short) is a triple

$$X^{(n+1)} = (A, f_{n+1}, X_n)$$

where X_n is an n-skeleton of a complex in $\mathfrak{X}$ and where

$$A = ((X_2, X_1), A_i, d_i),\ d_i: A_i \to \Sigma A_{i-1} \vee X_1$$

is a twisted chain complex compatible with $\mathfrak{X}$ which coincides with $\check{k}(X^n, 2)$ in degree $\leqq n$. Moreover, $f_{n+1} \in [A_{n+1}, X_n]$ satisfies

$$(1 \vee j)d_{n+1} = \nabla f_{n+1} \tag{1}$$

where $j: X_1 \subset X_n$ is the inclusion. Here d_{n+1} is the boundary in A and ∇ is the difference construction, compare (III.5a.1). A **map between** $(n+1)$**-systems** is a pair (ξ, η) which we write

$$(\xi, \eta): (A, f_{n+1}, X_n) \to (B, g_{n+1}, Y_n) = Y^{(n+1)}.$$

Here $\eta: X_n \to Y_n$ is a map in $\mathbf{Complex}(\mathfrak{X})/\overset{0}{\simeq}$, and $\xi: EA \to EB$ is a η_1-chain map in $\mathbf{Chain}^\vee$ where the restriction $\eta_1: X_1 \to Y_1 \subset Y_2$ of η yields the map $\eta_1:(X_2, X_1) \to (Y_2, Y_1)$ in **Coef**, see (III.4.3). Moreover, ξ coincides with the η_1-chain map $\check{K}(\eta)$ in degree $\leq n$ and for ξ_{n+1} there exists $\bar{\xi}_{n+1} \in [A_{n+1}, B_{n+1} \vee Y_1]_2$ such that $E\bar{\xi}_{n+1} = \xi_{n+1}$ and such that the following diagram commutes in $Ho(\mathbf{C})$.

$$\begin{array}{ccc} A_{n+1} & \xrightarrow{\bar{\xi}_{n+1}} & B_{n+1} \vee Y_1 \\ \downarrow{\scriptstyle f_{n+1}} & & \downarrow{\scriptstyle (g_{n+1}, j)} \\ X_n & \xrightarrow{\eta} & Y_n \end{array} \tag{2}$$

Let $\mathbf{TWIST}_{n+1}(\mathfrak{X})$ be the **category of (n + 1)-systems** and of maps as above. Composition is defined by $(\xi, \eta)(\bar{\xi}, \bar{\eta}) = (\xi\bar{\xi}, \eta\bar{\eta})$. Moreover, let $\mathbf{TWIST}^c_{n+1}(\mathfrak{X})$ be the full subcategory of $\mathbf{TWIST}_{n+1}(\mathfrak{X})$ consisting of all objects (A, f_{n+1}, X_n) which satisfy the following **cocycle condition**. There is $\bar{d}_{n+2} \in [A'_{n+2}, A_n \vee X_1]_2$ where $\Sigma A'_{n+2} = A_{n+2}$ with

$$\left. \begin{array}{l} E\bar{d}_{n+2} = d_{n+2} \text{ in } [A_{n+2}, \Sigma A_n \vee X_2]_2, \text{ and} \\ (f_{n+1}, j)_* \bar{d}_{n+2} = 0 \text{ in } [A'_{n+2}, X_n]. \end{array} \right\} \tag{3}$$

Here $j: X_1 \subset X_n$ is the inclusion. ||

Recall that $\mathbf{H}_n$ denotes the category of homotopy systems of order n in (VI.4.2) and (VI.5.7) respectively. The cocycle condition in (1.6) (3) corresponds exactly to the one in (VI.5.4).

(1.7) **Proposition.** *For the class $\mathfrak{X} = (\mathfrak{X}_0^1)^D$ of CW-complexes in* (1.2) *with $G = \pi_1 D$ we have canonical equivalences of categories*

$$N_2: \mathbf{TWIST}^c_2(\mathfrak{X}) \xrightarrow{\sim} \mathbf{H}_2(G) = \mathbf{H}_2$$

$$N_n: \mathbf{TWIST}_n(\mathfrak{X}) \xrightarrow{\sim} \mathbf{H}_n, \quad n \geq 3.$$

The same result holds for the subcategories of objects which satisfy the cocycle condition.

Proof of (1.7). We define N_2 by $N_2(A, f_2, X^1) = (C, \rho(X^2))$ where X^2 is the mapping cone of f_2 and where $\rho(X^2)$ is the crossed module of (X^2, X^1). Moreover, C is the $\pi_1(X^2)$-chain complex given by the twisted chain complex A via diagram (1) in the proof of (III.5.9). Similarly, we define the functor N_n by $N_n(A, f_n, X^{n-1}) = (C, f_n, X^{n-1})$. It is easy to check that N_n, $n \geq 2$, is an equivalence of categories. Indeed, (1.6)(1) corresponds to (VI.5.7)(1) and (1.6)(2) corresponds exactly to (VI.5.7)(2), $n \geq 2$. For $n \geq 2$ we know that $\bar{\xi}_{n+1}$ in (1.6)(2) is uniquely determined by ξ_{n+1} since we have (1.1)(2). This is not true for $\bar{\xi}_2$, in fact, the map $\bar{\xi}_2$ corresponds exactly to the map $\bar{\xi}_2$ in (VI.4.2)(2). □

As in (VI.5.8) we have obvious functors ($n \geqq 2$)

(1.8)
$$\begin{array}{ccc} \mathbf{Complex}(\mathfrak{X})/\overset{0}{\simeq} \xrightarrow[r_{n+1}]{} \mathbf{TWIST}_{n+1}(\mathfrak{X}) \xrightarrow[\lambda]{} & \mathbf{TWIST}_n(\mathfrak{X}) \\ & \downarrow \check{K} \\ \mathbf{Coef} \xleftarrow{C} & \mathbf{Chain}^{\vee}(\mathfrak{X}) \end{array}$$

The functor r_{n+1} carries a complex $X \in \mathfrak{X}$ to

$$r_{n+1}(X) = (\check{k}(X,2), f_{n+1}, X_n), \tag{1}$$

where f_{n+1} is the attaching map of $X_n \subset X_{n+1}$ and where $\check{k}(X,2)$ is the twisted chain complex in (III.5a.7). For a map $F: X \to Y$ in **Complex** ($\mathfrak{X}$) we obtain $r_{n+1}(F) = (\xi, \eta)$ by $\xi = \check{K}(F,2)$ and by $\eta = F|X_n$. Moreover $(\bar{\xi}_{n+1}, \eta)$ is associated to the twisted map F^{n+1} in (1.3), compare (V.3.12). One can see now that r_{n+1} is a well-defined functor. We define λ in (1.8) similarly by

$$\lambda(A, f_{n+1}, X_n) = (A, f_n, X_{n-1}), \tag{2}$$

where f_n is the attaching map of the principal cofibration $X_{n-1} \subset X_n$.

Lemma. *The full image of λ in* (1.8) *is the category* $\mathbf{TWIST}_n^c(\mathfrak{X})$ *of all objects which satisfy the cocycle condition.* (3)

Proof. If the cocycle condition is satisfied for $X^{(n)}$, then there exists a twisted map $f_{n+1}: A_{n+1} \to X_n = C_{f_n}$ associated to $\bar{d}_{n+1}$, see (II.11.7), such that (A, f_{n+1}, X_n) is an object in $\mathbf{TWIST}_{n+1}(\mathfrak{X})$ with $\lambda(A, f_{n+1}, X_n) = X^{(n)}$. On the other hand, if $\lambda(X^{(n+1)}) = X^{(n)}$ then we know by (1.1)(3) with $\varepsilon = 1$ that $X^{(n)}$ satisfies the cocycle condition since f_{n+1} is a twisted map associated to an element $\bar{d}_{n+1}$ by (1.1)(3). □

Finally, we obtain the functor $\check{K}$ in (1.8) by

$$\check{K}(A, f_n, X_{n-1}) = A, \quad \check{K}(\xi, \eta) = \xi. \tag{4}$$

The functor C is defined in (III.4.5)(5).

Recall that exact sequences for functors form 'towers of categories' as defined in (IV.4.15). We prove the following 'tower theorem'.

(1.9) **Theorem.** *Let $\mathfrak{X}$ be a good class of complexes as in* (1.1). *Then we have the following tower of categories which approximates the category* $\mathbf{Complex}(\mathfrak{X})/\overset{0}{\simeq}$ ($n \geqq 3$):

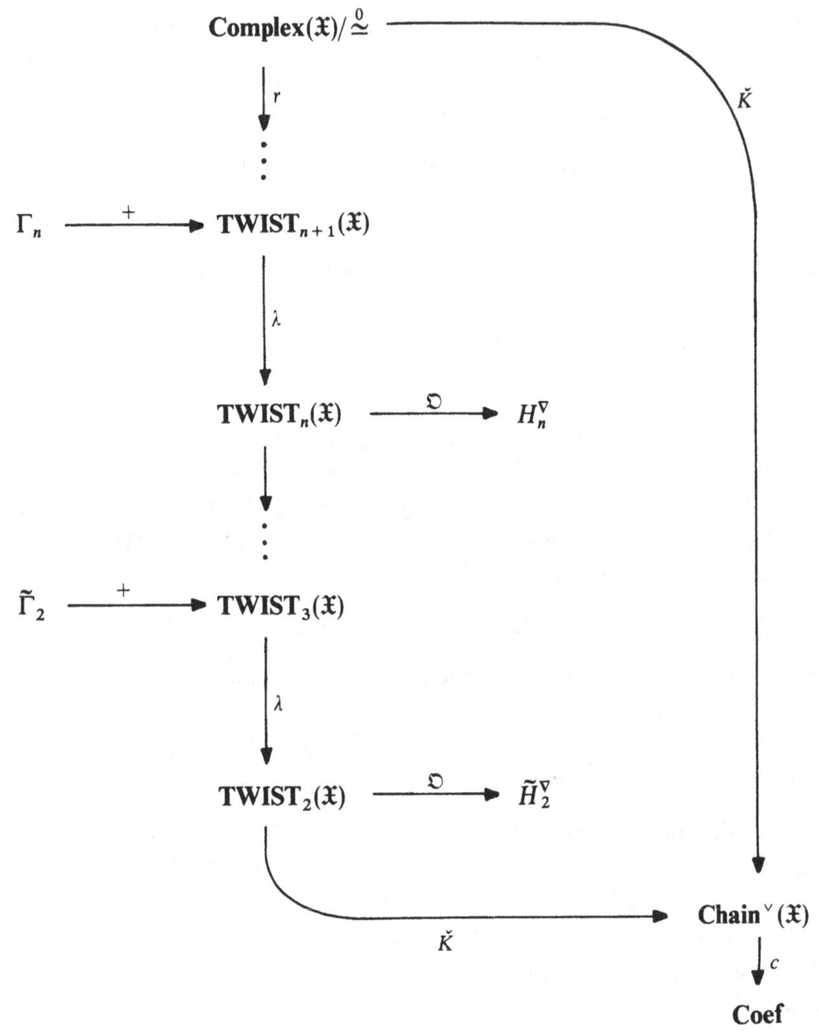

The natural systems of abelian groups in (1.9) are obtained by the groups $\tilde{E}_n$ and E_n which we define as follows.

Let $X^{(n)} = (A, f_n, X_{n-1})$ and $Y^{(n)} = (B, g_n, Y_{n-1})$ be objects in the category **TWIST**$_n(\mathfrak{X})$, $n \geqq 2$. Moreover, let $\varphi:(X_2, X_1) \to (Y_2, Y_1)$ be a map in **Coef**. Then we have the following diagram where C_g is the mapping cone of $g \in g_n$.

(1.10)
$$\begin{array}{ccc} [\Sigma A_n, Y_{n-1}] \xrightarrow{i_{g*}} & [\Sigma A_n, C_g] \supset \tilde{E}_n \supset E_n & \\ & \Big\downarrow d^*_{n+1}(\ldots,\varphi) & \\ & [A_{n+1}, C_g] \xrightarrow{\nabla} [A_{n+1}, \Sigma C_n \vee C_g]_2 & . \end{array}$$

Here $i_g: Y_{n-1} \subset C_g$ is the inclusion and ∇ is the difference operator (II.12.2). In case A_{n+1} is a double suspension, see (1.1), we know that ∇ is a homomorphism of groups. The map $d_{n+1}: A_{n+1} \to \Sigma A_n \vee X_1$ is the twisted boundary of A and we set

$$d^*_{n+1}(\beta, \varphi) = (\beta, j\varphi)d_{n+1} \tag{1}$$

for $\beta \in [\Sigma A_n, C_g]$. Here $j: Y_1 \subset Y_{n-1} \subset C_g$ is the inclusion. By (1.1)(2) we see that d_{n+1} is a partial suspension for $n \geqq 2$, therefore $d^*_{n+1}(\cdot, \varphi)$ in (1.10) is a homomorphism of abelian groups in this case.

We now define for any suspension ΣA (with $A = \Sigma A'$) the subgroups

$$E_n \subset \tilde{E}_n \subset [\Sigma A, C_g]. \tag{2}$$

For $A = A_n$ we get the groups in (1.10). Here E_n in just the image

$$E_n = \text{image}(i_{g*}: [\Sigma A, Y_{n-1}] \to [\Sigma A, C_g]). \tag{3}$$

For the definition of $\tilde{E}_n$ we use the functional suspension (II.11.7). The group $\tilde{E}_n$ is the subgroup of all elements $\beta \in [\Sigma A, C_g]$ for which there exists an element $\beta_n \in [A, B \vee Y_1]_2$ such that

$$\left.\begin{array}{l} \beta \in E_g(\beta_n), \\ E\beta_n = 0 \text{ in } [\Sigma A, \Sigma B_n \vee Y_2]_2, \text{ and} \\ (g, j)_* \beta_n = 0 \text{ in } [A, Y_{n-1}]. \end{array}\right\} \tag{4}$$

Here we use the convention in (III.5a.4). Clearly, E_n is contained in $\tilde{E}_n$ since $E_n = \text{image } (i_g)_* = E_g(0)$. Moreover, we define by (1.10) the abelian groups ($n \geqq 2$)

$$\left.\begin{array}{l} \Gamma_n(\varphi) = \Gamma_n(X^{(n)}, \varphi, Y^{(n)}) = E_n \cap \text{kernel } d^*_{n+1}(\cdot, \varphi), \\ H^\nabla_n(\varphi) = H^\nabla_n(X^{(n)}, \varphi, Y^{(n)}) = \text{kernel}(\nabla)/d^*_{n+1}(E_n, \varphi), \end{array}\right\} \tag{5}$$

where $d^*_{n+1}(E_n, \varphi) = \{d^*_{n+1}(x, \varphi): x \in E_n\}$. If we replace in (5) the groups E_n by $\tilde{E}_n$ we get the corresponding groups $\tilde{\Gamma}_n(\varphi)$ and $\tilde{H}^\nabla_n(\varphi)$ respectively. We clearly have the maps

$$\left.\begin{array}{l} \Gamma_n(\varphi) \subset \tilde{\Gamma}_n(\varphi), \text{ and} \\ H^\nabla_n(\varphi) \twoheadrightarrow \tilde{H}^\nabla_n(\varphi), \end{array}\right\} \tag{6}$$

which are injective and surjective respectively.

Remark. We have $E_n = \tilde{E}_n$ if the partial suspension E is injective on $[A_n, B_n \vee Y_2]_2$. Thus $\Gamma_n = \tilde{\Gamma}_n$ and $H^\nabla_n = \tilde{H}^\nabla_n$ in this case. Since $\mathfrak{X}$ is a good class this holds for $n \geqq 3$ by the assumption (1.1)(2). We say that the class $\mathfrak{X}$ **satisfies** $\Gamma = \tilde{\Gamma}$ **in degree** 2 if $\Gamma_2 = \tilde{\Gamma}_2$ and $H^\nabla_2 = \tilde{H}^\nabla_2$. (7)

Lemma. *$\tilde{H}_n^\nabla$ and $\tilde{\Gamma}_n$ are well-defined natural systems on the category* **TWIST**$_n(\mathfrak{X})$ $(n \geqq 2)$. (8)

Here we use the notation in (IV.3.1)(∗) for the natural systems in (5) above.

Proof of (8). Let $(\xi, \eta): X^{(n)} \to Y^{(n)}$, $(\xi', \eta'): X'^{(n)} \to X^{(n)}$, and $(\xi'', \eta''): Y^{(n)} \to Y'^{(n)}$ be maps in **TWIST**$_n(\mathfrak{X})$ which induce φ, φ', and φ'' respectively in **Coef**. Then we obtain the induced functions

$$(\xi', \eta')^*: \tilde{\Gamma}_n(\varphi) \to \tilde{\Gamma}_n(\varphi\varphi'), \qquad \beta \longmapsto (\xi'_n)^*(\beta, \varphi), \tag{9}$$

$$(\xi', \eta')^*: \tilde{H}_n^\nabla(\varphi) \to \tilde{H}_n^\nabla(\varphi\varphi'), \quad \{\alpha\} \longmapsto \{(E^{-1}\xi'_{n+1})^*(\alpha, \varphi)\}, \tag{10}$$

$$(\xi'', \eta'')_*: \tilde{\Gamma}_n(\varphi) \to \tilde{\Gamma}_n(\varphi''\varphi), \qquad \beta \longmapsto F_*(\beta), \tag{11}$$

$$(\xi'', \eta'')_*: \tilde{H}_n^\nabla(\varphi) \to \tilde{H}_n^\nabla(\varphi''\varphi), \quad \{\alpha\} \longmapsto \{F_*(\alpha)\}. \tag{12}$$

In (10) the map $E^{-1}\xi'_{n+1} = \bar{\xi}'_{n+1}$ is well defined by (1.1)(2). This shows that (9) and (10) are well defined. The map $F: Y_n \to Y'_n$ in (11), (12) is a twisted map associated to (ξ''_n, η''). The map in (11) does not depend on the choice of F since two such choices $F, \bar{F}$ satisfy $\bar{F} = F + \xi$, $\xi \in [\Sigma A_n, Y_{n-1}]$, see (V.3.12). Hence we have $(F + \xi)_*\beta = F\beta + (\xi, F)(\nabla\beta)$ with $\nabla\beta = E\beta_n = 0$, see (4). Also (12) does not depend on the choice of F since $(F + \xi)_*\alpha = F\alpha + (\xi, F)\nabla\alpha$ with $\nabla\alpha = 0$ by assumption on α. We leave it to the reader as an exercise to show that diagram (1.10) is natural with respect to the induced maps defined as in (9), (10) and (11), (12) respectively. □

(1.11) ***Definition of the action $\tilde{\Gamma}_n$.*** Let $(\xi, \eta): X^{(n+1)} \to Y^{(n+1)}$ be a morphism in **TWIST**$_{n+1}(\mathfrak{X})$ which induces φ in **Coef** and let $\alpha \in \tilde{\Gamma}_n(X^{(n)}, \varphi, Y^{(n)})$. Then we set

$$(\xi, \eta) + \alpha = (\xi, \eta + \alpha),$$

where $\eta + \alpha = \mu^*(\eta, \alpha)$ is determined by the cooperation $\mu: X_n \to X_n \vee \Sigma A_n$ which is a map in *Ho* **Fil(C)**. (Here we consider $X_{n-1} \subset X_n$ as a principal cofibration in the cofibration category **Fil(C)**, therefore μ is defined by (II.8.7).) ||

Remark. We can use the spectral sequence (III.4.18) in **Fil(C)** for the computation of the isotropy groups of the action $\tilde{\Gamma}_n +$, compare the method in (VI.5.16).

(1.12) **Lemma.** *$\tilde{\Gamma}_n +$ is a well-defined action on the functor λ:* **TWIST**$_{n+1}(\mathfrak{X}) \to$ **TWIST**$_n(\mathfrak{X})$ *and $\tilde{\Gamma}_n +$ has a linear distributivity law, $n \geqq 2$.*

Proof. We first check that $(\xi, \eta + \alpha)$ is again a morphism in **TWIST**$_{n+1}(\mathfrak{X})$. For this consider first diagram (1.6)(2) where we replace η by $\eta + \alpha$. We deduce from (1.6)(1) the formula

$$f^*_{n+1}(\eta+\alpha)=f^*_{n+1}\eta+d^*_{n+1}(\alpha,\varphi)=f^*_{n+1}\eta=(g_{n+1},j)\bar{\xi}_{n+1}. \tag{1}$$

Here we know $d^*_{n+1}(\alpha,\varphi)=0$ since $\alpha\in\tilde{\Gamma}_n(\varphi)$. Moreover, $\eta+\alpha$ is a map in **Complex**$/\overset{0}{\simeq}$ for which $\check{K}(\eta+\alpha)$ coincides with ξ in degree $\leq n$. In fact, $\eta+\alpha$ is a twisted map associated to $(\bar{\xi}_n+\beta_n,\eta_{n-1})$ where $\eta_{n-1}=\eta|X_{n-1}$. This follows since for $\alpha\in\tilde{E}_n$ we have β_n as in (1.10)(4). Now $\check{K}(\eta+\alpha)$ is given in degree n by

$$E(\bar{\xi}_n+\beta_n)=E\bar{\xi}_n=\xi_n \tag{2}$$

as follows from (II.12.5) and (II.11.8). This completes the proof that $(\xi,\eta+\alpha)$ is a well-defined morphism in $\mathbf{TWIST}_{n+1}(\mathfrak{X})$. Next we show that $\tilde{\Gamma}_n+$ has a linear distributivity law. For this we consider the composition

$$(\xi'',\eta''+\alpha'')(\xi,\eta+\alpha)=(\xi''\xi,\eta''\eta+\delta). \tag{3}$$

Here $(\xi'',\eta''):Y^{(n+1)}\to Y'^{(n+1)}$ induces φ'' in **Coef** and $\alpha''\in\tilde{\Gamma}_n(\varphi'')$. From (V.1.6) we deduce the formula

$$\delta=\xi^*_n(\alpha'',\varphi'')+\eta''_*(\alpha)+(\nabla\alpha)^*(\alpha'',\varphi'') \tag{4}$$

Here $\nabla\alpha=E\beta_n=0$ since $\alpha\in\tilde{E}_n$. Finally we show that $\tilde{\Gamma}_n+$ is an action on the functor λ, compare (IV.2.3). Clearly, we have by definition of λ in (1.8)(2)

$$\lambda(\xi,\eta+\alpha)=\lambda(\xi,\eta). \tag{5}$$

Now assume $\lambda(\xi,\eta)=\lambda(\xi',\eta')$. Then we have $\xi=\xi'$ and $\eta_{n-1}=\eta'_{n-1}$. Hence there is $\alpha\in[\Sigma A_n,Y_n]$ with $\eta+\alpha=\eta'$ in $Ho\,\mathbf{Fil}(\mathbf{C})$. Here we use the cofiber sequence in $\mathbf{Fil}(\mathbf{C})$, see (1.11). We have to show $\alpha\in\tilde{\Gamma}_n(\varphi)$. The assumption (1.1)(3) ($\varepsilon=0$) shows $\alpha\in E_g(\beta_n)$. As in (2) we see

$$E(\bar{\xi}_n+\beta_n)=\xi'_n=\xi_n=E\bar{\xi}_n \quad \text{on } Y_2 \tag{6}$$

and therefore $E\beta_n=0$ on Y_2, hence $\alpha\in\tilde{E}_n$. Moreover, we have for η and η' the formula (see (1.6)(2))

$$\begin{aligned} f^*_{n+1}\eta' &= (g_{n+1},j)(E^{-1}\xi'_{n+1})\\ &= (g_{n+1},j)(E^{-1}\xi_{n+1}) \\ &= f^*_{n+1}\eta \end{aligned} \tag{7}$$

Here we use the fact that by (1.1)(2) $E^{-1}\xi_{n+1}=\bar{\xi}_{n+1}$ is well defined for $n\geq 2$. Now (7) and (2) show $d^*_{n+1}(\alpha,\varphi)=0$. This completes the proof that $\alpha\in\tilde{\Gamma}_n(\varphi)$. □

(1.13) ***Definition of the obstruction.*** Let $(\xi,\eta):\lambda X^{(n+1)}=X^{(n)}\to Y^{(n)}=\lambda Y^{(n+1)}$ be a map in $\mathbf{TWIST}_n(\mathfrak{X})$, $\eta:X_{n-1}\to Y_{n-1}$. Then there is a twisted map $F:X_n\to Y_n$ associated to $(\bar{\xi}_n,\eta)$, compare (1.6)(2) where we replace n by $(n-1)$.

For F we have the diagram

$$\begin{array}{ccc} A_{n+1} & \xrightarrow[\bar{\xi}_{n+1}]{} & B_{n+1} \vee Y_2 \\ \downarrow{\scriptstyle f_{n+1}} & & \downarrow{\scriptstyle (g_{n+1}, j)} \\ X_n & \xrightarrow{F} & Y_n \end{array}, \tag{1}$$

where $\bar{\xi}_{n+1} = E^{-1}\xi_{n+1}$ is well defined by ξ_{n+1}, since $n \geqq 2$, see (1.1)(2). Diagram (1), however, needs not to be commutative. We define

$$\mathfrak{O}(F) = -(g_{n+1}, j)\bar{\xi}_{n+1} + F f_{n+1}. \tag{2}$$

We prove that this is an element in the kernel of ∇, see (1.10). Therefore it represents a coset ($X = X^{(n+1)}$, $Y = Y^{(n+1)}$)

$$\mathfrak{O}_{X,Y}(\xi, \eta) = \{\mathfrak{O}(F)\} \in \tilde{H}_n^{\nabla}(X^{(n)}, \varphi, Y^{(n)}). \tag{3}$$

This is the definition of the obstruction operator. ||

(1.14) **Lemma.** *The obstruction $\mathfrak{O}_{X,Y}(\xi, \eta)$ is well defined.*

Proof. First we check $\nabla\mathfrak{O}(F) = 0$. We have by (II.12.8)

$$\begin{aligned} \nabla(F f_{n+1}) &= -i_2 F f_{n+1} + (i_2 + i_1) F f_{n+1} \\ &= -i_2 F f_{n+1} + (\nabla_F, i_2 F)(i_2 + i_1) f_{n+1} \\ &= (\nabla_F, i_2 F)\nabla f_{n+1} \\ &= (E\bar{\xi}_n, i_2\varphi) d_{n+1}, \text{ see (V.3.12)(3)} \\ &= (\xi_n \odot \varphi) d_{n+1}, \text{ see (III.4.4)(1).} \end{aligned} \tag{1}$$

On the other hand, we have by (1.6)(1)

$$\nabla((g_{n+1}, j)\bar{\xi}_{n+1}) = (d_{n+1} \odot 1)\bar{\xi}_{n+1} \tag{2}$$

since $\bar{\xi}_{n+1}$ is a partial suspension by (1.1)(2). Since ξ is a twisted chain map we know that the partial suspension of (1) and (2), respectively, coincide on Y_2, hence (1) and (2) coincide since by (1.1)(2) the partial suspension is injective on $[A_{n+1}, \Sigma B_n \vee Y_2]_2$. This shows $\nabla\mathfrak{O}(F) = 0$ since ∇ is a homomorphism.

Next we consider the indeterminancy of the obstruction depending on the choice of $\bar{\xi}_n$ and F. Let G be a twisted map associated to $(\bar{\bar{\xi}}_n, \eta)$ with $E\bar{\bar{\xi}}_n = E\bar{\xi}_n = \xi_n$ on Y_2. Then there is an element $\alpha \in [\Sigma A_n, Y_n]$ with $F + \alpha = G$. As in (1.12)(5)(6) we see $\alpha \in \tilde{E}_n$. Now we get

$$\mathfrak{O}(G) = \mathfrak{O}(F + \alpha) = \mathfrak{O}(F) + (\alpha, F)\nabla f_{n+1} = \mathfrak{O}(F) + d_{n+1}^*(\alpha, \varphi). \tag{3}$$

This shows that the class $\{\mathfrak{O}(F)\}$ in $\tilde{H}_n^{\nabla}(\varphi)$ depends only on (ξ, η). Whence the obstruction (1.13)(3) is well defined. □

With the notation in (IV.4.10) we get

(1.15) **Lemma.** *The obstruction $\mathfrak{O}$ has the derivation property.*

Proof. Assume $(\xi,\eta):\lambda X^{(n+1)}\to\lambda Y^{(n+1)}$ and $(\xi',\eta'):\lambda Y^{(n+1)}\to\lambda Y'^{(n+1)}$ induce φ and φ' in **Coef** respectively. Let F and G be maps associated to $(\bar\xi_n,\eta)$ and $(\bar\xi',\eta')$ respectively. We have

$$\mathfrak{O}(GF) = -(g'_{n+1},j)(\xi'\xi)^-_{n+1} + GFf_{n+1}$$

with $(\xi'\xi)^-_{n+1} = (\bar\xi'_{n+1}, i_2\varphi')\bar\xi_{n+1}$. This shows

$$\begin{aligned}\mathfrak{O}(GF) = &-((g'_{n+1},j)\bar\xi'_{n+1}, j\varphi')\bar\xi_{n+1}\\ &+(Gg_{n+1}, j\varphi')\bar\xi_{n+1}\\ &-G(g_{n+1},j)\bar\xi_{n+1}\\ &+GFf_{n+1}.\end{aligned}$$

The second and third term cancel. Since $\bar\xi_{n+1}$ is a partial suspension we deduce

$$\mathfrak{O}(GF) = (\mathfrak{O}(G), j\varphi')\bar\xi_{n+1} + G_*\mathfrak{O}(F).$$

With (1.10)(10), (12) this proves the proposition of the lemma. □

In the next lemma we use the assumption (1.1)(4) on $\mathfrak{X}$.

(1.16) **Lemma.** $\mathfrak{O}$ *has the transitivity property.*
Compare (IV.4.10)(*d*).

Proof. Let $X = (A, f_{n+1}, X_n)$ be an object in $\mathbf{TWIST}_{n+1}(\mathfrak{X})$ and let $\{\alpha\}\in\tilde H_n^\nabla(\lambda X, 1, \lambda X)$. Then we get

$$X + \{\alpha\} = (A, f_{n+1} + \alpha, X_n)$$

with $\mathfrak{O}_{X+\{\alpha\},X}(1) = \{\alpha\}$. This is clear by (1.13). Since $\nabla\alpha = 0$ we see that (1.6)(1) is satisfied for $X + \{\alpha\}$. □

(1.17) **Lemma.** $\mathfrak{O}$ *has the obstruction property.*
Compare (IV.4.10)(*b*).

Proof. Assume $\mathfrak{O}_{X,Y}(\xi,\eta) = 0$. Then there is $\beta\in\tilde E_n$ with

$$\mathfrak{O}(F) = -(g_{n+1},j)\bar\xi_{n+1} + Ff_{n+1} = -(\beta,\varphi)d_{n+1} \tag{1}$$

Thus we get

$$\begin{aligned}\mathfrak{O}(F+\beta) &= -(g_{n+1},j)\bar\xi_{n+1} + (F+\beta)f_{n+1}\\ &= -(g_{n+1},j)\bar\xi_{n+1} + Ff_{n+1} + (\beta,\varphi)d_{n+1}\\ &= 0.\end{aligned} \tag{2}$$

Therefore $(\xi, F+\beta)$ is a map in $\mathbf{TWIST}_{n+1}(\mathfrak{X})$ with $\lambda(\xi, F+\beta) = (\xi,\eta)$. Here $F+\beta$ is a twisted map associated to $(\bar\xi_n + \beta_n, \eta)$ with β_n in (1.10)(4). Therefore the induced chain map in degree n is $E(\bar\xi_n + \beta_n) = E\bar\xi_n = \xi_n$, see (V.3.12)(3). Hence the map $(\xi, F+\beta)$ is well defined. □

(1.18) *Proof of* (1.9). By definition in (IV.4.10) and (IV.4.15) the theorem is a consequence of (1.10)(7), (1.12), (1.15), (1.16), and (1.17). □

As in (III.2.10) we have the following

(1.19) ***Naturality of the tower*** $\mathbf{TWIST}_*(\mathfrak{X})$ ***with respect to functors.*** Let $\alpha:\mathbf{C}\to\mathbf{K}$ be a based model functor, see (I.1.10). For a complex X in $\mathbf{C}$ we choose RMαX in **Fil(K)**. Then $RM\alpha X$ is a complex in $\mathbf{K}$. Assume that $\mathfrak{X}$ and $\mathfrak{X}_\alpha = RM\alpha(\mathfrak{X})$ are good classes of complexes in $\mathbf{C}$ and in $\mathbf{K}$ respectively. Then α induces a structure preserving map between towers of categories which carries the tower $\mathbf{TWIST}_*(\mathfrak{X})$ in $\mathbf{C}$ to the tower $\mathbf{TWIST}_*(\mathfrak{X}_\alpha)$ in $\mathbf{K}$. The maps induced by α on $\tilde{\Gamma}_n$ and $\tilde{H}_n^\nabla$, respectively, are defined by α_L in (II.6a.2). Hence we have for $n \geqq 2$ the commutative diagram of exact sequences (see (IV.4.13))

$$\begin{array}{ccccccc}
\tilde{\Gamma}_n & \xrightarrow{+} & \mathbf{TWIST}_{n+1}(\mathfrak{X}) & \xrightarrow{\lambda} & \mathbf{TWIST}_n(\mathfrak{X}) & \xrightarrow{\mathfrak{D}} & \tilde{H}_n^\nabla \\
\downarrow{\scriptstyle \alpha_L} & & \downarrow{\scriptstyle \alpha} & & \downarrow{\scriptstyle \alpha} & & \downarrow{\scriptstyle \alpha_L} \\
\tilde{\Gamma}_n & \xrightarrow{+} & \mathbf{TWIST}_{n+1}(\mathfrak{X}_\alpha) & \xrightarrow{\lambda} & \mathbf{TWIST}_n(\mathfrak{X}_\alpha) & \xrightarrow{\mathfrak{D}} & \tilde{H}_n^\nabla
\end{array}$$

Finally, we consider the special case of theorem (1.9) for the class of CW-complexes in (1.7).

(1.20) **Proposition.** *Let* $\mathfrak{X} = (\mathfrak{X}_0^1)^D$ *be the class of CW-complexes in* $\mathbf{CW}_0^D$. *Then we have a tower of categories which approximates* $\mathbf{CW}_0^D/\overset{0}{\simeq}$ *by the following equivalent exact sequences,* $n \geqq 2$.

$$\begin{array}{ccccccc}
Z^n\tilde{\Gamma}_n & \xrightarrow{+} & \mathbf{H}_{n+1} & \xrightarrow{\lambda} & \mathbf{H}_n & \xrightarrow{\mathfrak{D}} & H^{n+1}\tilde{\Gamma}_n^\nabla \\
{\scriptstyle N}\uparrow\cong & & \sim\uparrow{\scriptstyle N_{n+1}} & & \sim\uparrow{\scriptstyle N_n} & & \cong\uparrow{\scriptstyle N^\nabla} \\
\tilde{\Gamma}_n & \xrightarrow{+} & \mathbf{TWIST}_{n+1}(\mathfrak{X}) & \xrightarrow{\lambda} & \mathbf{TWIST}_n(\mathfrak{X}) & \xrightarrow{\mathfrak{D}} & \tilde{H}_n^\nabla
\end{array}$$

Here the natural systems $Z_n\tilde{\Gamma}_n$ *and* $H^{n+1}\tilde{\Gamma}_n^\nabla$ *are given by*

$$Z^n\tilde{\Gamma}_n(X,\varphi,Y) = \mathrm{Hom}_\varphi(\hat{C}_n X, \tilde{\Gamma}_n(Y,D)),$$
$$H^{n+1}\tilde{\Gamma}_n^\nabla(X,\varphi,Y) = \hat{H}^{n+1}(X^{n+1}, \varphi^*\tilde{\Gamma}_n(Y,D))$$

where X *and* Y *are objects in* $\mathbf{H}_n$ *and where* $\varphi:\pi_1 X \to \pi_1 Y$ *is a homomorphism.*

Proof of (1.20). We have the commutative diagram

$$\begin{array}{ccccc}
\pi_n X^{n-1} & \xrightarrow{i} & \pi_n(X^n) & \xrightarrow{\nabla} & \pi_n(\bigvee_{Z_n} S^n \vee X^n)_2 \\
 & & \downarrow{\scriptstyle j} & & \cong\downarrow \\
 & & \pi_n(X^n, X^{n-1}) & \xrightarrow[h]{} & \hat{C}_n X
\end{array}$$

which shows that

$$\text{kernel}(\nabla) = \text{kernel}(hj) = \tilde{\Gamma}_n(X, D).$$

Therefore we get the isomorphism N^∇ in (1.20). On the other hand, we get the isomorphism N for $n = 2$ by the last equation in (VI.2.9) which shows that $\tilde{E}_2 = \tilde{\Gamma}_2(X, D)$. □

(1.21) **Description of the obstruction operator by use of track addition.** We use the notation in (1.13). The map $f_{n+1}: \Sigma A'_{n+1} = A_{n+1} \to X_n$ is a twisted map associated to $\bar{d}_{n+1}: A'_{n+1} \to A_n \vee X_1$, see (1.6)(3). Hence we can find a track H_X as in the diagram ($n \geq 2$)

$$
\begin{array}{ccc}
& \Downarrow H_X & o \\
A'_{n+1} \xrightarrow{\bar{d}_{n+1}} & A_n \vee X_1 \xrightarrow{(f_n, j)} & X_{n-1}
\end{array}
\tag{1}
$$

Here $(f_n, j)\bar{d}_{n+1}$ is a map in **C**. The track H_X gives us the twisted map $f_{n+1} = C(\bar{d}_{n+1}, o, H_X, G_1)$ as in (V.2.4). In the same way the map g_{n+1} is a twisted map given by tracks H_Y, G_3. Now consider the following diagram associated to the diagram in (1.13) where $\varepsilon = 1$ for $n = 2$ and where $\varepsilon = 2$ for $n \geq 3$.

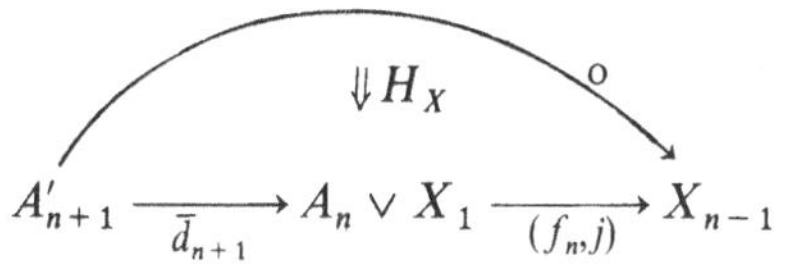

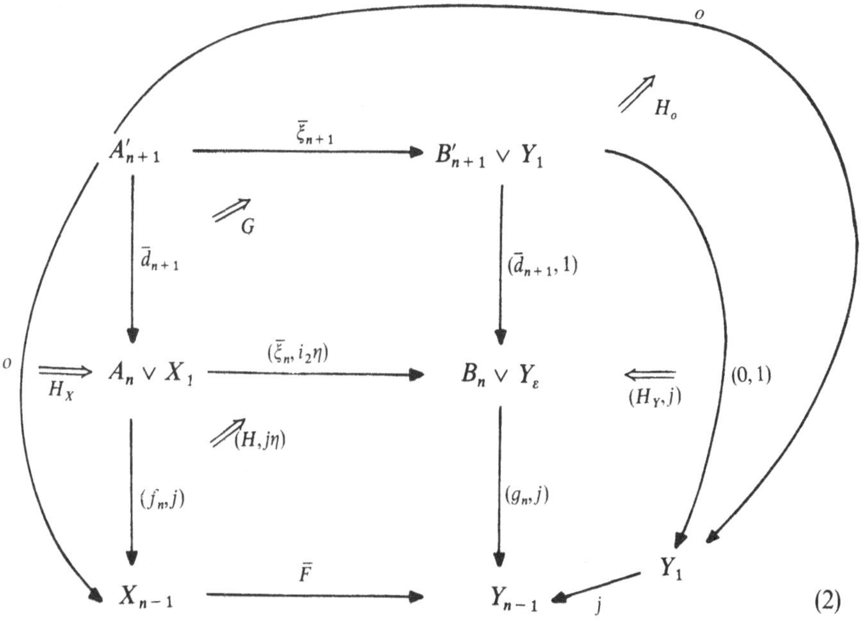

The maps $\eta:X_1 \to Y_1$ and $\bar{F}$ are restrictions of $F:X_n \to Y_n$. Since $\mathfrak{X}$ is a good class the map $\bar{\bar{\xi}}_{n+1}$ with $E\bar{\bar{\xi}}_{n+1} = \bar{\xi}_{n+1}$ exists. For $n \geqq 2$ the track H exists since F is a twisted map associated to $(\bar{\xi}_n, \bar{F})$. Since $\bar{\bar{\xi}}_{n+1}$ is trivial on Y_1 we know that there exists a track H_0.

Lemma. *Assume a track G as in (2) exists, $n \geqq 2$. Then any choice of H and a* **good choice** *of (G, H_0) yields, by track addition, the element*

$$\begin{aligned} &\alpha:\Sigma A'_{n+1} = A_{n+1} \to Y_{n-1} \subset Y_n \\ &\alpha = \bar{F}^* H_X + \bar{d}^*_{n+1}(H, j\eta) + (g_n, j)_* G - \bar{\bar{\xi}}^*_{n+1}(H_Y, j) + H_0 \end{aligned} \tag{3}$$

which represents the obstruction $\mathfrak{O}(F)$.

Remark. A track G in (2) exists for $n \geqq 3$ provided the partial suspension

$$E:[A'_{n+1}, B_n \vee Y_2]_2 \to [A_{n+1}, \Sigma B_n \vee Y_2]_2$$

is injective. (This, for example, is the case if $\mathfrak{X}$ is a very good class, see (2.2) below.) If, in addition, for $n = 2$ the map

$$[A'_{n+1}, B_n \vee Y_1]_2 \to [A'_{n+1}, B_n \vee Y_2]_2 \tag{4}$$

is injective then G also exists for $n = 2$.

Proof of lemma (3). Since A_{n+1} $(n \geqq 2)$ is a double suspension we have $\mathfrak{O}(F) = Ff_{n+1} - (g_{n+1}, j)\bar{\xi}_{n+1}$. The map Ff_{n+1} is represented by the addition of tracks in the diagram

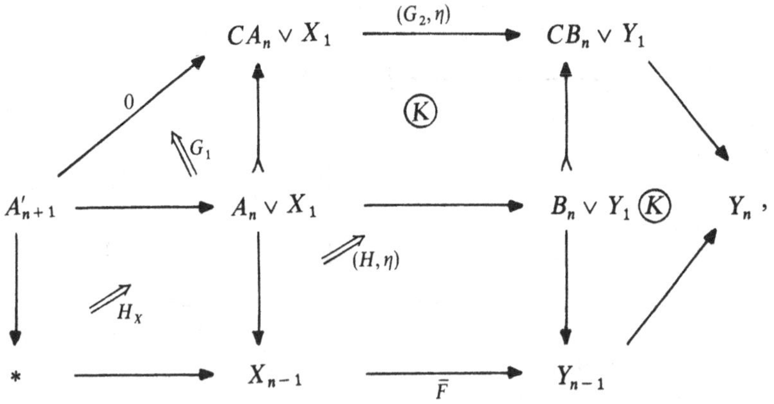

where Ⓚ denotes a commutative diagram, compare (V.2.3). On the other hand, $(g_{n+1}, j)\bar{\xi}_{n+1}$ is represented by the addition of tracks in the following diagram.

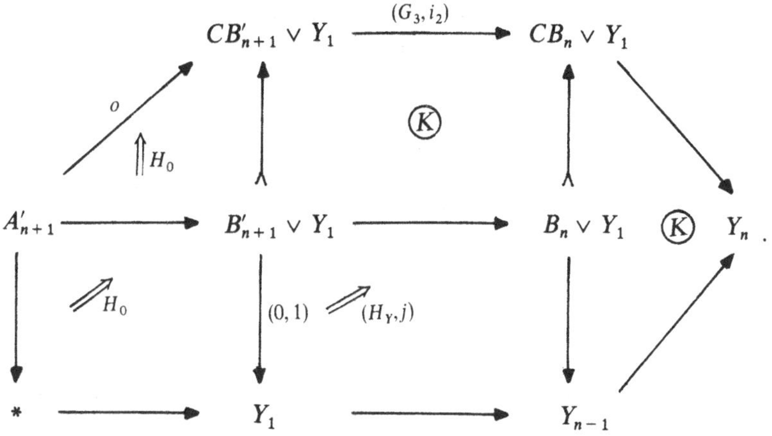

Here $-H_0$ is given by H_0. Now it is possible to choose G in (2) such that the following track addition represents the trivial homotopy class $0: \Sigma A'_{n+1} \to Y_\varepsilon$, (this defines a **good choice** of (G, H_0)).

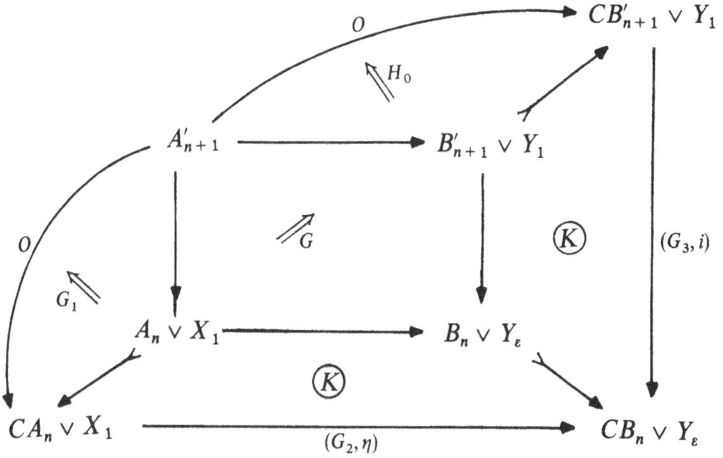

This implies the proposition in (3) when we consider the track addition corresponding to the sum of elements in $\mathfrak{O}(F)$ above. □

Lemma (3) above shows that the obstruction $\mathfrak{O}(F)$ is a kind of a 'Toda bracket'.

§ 2 Twisted maps and 1-homotopies

In this section we prove the theorem on the CW-tower in (VI.§ 5, § 6). We prove this result more generally in a cofibration category **C**. For this we introduce a homotopy relation, $\simeq$, on this category $\mathbf{TWIST}^c_{n+1}(\mathfrak{X})$ defined in § 1 above. Here $\mathfrak{X}$ is a 'very good class of complexes' in $\mathbf{C}$. Examples of such very good classes are described in § 3, in particular, the class $(\mathfrak{X}^1_0)^D$ of relative CW-complexes in $\mathbf{Top}^D$ is such a very good class. We describe an exact sequence for the functor $\lambda:\mathbf{TWIST}^c_{n+1}(\mathfrak{X})/\simeq \to \mathbf{TWIST}^c_n(\mathfrak{X})/\simeq$ which is isomorphic to the exact sequence for $\lambda:\mathbf{H}^c_{n+1}/\simeq \to \mathbf{H}^c_{n}/\simeq$ in case $\mathfrak{X}=(\mathfrak{X}^0_1)^D$.

Let $\mathbf{C}$ be a cofibration category with an initial object $*$. For the definition of a very good class of complexes in $\mathbf{C}$ we need the

(2.1) ***Definition.*** Let $A=\Sigma A'$ and $B=\Sigma B'$ be suspensions. We say that (A, B) is **E-stable** (with respect to the class $\mathfrak{X}$) if for all $Y\in\mathfrak{X}$ the partial suspension

$$E:[A', B' \vee Y]_2 \twoheadrightarrow [A, B \vee Y]_2$$

is surjective and if the iterated partial suspension

$$E^i:[A, B \vee Y]_2 \xrightarrow{\approx} [\Sigma^i A, \Sigma^i B \vee Y]_2$$

is bijective, $i \geqq 1$. ||

We assume that for $Y\in\mathfrak{X}$ also all skeleta Y_n are complexes in $\mathfrak{X}$, hence we can replace Y in (2.1) by Y_n for all $n\geqslant 0$

(2.2) ***Definition.*** Let $\mathfrak{X}$ be an admissible class of complexes as in (III.5a.5) with $X_0=*$ for $X\in\mathfrak{X}$. We say that $\mathfrak{X}$ is a **very good class** if for all $X, Y\in\mathfrak{X}$ with attaching maps $A_i\to X_{i-1}$ and $B_i\to Y_{i-1}$ respectively the following properties (1)...(4) are satisfied.

A_1 is a based object, $A_2=\Sigma A_2'$ is a suspension and $A_n=\Sigma A_n'=\Sigma^2 A_n''$ is a double suspension for $n\geqq 3$. (1)

The pair $(\Sigma A_2, \Sigma B_2)$ is E-stable and $(\Sigma^\varepsilon A_{m-\varepsilon}, B_m)$ is E-stable for $m\geqq 3$ and $\varepsilon\in\{-1,0,1\}$. Moreover, the pair $(A'_{m+1}, \Sigma A_{m-1})$ is injective on X_2 for $m\geqq 3$, see (III.5a.3). (2)

For $n\geqq 2$ and $\varepsilon\in\{-1,0,1\}$ the map

$$(\pi_g, j)_*:\pi_{1-\varepsilon}^{A_{n+\varepsilon}}(CB_n \vee Y_1, B_n \vee Y_1)\to \pi_{1-\varepsilon}^{A_{n+\varepsilon}}(Y_2, Y_{n-1}) \tag{3}$$

is surjective. Here $g:B_n\to Y_{n-1}$ is the attaching map of $C_g=Y_n$ and $j:Y_1\subset Y_{n-1}$ is the inclusion.

For $n\geqq 2$ all maps $f_{n+1}:A_{n+1}\to X_n$ are attaching maps of complexes in $\mathfrak{X}$. (4) ||

(2.3) ***Remark.*** A very good class is also a good class in the sense of (1.1).

(2.4) ***Example.*** The class $(\mathfrak{X}_0^1)^D$ of all CW-complexes in $\mathbf{CW}_0^D$ is a very good class of complexes in the cofibration category $\mathbf{Top}^D$, see (1.2). This again follows from the properties of CW-complexes described in (III.5.11).

We now introduce the notion of homotopy on the category $\mathbf{TWIST}_{n+1}(\mathfrak{X})$, for this compare also (III.4.5) where we define homotopies in the category $\mathbf{Chain}^{\vee}$.

(2.5) ***Definition.*** Let $\mathfrak{X}$ be very good and let $(\xi,\eta),(\xi',\eta'):X^{(n+1)}\to Y^{(n+1)}$ be maps in $\mathbf{TWIST}_{n+1}(\mathfrak{X})$ as in (1.6), $n\geqq 1$. Then $(\xi,\eta)\simeq(\xi',\eta')$ are **homotopic** if (a) and (b) hold.

(a) There is $\bar{\alpha}_{n+1}\in[\Sigma A_n, B_{n+1}\vee Y_1]_2$ such that

$$H:\eta+(g_{n+1},j)\bar{\alpha}_{n+1}\overset{1}{\simeq}\eta'$$

are 1-homotopic maps $X_n\to Y_n$, $j:Y_1\subset Y_n$. Here the action $+$ is defined by the cooperation $\mu:X_n\to X_n\vee\Sigma A_n$ as in (1.11). We set $\alpha_{n+1}=E\bar{\alpha}_{n+1}$.

(b) There are $\alpha_i\in[\Sigma^2 A_{i-1},\Sigma B_i\vee Y_1]_2$, $i\geqq n+2$, such that for $i\geqq n+1$

$$E\xi_i'-E\xi_i=(E\alpha_i\odot\varphi)(E^2 d_i)+(Ed_{i+1}\odot 1)\alpha_{i+1}$$

Here φ is the map in **Coef** induced by (ξ,η); by (a) we show that φ is also the map in **Coef** induced by (ξ',η'). The equation holds in $[\Sigma^2 A_i,\Sigma^2 B_{i+1}\vee Y_2]_2$.

We say that the pair (α,H) with $\alpha=(\alpha_i, i\geqq n+1)$ is a **homotopy** from (ξ,η) to (ξ',η'). ‖

(2.6) **Lemma.** (2.5) (a) *implies that there is a* 1-*homotopy* $H_1:i_g\eta\simeq i_g\eta'$ *of maps* $X_n\to Y_{n+1}=C_g$, $i_g:Y_n\subset Y_{n+1}$.

This proves that (ξ',η') induces φ in (2.5)(b).

Proof of (2.6). Let $H_0=H|IX_{n-1}$ be the restriction of the homotopy H. Then the difference

$$d(\eta+(g_{n+1},j)\bar{\alpha}_{n+1},H_0,\eta')=0 \tag{1}$$

is trivial since H exists, see (II.8.13). By (II.13.9) we get

$$d(\eta,H_0,\eta')=(g_{n+1},j)\bar{\alpha}_{n+1}. \tag{2}$$

This shows that

$$d(i_g\eta,i_gH_0,i_g\eta')=i_g(g_{n+1},j)\bar{\alpha}_{n+1}=0. \tag{3}$$

Hence a homotopy H_1 (which extends H_0) exists. Clearly, H_1 is a 1-homotopy since H_0 is one. We can choose H_1 as a twisted map associated to $(C_w = IX_n)$

$$\begin{array}{ccc} \Sigma A_n & \xrightarrow{\bar{\alpha}_{n+1}} & B_{n+1} \vee Y_1 \\ {\scriptstyle w}\downarrow & & \downarrow{\scriptstyle (g_{n+1}, j)} \\ I^{\cdot} X_n & \xrightarrow[(\eta, H_0, \eta')]{} & Y_n \end{array} \tag{4}$$

□

One can readily check that homotopy in (2.5) is a natural equivalence relation and that the functors in (1.8) induce functors

(2.7)
$$\begin{array}{ccccc} \mathbf{Complex}(\mathfrak{X})/\overset{1}{\simeq} & \xrightarrow[r]{} & \mathbf{TWIST}^c_{n+1}(\mathfrak{X})/\simeq & \xrightarrow[\lambda]{} & \mathbf{TWIST}^c_n(\mathfrak{X})/\simeq \\ & & & & \downarrow{\scriptstyle \check{K}} \\ & & & \mathbf{Coef} \xleftarrow{\;C\;} & \mathbf{Chain}^{\vee}(\mathfrak{X})/\simeq \end{array}$$

between homotopy categories. The following 'tower theorem' is the precise analogue of theorem (VI.5.11) and (VI.6.2) in a cofibration category.

(2.8) **Theorem.** *Let $\mathfrak{X}$ be a very good class of complexes and let $n \geqq 2$. Then we have a commutative diagram with exact rows*

$$\begin{array}{ccccccc} \tilde{\Gamma}_n & \xrightarrow{+} & \mathbf{TWIST}^c_{n+1}(\mathfrak{X}) & \xrightarrow{\lambda} & \mathbf{TWIST}^c_n(\mathfrak{X}) & \xrightarrow{\mathfrak{D}} & H^{n+1}_{-1}\tilde{\Gamma} \\ \downarrow{\scriptstyle q} & & \downarrow{\scriptstyle p} & & \downarrow{\scriptstyle p} & & \downarrow{\scriptstyle 1} \\ H^n_0\tilde{\Gamma} & \xrightarrow{+} & \mathbf{TWIST}^c_{n+1}(\mathfrak{X})/\simeq & \xrightarrow{\lambda} & \mathbf{TWIST}^c_n(\mathfrak{X})/\simeq & \xrightarrow{\mathfrak{D}} & H^{n+1}_{-1}\tilde{\Gamma} \end{array}$$

Here $H^{n+1}_{-1}\tilde{\Gamma} \subset \tilde{H}^{\nabla}_n$ is a natural subsystem of $\tilde{H}^{\nabla}_n$ in (1.10)(5) *and the top row is given by the corresponding exact sequence in* (1.9).

Clearly, this result yields a tower of categories $\mathbf{TWIST}^c_*(\mathfrak{X})/\simeq$ which approximates the category $\mathbf{Complex}(\mathfrak{X})/\overset{1}{\simeq}$. The tower of categories in (VI.6.2) for $(\mathbf{CW}^D_0)/\simeq$ is exactly the special case with $\mathfrak{X} = (\mathfrak{X}^1_0)^D$, see (2.4). Clearly all properties of the tower as described in (VI.6.5) and (VI.6.12) hold in the same way for the tower $\mathbf{TWIST}^c_*(\mathfrak{X})/\simeq$.

The natural systems of abelian groups $H^p_k\tilde{\Gamma}$ in (2.8) are defined on the homotopy category $\mathbf{TWIST}^c_n(\mathfrak{X})/\simeq$. Let $X^{(n)} = (A, f_n, X_{n-1})$ and $Y^{(n)} = (B, g_n, Y_{n-1})$ be objects in $\mathbf{TWIST}^c_n(\mathfrak{X})$ and let $\varphi: (X_2, X_1) \to (Y_2, Y_1)$ be a map in **Coef**. Then we define $H^p_k\tilde{\Gamma}(X^{(n)}, \varphi, Y^{(n)})$ by use of the left-hand column in the following diagram.

(2.9)

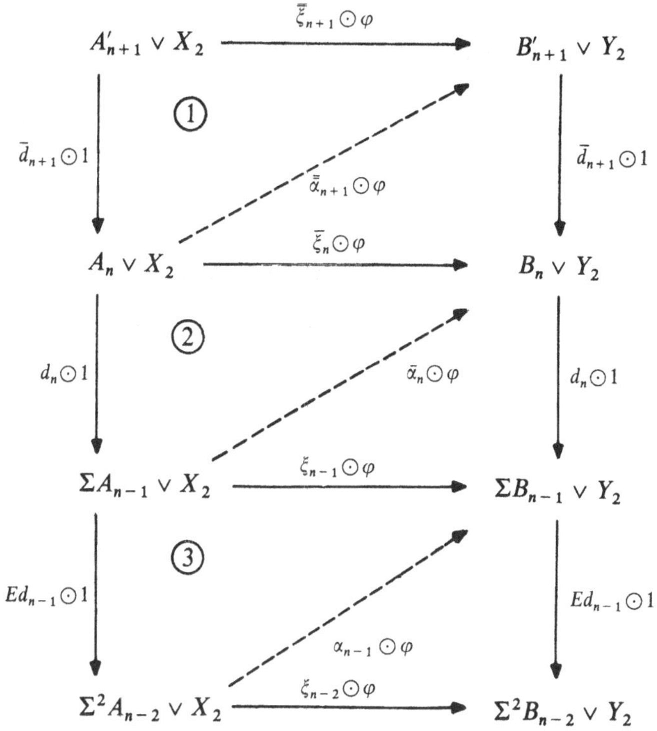

In ③ we set $A_0 = * = B_0$ for $n = 2$. This is a diagram in $Ho(\mathbf{C})$ which as well can be considered as a diagram in the category **Wedge** in (III.4.4). The boundaries d_n and d_{n-1} are given by the twisted chain complexes A and B respectively. Moreover, we obtain $\bar{d}_{n+1}$ for $n \geqq 3$ by stability of (A'_{n+1}, A_n) in (2.2)(2), $(\varepsilon = -1, B_m = A_m)$ and by the condition

$$E\bar{d}_{n+1} = d_{n+1} \quad \text{on } X_1. \tag{1}$$

Hence $\bar{d}_{n+1}$ is well defined by the condition in (1) for $n \geqq 3$. We get $\bar{d}_{n+1}$ for $n \geqq 2$ as well by the cocycle condition for $X^{(n)}$, see (1.6)(3). (For $n = 2$, however, the element $\bar{d}_3$ is not unique and needs not to be a partial suspension.) Moreover, the cocycle condition shows for $n \geqq 3$

$$(d_n \odot 1)\bar{d}_{n+1} = \nabla(f_n, j)\bar{d}_{n+1} = 0 \quad \text{on } X_{n-1}.$$

This implies $(d_n \odot 1)\bar{d}_{n+1} = 0$ on X_2 by the assumption that $(A'_{n+1}, \Sigma A_{n-1})$ is injective on X_2, see (2.2)(2).

Moreover, (2.2)(2) shows that for a φ-chain map $\xi: \Sigma A \to \Sigma B$ we get $\bar{\bar{\xi}}_{n+1}$ and $\bar{\xi}_n$ with $E^2\bar{\bar{\xi}}_{n+1} = E\bar{\xi}_{n+1} = \xi_{n+1}$ and $E\bar{\xi}_n = \xi_n$, respectively (as in (2.9)), where $\bar{\xi}_n$ is unique for $n \geqq 3$. The stability condition also shows that the diagram of unbroken arrows (2.9) commutes for $n \geqq 3$. In fact, the double partial suspension of this diagram commutes since ξ is a φ-chain map.

Next we consider a chain homotopy $\alpha: \xi \simeq \xi'$ for φ-chain maps $\xi, \xi': A \to B$. Hence elements $\alpha_i \in [\Sigma^2 A_{i-1}, \Sigma B_i \vee Y_1]_2$ are given such that (2.5)(b) holds on Y_2. Now (2.2)(2) ($\varepsilon = 1$) yields maps $\bar{\alpha}_n$ and $\bar{\bar{\alpha}}_{n+1}$ as in (2.9), ($n \geqq 2$), with $E\bar{\alpha}_n = \alpha_n$ and $E^2\bar{\bar{\alpha}}_{n+1} = E\bar{\alpha}_{n+1} = \alpha_{n+1}$. Moreover, we get for $n \geqq 3$ the equations

$$\xi'_{n-1} - \xi_{n-1} = (d_n \odot 1)\bar{\alpha}_n + (\alpha_n \odot \varphi)Ed_{n-1}, \tag{2}$$

$$\bar{\xi}'_n - \bar{\xi}_n = (\bar{d}_{n+1} \odot 1)\bar{\bar{\alpha}}_{n+1} + (\bar{\alpha}_n \odot \varphi)d_n, \tag{3}$$

in $[\Sigma A_{n-1}, \Sigma B_{n-1} \vee Y_2]_2$ and $[A_n, B_n \vee Y_2]_2$, respectively, since the partial suspension of (2) and the double partial suspension of (3) are equations as in (2.5)(b). Here we use the stability of $(\Sigma A_2, \Sigma B_2)$ for (2) with $n = 3$, see (2.2)(2).

The desuspension of A in the left-hand column of (2.9) and the attaching map g_n in $Y^{(n)}$ yield the following diagram where C_g is the mapping cone of $g: B_n \to Y_{n-1}$ with $g \in g_n$; let $p + k = n \geqq 2$ and $k \geqq -1$.

(2.10)
$$\begin{array}{ccccc}
E_n & \subset & \tilde{E}_n & \subset & [\Sigma^{k+2} A_{p-1}, C_g] \\
\downarrow & & \downarrow & & \downarrow \delta^{p-1}_{k+1} \\
E_n & \subset & \tilde{E}_n & \subset & [\Sigma^{k+1} A_p, C_g] \\
 & & & & \downarrow \delta^p_k \\
 & & & & [\Sigma^k A_{p+1}, C_g]
\end{array}$$

By the inclusion $j: Y_2 \subset Y_n = C_g$ we obtain the coboundary

$$\delta^p_k(\beta) = (E^k d_{p+1})^*(\beta, j\varphi) \tag{1}$$

as in (III.4.8). For $k = -1$, $p + 1 = n + 2 \geqq 4$, the object $\Sigma^{-1} A_{p+1} = A'_{p+1}$ is the desuspension of A_{p+1} in (2.2)(1); moreover, $E^{-1} d_{p+1} = \bar{d}_{p+1}$ is well defined by (2.9)(1). Since $\bar{d}_{p+1}$ is a partial suspension we obtain the homomorphism δ^{n+1}_{-1}, $n \geqq 2$, by (1).

The groups E_n and $\tilde{E}_n$ in (2.10) are defined by (1.10)(3) and (1.10)(4). Since the composition of twisted maps is twisted we see that diagram (2.10) commutes. Therefore we can define for $p + k = n \geqq 2$, $k \geqq -1$ the abelian groups

$$H^p_k\Gamma(\varphi) = H^p_k\Gamma(X^{(n)}, \varphi, Y^{(n)}) = \frac{E_n \cap \operatorname{kernel}(\delta^p_k)}{\delta^{p-1}_{k+1}(E_n)}. \tag{2}$$

When we replace E_n in this formula by $\tilde{E}_n$ we get $H^p_k\tilde{\Gamma}(\varphi)$. These groups are subquotients of the abelian group $[\Sigma^{k+1} A_p, C_g]$, see (2.2)(1). For the twisted cohomology in (III.4.8) we have the canonical maps

$$H_k^p\Gamma(\varphi)\to H_k^p\tilde{\Gamma}(\varphi)\to H_k^p(EA,\varphi) \tag{3}$$

which is the identity on cocycles. Moreover, for $\Gamma_n(\varphi)$ and $\tilde{\Gamma}_n(\varphi)$ in (1.10)(5) we have the quotient maps

$$\Gamma_n(\varphi)\twoheadrightarrow H_0^n\Gamma(\varphi),\quad \tilde{\Gamma}_n(\varphi)\twoheadrightarrow H_0^n\tilde{\Gamma}(\varphi). \tag{4}$$

We point out that by (1.10)(7) we have

$$H_{-1}^{n+1}\Gamma(\varphi)=H_{-1}^{n+1}\tilde{\Gamma}(\varphi),\quad H_0^n\Gamma(\varphi)=H_0^n\tilde{\Gamma}(\varphi)\quad \text{for } n\geqq 3. \tag{5}$$

We say that the class $\mathfrak{X}$ **satisfies** $\Gamma=\tilde{\Gamma}$ **in degree** 2 if (5) holds also for $n=2$. By (2) the natural systems of abelian groups in (2.8) are defined. The induced maps are given as follows where we use the notation in the proof of (1.10) (8). The homomorphism

$$(\xi',\eta')^*:H_k^p\tilde{\Gamma}(\varphi)\to H_k^p\tilde{\Gamma}(\varphi\varphi') \tag{6}$$

carries the class $\{\alpha\}$ to the class $\{(E^k\xi_p)^*\,(\alpha,j\varphi)\}$. Here the map $E^{-1}\xi'_p=\bar{\xi}'_p$ $(k=-1,\ p=n+1)$ is well defined as in (2.9). The homomorphism

$$(\xi'',\eta'')_*:H_k^p\tilde{\Gamma}(\varphi)\to H_k^p\tilde{\Gamma}(\varphi''\varphi) \tag{7}$$

carries the class $\{\alpha\}$ to $\{F_*\alpha\}$ where F is the map in (1.10)(11). One can check:

Lemma. *The induced maps* (6) *and* (7) *depend only on the homotopy class of* (ξ',η') *and* (ξ'',η''), *respectively, in* $\mathbf{TWIST}_n(\mathfrak{X})/\simeq$ *and hence* $H_k^p\tilde{\Gamma}$ *and* $H_k^p\Gamma$ *are well defined natural systems on this category.* (8)

For the proof we use the existence of $\bar{\alpha}$ and $\bar{\bar{\alpha}}$ in (2.9) and we use (2.9)(2), (3) for $(\xi',\eta')^*$. Moreover, we use (2.6) and the naturality of the functional suspension for twisted maps for $(\xi'',\eta'')_*$. In fact, diagram (2.10) is natural with respect to maps $(\xi'',\eta'')_*$ and $(\xi',\eta')^*$ respectively.

Next we prove theorem (2.8) by the following lemmas.

(2.11) **Lemma.** *The top row of* (2.8) *is an exact 'subsequence' of the corresponding exact sequence in* (1.9).

Proof. Let X and Y in (1.13) be objects in $\mathbf{TWIST}_{n+1}^c(\mathfrak{X})$. For $\{\mathfrak{O}(F)\}\in\tilde{H}_n^\nabla(\varphi)$ we show $\{\mathfrak{O}(F)\}\in H_{-1}^{n+1}\tilde{\Gamma}(\varphi)\subset\tilde{H}_n^\nabla(\varphi)$. Both groups are subquotients of $[A_{n+1},C_g]$, $C_g=Y_n$. We have to show (1) and (2):

$$\mathfrak{O}(F)\in\tilde{E}_n \tag{1}$$

$$\mathfrak{O}(F)\in\text{kernel } \delta_{-1}^{n+1}\quad\text{or equivalently } \bar{d}_{n+2}^*(\theta(F),\varphi)=0. \tag{2}$$

We first check (2). Since $\bar{d}_{n+2}$ is a partial suspension we get

$$(\mathfrak{D}(F),\varphi)\bar{d}_{n+2} = -((g_{n+1},j)\bar{\xi}_{n+1},\varphi)\bar{d}_{n+2} + (Ff_{n+1},\varphi)\bar{d}_{n+2}. \tag{3}$$

Commutativity of ① in (2.9) yields

$$\begin{aligned} &-((g_{n+1},j)\bar{\xi}_{n+1},\varphi)\bar{d}_{n+2} = -(g_{n+1},j)(\bar{\xi}_{n+1}\odot\varphi)\bar{d}_{n+2} = \\ &-(g_{n+1},j)(\bar{d}_{n+2}\odot 1)\bar{\bar{\xi}}_{n+2} \end{aligned} \tag{4}$$

On the other hand,

$$(Ff_{n+1},\varphi)\bar{d}_{n+2} = F(f_{n+1},j)\bar{d}_{n+2}. \tag{5}$$

We know by the cocycle condition

$$(f_{n+1},j)\bar{d}_{n+2} = 0, \quad (g_{n+1},j)\bar{d}_{n+2} = 0. \tag{6}$$

Therefore (4) and (5) and thus (3) are trivial and hence (2) is proved.

Next we check (1). Recall the definition of the functional suspension E_g, $g \in g_n$, for $(C_g, Y_{n-1}) = (Y_n, Y_{n-1})$. We know that $\bar{\xi}_{n+1} = E\bar{\bar{\xi}}_{n+1}$ is a partial suspension on Y_1. The attaching map f_{n+1} is a twisted map associated to $\bar{d}_{n+1}$, see (2.9)(1), and the map F in (1.13) is a twisted map associated to $(\bar{\xi}_n,\eta)$. This shows

$$(g_{n+1},j)\bar{\xi}_{n+1} \in E_g((\bar{d}_{n+1}\odot j)\bar{\bar{\xi}}_{n+1}) \text{ and} \tag{7}$$

$$Ff_{n+1} \in E_g((\bar{\xi}_n \odot j\varphi)\bar{d}_{n+1}). \tag{8}$$

Here we use the inclusion $j: Y_1 \subset Y_{n-1}$. By diagram ② in (2.9) (with n replaced by $n+1$) we get

$$\left.\begin{aligned} &E\beta_n = 0 \text{ on } Y_2 \quad \text{for} \\ &\beta_n = -(\bar{d}_{n+1}\odot j)\bar{\bar{\xi}}_{n+1} + (\bar{\xi}_n \odot j\varphi)\bar{d}_{n+1} \end{aligned}\right\} \tag{9}$$

Moreover (7), (8) and (9) show $\mathfrak{D}(F) \in E_g(\beta_n)$ and hence by (9) condition (1) is satisfied.

Now exactness of the sequences in (1.9) shows that the obstruction in the top row of (2.8) satisfies the obstruction property and the derivation property, see (IV.4.10). We still have to check the transitivity property of the obstruction. For X and $\{\alpha\} \in H^{n+1}_{-1}\tilde{\Gamma}(X,1,X)$ we get the object $X + \{\alpha\}$ as in (1.16). It is clear that $X + \{\alpha\}$ satisfies the cocycle condition since α is a cocycle and since X satisfies the cocycle condition. □

By lemma (2.11) we see that the top row in (2.8) is an exact sequence. Next we consider the bottom row of (2.8). All maps and operators in the bottom row are determined by use of the top row and the quotient functors. We have to show that the bottom row is well defined and that exactness holds.

(2.12) **Lemma.** $\tilde{\Gamma}_n +$ *induces an action of* $H^n_0\tilde{\Gamma}$ *on the functor* λ *between homotopy categories in the bottom row of* (2.8).

Proof of (2.12). Let (ξ,η), $(\xi',\eta'):X\to Y$ be maps in $\mathbf{TWIST}^c_{n+1}(\mathfrak{X})$, η,η': $X_n\to Y_n$, and let $\beta\in\tilde{\Gamma}_n(\lambda X,\varphi,\lambda Y)$ where φ is the map in **Coef** induced by (ξ,η). We have to show

$$p(\beta)=0 \text{ in } \tilde{H}^n_0\Gamma(\varphi) \text{ implies } (\xi,\eta)+\beta\simeq(\xi,\eta) \tag{1}$$

where p is the quotient map in (2.10)(4), and

$$\lambda(\xi,\eta)\simeq\lambda(\xi',\eta')\Rightarrow\exists\beta \quad \text{with } (\xi,\eta)+\beta\simeq(\xi',\eta') \tag{2}$$

We first check (1). The assumption $p(\beta)=0$ is equivalent to the existence of an element $\beta'\in\tilde{E}_n\subset[\Sigma^2A_{n-1},Y_n]$ such that β is the composite map

$$\beta:\Sigma A_n \xrightarrow[Ed_n]{} \Sigma^2A_{n-1}\vee X_1 \xrightarrow[(\beta',j\varphi)]{} Y_n \tag{3}$$

Let H be the trivial homotopy $H:\eta_{n-1}\simeq\eta_{n-1}$, with $\eta_{n-1}=\eta|X_{n-1}$. Then we have by (II.13.9) the equation

$$d(\eta,H+\beta',\eta+\beta)=d(\eta,H,\eta)+\beta-(\beta',j\varphi)Ed_n \tag{4}$$

Since, clearly, $d(\eta,H,\eta)=0$ we see by (3) that the difference in (4) is trivial. Hence there exists a 1-homotopy

$$\bar{H}:\eta\overset{1}{\simeq}\eta+\beta \text{ which extends } H+\beta'. \tag{5}$$

Now the definition in (2.5) shows that with $\alpha=0$ we have a homotopy $(\alpha,\bar{H}):(\xi,\eta)\simeq(\xi,\eta+\beta)$. This completes the proof of (1).

Next we proof (2). Assume we have a homotopy $(H,\alpha):\lambda(\xi,\eta)=(\xi,\eta_{n-1})\simeq\lambda(\xi',\eta')=(\xi',\eta'_{n-1})$ in $\mathbf{TWIST}^c_n(\mathfrak{X})$, compare (2.5). We have to construct an element $\beta\in\tilde{\Gamma}_n(\varphi)$ with $(\xi,\eta+\beta)\simeq(\xi',\eta')$ in $\mathbf{TWIST}^c_{n+1}(\mathfrak{X})$. By (H,α) we get a 1-homotopy

$$H_1:i_g\eta_{n-1}\overset{1}{\simeq}i_g\eta'_{n-1},\quad H_1:IX_{n-1}\to Y_n \tag{6}$$

as in (2.6), where we replace n by $n-1$. By (2.5)(b) we have for $i\geqq n$ the equation

$$E\xi'_i-E\xi_i=(E\alpha_i\odot\varphi)(E^2d_i)+(Ed_{i+1}\odot 1)\alpha_{i+1} \tag{7}$$

in $[\Sigma^2A_i,\Sigma^2B_i\vee Y_2]_2$. By (2.2)(2) we can find $\bar{\alpha}_{n+1}$ with

$$E\bar{\alpha}_{n+1}=\alpha_{n+1},\quad \bar{\alpha}_{n+1}\in[\Sigma A_n,B_{n+1}\vee Y_1]_2 \tag{8}$$

We define β in (2) by

$$\beta=d(\eta,H_1,\eta')-(g_{n+1},j)\bar{\alpha}_{n+1} \tag{9}$$

This is an element in $[\Sigma A_n,Y_n]$ defined by $\bar{\alpha}_{n+1}$ in (8) and by H_1 in (6). We have to check that there is a homotopy

$$(\alpha,G):(\xi,\eta+\beta)\simeq(\xi',\eta') \tag{10}$$

and that

$$\beta \in \tilde{\Gamma}_n(\varphi) \tag{11}$$

Then the proof of (2) (and whence the proof of (2.12)) is complete.

We first prove (10). We define (α, G) by α in (7) and by a 1-homotopy

$$G:(\eta + \beta) + (g_{n+1}, j)\bar{\alpha}_{n+1} \simeq \eta'. \tag{12}$$

This 1-homotopy exists by definition of β in (9), see (II.8.15); the homotopy G is an extension of H_1. Clearly (7) and (12) show that (α, G) is a well-defined homotopy as in (2.5).

Now we prove (11). Condition (11) is equivalent to

$$\beta \in \tilde{E}_n \subset [\Sigma A_n, Y_n] \tag{13}$$

and

$$(\beta, \varphi)d_{n+1} = 0. \tag{14}$$

We first consider (14). For $\Delta = d(\eta, H_1, \eta')$ in (9) we have a homotopy $\eta + \Delta \simeq \eta'$ which extends H_1. This shows

$$\eta' f_{n+1} = (\eta + \Delta) f_{n+1} = \eta f_{n+1} + (\Delta, \varphi)d_{n+1} \tag{15}$$

by (1.6)(1). On the other hand, (1.6)(2) yields

$$\eta f_{n+1} = (g_{n+1}, j)\bar{\xi}_{n+1}, \quad \eta' f_{n+1} = (g_{n+1}, j)\bar{\xi}'_{n+1}. \tag{16}$$

Now (15) and (16) imply

$$(\Delta, \varphi)d_{n+1} = (g_{n+1}, j)(\bar{\xi}'_{n+1} - \bar{\xi}_{n+1}) \tag{17}$$

Since d_{n+1} is a partial suspension we derive from (9)

$$\begin{aligned} (\beta, \varphi)d_{n+1} &= (\Delta, \varphi)d_{n+1} - (g_{n+1}, j)(\bar{\alpha}_{n+1} \odot \varphi)d_{n+1} \\ &= (g_{n+1}, j)(\bar{\xi}'_{n+1} - \bar{\xi}_{n+1} - (\bar{\alpha}_{n+1} \odot \varphi)d_{n+1}) \qquad (18) \\ &= (g_{n+1}, j)(\bar{d}_{n+2} \odot 1)\bar{\bar{\alpha}}_{n+2} \qquad (19) \end{aligned}$$

Here (19) follows from (2.9)(3) (where we replace n by $n+1$). Now (19) is the trivial element since $(g_{n+1}, j)\bar{d}_{n+2} = 0$. Here we use the assumption that Y satisfies the cocycle condition. This completes the proof of (14).

We now prove (13). The map $w:\Sigma A_n \to I^{\cdot}X_n$ with $I^{\cdot}X_n = X_n \cup IX_{n-1} \cup X_n$ is twisted by (II.13.7). Also the map

$$\bar{M} = (\eta, H_1, \eta'):(I^{\cdot}X_n, IX_{n-2}) \to (Y_n, Y_{n-1}) \tag{20}$$

is a twisted map between mapping cones. This follows since H_1 is twisted by construction in (2.6)(4) and since η and η' are twisted. Since $\lambda(\xi, \eta) = (\xi, \eta_{n-1})$ we know that η is associated to $(\bar{\xi}_n, \eta_{n-1})$, similarly η' is associated to $(\bar{\xi}'_n, \eta'_{n-1})$. Now it is clear that the composition

$$\Delta = d(\eta, H_1, \eta') = \bar{M}w \tag{21}$$

is twisted with $\Delta \in E_g(\delta)$ where δ is obtained 'by ξ in (II.13.8)' and by the map

$$(\bar{\xi}_n, \bar{\alpha}_n, \bar{\xi}'_n): A_n \vee \Sigma A_{n-1} \vee A_n \to B_n \vee Y_1 \tag{22}$$

which is associated to $\bar{M}$. Therefore we get

$$\delta = -\bar{\xi}_n + (\bar{\nabla} f_n)(-\alpha_n, j\varphi) + \bar{\xi}'_n \tag{23}$$

where φ is the map in **Coef** induced by (ξ, η) (and by (ξ', η')). We have $\bar{\nabla} f_n = d_n$ on Y_2, compare (II.13.8) and (1.6)(1). On the other hand, we have by the cocycle condition for g_{n+1}

$$(g_{n+1}, j)\bar{\alpha}_{n+1} \in E_g((\bar{d}_{n+1} \odot j)\bar{\bar{\alpha}}_{n+1}) \tag{24}$$

Since E_g is a homomorphism we deduce from (23) and (24)

$$\beta = \Delta - (g_{n+1}, j)\alpha_{n+1} \in E_g(\delta - (\bar{d}_{n+1} \odot j)\bar{\bar{\alpha}}_{n+1}) \tag{25}$$

We know by (2.9)(2)

$$E(\delta - (\bar{d}_{n+1} \odot j)\bar{\bar{\alpha}}_{n+1}) = 0 \quad \text{on } Y_2 \tag{26}$$

This shows by (25) that $\beta \in \tilde{E}_n$ and (13) is proved. □

(2.13) **Lemma.** *The obstruction* $\mathfrak{O}$ *in the bottom row of* (2.8) *is well defined.*

Proof. We have to show that the element

$$\{\mathfrak{O}(F)\} = \mathfrak{O}_{X,Y}(\xi, \eta) \in H^{n+1}_{-1}\tilde{\Gamma}(\varphi) \tag{1}$$

in the proof of (2.11) depends only on the homotopy class of $(\xi, \eta): \lambda X \to \lambda Y$ in $\mathbf{TWIST}^c_n(\mathfrak{X})/\simeq$, see (1.13). Here X and Y are $(n+1)$-systems which satisfy the cocycle condition. Now let $(H, \alpha): (\xi, \eta) \simeq (\xi', \eta')$ be a homotopy in $\mathbf{TWIST}_n(\mathfrak{X})$ and let F and G be maps associated to $(\bar{\xi}_n, \eta)$ and $(\bar{\xi}'_n, \eta')$ respectively. We have to show

$$\{\mathfrak{O}(F)\} = \{\mathfrak{O}(G)\} \quad \text{in } H^{n+1}_{-1}\tilde{\Gamma}(\varphi). \tag{2}$$

The n-skeleta X_n, Y_n of X and Y, respectively, are objects in $\mathfrak{X}$. Therefore we obtain by the functor r_{n+1} in (1.8) maps

$$U, V: r_{n+1}X_n = R \to r_{n+1}Y_n = Q \tag{3}$$

in $\mathbf{TWIST}^c_{n+1}(\mathfrak{X})$ by $U = r_{n+1}(F + g_{n+1}\bar{\alpha}_{n+1})$ and $V = r_{n+1}(G)$. Here the map $F + g_{n+1}\bar{\alpha}_{n+1}$ is associated to

$$\bar{\xi}_n + (\bar{d}_{n+1} \odot 1)\bar{\bar{\alpha}}_{n+1}: A_n \to B_n \vee Y_1. \tag{4}$$

This follows from the cocycle condition for g_{n+1} and from $E\bar{\bar{\alpha}}_{n+1} = E\bar{\alpha}_{n+1}$, see (2.9). Hence the homotopy (H, α) above yields a homotopy

$$\lambda(U) \simeq \lambda(V) \quad \text{in } \mathbf{Twist}^c_n(\mathfrak{X}). \tag{5}$$

Now (2.12) shows that there is $\beta \in \tilde{\Gamma}_n(\lambda R, \varphi, \lambda Q)$ with

$$U + \beta \simeq V \quad \text{in } \mathbf{TWIST}^c_{n+1}(\mathfrak{X}) \tag{6}$$

Since R is given by the n-skeleton X_n we see $\tilde{\Gamma}_n(\lambda R, \varphi, \lambda Q) = \tilde{E}_n$, compare (1.10)(5). Next (6) and the definition of U and V show that we have a 1-homotopy

$$F + g_{n+1}\bar{\alpha}_{n+1} + \beta \simeq G. \tag{7}$$

This shows

$$\begin{aligned}
Gf_{n+1} &= (F + g_{n+1}\bar{\alpha}_{n+1} + \beta)f_{n+1} \\
&= Ff_{n+1} + (g_{n+1}\bar{\alpha}_{n+1} + \beta, \varphi)d_{n+1}.
\end{aligned} \tag{8}$$

Therefore we get

$$\begin{aligned}
-\mathfrak{O}(F) + \mathfrak{O}(G) &= -Ff_{n+1} + (g_{n+1}, j)\bar{\xi}_{n+1} - (g_{n+1}, j)\bar{\xi}'_{n+1} + Gf_{n+1} \\
&= (g_{n+1}\bar{\alpha}_{n+1} + \beta, \varphi)d_{n+1} + (g_{n+1}, j)(\bar{\xi}_{n+1} - \bar{\xi}'_{n+1}) \\
&= (g_{n+1}, i)[(\bar{\alpha}_{n+1} \odot \varphi)d_{n+1} + \bar{\xi}_{n+1} - \bar{\xi}'_{n+1}] + (\beta, \varphi)d_{n+1} \\
&= (\beta, \varphi)d_{n+1}.
\end{aligned} \tag{9}$$

The last equation is a consequence of (2.9)(3) since $(g_{n+1}, j)\bar{d}_{n+2} = 0$ by the cocycle condition for Y. Now (7) shows that (2) is satisfied since $\beta \in \tilde{E}_n$, see (2.10)(2). $\qquad\square$

(2.14) *Proof of* (2.8). The theorem is a consequence of (2.11), (2.12) and (2.13) compare the definition of exact sequences for functors in (IV.4.10). $\qquad\square$

As in (1.19) we get the

(2.15) *Naturality of the tower* $\mathbf{TWIST}^c_*(\mathfrak{X})/\simeq$ *with respect to functors.* Let $\alpha: \mathbf{C} \to \mathbf{K}$ be a based model functor as in (I.1.10). Let $\mathfrak{X}$ be a very good class of complexes in $\mathbf{C}$ and assume that $\mathfrak{X}_\alpha = RM\alpha(\mathfrak{X})$ is a very good class of complexes in $\mathbf{K}$. Then the functor α induces a structure preserving map which carries the tower $\mathbf{TWIST}^c_*(\mathfrak{X})$ to the tower $\mathbf{TWIST}^c_*(\mathfrak{X}_\alpha)$ and the same holds when we divide out homotopies. In particular, we have for $n \geqslant 2$ the commutative diagram of exact sequences, see (IV.4.13),

$$
\begin{array}{ccccccc}
H^n\tilde{\Gamma}_n & \xrightarrow{+} & \mathbf{TWIST}^c_{n+1}(\mathfrak{X})/\simeq & \xrightarrow{\lambda} & \mathbf{TWIST}^c_n(\mathfrak{X})/\simeq & \xrightarrow{\mathfrak{O}} & H^{n+1}\tilde{\Gamma}_n \\
\downarrow{\scriptstyle\alpha_L} & & \downarrow{\scriptstyle\alpha} & & \downarrow{\scriptstyle\alpha} & & \downarrow{\scriptstyle\alpha_L} \\
H^n\tilde{\Gamma}_n & \xrightarrow{+} & \mathbf{TWIST}^c_{n+1}(\mathfrak{X}_\alpha)/\simeq & \xrightarrow{\lambda} & \mathbf{TWIST}^c_n(\mathfrak{X}_\alpha)/\simeq & \xrightarrow{\mathfrak{O}} & H^{n+1}\tilde{\Gamma}_n
\end{array}
$$

Here α_L is induced by α_L in (II.6a.2).

Finally, we consider the special case of theorem (2.8) for the class of CW-complexes in (2.4).

(2.16) **Proposition.** *Let* $\mathfrak{X} = (\mathfrak{X}_0^1)^D$ *be the class of CW-complexes in* $\mathbf{CW}_0^D$. *Then the diagram of exact sequences in* (2.8) *is equivalent to the diagram of exact sequences in* (IV.5.11). *In particular, we get by* (1.20) *the equivalent exact sequences* ($n \geqq 2$)

$$\begin{array}{ccccccc} H^n\tilde{\Gamma}_n & \xrightarrow{+} & \mathbf{H}_{n+1}^c/\simeq & \xrightarrow{\lambda} & \mathbf{H}_n^c/\simeq & \xrightarrow{\mathfrak{D}} & H^{n+1}\tilde{\Gamma}_n \\ N\uparrow\cong & & \sim\uparrow N_{n+1} & & \sim\uparrow N_n & & \cong\uparrow N \\ H_0^n\tilde{\Gamma} & \xrightarrow{+} & \mathbf{TWIST}_{n+1}^c(\mathfrak{X})/\simeq & \xrightarrow{\lambda} & \mathbf{TWIST}_n^c(\mathfrak{X})/\simeq & \xrightarrow{\mathfrak{D}} & H_{-1}^{n+1}\tilde{\Gamma} \end{array}$$

Here the isomorphism N is obtained as in (III.5.9), compare also the proof of (1.20). Clearly, (2.16) yields a *proof of* (VI.5.11) and hence of (VI.6.2).

§3 Examples of complexes in topology

We show that the 'principal reduction' of CW-complexes and of Postnikov towers yield classes of complexes which are very good and good respectively. This result shows that there are various kinds of towers of categories which approximate the homotopy category of spaces.

Recall the definition of a complex in (III.3.1).

(3.1) **Theorem** (***r-fold principal reduction of CW-complexes***). *Let D be a path connected CW-space and let $k \geqq r-1 \geqq 0$. A relative CW-complex (X, D) with k-skeleton $X^k = D$ has the structure of a complex $\{X_i\}$ in the cofibration category* $\mathbf{Top}^D$ *by setting*

$$\begin{cases} X_i = X^{ir+k},\ i \geqq 0 \\ D = X_0 \subset X_1 \subset X_2 \subset \cdots \subset \varinjlim X_n = X \end{cases}$$

The class $(\mathfrak{X}_k^r)^D$ of all such complexes is very good. For $k > r-1 \geqq 0$ this class satisfies $\Gamma = \tilde{\Gamma}$ in degree 2, *see* (2.10)(5).

(3.2) ***Remark.*** It is enough to consider $k < 2r-1$ in theorem (3.1) since $(\mathfrak{X}_{2r-1}^r)^D$ is the subclass of $(\mathfrak{X}_{r-1}^r)^D$ consisting of all X with $X_1 = D$.

Theorem (3.1) also shows that the class $(\mathfrak{X}_0^1)^D$ of complexes in $\mathbf{CW}_0^D$ is very good, see (2.4).

Proof. We have shown in 2.3.5 of Baues (1977) that for $n \geqq 0$ the inclusion $X_n \subset X_{n+1}$, with X_n defined in (3.1), is actually a principal cofibration in the cofibration category $\mathbf{Top}^D$. In fact there are attaching maps f_{n+1}

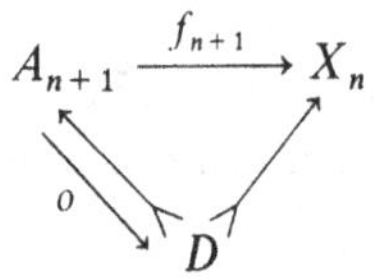

in $\mathbf{Top}^D$ where $A_{n+1} = (D \rightarrowtail A_{n+1} \xrightarrow{0} D)$ is a based object in $\mathbf{Top}^D$ such that $X_{n+1} \simeq C_{f_{n+1}}$ under X_n. Here the mapping cone $C_{f_{n+1}}$ is a mapping cone in the cofibration category $\mathbf{Top}^D$. The pair (A_{n+1}, D) is a relative CW-complex such that $A_{n+1} - D$ has only cells in dimension N with $rn + k \leqq N < r(n+1) + k$. Up to a shift in dimension by $+1$ these cells correspond exactly to the cells of $X_{n+1} - X_n$.

From the addendum of (V.7.6) we derive that $(\mathfrak{X}_k^r)^D$ is an admissible class of complexes in $\mathbf{Top}^D$, compare (III.§ 5a). Here we use the isomorphism

$$[A, B \vee X]_2 = \pi_1^A(CB \vee X, B \vee X)$$

in the cofibration category $\mathbf{Top}^D$ provided A is a suspension in $\mathbf{Top}^D$. Assume that (B, D) is $(b-1)$-connected then (2) and (V.7.6) (addendum) show that (A, B) is surjective on X_1 if $\dim(A - D) \leqq b + r + k - 1$ and (A, B) is injective on X_2 if $\dim(A - D) < b + 2r + k - 1$. This implies readily that $(\mathfrak{X}_k^r)^D$ is an admissible class of complexes and that $(\mathfrak{X}_k^r)^D$ satisfies the last condition in (2.2)(2). Moreover, all conditions on a very good class of complexes in (2.2) can be checked for $(\mathfrak{X}_k^r)^D$ by the general suspension theorem under D and by its addendum, see (V.7.6). For $k > r - 1$ this also shows that $\Gamma = \tilde{\Gamma}$ in degree 2. □

For $D = *$ we derive from (3.1) the

(3.3) **Corollary.** *Let X be a CW-complex with trivial k-skeleton $X^k = *$ and let $X_i = X^{ir+k}$ where $k \geqq r - 1 \geqq 0$. Then*

$$* = \mathbf{X}_0 \subset X_1 \subset X_2 \subset \cdots \subset \varinjlim X_n = X$$

is a based complex in the cofibration category $\mathbf{Top}^*$, *see* (III.3.3). *The class* $(\mathfrak{X}_k^r)^*$ *of all such complexes in* $\mathbf{Top}^*$ *is a very good class of complexes.*

The corollary shows that for $X^k = *$ the inclusion $X_n \subset X_{n+1}$ is a principal cofibration in $\mathbf{Top}^*$. In particular, X_1 is a suspension. In (3.1), however, the space X_1 in general is not a suspension in $\mathbf{Top}^D$.

Let $\mathbf{CW}_k^D$ be *the full subcategory of* $\mathbf{CW}_0^D$ *consisting of all complexes with* $X^k = D$. For $k \geqq r - 1 \geqq 0$ there is the canonical equivalence of categories

$$\theta_k^r : \mathbf{CW}_k^D/\simeq \; \cong \mathbf{Complex}\,(\mathfrak{X}_k^r)^D/\overset{1}{\simeq} \tag{3.4}$$

which carries X to the complex $\{X_i\}$ in (3.1). By (3.1) and (2.8) we have a tower of categories

$$\mathbf{TWIST}_n^c(\mathfrak{X}_k^r)^D/\simeq, \quad n \geqq 2, \tag{3.5}$$

which approximates the homotopy category $\mathbf{CW}_k^D/\simeq$. This tower satisfies $\Gamma = \tilde{\Gamma}$ in degree 2 if $k > r - 1 \geqslant 0$.

By (3.5) we actually obtain $k+1$ different towers of categories $(r = 1, \ldots, k+1)$ which approximate the homotopy category $\mathbf{CW}_k^D/\simeq$. There are even more such towers by restricting the towers for $\mathbf{CW}_{k-1}^D/\simeq$ to the subcategory $\mathbf{CW}_k^D/\simeq$. It remains to be seen how all these towers are connected with each other. In (VI.§6) we described in detail the properties of the tower $\mathbf{TWIST}_*^c(\mathfrak{X}_0^1)^D$. Similar properties are available for the towers (3.5). This yields many new results on the homotopy classification problems.

The next result allows the application of the tower of categories in section §1 in a fibration category. Recall that cocomplexes in a fibration category are by definition the strict duals of complexes in a cofibration category.

(3.6) **Theorem (*r-fold principal reduction of Postnikov-towers*).** *Let B be a path connected CW-space and let $k \geqq r \geqq 1$. Consider all fibrations $p: E \twoheadrightarrow B$ in* **Top** *for which E and the fiber F are CW-spaces and for which F is k-connected. Then the Postnikov decomposition $\{E^i\}$ of p in* (III.7.2) *yields the fibrations in* **Top**

$$\begin{cases} B = E_0 \twoheadleftarrow E_1 \twoheadleftarrow E_2 \ldots \twoheadleftarrow E_n \ldots \\ \text{with } E_i = E^{ir+k}, \; i \geqq 0, \end{cases}$$

which form a cocomplex in the category $\mathbf{Top}_B$. *The class $(\mathfrak{C}_k^r)_B$ of all such cocomplexes is a good class of cocomplexes in the fibration category* $\mathbf{Top}_B$. *This class satisfies $\Gamma = \tilde{\Gamma}$ in degree* 2, *see* (1.10)(7).

(3.7) ***Remark.*** For $k = r = 1$ the theorem shows that the Postnikov-tower of a fibration $p: E \to B$ with simply connected fiber is a cocomplex in $\mathbf{Top}_B$. We proved this in (III.7.4)(*b*). The classifying maps are the twisted k-invariants.

Proof of (3.6). The principal reduction in 2.3 of Baues (1977) shows that $E_n \to E_{n-1}$ is actually a principal fibration in $\mathbf{Top}_B$. For $r = 1$ the classifying maps are described by the k-invariants (III.7.4). For $r \geqq 1$ the classifying map f_n of $E_n \to E_{n-1}$ is a map over B as in the diagram

$$\begin{array}{ccc} E_{n-1} & \xrightarrow{f_n} & A_n \\ & \searrow & \swarrow\!\!\nearrow o \\ & B & \end{array}.$$

Here A_n is an appropriate based object in $\mathbf{Top}_B$. The fiber F_n of $A_n \twoheadrightarrow B$ has the homotopy groups

$$\pi_j(\Omega F_n) = \begin{cases} \pi_j F & \text{for } (n-1)r + k < j \leqq nr + k \\ 0 & \text{otherwise} \end{cases}$$

Here F is the fiber of $E \twoheadrightarrow B$ and ΩF_n is the fiber of $E_n \to E_{n-1}$. The construction of f_n is originally due to MC–Clendon (1974). From the addendum of (V.10.6) we derive that $(\mathfrak{E}_k^r)^B$, for $k \geqq r \geqq 1$, is actually an admissible class of cocomplexes. Here we use the equation

$$[Y \times X, A]_2 = \pi_A^1(WY \times X \mid Y \times X)$$

which is satisfied for a loop space A in $\mathbf{Top}_B$. Also the conditions on a good class can be checked by the general loop theorem (V.10.6). Moreover, $\Gamma = \tilde{\Gamma}$ in degree 2 is a consequence of this result. □

Next we consider the important special case of (3.6) with $B = *$.

(3.8) **Corollary.** *Let E be a k-connected CW-space with basepoint, $k \geqq r \geqq 1$. Then the Postnikov decomposition of E yields the based cocomplex $\{E_i\}$ with*

$$\begin{cases} E_i = E^{ir+k} \\ * = E_0 \twoheadleftarrow E_1 \twoheadleftarrow E_2 \twoheadleftarrow \cdots. \end{cases}$$

The class $(\mathfrak{E}_k^r)_*$ of all these cocomplexes is a good class of cocomplexes in the fibration category **Top**.

In particular, the fibration $E_n \to E_{n-1}$ is a principal fibration in **Top** and E_1 is a loop space in **Top**. For the cocomplexes $\{E_i\}$ in (3.6), however, the space E_1 in general is not a loop space in $\mathbf{Top}_B$.

Let $\mathbf{CW}_B^k$ be the *full subcategory of* $\mathbf{Top}_B$ *consisting of fibrations $E \twoheadrightarrow B$ for which E and the fiber F are CW-spaces and for which F is k-connected.* For $k \geqq r \geqq 1$ we have the canonical functor of homotopy categories

(3.9) $$\theta_k^r : \mathbf{CW}_B^k/\simeq \;\to \mathbf{Cocomplex}\,(\mathfrak{E}_k^r)_B/\overset{0}{\simeq}$$

which carries $E \twoheadrightarrow B$ to the cocomplex $\{E_i\}$ in (3.6). The functor θ_k^r is well defined by naturality of the Postnikov decomposition, compare for example 5.3 in Baues (1977). The functor (3.9) is full and faithful on the subcategory consisting of objects $E \twoheadrightarrow B$ for which E is a finite dimensional CW-space or for which $E \twoheadrightarrow B$ has a finite Postnikov decomposition. By (3.6) and (1.9) we obtain for $k \geqq r \geqq 1$ the tower of categories

(3.10) $$\mathbf{TWIST}_n(\mathfrak{E}_k^r)_B, \quad n \geqq 2,$$

which approximates the homotopy category $\mathbf{CW}_{B}^{k}/\simeq$ via θ_k^r in (3.9). This tower satisfies $\Gamma = \tilde{\Gamma}$ in degree 2.

There are k-different towers (3.10) which approximate the category $\mathbf{CW}_{B}^{k}/\simeq$. Moreover, we get such towers by restricting the towers for $\mathbf{CW}_{B}^{k-1}/\simeq$ to the subcategory $\mathbf{CW}_{B}^{k}/\simeq$. The connections between these towers are helpful for explicit computations.

The tower of categories (3.10) yields many new results on the homotopy classification problems in the category $\mathbf{Top}_B/\simeq$. In the next chapter we discuss in detail the tower $\mathbf{TWIST}_*(\mathfrak{C}_1^1)_B$ which approximates the homotopy category $\mathbf{CW}_{B}^{1}/\simeq$.

§4 Complexes in the category of chain algebras

We show that a cofibration in the category **DA** of chain algebras has in a canonical way the structure of a complex in **DA**. Moreover, we describe the 'principal reduction' of such complexes. This is an algebraic illustration of the result for CW-complexes in (3.1). In the category of chain Lie algebras **DL** each cofibration as well is a complex in **DL** and one can study the principal reduction of such complexes too; we leave it, however, to the reader to formulate (and to prove) the results of this section for chain Lie algebras. Complexes in the category of commutative cochain algebras $\mathbf{CDA}_*^0$ are discussed in (VIII.§4) below. All these complexes yield towers of categories as described in §1 and §2 above. These towers have properties similar to those discussed in Chapter VI for CW-complexes. Unfortunately it is beyond the limits of this book to describe this in detail.

Let R be a commutative ring of coefficients. We show that the objects in $\mathbf{DA}_c^D$ are complexes in the cofibration category $\mathbf{DA}_c^D$. Here we derive the structure from the I-category $\mathbf{DA}_c^D$ in (I.3.3). The homotopy category $\mathbf{DA}_c^D/\simeq$ plays an important role since we have by (II.3.6) the following result:

(4.1) **Theorem.** *Let R be a principal ideal domain and let D be flat as a module. Then the inclusion*

$$\mathbf{DA}_c^D/\simeq \xrightarrow{\sim} Ho(\mathbf{DA}(\text{flat})^D)$$

is an equivalence of categories.

Here the right-hand side is the localization with respect to weak equivalences and the left-hand side is the homotopy category of cofibrant objects in $\mathbf{DA}^D$. An object in $\mathbf{DA}_c^D$ is a cofibration

(4.2) $$D \rightarrowtail X = (D \amalg T(V), d)$$

in **DA**. This is a **complex** with (relative) skeleta

$$X^n = (D \amalg T(V_{\leqq n}), d) \subset X \tag{1}$$

where $V_{\leqq n} = \{x \in V : |X| \leqq n\}$. One readily checks that X^n is a chain algebra with the differential given by the one of X. The 0-skeleton

$$X^0 = D \vee T(V_0) \tag{2}$$

is just given by the chain algebra D and by the free R-module V_0 concentrated in degree 0. Clearly, we have the filtration

$$D \rightarrowtail X^0 \rightarrowtail X^1 \rightarrowtail \cdots \rightarrowtail X = \lim_{\rightarrow} X^n. \tag{3}$$

Here $X^n \rightarrowtail X^{n+1}$ $(n \geqq 0)$ is a principal cofibration since we have the push out diagram

$$\begin{array}{ccc} CT(s^{-1}V_{n+1}) & \xrightarrow{\pi_f} & X^{n+1} \\ \cup & \text{push} & \cup \\ T(s^{-1}V_{n+1}) & \xrightarrow[f_{n+1}]{} & X^n \end{array} \tag{4}$$

The **attaching map** f_{n+1} is given by the differential d in X, namely

$$f_{n+1}(s^{-1}v) = dv, \quad v \in V_{n+1}, \tag{5}$$

Moreover, $T(s^{-1}V_{n+1})$ is the tensor algebra with trivial differential so that f_{n+1} is a well defined map in **DA** since $dd = 0$. We have the **cone** in (4) by

$$\left.\begin{array}{r} CT(s^{-1}V_{n+1}) = (T(s^{-1}V_{n+1} \oplus V_{n+1}), d) \\ d(s^{-1}v) = 0, \quad dv = s^{-1}v, \quad v \in V_{n-1} \end{array}\right\} \tag{6}$$

By (I.7.19) we know that this is actually the cone given by the cylinder I in (I.7.11); here $T(s^{-1}V_{n+1})$ is a based object with the augmentation ε, $\varepsilon(s^{-1}V_{n+1}) = 0$. By the definition of f_{n+1} it is immediately clear that (4) is a push out diagram with π_f being the identity on V_{n+1}.

(4.3) ***Remark.*** The inclusion $D \rightarrowtail X^0 = D \vee T(V_0)$ is not a principal cofibration since we have no attaching map. This inclusion, however, is a *weak* principal cofibration in the sense that we have the **coaction**

$$\mu : X^0 \to X^0 \vee T(V_0') = D \vee T(V_0 \oplus V_0') \tag{1}$$

(with $V_0 = V_0'$) which is the identity on D and which carries v to $v + v'$ for $v \in V_0$. We now define the structure of a (weak) complex $\{X_n\}$ for $D \rightarrowtail X$ by setting

$$X_n = \left\{\begin{array}{ll} D \quad , & \text{for } n = 0 \\ X^{n-1}, & \text{for } n \geq 1 \end{array}\right\} \tag{2}$$

This complex is the analogue of a CW-complex in $\mathbf{CW}_0^D$ with skeletal filtration, see (VI.1.1)(1). As in (VI.1.1)(4) we see that $\{X_n\}$ is a complex in $\mathbf{DA}_c^D$ with attaching maps

$$(f_{n+1}, i): T(s^{-1}V_{n+1}) \vee D \to X^n \tag{3}$$

where $i: D \subset X^n$ is the inclusion.

Let $(\mathfrak{X}_0^1)^D$ be the class of all complexes $\{X_n\}$ as in (4.3)(2). Then we have the equality of categories

(4.4) $$\mathbf{Complex}(\mathfrak{X}_0^1)^D = \mathbf{DA}_c^D$$

since each map in $\mathbf{DA}_c^D$ is filtration preserving. Moreover, for degree reasons we get

(4.5) $$\mathbf{Complex}(\mathfrak{X}_0^1)^D/\overset{1}{\simeq} = \mathbf{DA}_c^D/\overset{1}{\simeq} = \mathbf{DA}_c^D/\simeq$$

since a homotopy is also a 1-homotopy, see (III.1.5).

As in (3.1) there is a principal reduction theorem for the complexes above.

(4.6) **Theorem.** *Let D be a chain algebra and let $k \geqq r - 1 \geqq 0$. A cofibration $D \rightarrowtail X = (D \coprod T(V), d)$ in* **DA** *with $V_i = 0$ for $0 \leqq i \leqq k - 1$ has the structure of a complex $\{X_i\}$ in $\mathbf{DA}_c^D$ by defining*

$$X_i = D \coprod T(V_0, \ldots, V_{ir+k-1}), \quad i \geqq 0,$$

with $X_0 = D$. The class $(\mathfrak{X}_k^r)^D$ of all such complexes is very good.

We have similar consequences of this theorem as in (3.4) and (3.5), in particular, we obtain the tower of categories $\mathbf{TWIST}_*^c(\mathfrak{X}_k^r)^D$ which approximates $\mathbf{DA}_k^D/\simeq$ (the 'weakness' of a cofibration $D \rightarrowtail X_0$ for $\{X_i\} \in (\mathfrak{X}_k^r)^D$ mentioned in (4.3) is not essential in the proofs). For $k = 0, r = 1$ the class $(\mathfrak{X}_0^1)^D$ yields a tower of categories

(4.7) $$\mathbf{K}_n = \mathbf{TWIST}_n^c(\mathfrak{X}_0^1)^D, \quad n \geqq 2,$$

with almost the same properties as the tower $\{\mathbf{H}_n, n \geqq 2\}$ in (VI.6.2). We will discuss the subtower $\mathbf{K}_n = \mathbf{TWIST}_n^c(\mathfrak{X}_1^1)^*$, $n \geqq 2$, in Chapter IX.

VIII

Homotopy theory of Postnikov towers and the Sullivan–de Rham equivalence of rational homotopy categories

In this chapter we consider topological fibrations and fiber preserving maps over a fixed base space D. The results are also of interest when $D = *$ is a point. We approximate the homotopy category of fiber preserving maps by a tower of categories which is deduced from the Postnikov-decomposition of the fibrations. This tower is a special case of TWIST_* in (VII.1.9). Here our axiomatic approach saves a lot of work since we can use the results in a fibration category dual to results already proved in a cofibration category.

We apply the towers of categories in Chapter VII in rational homotopy theory. For this we first describe a de Rham theorem for cohomology groups with local coefficients. Then we show that in the cofibration category of commutative cochain algebras one has towers of categories which are strictly analogous to the towers of categories for Postnikov decompositions. Moreover, the Sullivan–de Rham functor yields a structure preserving map between these towers. This fact leads to elementary proofs, and to generalizations, of some fundamental results in rational homotopy theory.

§1 The tower of categories for Postnikov decompositions

We work in the fibration category **Top** of topological spaces. The category $\mathbf{Top}_D$ of spaces over D is a fibration category by (II.1.4). We consider fibrations in **Top**

$$FX \subset X \overset{p}{\twoheadrightarrow} D$$

where we assume that X, the fiber $FX = p^{-1}(*)$, and D are pathconnected CW-spaces and that FX is simply connected. Let $\mathbf{CW}_D^1$ be the full subcategory of $\mathbf{Top}_D$ consisting of such fibrations, see (VII.3.9). We thus have the full subcategory

$$\mathbf{CW}^1_D/\simeq \;\subset \mathbf{Top}_D/\simeq \tag{1.1}$$

where $\simeq$ denotes the homotopy relation over D. In this section we describe a tower of categories which approximates the category $\mathbf{CW}^1_D/\simeq$.

Let $\pi = \pi_1(D)$ and let $\mathbf{Mod}_\pi$ be the category of right π-modules. There are functors ($n \geqq 2$)

$$\pi_n : \mathbf{CW}^1_D \to \mathbf{Mod}_\pi \tag{1.2}$$

which carry the fibration $X \twoheadrightarrow D$ to the nth homotopy group of the fiber FX, $\pi_n(FX)$, which is a π-module. Moreover, π_n carries a map $f: Y \to X$ over D to the induced map $\pi_n(Ff): \pi_n(FY) \to \pi_n(FX)$ where $Ff: FY \to FX$ is the restriction of f to the fibers. We need no basepoints for FX and FY since these spaces are simply connected.

We now define the category of coefficients which will be used for the tower of categories below.

(1.3) ***Definition.*** Let $\mathbf{Coef} = \mathbf{K}^2_D$ be the full subcategory of $\mathbf{CW}^1_D/\simeq$ consisting of fibrations $X_1 \twoheadrightarrow D$ for which the fiber is an Eilenberg–Mac Lane space $K(\pi_2, 2)$. We discuss properties of this category in § 2 below. ||

The Postnikov decomposition $\{X^i\}$ of $X \twoheadrightarrow D$ yields by (III.7.4) or (VII.3.6) the cocomplex

$$\begin{cases} D = X_0 \twoheadleftarrow X_1 \twoheadleftarrow X_2 \twoheadleftarrow \cdots \\ \text{with } X_i = X^{i+1} \text{ and with classifying} \\ \text{maps } f_n : X_{n-1} \to A_n = L(\pi_{n+1}, n+2). \end{cases} \tag{1.4}$$

Here A_n is the based object in $\mathbf{Top}_D$ defined in (III.6.5). By naturality of the Postnikov decomposition we have the functor E_1 as in the commutative diagram

$$\begin{array}{ccc} \mathbf{CW}^1_D/\simeq & \xrightarrow{E_1} & \mathbf{Coef} \\ & {\scriptstyle \pi_2}\searrow \quad \swarrow{\scriptstyle \pi_2} & \\ & \mathbf{Mod}_\pi & \end{array} \tag{1.5}$$

Here E_1 carries $X \twoheadrightarrow D$ to $X_1 \twoheadrightarrow D$. In the following theorem we use the category of graded π-modules $\mathbf{Mod}^{\geqq k}_\pi$ which consists of sequences of π-modules $(\pi_k, \pi_{k+1}, \ldots)$ and of π-linear maps of degree 0.

(1.6) **Theorem.** *There is a tower of categories which approximates the homotopy* *category* $\mathbf{CW}_D^1/\simeq$, $n \geqq 2$,

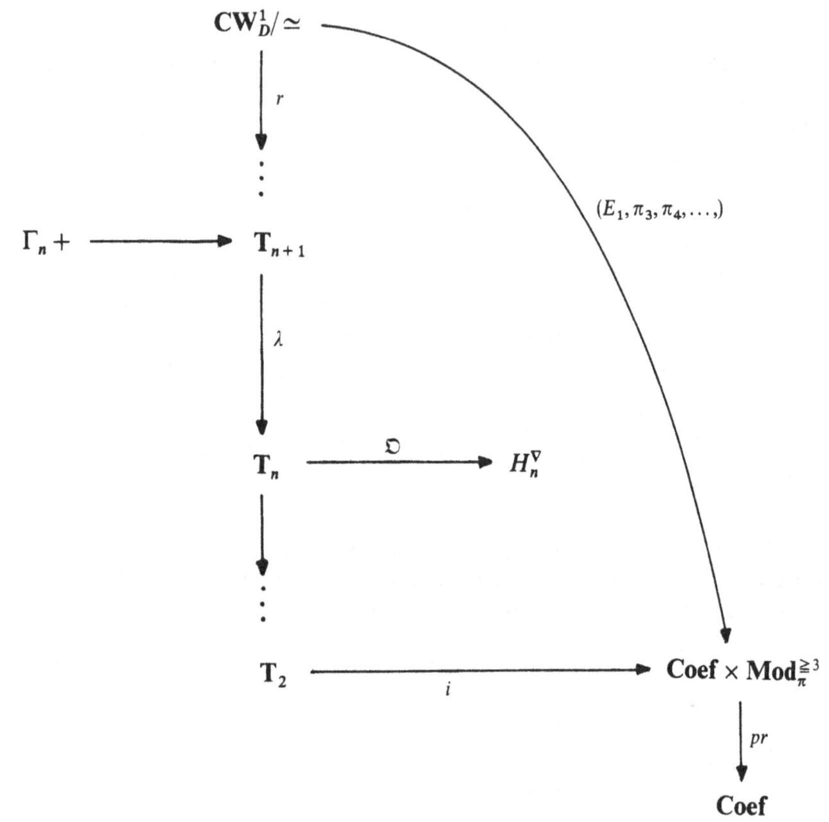

Here i is a faithful functor and the composition of functors i, λ, and r is $(E_1, \pi_3, \pi_4, \ldots)$.

Proof of (1.6). We use the good class of cocomplexes $(\mathfrak{C}_1^1)_D$ in (VII.3.6) and we use the functor θ_1^1 in (VII.3.9). Then (VII.1.9) gives us the tower of categories

$$\mathbf{T}_n = \mathbf{TWIST}_n(\mathfrak{C}_1^1)_D, \quad n \geqq 2,$$

where we dualize the definition in (VII.1.6). Below we describe the category $\mathbf{T}_2$ more explicitly. This shows that i in the theorem is a faithful functor.

□

Recall that we have (for each π-module A) the Eilenberg–Mac Lane object $L(A, n)$ in $\mathbf{Top}_D$, see (III.6.5), which is a based object

(1.7) $$K(A, n) \subset L(A, n) \underset{0}{\overset{p}{\rightleftarrows}} D$$

in $\mathbf{Top}_D$.

(1.8) *Lemma. For π-modules A, A' there is the canonical isomorphism*

$$\pi_n: [L(A,n), L(A',n)]_0 = \mathrm{Hom}_\pi(A, A').$$

Here the left-hand side denotes the set of homotopy classes in $\mathbf{Top}_D/\simeq$ *which are based up to homotopy*, see (II.9.8). The right-hand side is the group of π-linear maps $A \to A'$.

We define the isomorphism π_n in (1.8) by the functor in (1.2). This shows that the isomorphism is compatible with the loop operation Ω in the fibration category $\mathbf{Top}_D$, dual to (II.9.8), so that the diagram

(1.9)
$$\begin{array}{ccc} [L(A,n), L(A',n)]_0 & & \\ & \overset{\pi_n}{\underset{\cong}{\searrow}} & \\ \Omega \downarrow & & \mathrm{Hom}_\pi(A, A') \\ & \overset{\cong}{\underset{\pi_{n-1}}{\nearrow}} & \\ [L(A,n-1), L(A',n-1)]_0 & & \end{array}$$

commutes. Here we identify (see (III.6.6))

(1.10)
$$\Omega L(A,n) = L(A,n-1)$$

as usual. The Postnikov decomposition in (1.4) is a cocomplex for which the principal fibration $X_n \to X_{n-1}$ in $\mathbf{Top}_D$ has the classifying map

(1.11)
$$f_n: X_{n-1} \to A_n = L(\pi_{n+1}, n+2).$$

Here $\pi_{n+1} = \pi_{n+1}FX$ denotes the homotopy group of the fiber FX of $X \twoheadrightarrow D$.

For the difference ∇f_n we obtain a unique map d_n, for which the following diagram commutes up to homotopy over D, and which is trivial on X_1, ($n \geqq 2$):

(1.12)
$$\begin{array}{ccc} L(\pi_n, n) \times {}_D X_{n-1} & \xrightarrow{\nabla f_n} & L(\pi_{n+1}, n+2) \\ \downarrow 1 \times p & \overset{d_n}{\nearrow} & \\ L(\pi_n, n) \times {}_D X_1 & & \end{array}$$

This follows from the addendum of the general loop theorem (V.10.6). Clearly $p: X_{n-1} \twoheadrightarrow X_1$ in the diagram is given by (1.4). Moreover, we use (1.10) for the definition of ∇f_n, see (II.12.2). The product $X \times_D Y$ in (1.12) denotes the product in the category $\mathbf{Top}_D$.

(1.13) ***Remark.*** We cannot expect that d_n factors over $L(\pi_n, n) \times {}_D X_0 = L(\pi_n, n)$ since in general there exist non trivial cup product maps $K(\pi_n, n) \times K(\pi_2, 2) \to K(\pi_{n+1}, n+2)$. ||

We now are ready for the definition of the bottom category $\mathbf{T}_2$ in (1.6).

(1.14) ***Definition.*** Objects of $\mathbf{T}_2$ are **twisted 2-systems** $X^{(2)}$ which are given by tuples

$$X^{(2)} = \{X_1, f_2, \pi_n, d_n, n \geqq 3\}. \tag{1}$$

Here X_1 is an object in **Coef** and π_n is *a* π-module. Moreover,

$$\left.\begin{array}{l} f_2: X_1 \to L(\pi_3, 4), \text{ and} \\ d_n: L(\pi_n, n) \times_D X_1 \to L(\pi_{n+1}, n+2) \end{array}\right\} \tag{2}$$

are maps in $\mathbf{Top}_D/\simeq$ and d_n is trivial on X_1. These maps satisfy

$$d_n(Ld_{n-1} \odot 1) = 0 \tag{3}$$

for $n \geqq 3$ where we set $d_2 = \nabla f_2$. Recall that the trivial map 0 in (3) is given by the section 0 in (1.7). Next we define the **maps in $\mathbf{T}_2$** by a tuple

$$(\varphi, \xi_n, n \geqq 3): Y^{(2)} \to X^{(2)} \tag{4}$$

where $Y^{(2)} = \{Y_1, g_2, \pi'_n, d'_n, n \geqq 3\}$.

The tuple in (4) is given by a map $\varphi: Y_1 \to X_1$ in **Coef** and by π-linear maps $\xi_n: \pi'_n \to \pi_n$ in $\mathbf{Mod}_\pi$. For these maps the following diagrams (5) and (6) commute in $\mathbf{Top}_D/\simeq$ $(n \geqq 3)$

$$\begin{array}{ccc} Y_1 & \xrightarrow{\varphi} & X_1 \\ \downarrow{\scriptstyle g_2} & & \downarrow{\scriptstyle f_2} \\ L(\pi'_3, 4) & \xrightarrow{\xi_3} & L(\pi_3, 4) \end{array} \tag{5}$$

$$\begin{array}{ccc} L(\pi'_n, n) \times_D Y_1 & \xrightarrow{\xi_n \odot \varphi} & L(\pi_n, n) \times_D X_1 \\ \downarrow{\scriptstyle d'_n} & & \downarrow{\scriptstyle d_n} \\ L(\pi'_{n+1}, n+2) & \xrightarrow{\xi_{n+1}} & L(\pi_{n+1}, n+2) \end{array} \tag{6}$$

Here we use (1.8). ||

Remark. The category $\mathbf{T}_2$ depends on the category **Coef** and thus the morphisms are not purely algebraic. For $D = *$, however, we have the equivalence **Coef** = **Ab** and in this case the category $\mathbf{T}_2$ can be considered as being a purely algebraic category. ||

There is an obvious forgetful functor *i*, see (1.6), which is faithful. Moreover, we have the functor

(1.15) $$\mathbf{CW}^1_D/\simeq \to \mathbf{T}_2$$

which carries $X \twoheadrightarrow D$ to $X^{(2)}$ where $X^{(2)}$ is defined by (1.11) and (1.12). In a similar way as in (1.14) one defines the categories $\mathbf{T}_n$ ($n \geqq 2$) in (1.6) with objects

(1.16) $$X^{(n)} = \{X_{n-1}, f_n, \pi_i, d_i, i \geqq n+1\}.$$

The addendum of (V.10.6) implies that

(1.17) $$\begin{aligned}[L(A',n)X_D Y_1, L(A,n)]_2 &= [L(A',n), L(A,n)]_0 \\ &= \mathrm{Hom}_\pi(A', A).\end{aligned}$$

This shows by (1.8) that the definition of $\mathbf{T}_2$ above is actually consistent with the definition in the proof of (1.6), compare (VII.1.6).

For the definition of the **natural systems** Γ_{n-1} and H^∇_{n-1} on the category $\mathbf{T}_{n-1}$ ($n \geqq 3$) we use the diagram (1.18) below which is dual to diagram (VII.1.10) (where we replace n by $n-1$). Let

$$X = X^{(n-1)} = \{X_{n-2}, f_{n-1}, \pi_i, d_i, i \geqq n\},$$
$$Y = Y^{(n-1)} = \{Y_{n-2}, g_{n-1}, \pi'_i, d'_i, i \geqq n\}$$

be objects in $\mathbf{T}_{n-1}$ and let $p_g: P_g = Y_{n-1} \twoheadrightarrow Y_{n-2}$ be the principal fibration induced by $g \in g_{n-1}$. For a map $\varphi: Y_1 \to X_1$ in **Coef**, with Y_1 and X_1 given by Y_{n-2} and X_{n-2} respectively, we have the diagram (of homotopy groups in $\mathbf{Top}_D/\simeq$)

(1.18) $$\begin{array}{ccc} [Y_{n-2}, L(\pi_n, n)] & \xrightarrow{p_g^*} & [Y_{n-1}, L(\pi_n, n)] \\ & \swarrow{\scriptstyle (d_n)_*(.,\varphi)} & \\ [Y_{n-1}, L(\pi_{n+1}, n+2)] & \xrightarrow{\nabla} & [L(\pi'_n, n) \times_D Y_{n-1}, L(\pi_{n+1}, n+2)] \end{array}$$

Here ∇ is the difference construction for P_g. Moreover, $(d_n)_*$ (α, φ) is the composition

$$Y_{n-1} \xrightarrow{(\alpha, \varphi p)} L(\pi_n, n) \times_D X_1 \xrightarrow{d_n} L(\pi_{n+1}, n+2)$$

where $p: Y_{n-1} \to Y_1$ is the projection.

Recall that we have the cohomology with local coefficients (III.6.7) which for $p: X \twoheadrightarrow D$ in $\mathbf{CW}^1_D$ satisfies

(1.19) $$[X, L(A,n)] = \hat{H}^n(X, A).$$

Here A is a $\pi_1(X) = \pi_1(D)$-module. This shows that diagram (1.18) is equivalent to the following diagram of cohomology groups with $\pi_n = \pi_n(FX)$:

(1.20) $$\begin{array}{ccc} \hat{H}^n(Y_{n-2}, \pi_n) & \rightarrowtail^{p^*} & \hat{H}^n(Y_{n-1}, \pi_n) \\ & \swarrow{\scriptstyle (d_n)_*(.,\varphi)} & \\ \hat{H}^{n+2}(Y_{n-1}, \pi_{n+1}) & \xrightarrow{\nabla} & \hat{H}^{n+2}(L(\pi'_n, n) \times_D Y_{n-1}, \pi_{n+1}) \end{array}$$

As in (VII.1.10)(5) we now define the groups

$$\left\{\begin{array}{l}\Gamma_{n-1}(Y,\varphi,X)=E_{n-1}\cap\text{kernel}\,(d_n)_*(.,\varphi),\\ H^{\nabla}_{n-1}(Y,\varphi,X)=\text{kernel}\,(\nabla)/(d_n)_*(E_{n-1},\varphi),\end{array}\right. \tag{1.21}$$

where $E_{n-1}=p_g^*\hat{H}^n(Y_{n-2},\pi_n)\subset\hat{H}^n(Y_{n-1},\pi_n)$. Here p_g^* is injective since p_g is an n-connected map. In (1.21) the objects Y and X are objects in $\mathbf{CW}^1_D$ or objects in $\mathbf{T}_{n-1}$.

In (VII.1.10)(8) we proved that the groups Γ_{n-1}, H^{∇}_{n-1} above are natural systems of abelian groups on the category $\mathbf{T}_{n-1}$, $n\geqq 3$.

We now describe more explicitly some properties of the tower (1.6). First we have the following well-known Whitehead theorem which we derive from (IV.4.11).

(1.22) **Proposition.** *The functor* (see (1.2))

$$(\pi_2,\pi_3,\dots)\colon \mathbf{CW}^1_D/\simeq\;\to\mathbf{Mod}^{\geqq 2}_{\pi}$$

satisfies the sufficiency condition, that is: A map $f\colon Y\to X$ over D is a homotopy equivalence in $\mathbf{CW}^1_D/\simeq$ if and only if the induced maps $\pi_n(Ff)\colon\pi_n(FY)\to\pi_n(FX)$ are isomorphisms for all $n\geqq 2$.

Next we consider the homotopy classification problem for maps in $\mathbf{CW}^1_D/\simeq$. Let $Y\twoheadrightarrow D$, $X\twoheadrightarrow D$ be objects in $\mathbf{CW}^1_D/\simeq$ and let $\varphi\colon Y_1\to X_1$ be a map in **Coef**. We denote by $[Y,X]^{\varphi}_D$ the set of all homotopy classes of maps $Y\to X$ over D which induce φ on Y_1. Moreover, $[Y,X]^{\varphi}_n$ is the set of all morphisms $rY\to rX$ in $\mathbf{T}_n$ which induce φ. Then (1.6) yields the following diagram of sets and of group actions with $n\geqq 2$.

$$\begin{array}{ccccc}
 & & [Y,X]^{\varphi}_D & & \\
 & & \downarrow & & \\
 & & \vdots & & \\
\Gamma_n(Y,\varphi,X) & \xrightarrow{\;+\;} & [Y,X]^{\varphi}_{n+1} & & \\
 & & \downarrow{\scriptstyle\lambda} & & \\
 & & [Y,X]^{\varphi}_n & \xrightarrow{\;\mathfrak{D}\;} & H^{\nabla}_n(Y,\varphi,X)\\
 & & \downarrow & &
\end{array}\tag{1.23}$$

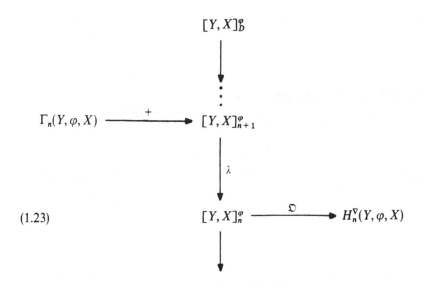

$$\begin{array}{ccccc} & & \vdots & & \\ \Gamma_2(Y, \varphi, X) & \xrightarrow{\ +\ } & [Y, X]_3^\varphi & & \\ & & \downarrow & & \\ & & [Y, X]_2^\varphi & \xrightarrow{\ \mathfrak{D}\ } & H_2^\nabla(Y, \varphi, X) \\ & & \cap & & \\ & & \underset{n \geqq 3}{\times} \mathrm{Hom}_\pi(\pi'_n, \pi_n) & & \end{array}$$

Here $\pi_n = \pi_n(FX)$, $\pi'_n = \pi_n(FY)$ denote the homotopy groups of the fibers of $X \twoheadrightarrow D$ and $Y \twoheadrightarrow D$ respectively. The composition of all maps in the column is given by the functor $(\pi_3, \pi_4, \ldots)$, see (1.2). The abelian groups Γ_n and H_n^∇ are defined in (1.21). As in (VI.6.5) we have the following properties of diagram (1.23):

(a) $\mathfrak{D}$ is a function with image $(\lambda) =$ kernel $(\mathfrak{D})$ and $\mathfrak{D}$ is a derivation, that is,

$$\mathfrak{D}(fg) = f_* \mathfrak{D}(g) + g^* \mathfrak{D}(f).$$

Here fg denotes the composition in $\mathbf{T}_n$.

(b) For all $f \in [Y, X]_n^\varphi$ with $\mathfrak{D}(f) = 0$ the group at the left-hand side, $\Gamma_n(Y, \varphi, X)$, in (1.23) acts transitively on the subset $\lambda^{-1}(f)$ of $[Y, X]_{n+1}^\varphi$. Moreover, the action is linear, this means for $f_0 \in \lambda^{-1}(f)$, $g_0 \in \lambda^{-1}(g)$:

$$(f_0 + \alpha)(g_0 + \beta) = f_0 g_0 + f_* \beta + g^* \alpha.$$

(c) If the homotopy groups $\pi_k(FX)$ vanish for $k \geqq N$ then the function

$$r: [Y, X]_D^\varphi \to [Y, X]_n^\varphi$$

is bijective for $n = N - 1$ and surjective for $n = N - 2$.

(d) The isotropy groups of the action in (b) can be computed by the spectral sequence $\{\bar{E}_r^{s,t}\{M^p\}\}$ in (III.9.8) with

$$M^p = \mathrm{Map}(Y_n, X_{p-1})_D^\phi, \quad p \geqq 1.$$

Here ϕ denotes the empty space. We use the result dual to (III.4.18) in the fibration category $\mathbf{Top}_D$. For

$$\beta \in \Gamma_n(Y, \varphi, X) \subset \hat{H}^{n+1}(Y_n, \pi_{n+1} FX) = \bar{E}_1^{n+1, n+1},$$

and for $(\xi, \eta) \in [Y, X]_{n+1}^\varphi$ we have $(\xi, \eta) + \beta = (\xi, \eta + \beta) = (\xi, \eta)$ if and only if

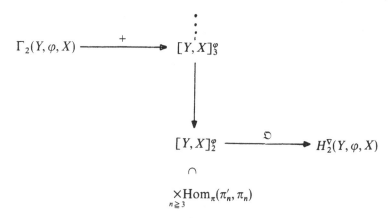

$\eta + \beta = \eta$ in $[Y_n, X_n]^{\varphi}_D$. This is the case if and only if β is in the image of one of the differentials $\partial_1, \ldots, \partial_{n-1}$ as pictured in the diagram

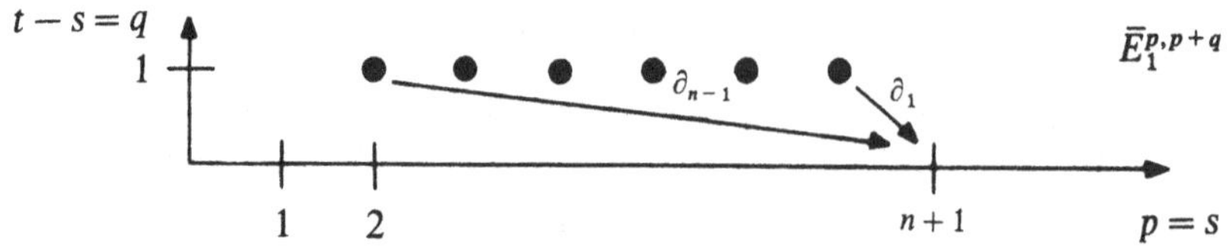

of the $\bar{E}_1$-term with

$$\bar{E}_1^{p,p+q} = [Y_n, \Omega^{q+1} A_{p-1}]_D = \hat{H}^{p-q}(Y_n, \pi_p FX).$$

Here $\partial_n = 0$ since we assume $\pi_1 FX = 0$. Moreover, the differential ∂_1 is given by d_n in (1.18)

$$\begin{cases} \partial_1 : \bar{E}_1^{n,n+1} = \hat{H}^{n-1}(Y_n, \pi_n) \to \bar{E}_1^{n+1,n+1} = \hat{H}^{n+1}(Y_n, \pi_{n+1}), \\ \partial_1 = (Ld_n)^*(., \varphi p). \end{cases}$$

Here we set $\pi_n = \pi_n FX$ and Ld_n denotes the partial loop operation (in $\mathbf{Top}_D$) applied to d_n in (1.18). Then $(Ld_n)^*(., \varphi p)$ is defined as in (1.20).

Diagram (1.23) shows that the enumeration of the set $[Y, X]_D$ can be achieved by the inductive computation of the obstruction operator in (a), and of the isotropy groups of the action in (b). Here we assume that the fiber FX has finite homotopy dimension or that Y is homotopy equivalent to a CW-complex of finite dimension. Clearly, in general, this program is very hard; but there are nice examples, in particular if $D = *$ is a point. As in (VI.6.6) diagram (1.23) yields **'higher order obstructions'** for the realizability of a homomorphism $\xi: \pi_*(FY) \to \pi_*(FX)$ of $\pi_1(D)$-modules via a map $Y \to X$ over D. This result improves equation 2 in Adams (1956).

Next we derive from the tower in (1.6) the following structure of the **group of homotopy equivalences** over D of the fibration $X \twoheadrightarrow D$ in $\mathbf{CW}^1_D/\simeq$. We denote this group by

(1.24) $$\mathrm{Aut}(X)_D \subset [X, X]_D.$$

This is the group of units in the monoid $[X, X]_D$. Let $X_1 \twoheadrightarrow D$ be given by X as in (1.4) and let

(1.25) $$G = \mathrm{Aut}(X_1)_D.$$

Thus G is the group of automorphisms in the category **Coef**. As in (VI.6.12) the tower of categories for Postnikov decompositions in (1.6) yields the following tower of groups with $n \geq 2$:

$$\begin{array}{ccccc}
 & & \mathrm{Aut}(X)_D & & \\
 & & \downarrow & & \\
 & & \vdots & & \\
\Gamma_n(X,1,X) & \xrightarrow{1^+} & E^{n+1}(X) & & \\
 & & \downarrow \lambda & & \\
(1.26) \qquad & & E^n(X) & \xrightarrow{\bar{\mathfrak{O}}} & H_n^{\nabla}(X,1,X). \\
 & & \downarrow & & \\
 & & \vdots & & \\
 & & E^2(X) & & \\
 & & \downarrow & & \\
 & & \mathrm{Aut}_\pi(\pi_*(FX)) & &
\end{array}$$

Here $E^n(X)$ is the group of equivalences of the object $r(X)$ in $\mathbf{T}_n$. Thus $E^n(X)$ is a subset:

$$E^n(X) \subset \bigcup_{\varphi \in G} [X,X]_n^{\varphi}.$$

The composition of all homomorphisms of groups in the column of (1.26) is given by the functors $\pi_n, n \geqq 2$, in (1.2). The obstruction operator $\bar{\mathfrak{O}}$ is a derivation defined as in (IV.4.11) by the corresponding obstruction in (1.23). This shows that kernel $(\bar{\mathfrak{O}})$ is a subgroup of $E^n(X)$ and by sufficiency of the functor λ we have

(a) $$\lambda E^{n+1}(X) = E^n(X) \cap \mathrm{kernel}\,(\bar{\mathfrak{O}}).$$

The action $\Gamma_n +$ in (1.6) and (1.23) respectively yields the homomorphism 1^+ in (1.26) with $1^+(\alpha) = 1 + \alpha$ where 1 is the identity in $E^{n+1}(X)$. From (1.23)(b) we deduce

(b) $$\mathrm{image}\,(1^+) = \lambda^{-1}(1) \cap E^{n+1}(X).$$

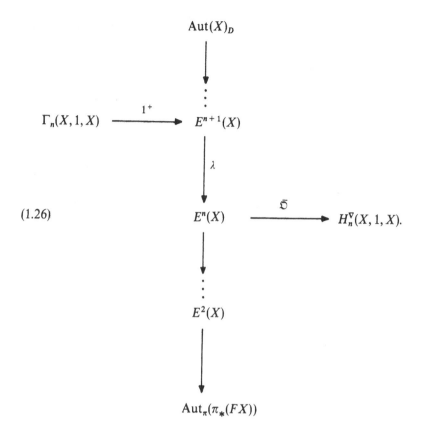

By (a) and (b) we have the short exact sequence of groups ($n \geqq 2$)

(c) $$\frac{\Gamma_n(X,1,X)}{\text{kernel } 1^+} \rightarrowtail E^{n+1}(X) \twoheadrightarrow E^n(X) \cap \text{kernel } \bar{\mathfrak{O}}.$$

The associated homomorphism, see (IV.3.9), is induced by

(d) $$\begin{cases} h: E^n(X) \to \operatorname{Aut} \Gamma_n(X,1,X), \\ \text{with } h(u)(\alpha) = u_*(u^{-1})^*(\alpha) = (u^{-1})_* u^*(\alpha). \end{cases}$$

This follows from the linear distributivity law in (1.23)(b). Moreover, by (1.23)(c) we know

(e) If FX has trivial homotopy groups in degree $\geqq N$ then the homomorphism $r: \operatorname{Aut}(X)_D \to E^n(X)$ is bijective for $n = N-1$ and surjective for $N-2$.

(f) kernel (1^+) is the isotropy group of the action $\Gamma_n +$ on $1 \in [X,X]^{\varphi}_{n+1}$ where $\varphi = 1$. This group can be computed by the spectral sequence (1.23)(d).

(1.27) ***Remark.*** The group $\operatorname{Aut}(X)_D$ is as well studied by Didierjean (1981) and Shih (1964). Both authors define a spectral sequence for $\operatorname{Aut}(X)_D$ by use of the Postnikov-tower of $X \twoheadrightarrow D$. This spectral sequence, however, consists partially of non-abelian groups. The method in (1.26) has the advantage that the groups Γ_n and H^{∇}_n are always abelian and that the isotropy groups of Γ_n are given by a spectral sequence of abelian groups, see (1.26)(f). In addition the result on $\operatorname{Aut}(X)_D$ is just a detail of the tower of categories in (1.6). We leave it as an exercise to deduce from (1.26) results as in (VI.6.13) on solvability and nilpotency of certain subgroups of $\operatorname{Aut}(X)_D$, compare Dror–Zabrodsky (1979).

§ 2. The category of $K(A, n)$-fibrations over D

Let D be a path connected CW-space with basepoint and let $\pi = \pi_1(D)$. A **$K(A, n)$-fibration over D** is a fibration

(2.1) $$K(A,n) \subset X \twoheadrightarrow D,$$

for which the fiber is the Eilenberg–Mac Lane space $K(A,n)$. Such a fibration yields for $n \geqq 2$ the structure of a π-module on $A = \pi_n(FX)$. By (III.7.3) we know that a $K(A,n)$-fibration is a principal fibration classified by the twisted k-invariant

(2.2) $$k(X) \in \hat{H}^{n+1}(D, A).$$

(2.3) ***Definition.*** Let $n \geqq 2$. We denote by $\mathbf{K}_D^n$ the full subcategory of $\mathbf{CW}_D^1/\simeq$ consisting of $K(A,n)$-fibrations where A ranges over all π-modules. ||

For $n=2$ the category $\mathbf{K}_D^2 = \mathbf{Coef}$ is the coefficient category used in (1.3). We can apply the general results in (1.6) to the full subcategory $\mathbf{K}_D^n$ of $\mathbf{CW}_D^1/\simeq$. To this end we introduce the functor

$$\pi_n : \mathbf{K}_D^n \to \mathbf{k}_D^{n+1} \subset \mathbf{Mod}_\pi, \tag{2.3}$$

where $\mathbf{k}_D^{n+1}$ is the following **category of k-invariants**: objects are pairs (A,k) where A is a π-module and where $k \in \hat{H}^{n+1}(D,A)$. Morphisms $\xi:(A',k') \to (A,k)$ are π-linear maps $\xi: A' \to A$ which satisfy $\xi_*(k') = k$. This yields the faithful inclusion into the category $\mathbf{Mod}_\pi$ of π-modules. There is a natural system of abelian groups on the category $\mathbf{k}_D^{n+1}$ by the groups

$$\hat{H}^n(\xi) = \hat{H}^n(D,A), \tag{2.4}$$

which is a covariant functor in A (and thus a $\mathbf{k}_D^{n+1}$-module, see (IV.5.13)). The natural system (2.4) is used in the following result which is an application of (1.6), compare the notation in (IV.3.2).

(2.5) **Theorem.** *There is a linear extension of categories* ($n \geqq 2$):

$$\hat{H}^n + \to \mathbf{K}_D^n \xrightarrow{\pi_n} \mathbf{k}_D^{n+1}.$$

This implies that the functor π_n is full and that each object in $\mathbf{k}_D^{n+1}$ is realizable. Moreover, the action of $\hat{H}^n$ has only trivial isotropy groups. The extension (2.5) is an example for the extension **PRIN** in (V.3.12).

Proof of (2.5). If we restrict the tower in (1.6) to the full subcategory $\mathbf{K}_D^n$ ($n \geqq 3$) we see that all H_i^∇ are trivial and that the single action $\Gamma_i +$, which is not trivial, is the action $\Gamma_{n-1} +$ where $\Gamma_{n-1} + = \hat{H}^n +$. Moreover, for $n \geqq 3$ the full subcategory of $\mathbf{T}_2$, which corresponds to $\mathbf{K}_D^n$, is $\mathbf{k}_D^{n+1}$. For $n=2$ we leave the proof to the reader. The isotropy groups of the action $\Gamma_{n-1} +$ are trivial by (1.23)(d). □

Clearly, (2.5) implies the following corolaries:

(2.6) **Corollary.** *Let $X \twoheadrightarrow D$ and $Y \twoheadrightarrow D$ be fibrations with fiber $K(A,n)$ and $K(B,n)$ respectively, $n \geqq 2$. Then the group $\hat{H}_n(D,A)$ acts freely on the set $[Y,X]_D$ and the set of orbits is $\{\alpha \in \mathrm{Hom}_\pi(B,A): \alpha_* k(Y) = k(X)\}$.*

(2.7) **Corollary.** *Let $X \twoheadrightarrow D$ be a fibration with fiber $K(A,n)$, $n \geqq 2$. Then there is a short exact sequence of groups*

$$\hat{H}^n(D,A) \rightarrowtail \mathrm{Aut}(X)_D \twoheadrightarrow \{\alpha \in \mathrm{Aut}_\pi(A): \alpha_* k(X) = k(X)\}.$$

This corresponds to a result of Didierjean (1981) and Tsukiyama (1980). The associated action in the extension of groups (2.7) is given by $\alpha \longmapsto \alpha_* \in \mathrm{Aut}(\hat{H}^n(D, A))$, compare (1.26)(d). If $k(X) = 0$, the extension in (2.7) splits since the fibration $X \longrightarrow\!\!\!\!\!\rightarrow D$ admits a section in this case. We can use (1.8) for the definition of the splitting. More generally, this yields the following fact:

(2.8) **Proposition.** *The restriction of the extension in* (2.5) *to the full subcategory of objects X with $k(X) = 0$ is a split extension of categories.*

By (IV.6.1) the extension of categories in (2.5) yields the cohomology class.

$$\{\mathbf{K}_D^n\} \in H^2(\mathbf{k}_D^{n+1}, \hat{H}^n). \tag{2.8}$$

Here H^2 is the cohomology of the category $\mathbf{k}_D^{n+1}$ and $\{\mathbf{K}_D^n\}$ is the cohomology class represented by the extension in (2.5). We consider the cohomology class (2.8) as a '**characteristic cohomology class**' of homotopy theory and we expect that there is actually a nice homological algebra of such **universal** cohomology classes which can be exploited for the homotopy classification problems.

§ 3 The de Rham theorem for cohomology groups with local coefficients

In this section we study commutative cochain algebras as in (I.§ 8). We describe various properties which we derive from the general homotopy theory in a cofibration category. We make use of the analogy between the fibration category of spaces (over D) and the cofibration category of commutative cochain algebras (under αD). These two categories have similar properties which can be compared by using the Sullivan–de Rham functor.

Let R be a field of characteristic zero and let $\mathbf{C} = \mathbf{CDA}_*^0$ be the cofibration category of commutative cochain algebras (connected with augmentation) which is described in (I.8.17). We have the initial object $* = R$ in $\mathbf{C}$ which is also the final object so that each cofibrant object in $\mathbf{C}$ is also a based object in $\mathbf{C}$.

We first discuss some simple illustrations of the homotopy theory in Chapter II.

(3.1) ***Definition.*** Let V be an R-module (non graded) and let $n \geqq 1$. We define the Eilenberg–Mac Lane object $\Lambda(V; n)$ in $\mathbf{C}$ by the free object $\Lambda(V)$ where V is concentrated in degree n with $dV = 0$, compare (I.8.2). ||

One can check that for $n \geqq 1$

$$[\Lambda(V; n), X] = \mathrm{Hom}(V, H^n X), \tag{3.2}$$

where H^nX is the cohomology of the underlying cochain complex of X. The left-hand side in (3.2) is the set of homotopy classes. This set is actually an abelian group since

$$\Sigma\Lambda(V; n+1) = \Lambda(V; n), \tag{3.3}$$

and thus for $k \geqq 0$

$$[\Lambda(V; n), X] = [\Sigma^k\Lambda(V; n+k), X] = \pi_k^{\Lambda(V;n+k)}(X). \tag{3.4}$$

For these homotopy groups we have the exact sequence in (II.7.8) for a pair (U, V), $U \rightarrowtail V$. This sequence can be identified with the long exact cohomology sequence of the pair of cochain complexes (U, V) via the commutative diagram ($A = \Lambda(V, n+k)$, $k \geqq 1$)

$$\begin{array}{ccccccccc} H^n(V) & \longrightarrow & H^n(U) & \longrightarrow & H^n(U, V) & \longrightarrow & H^{n+1}(V) & \longrightarrow \\ \| & & \| & & \| & & \| \\ \pi_k^A(V) & \longrightarrow & \pi_k^A(U) & \longrightarrow & \pi_k^A(U, V) & \longrightarrow & \pi_{k-1}^A(V) & \longrightarrow \end{array} . \tag{3.5}$$

For the proof of these results consider the cylinder $I_*\Lambda(V; n)$ defined in (I.8.19) which is given by

$$\begin{cases} I_*\Lambda(V; n) = (\Lambda(V' \oplus V'' \oplus sV), d) \\ dsv = -v' + v'' \quad \text{for } v \in V = V^n. \end{cases} \tag{3.6}$$

Next we define the analogue of a sphere in the category **C**.

(3.7) ***Definition.*** A **sphere** S_n $(n \geqq 1)$ in **C** is an object S_n together with isomorphisms

$$H^k(S_n) = \begin{cases} R & \text{if } k = n, \quad k = 0 \\ 0 & \text{otherwise.} \end{cases}$$ ||

Remark. A sphere S_n exists since we can take for S_n the unique object A in **C** for which the underlying graded module is given by $A^k = R$ for $k = n, k = 0$ and $A^k = 0$ otherwise, $dA = 0$, $A = HA = HS_n$. Furthermore we can take for S_n the object $A_R(S^n)$ given by the Sullivan–de Rham functor in (I.8.23). One can check that a sphere S_n is unique up to a canonical isomorphism in $Ho(\mathbf{C})$.

A minimal model of a sphere is given by

$$S_n = \begin{cases} \Lambda(t), & |t| = n \text{ odd}, \quad dt = 0, \\ \Lambda(t, x), & |t| = n \text{ even}, \quad |x| = 2n-1, \quad dx = t^2. \end{cases} \tag{3.8}$$

For $n \geqq 2$ there is a canonical isomorphism of groups (compare (I.8.21))

$$\mathrm{Hom}_R(\pi_\psi^n(A), R) = [A, S_n], \tag{3.9}$$

where A is a cofibrant object in **C**. The right-hand side has for $n \geqq 2$ a group structure since

$$[A, S_n] = [\Sigma A, S_{n-1}]. \tag{3.10}$$

Proof of (3.9). We can assume that $A = (\Lambda(V), d)$ is minimal with $V^0 = 0$. Let $A(n) = (\Lambda(V^{\leqq n}), d)$. Cofiber sequences show for $n \geqq 2$

$$\begin{aligned}[A, S_n] &= [A(n), S_n] \\ &= [A(n)/A(n-1), S_n] \\ &= [\Lambda(V^n; n), S_n] \\ &= \mathrm{Hom}(V^n, R).\end{aligned}$$

The assumption $n \geqq 2$ is needed for

$$A(n)/A(n-1) = \Lambda(V^n; n).$$

For $n = 1$ the result is still true if $A(1) = \Lambda(V^1; 1)$. Compare also 8.13 in Bousfield–Gugenheim. □

(3.11) ***Remark.*** Let $* \rightarrowtail B \rightarrowtail A$ be a cofibrant pair in **C**. Then the short exact sequence of cochain complexes

$$0 \to QB \to QA \to Q(A/B) \to 0$$

induces a long exact sequence of ψ-homotopy groups (see (I.8.21)(1)):

$$\pi_\psi^n(B) \to \pi_\psi^n(A) \to \pi_\psi^n(A, B) \to \pi_\psi^{n+1}(B).$$

If we apply the functor Hom$(-, R)$ to this sequence we obtain via (3.9) and (3.10) an exact sequence which can be identified with the cofiber sequence $(n \geqq 2)$

$$[B, S_n] \leftarrow [A, S_n] \leftarrow [(A/B)_0, S_n] \leftarrow [\Sigma B, S_n].$$

Recall that $\mathbf{F} = \mathbf{Top}_0^*$ is a *fibration category* with the structure in (I.5.5) and recall that we have the Sullivan–de Rham functor

$$\alpha = A_R : (\mathbf{Top}_0^*)^{op} = \mathbf{F}^{op} \to \mathbf{C} = \mathbf{CDA}_*^0 \tag{3.12}$$

in (I.8.23)(3). The objects $\Lambda(V; n)$ and S_n, respectively, are minimal models $(n \geqq 1)$

$$* \rightarrowtail \Lambda(R; n) \xrightarrow{\sim} \alpha K(R, n) \quad \text{and} \tag{3.13}$$

$$* \rightarrowtail S_n \xrightarrow{\sim} \alpha S^n \tag{3.14}$$

where $K(R, n)$ is the Eilenberg–Mac Lane space of R and where S^n is the n-sphere. Let X be a CW-space in **F**. Then the natural de Rham isomorphism in (I.8.23)(2) can be expressed by the commutative diagram ($n \geqq 1$)

$$\begin{array}{ccc} H^n(X,R) & \xrightarrow{\simeq} & H^n\alpha X \\ \| & & \| \\ [X,K(R,n)] & \xrightarrow[\alpha]{\cong} & [\alpha K(R,n),\alpha X] = [\Lambda(R;n),\alpha X] \end{array} \tag{3.15}$$

Moreover for homotopy groups we get the natural homorphism of R-modules ($n \geqq 2$)

$$\begin{array}{ccc} \pi_n(X)\otimes R & \xrightarrow{\alpha} & \mathrm{Hom}_R(\pi_\psi^n M\alpha X, R) \\ \| & & \| \\ [S^n, X]\otimes R & \xrightarrow[\alpha]{} & [\alpha X, \alpha S^n] = [M\alpha X, S_n] \end{array} \tag{3.16}$$

This actually is an isomorphism provided X is a nilpotent space for which $H_*(X,R)$ is an R-module of finite type. This follows from the equivalence of categories in (I.8.27).

Next we consider the **cohomology with local coefficients**. Let D be a CW-space in $\mathbf{F} = \mathbf{Top}_0^*$ and let A be a $\pi_1(D)$-module. Then we have as in (1.7) the Eilenberg–Mac Lane object which is a based object

$$K(A,n) \xrightarrow{i} L(A,n) \underset{0}{\overset{p}{\rightleftarrows}} D \tag{3.17}$$

in $\mathbf{F}_D$ (here we choose a base point $*\in D$). Similarly we define in the cofibration category $\mathbf{C}$:

(3.18) ***Definition.*** Let B be an object in $\mathbf{C}$ and let V be an R-module (non graded). An **Eilenberg–Mac Lane object in $\mathbf{C}^B$** is a minimal cofibration i,

$$\Lambda(V;n) \xleftarrow{q} L(V,n) \underset{0}{\overset{i}{\leftrightarrows}} B \quad (n \geqq 1)$$

with cofiber $\Lambda(V;n)$ and with retraction 0. This is a based object in $\mathbf{C}^B$. ||

(1.19) **Lemma.** *The generators V of*

$$L(V,n) = (B\otimes\Lambda(V;n), d)$$

can be chosen such that the retraction 0 *is trivial on V, that is, $0 = 1\otimes\varepsilon$.*

Proof. Let r be a retraction of i in (3.18). Then we obtain an isomorphism

$$\tau\colon L(V,n) = B\otimes\Lambda(V,n) \to B\otimes\Lambda(V,n)$$

of graded algebras by $\tau(b) = b$ for $b\in B$ and $\tau(v) = r(v) + v$ for $v\in V$. We have $(1\otimes\varepsilon)\tau = r$. □

(3.20) ***Definition.*** Let V be a (non-graded) R-module and let B be an object in $\mathbf{C}$. *A* **coaction** (V, ∂_V) of B on V is a homomorphism

$$\partial_V: V \to B^1 \otimes V$$

with the properties (a) and (b):

(a) The following diagram commutes where d is the differential of B and where μ is the multiplication in B:

$$\begin{array}{ccc} V & \xrightarrow{\quad\quad\quad\partial_V\quad\quad\quad} & B^1 \otimes V \\ \downarrow{\scriptstyle \partial_V} & & \downarrow{\scriptstyle d\otimes 1} \\ B^1 \otimes V & \xrightarrow{1\otimes\partial_V} B^1 \otimes B^1 \otimes V \xrightarrow{\mu\otimes 1} & B^2 \otimes V \end{array}$$

(b) There is a well-ordered basis J_V of V such that $\partial_V(\alpha) \in B^1 \otimes V_{<\alpha}$ for $\alpha \in J_V$. Compare the notation in (I.8.5).

Now let (V, ∂_V) and (W, ∂_W) be coaction of B on V and W respectively. A **map** $\alpha:(V, \partial_V) \to (W, \partial_W)$ is given by a commutative diagram

$$\text{(c)} \qquad \begin{array}{ccc} V & \xrightarrow{\alpha} & W \\ \downarrow{\scriptstyle \partial_V} & & \downarrow{\scriptstyle \partial_w} \\ B^1 \otimes V & \xrightarrow{1\otimes\alpha} & B^1 \otimes W \end{array}$$

in the category of R-modules. This shows that the set of maps $(V, \partial_V) \to (W, \partial_W)$ is a submodule of $\mathrm{Hom}_R(V, W)$. Let $\mathbf{Co}_B$ be the *category of all coactions* and of maps between coactions. For a map $f: B \to B'$ in $\mathbf{C}$ we get the coaction

(d) $$f_*(V, \partial_V) = (V, (f \otimes 1)\partial_V)$$

of B' on V. Thus f_* is a functor from $\mathbf{Co}_B$ to $\mathbf{Co}_{B'}$. ||

The coaction (V, ∂_V) is the analogue of the $\pi_1(D)$-module A in (3.17), in fact we have:

(3.21) **Lemma.** *An Eilenberg–Mac Lane object* $L(V, n)$, $n \geqq 1$, *in* $\mathbf{C}^B$ *is determined by a coaction* $V = (V, \partial_V)$.

Proof. We choose generators V of $L(V, n)$ as in (3.19). Then the differential d of $L(V, n) = (B \otimes \Lambda(V; n), d)$ is given by a map

$$d = (\partial_V, \delta): V \to L(V, n)^{n+1} = B^1 \otimes V \oplus B^{n+1}$$

Since the retraction $1 \otimes \varepsilon = 0$ is a chain map we see that $\delta = 0$. Moreover

$dd = 0$ shows that ∂_V is a coaction. Vice versa each coaction yields via $d = (\partial_V, 0)$ an object $L(V,n)$. □

Addendum. *Let $f: B' \xrightarrow{\sim} B$ be a weak equivalence in* **C** *and let (V, ∂_V) be a coaction of B on V. Then there exists a coaction (V, ∂'_V) of B' on V, unique up to isomorphism, such that $f_*(V, \partial'_V) \cong (V, \partial_V)$ in* $\mathbf{Co}_B$. *We also write $(V, \partial'_V) \in f_*^{-1}(V, \partial_V)$.*

Proof. For the Eilenberg–Mac Lane object $L(V,n)$ in $\mathbf{C}^B$ we construct a commutative diagram in **C**

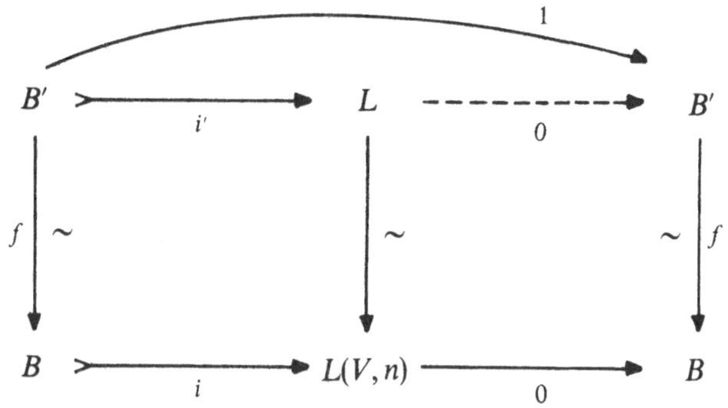

Here i' is a minimal model of if and $0'$ is a lifting as in (II.1.11)(b). It is clear that L is an Eilenberg–Mac Lane object in $\mathbf{C}^{B'}$ for which $f_* L \cong L(V,n)$. Now let (V, ∂'_V) be given by L as in (3.21). Then we get the addendum. □

In $\mathbf{F}_D$, respectively in $\mathbf{C}^B$, we have

(3.22)
$$\begin{cases} \Omega L(A, n+1) = L(A, n) \quad \text{and} \\ \Sigma L(V, n+1) = L(V, n) \end{cases}$$

Compare (III.6.6); for the second equation consider the suspension in $\mathbf{C}^B$ which is defined by the cylinder in (I.8.19). For $V = (V, \partial_V)$ the cylinder

$$I_B L(V,n) = (B \otimes \Lambda(V' \oplus V'' \oplus sV), d) \tag{1}$$

has the differential

$$d(sv) = -v' + v'' + (1 \otimes s)\partial_V(v) \tag{2}$$

where $(1 \otimes s)\,(b \otimes v) = -b \otimes sv$ for $b \in B^1$, $v \in V$. The formula for d is easily checked by (I.8.19)(14). By (1) we get for $V = V^n$

$$\left.\begin{aligned} \Sigma L(V; n) = (B \otimes \Lambda(sV), d) \quad \text{with} \\ ds(v) = (1 \otimes s)\partial_V \end{aligned}\right\} \tag{3}$$

Therefore the isomorphism in (3.22) is given by $sv \longmapsto -v$.

Now let X be a CW-space for which $* \rightarrowtail X$ is a closed cofibration and let $f: X \to D$ be a map in $\mathbf{Top}_0^*$. Then the **reduced cohomology with local coefficients** is given by the homotopy set

$$\hat{H}^n(X, *; f^*A) = [X, L(A, n)]_D^* \quad (n \geqq 1). \tag{3.23}$$

Here the right-hand side is the set of homotopy classes of basepoint preserving maps over D. This is a set of homotopy classes in $Ho(\mathbf{F}_D)$ which by (3.22) is an abelian group.

Remark. For $n \geqq 2$ the reduced cohomology coincides with the cohomology since in this case

$$[X, L(A, n)]_D^* = [X, L(A, n)]_D^\phi$$

by (II.5.19). This is not true for $n = 1$.

(3.24) ***Definition.*** *Let* $V = (V, \partial_V)$ be a coaction of B on V and let $f: B \to X$ be an object in $\mathbf{C}^B$. Then we define the **reduced cohomology of X with local coefficients in $f_* V$** by $(n \geqq 1)$

$$\hat{H}^n(X; f_* V) = [L(V, n), X]^B.$$

Here the right-hand side is a set of homotopy classes in $Ho(\mathbf{C}^B)$; this set is actually an abelian group by (3.22). The cohomology is a functor which is covariant in X and contravariant in V. ||

Remark. One can define the non reduced cohomology by the homotopy set of maps $L(V, n) \to X$ under B which are not augmentation preserving. Here homotopies are given as well by the cylinder in (I.8.19).

There is a de Rham theorem for the cohomology groups above. For this we use the functor

$$\alpha: Ho(\mathbf{F}_D)^{op} \to Ho(\mathbf{C}^{\alpha D}) \tag{3.25}$$

which is given by α in (3.12) and by (II.5a.2).

(3.26) **Proposition.** *Let A be a $\pi_1(D)$-module and suppose that $A \otimes_{\mathbb{Z}} R$ is a finite dimensional R-module and that the action of $\pi_1(D)$ on $A \otimes_{\mathbb{Z}} R$ is nilpotent. Then the $\pi_1 D$-module A determines a coaction $\alpha A = (V, \partial_V)$ of $\alpha(D)$ on $V = \operatorname{Hom}_{\mathbb{Z}}(A, R)$. Moreover, for $n \geqq 1$ one has the natural isomorphism of groups*

$$\alpha_*: \hat{H}^n(X, *; f^*A) \underset{\mathbb{Z}}{\otimes} R \cong \hat{H}^n(\alpha X, (\alpha f)_*(\alpha A)).$$

This isomorphism generalizes the de Rham isomorphism for cohomology groups in (3.15) *above.*

We point out that in the proposition we do not assume that A is a nilpotent $\pi_1(D)$-module, only $A \bigotimes_{\mathbb{Z}} R$ is nilpotent.

Proof. The based object $L(A,n)$ in $\mathbf{F}_D$ yields the based object (in $\mathbf{C}^{\alpha D}$)

$$\alpha D \rightarrowtail M\alpha L(A,n) \xrightarrow{\sim} \alpha L(A,n) \xrightarrow{\alpha(0)} \alpha D \tag{1}$$

where we can assume that the cofibration is minimal. From (I.8.24) we derive that the cofiber is actually a minimal model of $\alpha K(A,n)$, namely $\Lambda(V;n)$. Therefore (1) is an Eilenberg–Mac Lane object $L(V,n) = M\alpha L(A,n)$ in $\mathbf{C}^{\alpha D}$ and thus (1) is determined by a coaction $\alpha(A) = (V, \partial_V)$ as in (3.21). Moreover, the functor α in (3.25) induces the homomorphism of groups

$$[X, L(A,n)] \xrightarrow{\alpha} [L(V,n), \alpha X] \tag{2}$$

which induces the isomorphism in (3.24). This can be seen inductively by (3.15) since nilpotency of $A \bigotimes_{\mathbb{Z}} R$ implies that $L(A \bigotimes_{\mathbb{Z}} R, n)$ is a finite cocomplex $\{L_i\}$ in $\mathbf{F}$ with classifying maps $L_{i-1} \to K(A_i, n)$ where A_i are R-modules with trivial action of $\pi_1(D)$, see (III.7.4). Now cofibration sequences and (3.15) show that α in (2) is an isomorphisms. □

(3.27) **Addendum.** *For a map $f: D' \to D$ in $\mathbf{F}$ we have the isomorphism in $\mathbf{Co}_{\alpha D'}$*

$$\alpha(f^*A) = (\alpha f)_*(\alpha(A)).$$

Moreover, if $R = \mathbb{Q}$, each coaction (W, ∂_W) of αD on W is realizable by a $\pi_1(D)$-module A with $\alpha(A) \cong (W, \partial_W)$ in $\mathbf{Co}_{\alpha D}$.

Realizability of (W, ∂_W) is seen by realizing inductively the elementary cofibrations for $\alpha D \rightarrowtail L(W,n)$.

(3.28) **Lemma.** *For coactions V, W in $\mathbf{Co}_B$ there is a canonical isomorphism of abelian groups ($n \geqq 1$):*

$$\pi^n: [L(V,n), L(W,n)]_0 = \mathbf{Co}_B(V, W).$$

Here the left-hand side denotes the set of homotopy classes in $Ho(\mathbf{C}^B)$ which are based up to homotopy, see (II.9.8), *and the right-hand side is the set of morphisms in $\mathbf{Co}_B$.*

Proof. Since B is connected we can assume $B^0 = R$ (otherwise prove the result first for a minimal model of B). A map $F: L(V,n) \to L(W,n)$ under B is given by a commutative diagram

$$\begin{array}{ccc} V & \xrightarrow{(\alpha, a)} & W \oplus B^n = L(W,n)^n \\ \downarrow{\scriptstyle \partial_V} & & \downarrow{\scriptstyle \partial_w \oplus d} \\ B^1 \otimes V & \xrightarrow[1 \otimes \alpha,\, \mu(1 \otimes a)]{} & B^1 \otimes W \oplus B^{n+1} = L(W,n)^{n+1} \end{array}$$

This shows that a yields a map $a: L(V,n) \to B$ which is homotopic to $1 \otimes \varepsilon$ since F is based up to homotopy. Clearly π^n in (3.28) carries F to α. □

As a special case of (3.26) we get:

(3.29) **Lemma.** *Let A, B be nilpotent $\pi_1(D)$-modules and assume $A \otimes R$ and $B \otimes R$ are finite dimensional R-modules. Then we have the canonical isomorphism*

$$\mathrm{Hom}_\pi(A, B) \otimes R = \mathbf{Co}_{\alpha D}(\alpha A, \alpha B).$$

For the proof we use (3.28), (1.8), (3.24) and (3.23).

§ 4 The tower of categories for commutative cochain algebras under D

As in § 3 let R be a field of characteristic zero and let $\mathbf{C} = \mathbf{CDA}_*^0$ be the cofibration category of commutative cochain algebras (connected with augmentation), see (I.§8). In this section we show that for an object D in $\mathbf{C}$ the homotopy category $Ho\,(\mathbf{C}^D)$ can be approximated by a tower of categories. The tower is analogous to the tower of categories for Postnikov decompositions in § 2. The tower of categories here, however, is given in terms of the purely algebraic category $\mathbf{CDA}_*^0$ and thus it is a kind of an algebraic illustration of the tower of categories for Postnikov decompositions in § 2.

We first consider the analogue of a $K(A,n)$-fibration in the category $\mathbf{C}$, see (2.1). *A* $\Lambda(V,n)$**-cofibration under** B is a cofibration

(4.1) $$B \rightarrowtail A \xrightarrow{q} \Lambda(V,n) \quad \text{in } \mathbf{C}$$

for which the cofiber is $A/B = \Lambda(V,n)$. As in (III.7.3) we get

(4.2) **Proposition.** *Let $n \geqq 2$ and let $b: D \rightarrowtail B$ be a cofibration in $\mathbf{C}$ with simply connected cofiber,* i.e. $H^0(B/D) = R$, $H^1(B/D) = 0$. *Then each $\Lambda(V,n)$-cofibration $B \rightarrowtail A$ determines a coaction $V = (V, \partial_V)$ of D on V and there exists a map $f: L(V, n+1) \to B$ under D such that $B \rightarrowtail A$ is a principal cofibration in the cofibration category $\mathbf{C}^D$ with classifying map f.*

We call the element f,

(4.3) $$f \in [L(V, n+1), B]^D = \hat{H}^{n+1}(B, b_* V),$$

which is uniquely determined by $i: B \rightarrowtail A$ the (twisted) **k-invariant** of the $\Lambda(V, n)$-cofibration i.

Proof of (4.2). It is enough to prove the lemma for the case that $D \rightarrowtail B$ is a minimal cofibration. By the assumption in (4.2) we have $B^1 = D^1$. For the cofibration $B \rightarrowtail A$ we choose an isomorphism

$$A = (B \otimes \Lambda(V; n), d).$$

Then the differential d on generators V is given by

$$d = (\partial_V, f)\colon V \to A^{n+1} = (D^1 \otimes V) \oplus B^{n+1}.$$

This gives us the coaction (V, ∂_V) and f in (4.3) is represented by $f\colon V \to B^{n+1}$. In fact, the equation $dd = 0$ in A shows that ∂_V is a coaction and that $f\colon L(V, n+1) \to B$ with $v \mapsto f(v)$, $v \in V$, is well defined. Moreover, we have the push out

$$\begin{array}{ccc} CL(V, n+1) = (D \otimes \Lambda(V; n+1) \otimes \Lambda(sV; n), d) & \xrightarrow{\bar{f}} & A \\ \uparrow & & \uparrow \\ L(V, n+1) & \xrightarrow[f]{} & B \end{array}$$

with $d(sv) = +v + (1 \otimes s)\partial_V(v)$, see (3.22)(2), and with $\bar{f}(sv) = -v$. □

Now let $D \rightarrowtail A$ be a minimal cofibration with $A = (D \otimes \Lambda(V), d)$. Then we have subalgebras

(4.4) $$D \overset{i}{\rightarrowtail} A(n) = (D \otimes \Lambda(V^{\leqq n}), d) \rightarrowtail A$$

such that $A(n) \rightarrowtail A(n+1)$ is a $\Lambda(V^n; n)$-cofibration. Therefore (4.2) shows that for $V^0 = V^1 = 0$ there are coactions $V^n = (V^n, \partial_{V^n})$ of D on V^n and **k-invariants**

(4.5) $$k_n(A, D) \in \hat{H}^{n+1}(A(n-1), i_* V^n)$$

such that $A(n-1) \rightarrowtail A(n)$ is a principal cofibration in $\mathbf{C}^D$ with attaching map $k_n(A, D)$, $n \geqq 2$. This corresponds exactly to the k-invariants of a Postnikov-tower in (III.7.2). Moreover, as in (VII.3.6) we have

(4.6) **Theorem** (*r-fold principal reduction*). *Let D be an object in* $\mathbf{C}$ *and let $k \geqq r \geqq 1$. Consider all minimal cofibrations $D \rightarrowtail A$ for which the cofiber is k-connected. Then the subalgebras $A(n)$ of A in* (4.4) *yield the cofibrations*

$$\begin{cases} D = A_0 \rightarrowtail A_1 \rightarrowtail \cdots \rightarrowtail A_n \cdots \rightarrowtail A \\ \text{with } A_i = A(ir + k), \quad i \geqslant 0, \end{cases}$$

which form a complex in the category $\mathbf{C}^D$. *The class* $(\mathfrak{E}_k^r)^D$ *of all such complexes is admissible and good and this class satisfies* $\Gamma=\tilde{\Gamma}$ *in degree 2, compare* (VII.1.10) (7).

Proof. We use a general suspension theorem in $\mathbf{C}^D$ which is the analogue of the general loop theorem in $\mathbf{Top}_D$ in (V.10a). Then the same arguments as in (VII.3.6) yield the result. □

As in (VII.3.9) we get the functor

$$\theta_k^r:(\mathbf{CDA}_*^0)_r^D/\simeq\;\to\mathbf{Complex}\,(\mathfrak{E}_k^r)^D/\overset{0}{\simeq} \tag{4.7}$$

where $(\mathbf{CDA}_*^0)_k^D$ is the full subcategory of $\mathbf{C}^D$ consisting of cofibrations $D\rightarrowtail A$ with k-connected cofiber. The functor θ_k^r carries $D\rightarrowtail A$ to the complex $\{A_i\}$ in (4.6). The functor is full and faithful on the subcategory consisting of objects $D\rightarrowtail A$ with $A=A(n)$ for n sufficiently large, see (4.4).

By (VII.1.9) we obtain for $k\geqq r\geqq 1$ the **tower of categories**

$$\mathbf{TWIST}_n(\mathfrak{E}_k^r)^D,\quad n\geqslant 2, \tag{4.8}$$

which approximates the homotopy category $(\mathbf{CDA}_*^0)_k^D/\simeq$ via θ_k^r in (4.7). This tower is an algebraic analogue of the corresponding tower of categories in (VII.3.10). In particular, for $k=r=1$ one obtains a tower $\mathbf{T}_*$ of categories with properties exactly corresponding to the properties of the tower of categories for Postnikov-decompositions in §2, see (6.13) below. We leave it the reader to formulate explicitly the results on the tower (4.8) as described in §2, see also §6. For example, (3.28) is exactly the analogue of (1.8); this is used for the definition of the category $\mathbf{T}_2$, see (1.14).

§5 The category of $\Lambda(V,n)$-cofibrations under D

We describe the results on the homotopy category of $\Lambda(V,n)$-cofibrations under D which correspond exactly to the results on the homotopy category of $K(A,n)$-fibrations in §2.

Let $\mathbf{K}_n^D$ be the full subcategory of $Ho(\mathbf{C}^D)$, $\mathbf{C}=\mathbf{CDA}_*^0$, consisting of $\Lambda(V,n)$-cofibrations $D\rightarrowtail A\to\Lambda(V,n)$ where V ranges over all R-modules. As in (2.3) we have the functor ($n\geqslant 2$):

$$\pi_\psi^n:\mathbf{K}_n^D\to\mathbf{k}_{n+1}^D\subset\mathbf{Co}_D. \tag{5.1}$$

Here the category of k-invariants $\mathbf{k}_{n+1}^D$ is a subcategory of the category of coactions (3.20). Objects are pairs (V,k) where $V=(V,\partial_V)$ is a coaction of D on V and where $k\in\hat{H}^{n+1}(D;V)$. Morphisms $\xi:(V,k)\to(W,k')$ are morphisms $\xi: V\to W$ in $\mathbf{Co}_D$ with $\xi^*(k')=k$.

The functor π_ψ^n in (5.1) carries the $\Lambda(V,n)$-cofibrations $D\rightarrowtail A$ to the coaction (V, ∂_V) described in (4.2) and carries a map $F: A \to A'$ in $\mathbf{K}_n^D$ to $\pi_\psi^n(F)$: $V \to V'$, see (I.8.21). As in (2.4) we obtain a natural system of abelian groups on the category $\mathbf{k}_{n+1}^D$ by the groups

$$(5.2) \qquad \hat{H}^n(\xi) = \hat{H}^n(D, V)$$

for $\xi:(V,k) \to (W,k') \in \mathbf{Co}_D$. This is a contravariant functor in V. Now the linear extension in (2.5) corresponds exactly to

(5.3) **Theorem.** *There is a linear extension of categories* $(n \geqslant 2)$

$$\hat{H}^n + \to \mathbf{K}_n^D \to \mathbf{k}_{n+1}^D.$$

Proof. We use the same arguments as in (2.5). Here we use the tower of categories for **Complex** $(\mathfrak{C}_1^1)^D$ in (4.8). □

The theorem implies that the functor π_ψ^n is full and that each object in $\mathbf{k}_{n+1}^D$ is realizable (as follows from (4.2)). Moreover, the action of $\hat{H}^n$ has only trivial isotropy groups. Again, the extension (5.3) is actually an example for the extension **PRIN** in (V.3.12). Clearly, we have corollaries of (5.3) as in (2.6) and (2.7), see § 6. For example (2.7) corresponds to

(5.4) **Corollary.** *Let $D \rightarrowtail X$ be a $\Lambda(V, n)$-cofibration with coaction V and k-invariant k, $n \geqslant 2$. Then there is a short exact sequence of groups*

$$\hat{H}^n(D, V) \rightarrowtail \mathrm{Aut}(X)^D \twoheadrightarrow \{\psi \in \mathrm{Aut}(V): \psi^* k = k\}.$$

Here $\mathrm{Aut}(X)^D$ *is the group of homotopy equivalances of* X *in* $Ho(\mathbf{C}^D)$, $\mathbf{C} = \mathbf{CDA}_*^0$, *and* $\mathrm{Aut}(V)$ *is the group of automorphisms in the category* $\mathbf{Co}_D$.

We point out that this corollary is a special case of the **tower of groups**, corresponding to (1.26), which is available for any cofibration $D \rightarrowtail X$ in $\mathbf{CDA}_*^0$ and which approximates the group $\mathrm{Aut}(X)^D$, see § 6 below.

§ 6 Integral homotopy theory of spaces over D and minimal models

We here compare the 'integral' tower of categories for Postnikov decompositions with the 'rational' tower of categories for commutative cochain algebras. This yields elementary proofs, and also generalizations, of various fundamental results of Sullivan.

In this section the ring of coefficients is the ring $R = \mathbb{Q}$ of rational numbers. Let D be a path connected CW-space with base point as in § 1. We consider the following class of fibrations over D, compare (I.8.26).

(6.1) Definition. Let $(fn\mathbb{Z})_D$ be the class of all fibrations $p: X \twoheadrightarrow D$ in **Top** such that (a) and (b) hold:

(a) X and the fiber $F = p^{-1}(*)$ are very well pointed CW-spaces and F is simply connected.
(b) The group $\pi_1 D$ acts nilpotently on $\pi_*(F) \otimes \mathbb{Q}$ (or equivalently $\pi_1 D$ acts nilpotently on $H_*(F, \mathbb{Q})$) and $\pi_*(F) \otimes \mathbb{Q}$ or $H_*(F, \mathbb{Q})$ are rational vector spaces of finite type.

Let **Top** $(fn\mathbb{Z})_D$ be the full subcategory of $\mathbf{Top}_D$ consisting of objects in the class $(fn\,\mathbb{Z})_D$. Homotopies in this category are homotopies over D. Moreover, let $(ffn\,\mathbb{Z})_D$ be the subclass of all fibrations in $(fn\,\mathbb{Z})_D$ for which the fibers have only *finitely* many non trivial homotopy groups. Let $(fn\,\mathbb{Q})_D$ and $(ffn\,\mathbb{Q})_D$ be the corresponding subclasses consisting of fibrations with *rational fiber*, that is, $\pi_*(F) = \pi_*(F) \otimes \mathbb{Q}$. ||

Recall that we have the Sullivan–de Rham functor (I.8.23)

(6.2) $$\alpha = A_{\mathbb{Q}}: \mathbf{F} = \mathbf{Top}_0^* \to \mathbf{C} = \mathbf{CDA}_*^0$$

which maps the fibration category $\mathbf{Top}_0^*$ to the cofibration category $\mathbf{CDA}_*^0$ and which is a model functor on an appropriate subcategory of $\mathbf{Top}_0^*$ by the result of Halperin (I.8.24). This result shows that for a fibration $p: X \twoheadrightarrow D$ in $(fn\,\mathbb{Z})_D$ the minimal model

(6.3) $$\alpha(p): \alpha D \rightarrowtail M\alpha X \xrightarrow{\sim} \alpha X$$

gives us the minimal model $M\alpha X/\alpha D$ of the fiber $\alpha(F)$. Therefore we have

$$\begin{cases} M\alpha X = (\alpha(D) \otimes \Lambda(V), d) \text{ with} \\ V = \operatorname{Hom}(\pi_* F, \mathbb{Q}). \end{cases}$$

We now describe the class of all cofibrations in $\mathbf{C} = \mathbf{CDA}_*^0$ which can be obtained by minimal models of fibrations in $(fn\,\mathbb{Z})_D$.

(6.4) Definition. Let $\bar{D}$ be an object in **C**. Then $(fm\,\mathbb{Q})^{\bar{D}}$ denotes the class of all minimal cofibrations $\bar{D} \rightarrowtail X = (\bar{D} \otimes \Lambda(V), d)$ in **C** such that

(a) $V^0 = V^1 = 0$ and
(b) V is a rational vector space of finite type.

Let $\mathbf{CDA}_*^0\,(fm\,\mathbb{Q})^{\bar{D}}$ be the full subcategory of $\mathbf{C}^{\bar{D}}$ consisting of objects in the class $(fm\,\mathbb{Q})^{\bar{D}}$. Homotopies in this category are homotopies relative $\bar{D}$. Let $(ffm\,\mathbb{Q})^{\bar{D}}$ be the subclass of all objects in $(fm\,\mathbb{Q})^{\bar{D}}$ for which V is finitely generated.

Clearly a fibration in $(fn\,\mathbb{Z})_D$ yields via (6.3) a cofibration in $(fm\,\mathbb{Q})^{\alpha D}$. Let $\varphi: \overline{\alpha D} \cong \alpha D$ be an isomorphism in $Ho(\mathbf{C})$ which is given by

$$\text{(6.5)} \qquad \varphi: \overline{\alpha D} \underset{n}{\overset{\sim}{\longleftarrow}} M\alpha D \underset{m}{\overset{\sim}{\longrightarrow}} \alpha D.$$

We now define the functor

$$\text{(6.6)} \qquad M\alpha: \mathbf{Top}(fn\,\mathbb{Z})_D/\simeq \;\to \mathbf{CDA}^D_*(fm\,\mathbb{Q})^{\overline{\alpha D}}/\simeq$$

as follows. For objects X, Y in $(fn\,\mathbb{Z})_D$ we know

$$[X, Y]_D = [X, Y]^*_D, \tag{1}$$

by (II.5.19). Therefore we have the isomorphism of categories

$$\mathbf{Top}(fn\,\mathbb{Z})_D/\simeq \; = \mathbf{Top}^*_0(fn\,\mathbb{Z})_D/\simeq. \tag{2}$$

Now α in (6.2) yields the functor

$$M\alpha: \mathbf{Top}^*_0(fn\,\mathbb{Z})_D/\simeq \;\to \mathbf{CDA}^0_*(fm\,\mathbb{Q})^{\alpha D}/\simeq \tag{3}$$

which carries the object $(X \twoheadrightarrow D) \in (fn\,\mathbb{Z})_D$ to the minimal model (6.3) and which is defined on morphisms as in (II.3.11). Moreover, by (6.5) we have equivalences of categories ($\mathbf{C} = \mathbf{CDA}^0_*$)

$$Ho(\mathbf{C}^{\overline{\alpha D}})_c \underset{n_*}{\overset{\sim}{\longleftarrow}} Ho(\mathbf{C}^{M\alpha D})_c \underset{m_*}{\overset{\sim}{\longrightarrow}} Ho(\mathbf{C}^{\alpha D})_c. \tag{4}$$

Compare (II.4.5). By use of (2), (3) and (4) we get the functor $M\alpha$ in (6.6). This functor fits into the commutative diagram of functors

$$\begin{array}{ccc} \mathbf{Top}(fn\mathbb{Z})_D/\simeq & \xrightarrow{\;M\alpha\;} & \mathbf{CDA}^0_*(fm\mathbb{Q})^{\overline{\alpha D}}/\simeq \\ & {\scriptstyle F_{\mathbb{Q}}}\searrow \qquad \nearrow{\scriptstyle \sim\; M\alpha_{\mathbb{Q}}} & \\ & \mathbf{Top}(fn\mathbb{Q})_D/\simeq & \end{array} \tag{5}$$

where $F_{\mathbb{Q}}$ is the functor obtained by fiberwise $\mathbb{Q}$-localization as in Bousfield–Kan. Clearly $F_{\mathbb{Q}}$ is the identity on $\mathbf{Top}\,(fn\,\mathbb{Q})_D$ and $M\alpha_{\mathbb{Q}}$ is the restriction of $M\alpha$. We claim that $M\alpha_{\mathbb{Q}}$ is an equivalence of categories. We prove this for the subclass $(ffn\mathbb{Q})_D$, see (6.15):

(6.7) **Theorem.** *The functor*

$$M\alpha_{\mathbb{Q}}: \mathbf{Top}(ffn\mathbb{Q})_D/\simeq \;\longrightarrow \mathbf{CDA}^0_*(ffm\mathbb{Q})^{\overline{\alpha D}}/\simeq$$

is an equivalence of categories.

If $D = *$ is a point this result yields part of the Sullivan–de Rham equivalence of categories in (I.8.27). We point out that the theorem does not use any nilpotency condition on the space D. In case D is nilpotent the theorem corresponds to a result of Grivel who, however, only considers the subclass of $(fn\mathbb{Q})_D$ consisting of fibrations for which $\pi_1 D$ acts trivially on the homotopy groups of the fiber, compare also Silveira. Also Scheerer (1980, 1983) obtained the equivalence of categories in (6.7) by generalizing the method of Bousfield–Gugenheim. We give a new and elementary proof of (6.7) which relies only on the Sullivan–de Rham isomorphism for cohomology groups (I.8.23)(2) and on the compatibility of α with pull backs in (I.8.24).

As an easy illustration we first prove theorem (6.7) for the full subcategory of Eilenberg–Mac Lane fibrations over D. Let

$$\begin{cases} \mathbf{K}^n(fn\mathbb{Z})_D = \mathbf{K}^n_D \cap \mathbf{Top}(fn\mathbb{Z})_D, & \text{and} \\ \mathbf{K}_n(fm\mathbb{Q})^{\alpha D} = \mathbf{K}^{\alpha D}_n \cap \mathbf{DGA}^0_*(fm\mathbb{Q})^{\alpha D} \end{cases}$$

be the intersections of the categories defined in (2.3), (6.1) and (5.1), (6.4) respectively.

(6.8) **Proposition.** *There is a commutative diagram of linear extensions of categories* ($n \geqq 2$)

$$\begin{array}{ccccc} \hat{H}^n & \xrightarrow{+} & \mathbf{K}^n(fn\mathbb{Z})_D & \longrightarrow & \mathbf{k}^{n+1}(fn\mathbb{Z})_D \subset \mathbf{Mod}_\pi \\ \downarrow{\scriptstyle \alpha_*} & & \downarrow{\scriptstyle M\alpha} & & \downarrow{\scriptstyle \alpha} \\ \hat{H}^n & \xrightarrow{+} & \mathbf{K}_n(fm\mathbb{Q})^{\alpha D} & \longrightarrow & \mathbf{k}_{n+1}(fm\mathbb{Q})^{\alpha D} \subset \mathbf{Co}_{\alpha D} \end{array}$$

Here the top row is a subextension of (2.5) and the bottom row is a subextension of (5.3).

Proof of (6.8). The proposition follows from (II.8.24) and (V.3.16) since $M\alpha$ carries principal maps to principal maps. □

For morphism sets diagram (6.8) yields the following commutative diagram where X and Y are fibrations in $(fn\mathbb{Z})_D$ with fiber $K(A,n)$ and $K(B,n)$ respectively.

(6.9)
$$\begin{array}{ccccc} \hat{H}^n(D, A) & \overset{+}{\rightarrowtail} & [Y, X]_D & \twoheadrightarrow & \{\varphi \in \mathrm{Hom}_\pi(B, A) \mid \varphi_*(kY) = kX\} \\ {\scriptstyle \alpha_*}\downarrow & & \downarrow{\scriptstyle M\alpha} & & \downarrow{\scriptstyle \alpha} \\ \hat{H}^n(\alpha D; \alpha A) & \overset{+}{\rightarrowtail} & [\bar{X}, \bar{Y}]^{\alpha D} & \twoheadrightarrow & \{\psi \in \mathrm{Hom}(\alpha A, \alpha B) \mid \psi^*(k\bar{Y}) = k\bar{X}\} \end{array}$$

Here $\bar{X} = M\alpha X$, $\bar{Y} = M\alpha Y$ are minimal models under αD. The top row is the exact sequence described in (2.6); the bottom row is the corresponding exact sequence derived from (5.3). For the groups of homotopy equivalences we have the commutative diagram of short exact sequences of groups:

$$(6.10)\qquad \begin{array}{ccccc} \hat{H}^n(D, A) & \rightarrowtail & \mathrm{Aut}(X)_D & \twoheadrightarrow & \{\varphi \in \mathrm{Aut}_\pi(A) \mid \varphi_* kX = kX\} \\ \downarrow{\scriptstyle \alpha_*} & & \downarrow{\scriptstyle M\alpha} & & \downarrow{\scriptstyle \alpha} \\ \hat{H}^n(\alpha D, \alpha A) & \rightarrowtail & \mathrm{Aut}(\bar{X})^{\alpha D} & \twoheadrightarrow & \{\psi \in \mathrm{Aut}(\alpha A) \mid \psi^* k\bar{X} = k\bar{X}\} \end{array}$$

Compare (2.7) and (5.4). As a special case of (6.7) we now prove:

(6.11) **Proposition.** *The restriction*

$$M\alpha_{\mathbb{Q}} : \mathbf{K}^n(fn\mathbb{Q})_D \xrightarrow{\simeq} \mathbf{K}_n(fm\,\mathbb{Q})^{\alpha D}$$

of $M\alpha$ *in* (6.8) *is an equivalence of categories, compare* (6.6)(5).

Proof. Assume that Y, X in (6.9) are fibrations in $(fn\mathbb{Q})_D$. Then (3.29) shows that α in (6.9) is a bijection. Moreover, the de Rham isomophism (3.26) shows that α_* is an isomorphism. Thus $M\alpha$ in (6.9) is a bijection and therefore the functor $M\alpha_{\mathbb{Q}}$ in (6.11) is full and faithful. Also, each object (V, k) in $\mathbf{k}_{n+1}(fm\mathbb{Q})^{\alpha D}$ is realizable since V is realizable by (3.27) and since the k-invariant k is realizable by the de Rham isomorphism (3.26). □

The proof shows that $M\alpha_{\mathbb{Q}}$ in (6.11) yields an **isomorphism of linear extensions** of categories by restriction of the map $M\alpha$ in (6.8). Essentially the same arguments as in the proof of (6.11) yield the result in (6.7). We only replace linear extensions by 'towers of categories'.

Let $\mathfrak{X}$ and $\mathfrak{X}_{\mathbb{Q}}$ be the subclasses of $(\mathfrak{C}_1^1)_D$ in (VII.3.6) corresponding to $(fn\mathbb{Z})_D$ and $(fn\mathbb{Q})_D$ respectively. Moreover, let $\mathfrak{X}_\alpha = M\alpha(\mathfrak{X})$ be the subclass of $(\mathfrak{C}_1^1)^{\alpha D}$ in (4.6) given by $(fm\mathbb{Q})^{\alpha D}$, compare (VII.1.19). Here $M\alpha(\mathfrak{X})$ is a class of complexes by the result of Halperin (I.8.24). Diagram (6.6)(5) above now corresponds to the following diagram of functors

$$(6.12)\qquad \begin{array}{ccc} \mathbf{Cocomplex}(\mathfrak{X})/\overset{0}{\simeq} & \xrightarrow{M\alpha} & \mathbf{Complex}(\mathfrak{X}_\alpha)/\overset{0}{\simeq} \\ & {\scriptstyle F_{\mathbb{Q}}}\searrow \quad \nearrow{\scriptstyle M\alpha_{\mathbb{Q}}} & \\ & \mathbf{Cocomplex}(\mathfrak{X}_{\mathbb{Q}})/\overset{0}{\simeq} & \end{array}$$

Here $M\alpha_{\mathbb{Q}}$ is the restriction of $M\alpha$ as in (6.6)(5). By (VII.1.19) these functors induce maps between towers of categories. The map $M\alpha$ between towers of categories is described by the following diagram.

(6.13)

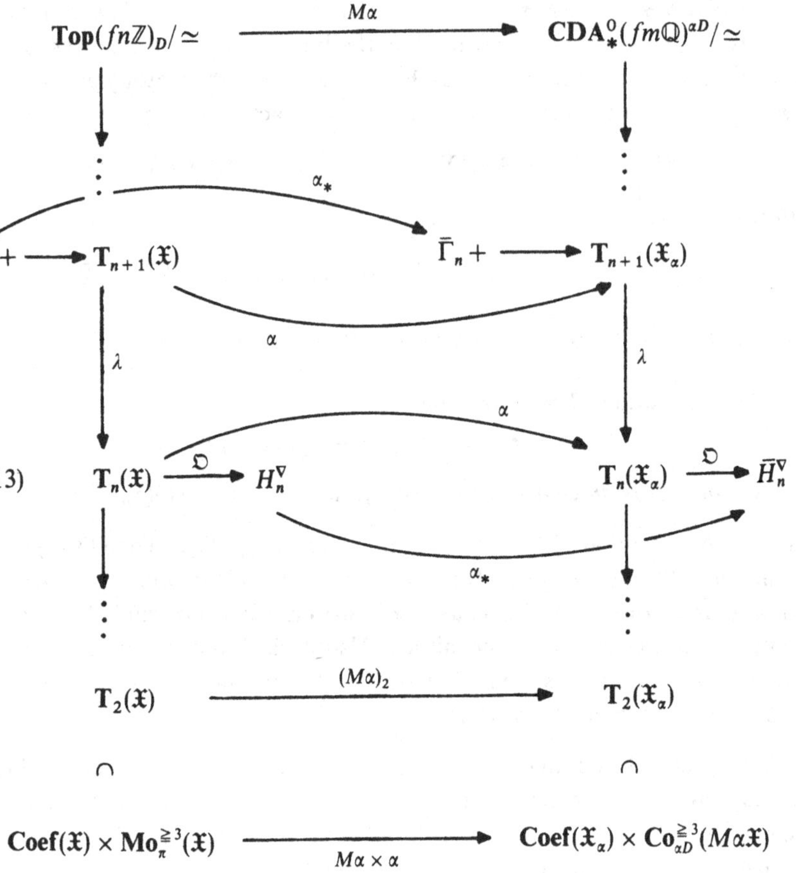

The left-hand side is a subtower of (1.6), the right-hand side is the corresponding subtower of (4.8) with $\mathbf{T}_n(\mathfrak{X}_\alpha) = \mathbf{TWIST}_n(\mathfrak{X}_\alpha)$.

Next we describe diagram (6.13) on the level of morphism sets. Let $Y \twoheadrightarrow D$ and $X \twoheadrightarrow D$ be objects in $(fn\mathbb{Z})_D$ and let $\varphi\colon Y_1 \to X_1$ be a map in $\mathbf{Coef}(\mathfrak{X})$. By (6.3) we obtain the objects $\bar{Y} = M\alpha Y$ and $\bar{X} = M\alpha X$ in $(fm\mathbb{Q})^{\alpha D}$ and the induced map $\bar{\varphi} = M\alpha(\varphi)\colon \bar{X}_1 \to \bar{Y}_1$ in $\mathbf{Coef}(\mathfrak{X}_\alpha)$. With these notations diagram (6.13) yields the commutative diagram of sets

(6.14)
$$\begin{array}{ccccccc}
\Gamma_n(Y,\varphi,X) & \xrightarrow{+} & [Y,X]^{\varphi}_{n+1} & \xrightarrow{\lambda} & [Y,X]^{\varphi}_n & \longrightarrow & H^{\nabla}_n(Y,\varphi,X) \\
\downarrow{\scriptstyle \alpha_*} & & \downarrow{\scriptstyle (M\alpha)_{n+1}} & & \downarrow{\scriptstyle (M\alpha)_n} & & \downarrow{\scriptstyle \alpha_*} \\
\bar{\Gamma}_n(\bar{X},\bar{\varphi},\bar{Y}) & \xrightarrow{+} & [\bar{X},\bar{Y}]^{\varphi}_{n+1} & \xrightarrow{\lambda} & [\bar{X},\bar{Y}]^{\varphi}_n & \longrightarrow & \bar{H}^{\nabla}_n(\bar{X},\bar{\varphi},\bar{Y})
\end{array},$$

the top row of which is the exact sequence in (1.23). Clearly, diagram (6.9) above is a special case of (6.14). The isotropy groups of the action at the left hand side of the diagram can be computed by spectral sequences as in (1.23)(d).

(6.15) *Proof of* (6.7). We show that $M\alpha_{\mathbb{Q}}$ in (6.12) induces an isomorphism of towers of categories

$$\mathbf{T}_*(\mathfrak{X}_{\mathbb{Q}}) \cong \mathbf{T}_*(\mathfrak{X}_\alpha).$$

For this we generalize the argument in the proof of (6.9). As in this proof it is readily seen that each object in $(fm\,\mathbb{Q})^{\alpha D}$ is realizable by an object in $(fn\,\mathbb{Q})_D$. Moreover, (6.11) and (3.29) show that $(M\alpha) \times \alpha$ in the bottom row of (6.13), restricted to $\mathfrak{X}_{\mathbb{Q}}$, is an equivalence of categories. This implies that $M\alpha_{\mathbb{Q}}: \mathbf{T}_2(\mathfrak{X}_{\mathbb{Q}}) \to \mathbf{T}_2(\mathfrak{X}_\alpha)$ is an equivalence of categories, compare (1.14). The de Rham theorem (3.26) shows that the maps α_* in (6.14) both are isomorphisms provided X and Y are in $(fn\mathbb{Q})_D$. Moreover, α_* induces an isomorphism of the isotropy groups of Γ_n+ and $\bar{\Gamma}_n+$ respectively. This follows from the de Rham theorem (3.26) by use of the spectral sequences (1.23)(d) which are natural with respect to α by (III.2.10). (In this spectral sequence we are allowed to use the reduced cohomology groups, see (3.23).) Inductively, we now derive from (6.14) that $(M\alpha)_n$ is a bijection for all n. Therefore, $M\alpha_{\mathbb{Q}}$ in (6.7) yields an isomorphism of towers of categories, compare also (IV.4.14). □

In addition to (6.14) we have for $(X \twoheadrightarrow D) \in (fn\mathbb{Z})_D$ the following commutative diagram in which the top row is the exact sequence (1.26), (moreover, diagram (6.10) is a special case of this diagram).

$$\begin{array}{ccccccc}
\Gamma_n(X,1,X) & \xrightarrow{+} & E^{n+1}(X) & \xrightarrow{\lambda} & E^n(X) & \xrightarrow{\bar{\mathfrak{D}}} & H_n^\nabla(X,1,X) \\
\downarrow{\scriptstyle \alpha_*} & & \downarrow{\scriptstyle (M\alpha)_{n+1}} & & \downarrow{\scriptstyle (M\alpha)_n} & & \downarrow{\scriptstyle \alpha_*} \\
\bar{\Gamma}_n(\bar{X},1,\bar{X}) & \xrightarrow{+} & E^{n+1}(\bar{X}) & \longrightarrow & E^n(\bar{X}) & \longrightarrow & \bar{H}_n^\nabla(\bar{X},1,\bar{X})
\end{array}$$

The bottom row is an exact sequence which depends purely algebraicly on the minimal model $\alpha D \rightarrowtail \bar{X} \xrightarrow{\sim} \alpha X$ of $\alpha(X \twoheadrightarrow D)$.

We consider various applications of the fundamental diagram (6.13). For the homotopy groups π_ψ^n defined in (I.8.21) we get the

(6.16) **Proposition.** *For a fibration* $(p: X \twoheadrightarrow D) \in (fn\mathbb{Z})_D$ *with fiber* F *we have the natural isomorphism of groups*

$$\pi_\psi^n(M\alpha X, \alpha D) \cong \mathrm{Hom}_{\mathbb{Z}}(\pi_n F, \mathbb{Q}).$$

Here $M\alpha X$ is a minimal model of $\alpha(p)$ as in (6.3).

Proof. We know that $M\alpha X/\alpha D$ is a minimal model of αF. Therefore we can apply (6.7) with $D = *$ since F is simply connected. Hence $\alpha_*: \pi_n(F) \otimes \mathbb{Q} = [M\alpha X/\alpha D, S_n]$. Moreover, $\pi_*(F) \otimes \mathbb{Q}$ is of finite type and we can use (3.9). □

Moreover, we compute rational homotopy groups of function spaces, see (III.8.2) where we set $Y = *$.

(6.17) **Proposition.** *Let $p: E \twoheadrightarrow D$ be a fibration in $(ffn\mathbb{Z})_D$ and let X be a path connected CW-space for which the inclusion $* \to X$ is a closed cofibration in* **Top**. *Then we have the isomorphism of groups* ($n \geqq 1$)

$$\pi_n(\mathrm{Map}(X, E)^*_D, u) \otimes_{\mathbb{Z}} \mathbb{Q} \cong [\Sigma^n_{\alpha D} M\alpha E, \alpha X]^{\bar{u}}.$$

Here $\alpha D \rightarrowtail M\alpha E \xrightarrow{\sim} \alpha E$ is a minimal model of $\alpha(p)$ and $\bar{u}: M\alpha E \xrightarrow{\sim} \alpha E \to \alpha X$ is given by $\alpha(u)$. For $n = 1$ the left-hand side is the $\mathbb{Q}$-localization of the nilpotent group π_1, see (III.8.3).

Proof. The isomorphism is induced by α as in (II.10a.1). The spectral sequence (III.9.8)(ii) and (III.2.10) show that we obtain an isomorphism since α induces an isomorphism on the E_1-term by the de Rham theorem (3.26). □

We point out that the group in (6.17) can be described in explicit algebraic terms by employing the cylinder construction in (I.8.19).

Remark. Though we have by (6.17) a good description of the rational homotopy groups of function spaces we do not know a minimal model of the function space Map $(X, E)^*_D$ in terms of $M\alpha E$ and αX. By (6.17) the generators of this minimal model are given by the homotopy groups (6.17). A special case of this problem is solved by Haefliger, see also Silveira.

The following result is the analogue of (6.17) for $n = 0$; it counts the path components of the space $\mathrm{Map}(X, E)^*_D$. Recall that a CW-space X is of **finite type** if X is homotopy equivalent to a CW-complex with finitely many cells in each dimension $\geqq 0$. If X is simply connected this is equivalent to the condition that the abelian groups $H_*(X, \mathbb{Z})$ or $\pi_*(X)$ are of finite type.

(6.18) **Proposition.** *Let $u: X \to D$ be given and let $E \twoheadrightarrow D$ be a fibration in $(ffn\mathbb{Z})_D$ with fiber F. Assume X and F are CW-spaces of finite type. Then the map*

$$\alpha: [X, E]_D \to [M\alpha E, \alpha X]^{\bar{u}}$$

is finite to one. Here we use the same notation as in (6.17).

Proof. We apply (III.4.18) and the fact that the cohomology groups $\hat{H}^n(X, u^*\pi_m F)$ are finitely generated abelian groups. Therefore the result follows as in (6.17) by (3.26). □

Finally, we obtain the following result on groups of homotopy equivalences (over a space D).

(6.19) **Proposition.** *Let $E \twoheadrightarrow D$ be a fibration in $(ffn\mathbb{Z})_D$ with fiber F and assume F and D are CW-spaces of finite type. Moreover, assume that there exists an isomorphism*

$$\overline{\alpha D} \cong \alpha D$$

in $Ho(\mathbf{CDA}_*^0)$ such that $\overline{\alpha D}$ is a vector space of finite type. Then the homomorphism

$$M\alpha\colon \mathrm{Aut}(E)_D \to \mathrm{Aut}(M\alpha E)^{\overline{\alpha D}},$$

given by (6.6), *has finite kernel and the image is an arithmetic group. This shows* $\mathrm{Aut}(E)_D$ *is a finitely presented group.*

This result is more general than the result of Scheerer (1980) since $(E \twoheadrightarrow D) \in (ffn\,\mathbb{Z})_D$ does not imply that $E \twoheadrightarrow D$ is a nilpotent fibration, see (6.1). We can still employ the inductive method in Scheerer (1980) or in Sullivan since we have the commutative diagram of groups in (6.16) and since we have the de Rham isomorphism (3.26) for cohomology groups with local coefficients. Here, in fact, the tower (1.6) is a crucial tool since this tower does not need the nilpotency of Postnikov decompositions. For $D = *$ the result corresponds to the original case considered by Sullivan, a different approach is due to Wilkerson.

IX

Homotopy theory of reduced complexes

In general, towers of categories are constructed by use of twisted chain complexes. In this chapter, we consider examples of towers of categories for which the underlying chain complexes are not twisted. This relies on the condition of 'simply connectedness' for reduced complexes (the non-simply connected and relative theory is discussed in Chapter VI). Hence in this chapter we describe the simplest examples of towers of categories from which, nevertheless, fundamental and classical results of homotopy theory can be deduced immediately. We also obtain some new results.

§ 1 Reduced complexes

We consider simultaneously four classes of 'reduced' complexes given by CW-complexes, localized CW-complexes, free chain algebras, and free chain Lie algebras respectively.

(A) ***Reduced CW-complexes.*** Let $\mathbf{C} = \mathbf{Top}^*$ be the cofibration category of pointed topological spaces in (I.5.4). A **reduced complex** in **C** is a CW-complex X with a trivial 1-skeleton $X^1 = *$. Such a complex determines the filtered object $X = \{X_n\}$ in **C** where $X_n = X^n$ is the n-skeleton. The 1-*sphere in* **C** is $S = S^1$. The ring of coefficients for **C** is the ring $R = \mathbb{Z}$ of integers.

(B) ***Localized reduced CW-complexes.*** Let R be a subring of $\mathbb{Q}$ and let $\mathbf{C} = \mathbf{CW\text{-}spaces}^*(R)$ be the cofibration category of pointed CW-spaces given by (I.5.10) where we assume $h_* = H_*(-, R)$ to be the singular homology with coefficients in R. A **reduced complex in C** is a filtered object $X = \{X_n\}$ in **C** with $X_1 = *$ such that $X_{n-1} \subset X_n$ is a principal cofibration in **C** with an attaching map $A_n \to X_{n-1}$ where A_n is a one point union of R-local $(n-1)$-spheres $S_R^{n-1}, n \geqq 2$. This implies that all X_n are R-local spaces. The **1-sphere in C** is the R-local 1-sphere $S = S_R^1$. For $R = \mathbb{Z}$ the reduced complexes here coincide with those in case (A) above.

(C) ***Reduced chain algebras.*** Let R be a principal ideal domain and let $\mathbf{C} = \mathbf{DA}_*$ (flat) be the cofibration category of augmented chain algebras over R in (I.7.10). A **reduced complex** in **C** is a free chain algebra $X = (T(V), d)$ with $V_0 = 0$. We derive from X the filtered object $X = \{X_n\}$ in **C** where $X_{n+1} = (T(V_{\leqq n}), d)$ is given by $V_{\leqq n} = \{x \in V : |x| \leqq n\}$. The **1-sphere in C** is $S = T = (T(t), d = 0)$ where t is a generator of degree 0.

(D) ***Reduced chain Lie algebras.*** Let $\mathbf{C} = \mathbf{DL}$ be the cofibration category of chain Lie algebras over a field R of characteristic 0, see (I.9.13). A **reduced complex in C** is a free chain Lie algebra $X = (L(V), d)$ with $V_0 = 0$. As in (C) above we obtain the filtered object $X = \{X_n\}$ where $X_{n+1} = (L(V_{\leqq n}), d)$. The **1-sphere in C** is $S = (L(t), d = 0)$ where t is a generator of degree 0.

In each case $*$ is the initial and final object of **C**. Now assume that **C** is one of the categories as described in (A), (B), (C), and (D) above. For a reduced complex X in **C** we obtain the **cellular chain complex** $\tilde{C}_* X$ in $\mathbf{Chain}_R$ by $\tilde{C}_n X = 0$ for $n \leqq 1$ and by

(1.1) $$\tilde{C}_n X = \pi_{n-1}^S (X_n, X_{n-1}) \quad \text{for } n \geqq 2$$

where we use the boundary in (III.10.6). Hence we get for an R-module M the **cellular (co) homology**

(1.2) $$\begin{cases} \tilde{H}_*(X; M) = H_*(\tilde{C}_* X \otimes_R M) \\ \tilde{H}^*(X; M) = H^* \mathrm{Hom}_R(\tilde{C}_* X, M). \end{cases}$$

(1.3) ***Remark.*** With the notation in (III.10.6)(3) we have $\tilde{H}_n(X, R) = H_n(\tilde{C}_* X) = H_{n-1}^S(s^{-1} X)$ where $s^{-1} X$ is the filtered object with $(s^{-1} X)_n = X_{n+1}$. In case (A) and (B) the cellular (co)-homology (1.2) coincides with the reduced singular (co)-homology of the space X. In case (C) and (D) we have an isomorphism of chain complexes

$$\tilde{C}_* X = s(QX, \bar{d}) \tag{1}$$

where $QX = V$ is the module of indecomposables and where $\bar{d}$ is the differential induced by the differential on X. The suspension s in $\mathbf{Chain}_R$ is given by $dsv = (-1)^n s dv, v \in V_n$. One can check by (1) that in case (C)

$$\tilde{H}_n(X, M) = \mathrm{Tor}_n^X(R, M) \quad (n \geqq 1) \tag{2}$$

is a 'differential Tor' and that in case (D)

$$\tilde{H}_n(X, M) = \mathrm{Tor}_n^{UX}(R, M) \quad (n \geqq 1), \tag{3}$$

compare (I.9.3)(5).

As in (III.10.7)(3) we define the Γ-groups of a reduced complex X by

(1.4) $$\Gamma_n^S(s^{-1}X) = \mathrm{image}\,(\pi_n^S(X_n) \to \pi_n^S(X_{n+1})).$$

It is easy to check that a reduced complex satisfies $\pi_q^S(X_n, X_{n-1}) = 0$ for $q < n-1$. Whence by (III.10.7) we have Whitehead's long exact sequence (of R-modules)

(1.5) $$\cdots \xrightarrow{h} \tilde{H}_4(X, R) \xrightarrow{b} \Gamma_2^S(s^{-1}X) \xrightarrow{i} \pi_2^S(X) \xrightarrow{h} \tilde{H}_3(X, R)$$

for each reduced complex X in **C**. Here $\pi_n^S(x) = [\Sigma^n S, X]$ is a homotopy group in **C**, compare (III.10.7)(1). Clearly in case (A) the sequence (1.5) is the classical one for a simply connected space, see (III.§11). We point out that the sequence (1.5) is an invariant of the homotopy type of X in **C** (this, in fact, holds in all cases (A), (B), (C) and (D)).

(1.6) ***Remark***: In case (A) and (B) the homotopy group

$$\pi_n^S(X) = [\Sigma^n S_R^1, X] = \pi_{n+1}(X) \tag{1}$$

is a usual homotopy group in **Top** since X is R-local. In case (C) and (D) the homotopy group

$$\pi_n^S(X) = H_n(X) \tag{2}$$

is the homology of the underlying chain complex of X, see (II.17.20) for case (C).

§2 The tower of categories for reduced complexes

Let **C** be a category as described in §1. We first show that reduced complexes in **C** are complexes in the sense of (III.§4). In fact they form a very good class of complexes so that we obtain a tower of categories which approximates the homotopy category of reduced complexes.

Let C be a free R-module with basis B. Then we call the sum

(2.1) $$M(C, n) = \bigvee_B \Sigma^{n-1} S, \quad n \geqq 2,$$

a '**Moore space**' of C in degree n and we call $M(R, n) = \Sigma^{n-1} S$ an ***n*-sphere in C**. Clearly, we have an obvious isomorphism

(2.2) $$[M(C, n), X] = \mathrm{Hom}_R(C, \pi_{n-1}^S X)$$

where the left-hand side denotes the set of homotopy classes in **C**. In addition we get the isomorphism

(2.3) $$\pi^S_{n-1}(M(C,n)) = C,$$

which is compatible with suspension, $n \geqq 2$. Let $X = \{X_n\}$ be a reduced complex in **C**. Then the inclusion $X_{n-1} \rightarrowtail X_n$ is a principal cofibration with an attaching map

(2.4) $$f_n: M(C_n, n-1) \to X_{n-1}.$$

Here $C_n = \tilde{C}_n X$ is given by the cellular chain complex of X. This shows that a reduced complex in **C** is actually a based complex in **C** in the sense of (III.3.1). The boundary $d_n: C_n \to C_{n-1}$ of $\tilde{C}_* X$ in (1.1) as well can be obtained by the composition ∂f_n:

(2.5) $$M(C_n, n-1) \to X_{n-1} \to X_{n-1}/X_{n-2} = M(C_{n-1}, n-1),$$

compare (III.3.9). This shows that the chain complex $K(X)$ in (III.3.5) is actually determined by the cellular chain complex $\tilde{C}_* X$. There is a coaction

(2.6) $$\mu: X_n \to X_n \vee M(C_n, n),$$

given by the principal cofibration $X_{n-1} \subset X_n$. In case (A) and (B) this is the usual coaction on a mapping cone in **Top**. In case (C) and (D) one can check that μ carries a generator $v \in V_{n-1} = C_n$ to the sum $v + v'$ where $v' \in C_n$ is the element corresponding to v.

(2.7) ***Remark.*** Let $\mathfrak{X} = \mathfrak{X}_{\mathrm{C}}$ be the class of all reduced complexes in **C**. Then $\mathfrak{X}$ is a very good class of complexes. In case (A) and (C) this follows from (VII.3.1) and (VII.4.6) since $\mathfrak{X} = (\mathfrak{X}^1_1)^*$. We leave it as an exercise to check this for case (B) and (D). As in (VII.§2) we thus have the tower $\mathbf{TWIST}^c_*(\mathfrak{X})/\simeq$ of categories which in case (A) is the subtower for 1-reduced CW-complexes of the tower in (VI.6.2).

As in (VI.5.7) we obtain the following definition:

(2.8) ***Definition.*** Let $n \geqq 2$. A **reduced homotopy system** in **C** **of order** $(n+1)$ is a triple (C, f_{n+1}, X_n) where X_n is an n-dimensional reduced complex in **C** and where $C = (C_*, d)$ is a free chain complex in $\mathbf{Chain}_R$ which coincides with $\tilde{C}_*(X_n)$ in degree $\leqq n$. Moreover, $f_{n+1}: C_{n+1} \to \pi^S_{n-1}(X_n)$ is a homomorphism of R-modules such that

$$d_{n+1} = \partial f_{n+1} \tag{1}$$

is given by the composition (2.5) and such that the **cocycle condition** $f_{n+1} d_{n+2} = 0$ is satisfied. A **morphism** between reduced homotopy systems of order $(n+1)$ is a pair (ξ, η) which we write

$$(\xi, \eta):(C, f_{n+1}, X_n) \to (C', g_{n+1}, Y_n).$$

Here $\eta: X_n \to Y_n$ is an element in $\mathbf{Fil(C)}(X_n, Y_n)/\overset{0}{\simeq}$ and $\xi: C \to C'$ is a chain map in $\mathbf{Chain}_R$ which coincides with $\tilde{C}_*(\eta)$ in degree $\leqq n$ and for which the following diagram commutes

$$\begin{array}{ccc} C_{n+1} & \xrightarrow{\xi_{n+1}} & C'_{n+1} \\ \Big\downarrow{\scriptstyle f_{n+1}} & & \Big\downarrow{\scriptstyle g_{n+1}} \\ \pi^S_{n-1} X_n & \xrightarrow[\eta_*]{} & \pi^S_{n-1} Y_n \end{array} \quad . \tag{2}$$

Let $\mathbf{rH}^c_{n+1} = \mathbf{rH}^c_{n+1}(\mathbf{C})$ be the category of reduced homotopy systems of order $(n+1)$ in $\mathbf{C}$. Clearly, composition is defined by $(\xi, \eta)(\bar{\xi}, \bar{\eta}) = (\xi, \bar{\xi}, \eta\bar{\eta})$. Next we define **homotopies** for morphisms in $\mathbf{rH}^c_{n+1}$ as in (VI.5.10). We set $(\xi, \eta) \simeq (\xi', \eta')$ if there exist homomorphisms $\alpha_{j+1}: C_j \to C'_{j+1}$ $(j \geqq n)$ of R-modules such that

(a) $\{\eta\} + g_{n+1}\alpha_{n+1} = \{\eta'\}$ in $[X_n, Y_n]$, and
(b) $\xi'_k - \xi_\lambda = \alpha_k d_k + d_{k+1}\alpha_{k+1}, k \geqq n+1$.

The action $+$ in (a) is defined by the coaction (2.6), $\{\eta\}$ denotes the homotopy class of η. ||

(2.9) ***Remark.*** The category $\mathbf{rH}^c_3/\simeq$ is the full subcategory of $\mathbf{Chain}_R/\simeq$ consisting of free chain complexes C with $C_i = 0$ for $i \leqq 1$.

The Γ-groups (1.4) and the cohomology groups (1.2) give us a bifunctor $H^k\Gamma_n$ on the category $\mathbf{rH}^c_n/\simeq$ which carries a pair (X, Y) of objects in $\mathbf{rH}^c_n$ to the R-module

(2.10) $$H^k\Gamma_n(X, Y) = \tilde{H}^k(X, \Gamma^S_{n-1}(s^{-1}Y)).$$

compare (VI.5.11)(2).

(2.11) **Theorem.** *Let* $\mathbf{C}$ *be a category as in* §1 *and let* $\mathbf{rC}/\simeq$ *be the full subcategory in* $Ho(\mathbf{C})$ *consisting of reduced complexes. The category* $\mathbf{rC}/\simeq$ *is approximated by the following tower of categories,* $n \geqq 3$.

$$\begin{array}{ccc} & & \mathbf{rC}/\simeq \\ & & \Big\downarrow{\scriptstyle r} \\ & & \vdots \\ H^n\Gamma_n & \xrightarrow{+} & \mathbf{rH}^c_{n+1}/\simeq \end{array}$$

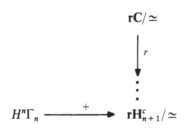

$$\begin{array}{ccc}
\Big\downarrow \lambda & & \\
\mathbf{rH}_n^c/\simeq & \xrightarrow{\mathfrak{O}} & H^{n+1}\Gamma_n\,. \\
\Big\downarrow & & \\
\vdots & & \\
\mathbf{rH}_3^c/\simeq & \xrightarrow{\mathfrak{O}} & H^4\Gamma_3 \\
\cap & & \\
\mathbf{Chain}_R/\simeq & &
\end{array}$$

In case (A) ($\mathbf{C} = \mathbf{Top}$) this tower is the subtower of (VI.6.2) given by CW-complexes with trivial 1-skeleton. The functors r and λ in (2.11) are defined in the same way as in (VI.5.8). Moreover, the obstruction $\mathfrak{O}$ and the action $+$ in (2.11) are given as in (VI.5.14) and (VI.5.15) respectively. The composition of all functors in the column of (2.11) carries a reduced complex X to its cellular chain complex $\tilde{C}_*X$, see (1.1).

(2.12) **Addendum.** *The isotropy groups of the action* $H^n\Gamma_n +$ *in* (2.11) *can be computed as in* (VI.5.16) *by use of a spectral sequence similar to the one in* (VI.5.9)(4).

Proof of (2.11). We observe that for the class $\mathfrak{X} = \mathfrak{X}_C = (\mathfrak{X}_1^1)^*$ in (2.7) we have an equivalence of categories

$$\mathbf{Complex}(\mathfrak{X})/\overset{1}{\simeq} = r\mathbf{C}/\simeq. \tag{1}$$

Hence, since $\mathfrak{X}$ is very good, we derive the tower in (2.11) from (VII.2.8). In fact, as in (VII.2.16) we can identify

$$\mathbf{rH}_n^c/\simeq = \mathbf{TWIST}_n^c(\mathfrak{X})/\simeq. \tag{2}$$

It is a good exercise to prove (2.11) directly along the lines of the proof of (VII.2.8). □

As in (VI.7.4) we derive from (2.11) the

(2.13) **Theorem.** *The functors* λ *in* (2.11) *and also* $\tilde{C}_*: \mathbf{rC}/\simeq \to \mathbf{rH}_3^c/\simeq$ *satisfy the strong sufficiency condition.*

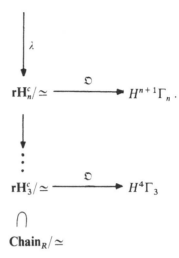

This result implies various important properties of reduced complexes which we describe in the rest of this section. By (I.6.11) the sufficiency condition for $\tilde{C}_*$ is equivalent to the

(2.14) **Whitehead-theorem for reduced complexes.** *A map $f:X \to Y$ in* **C** *between reduced complexes is a homotopy equivalence in* **C** *if and only if f induces an isomorphism $f_*:\tilde{H}_*(X,R) \cong \tilde{H}_*(Y,R)$ of homology groups.*

In case (A) this is the classical Whitehead-theorem for simply connected CW-complexes. In case (C) the result (2.14) is equivalent to the following theorem on chain algebras due to Moore (for this we use (1.3)(2) and (I.7.10)).

(2.15) **Theorem.** *Let R be a principal ideal domain and let $f:A \to B$ be a map in* $\mathbf{DA}_*$(flat) *with $H_0 A = R = H_0 B$. Then f is a weak equivalence if and only if the induced map*

$$f_*:\mathrm{Tor}_*^A(R,R) \cong \mathrm{Tor}_*^B(R,R)$$

is an isomorphism.

Moreover, we derive from (2.13) as in (VI.7.14) the following generalization of a result of Milnor.

(2.16) **Theorem.** *Let X be a reduced complex in* **C** *and suppose that a presentation*

$$0 \to R^{r_k} \to R^{b_k} \to \tilde{H}_k(X,R) \to 0$$

is given for each $k \geqq 0$, $b_1 = r_1 = 0$. Then there is a homotopy equivalence $K \simeq X$ in **C** *where K is a reduced complex for which $\tilde{C}_k(K)$ is a free R-module of $b_k + r_{k-1}$ generators.*

Clearly, in case (A) this is a consequence of (VI.7.14) where we set $D = *$. The theorem shows (for $b_k = r_k = 0$, $k \leqq n$):

(2.17) **Corollary.** *Assume X is a reduced complex in* **C** *with $\tilde{H}_k(X,R) = 0$ for $k \leqq n$. Then we have $\Gamma_k^S(s^{-1}X) = 0$ for $k \leqq n$.*

Whence exactness of the Γ-sequence (1.5) yields the

(2.18) **Hurewicz-theorem for reduced complexes.** *Assume X is a reduced complex in* **C** *with $\tilde{H}_k(X,R) = 0$ for $k \leqq n$. Then $h:\pi_r^S(X) \to \tilde{H}_{r+1}(X,R)$ is an isomorphism for $r = n$ and is surjective for $r = n+1$.*

In case (A) the proposition in (2.18) yields the classical Hurewicz-theorem for simply connected spaces, see (III.11.9). In case (C) we get by (1.3)(2) and (1.6)(2):

(2.19) **Hurewicz-theorem for chain algebras.** *Let A be a chain algebra in* $\mathbf{DA}_*$*(flat) with $H_0 A = R$ (R a principal ideal domain). Assume* $\mathrm{Tor}_k^A(R,R) = 0$ *for $k \leqq n$. Then*

$$h: H_r(A) \to \mathrm{Tor}_{r+1}^A(R,R)$$

is an isomorphism for $r = n$ and is surjective for $r = n + 1$.

Moreover, we derive from (2.16) the next result (compare (VI.7.15)).

(2.20) ***Minimal models of reduced complexes.*** Assume X is a reduced complex in $\mathbf{C}$ for which $\tilde{H}_*(X,R)$ is a free R-module. Then there exists a homotopy equivalence $K \simeq X$ in $\mathbf{C}$ for which $\tilde{C}_* K$ has trivial differential. Whence $\tilde{C}_* K = \tilde{H}_*(X,R)$. We call K a minimal model of X.

(2.21) ***Remark.*** If R is a field we obtain by (2.20) the minimal models of chain algebras A with $H_0 A = R$ and of chain Lie algebras L with $H_0 L = 0$. These as well are constructed in Baues–Lemaire, see also Neisendorfer.

(2.22) ***Definition.*** We say that a reduced complex X is **homologically (n − 1)-connected** if $\tilde{H}_r(X,R) = 0$ for $r \leqq n-1$. Moreover, we say that X has **homological dimension** $\leqq N$ if $\tilde{H}_r(X,R) = 0$ for $r > N$ and if $\tilde{H}_N(X,R)$ is a free R-module. ‖

(2.23) **Proposition.** *A reduced complex X is homologically $(n-1)$-connected and has homological dimension $\leqq N$ if and only if there is a homotopy equivalence $K \simeq X$ in $\mathbf{C}$ where K is a reduced complex with $K_{n-1} = *$ and $K = K_N$.*

Proof. This is as well an easy consequence of (2.16) since we can take $r_k = b_k = 0$ for $k \leqq n-1$, $k \geqq N$ and since we can take $r_N = 0$. □

§3 Functors on reduced complexes and the Quillen equivalence of rational homotopy categories

Assume that $\mathbf{C}$ and $\mathbf{C}'$ are categories as described in § 1 (A), (B), (C), and (D) respectively. Let S' be the 1-sphere of $\mathbf{C}'$ and let R' be the ring of coefficients of $\mathbf{C}'$ and assume a functor

(3.1) $$\alpha: \mathbf{C}' \to \mathbf{C}$$

is given. We consider the following properties of such a functor:

(i) α is a model functor on a subcategory of $\mathbf{C}'$ which contains all reduced complexes of $\mathbf{C}'$ and α carries objects in $\mathbf{C}'$ to fibrant objects in $\mathbf{C}$.

(ii) There is a weak equivalence $\Sigma S \xrightarrow{\sim} \alpha(\Sigma S')$ in $\mathbf{C}$ which induces a homomorphism of rings

$$a: R' = [\Sigma S', \Sigma S'] \xrightarrow{\alpha} [\alpha\Sigma S', \alpha\Sigma S'] = [\Sigma S, \Sigma S] = R.$$

(iii) α induces a bijection $(n \geqq 1)$

$$\alpha: [\Sigma^n S', X] \xrightarrow{\approx} [\alpha\Sigma^n S', \alpha X]$$

for all reduced complexes X in $\mathbf{C}'$.

(3.2) **Theorem.** *Suppose* α *satisfies* (i) *and* (ii). *Then we can choose for each reduced complex* X *in* $\mathbf{C}'$ *a weak equivalence* $M\alpha X \xrightarrow{\sim} \alpha X$ *in* $\mathbf{Fil(C)}$ *such that* $M\alpha X$ *is a reduced complex in* $\mathbf{C}$ *with*

$$\alpha: (\tilde{C}_* X) \otimes a^* R \cong \tilde{C}_*(M\alpha X) \quad \text{in } \mathbf{Chain}_R.$$

Moreover, α *induces a map between towers of categories as in* (3.4) *below and a map between spectral sequences as in* (2.12).

In particular, we derive from (3.2) the isomorphism

$$\tilde{H}_*(X, a^* R) = \tilde{H}_*(M\alpha X). \tag{3.3}$$

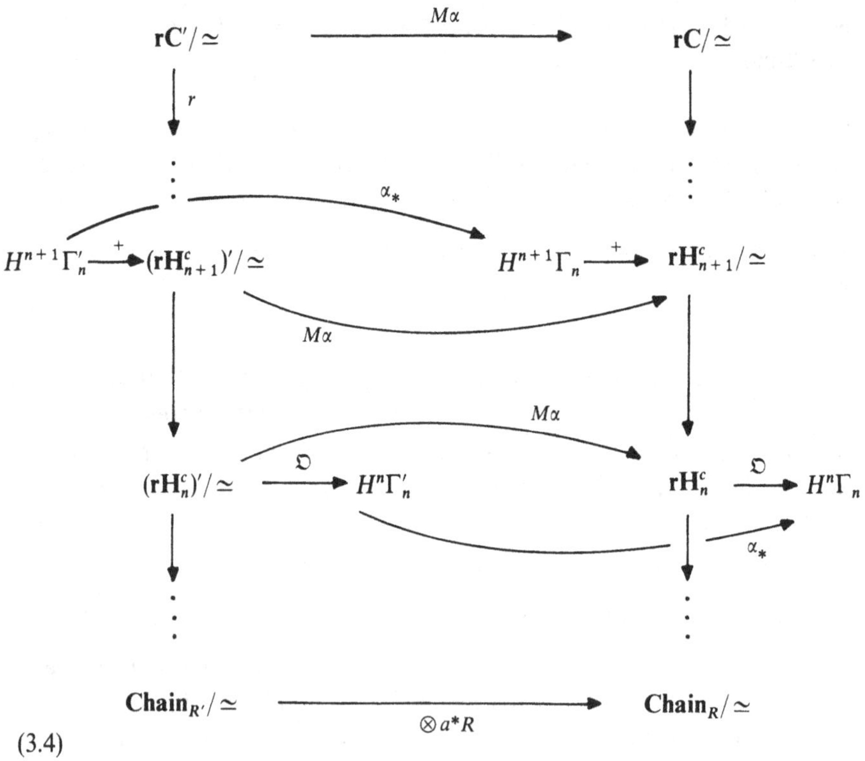

(3.4)

The R'-module a^*R is given by $a:R' \to R$ in (ii). The functor $M\alpha$ is determined on objects by the choice of $M\alpha X$ in (3.2). Moreover, α_* is the natural transformation on bifunctors, see (2.10), determined by the map α between Γ-groups in the following commutative diagram.

$$(3.5)\quad \begin{array}{ccccccccc} \cdots & \longrightarrow & \Gamma^{S'}_{n-1}(X) & \longrightarrow & \pi^{S'}_{n-1}(X) & \longrightarrow & \tilde{H}_n(X, R') & \longrightarrow & \cdots \\ & & \downarrow \alpha & & \downarrow \alpha & & \downarrow a_* & & \\ & \longrightarrow & \Gamma^{S}_{n-1}(\alpha X) & \longrightarrow & \pi^{S}_{n-1}(\alpha X) & \longrightarrow & \tilde{H}_n(X, a^*R) & \longrightarrow & \end{array}$$

The rows of this diagram are the exact Γ-sequences for X and $M\alpha X$ respectively, compare (1.5). The maps α are a-equivariant homomorphisms between modules.

(3.6) *Proof of* (3.2) *and* (3.5). We use the naturality of the tower $\mathbf{TWIST}_*$ with respect to functors, see (VII.1.19), and we use the naturality of the spectral sequences in (III.2.10). In addition we can derive from (2.5) and assumption (i), (ii) the isomorphism of chain complexes in (3.2). For (3.5) we use the naturality in (III.10.13)(3). □

(3.7) **Corollary.** *Suppose* α *satisfies* (i), (ii) *and* (iii). *Then* α *induces an isomorphism of towers of categories. In particular, the functor*

$$M\alpha:\mathbf{rC}'/\simeq \to \mathbf{rC}/\simeq,$$

restricted to reduced complexes of finite homological dimension, is an equivalence of categories, see (2.22).

Proof. We apply (IV.4.14) inductively. In fact, by (iii) the maps α in (3.5) are isomorphisms and the isotropy groups in (3.4) are isomorphic since the corresponding spectral sequences are isomorphic. □

We now consider various well known examples of functors α.

(3.8) ***Example of Quillen.*** Let

$$\lambda:\mathbf{Top}^*_1 \to \mathbf{DL}_1 \tag{1}$$

be the functor of Quillen in (I.9.19). This functor restricted to $\mathbf{CW\text{-}spaces}^*(\mathbb{Q}) \cap \mathbf{Top}^*_1$ satisfies by (I.9.21) the conditions (i), (ii) and (iii). Whence we see by (3.7) that the induced functor

$$Ho(\lambda):Ho_{\mathbb{Q}}(\mathbf{Top}^*_1) \to Ho(\mathbf{DL}_1) \tag{2}$$

is an equivalence on subcategories consisting of objects of finite homological dimension. In addition, we know by (3.7) that λ induces an isomorphism of towers of categories. This has similar consequences as in (VIII.6.14),..., (VIII.6.19). Quillen proved that (2) is an equivalence of categories, compare (I.9.20).

(3.9) ***Example of Adams–Hilton.*** Let R' be a subring of $\mathbb{Q}$ and let R be a principal ideal domain and suppose a ring homomorphism $a:R' \to R$ is given. We consider the functor

$$\alpha = SC_*\Omega(.)\otimes R:\mathbf{CW\text{-}spaces}^*(R') \to \mathbf{DA}_*(\text{flat}), \tag{1}$$

which carries a space X to the chain algebra $\alpha(X) = SC_*\Omega(X)\otimes_{\mathbb{Z}} R$ with coefficients in R. By (I.7.29) this functor satisfies the conditions (i) and (ii) above. Whence, by (3.2), we can find for each reduced complex X in **CW-spaces**(R') a model in $\mathbf{Fil}(\mathbf{DA}_*(\text{flat}))$

$$\left.\begin{aligned} &M\alpha X = (T(V), d) \xrightarrow{\sim} \alpha X, \quad \text{with} \\ &\tilde{C}_*(M\alpha X) = \tilde{C}_*(X)\otimes a^*R. \end{aligned}\right\} \tag{2}$$

Here V coincides as a module with $s^{-1}\tilde{C}_*X\otimes a^*R$. For $R' = R = \mathbb{Z}$ the construction of $M\alpha X$ in (2) is the main result in Adams–Hilton. From (2) we derive the following natural isomorphism ($R' = \mathbb{Z}$, see (3.3))

$$\operatorname{Tor}_n^{SC_*\Omega(X)\otimes R}(R, R) = H_n(X, R), \tag{3}$$

where R is any principal ideal domain and where X is any simply connected space in **Top***. For the left-hand side of (3) see (1.3)(2). The right-hand side of (3) is the singular homology of X with coefficients in R. In addition to the models (2) we derive from (3.2) a map between towers of categories which is very useful for computations.

(3.10) ***Example.*** The universal enveloping functor $U:\mathbf{DL} \to \mathbf{DA}_*$ in (I.9.13) satisfies the conditions (i) and (ii).

(3.11) ***Example.*** Let $R' \subset R \subset \mathbb{Q}$ be subrings and let

$$\alpha:\mathbf{CW\text{-}spaces}(R') \to \mathbf{CW\text{-}spaces}(R)$$

be the localization functor (II.4.4)(2) which carries X to its R-localization $\alpha(X) = X_R$. Then α satisfies (i) and (ii).

Clearly, for the examples (3.10), (3.11) we can apply (3.2) as well.

(3.12) ***Example.*** *Let* $\Sigma:\text{Top}^* \to \text{Top}^*$ be the suspension functor. Then Σ satisfies (i) and (ii).

§ 4 Whitehead's classification of simply connected 4-dimensional CW-complexes and the corresponding result for reduced complexes

J.H.C. Whitehead (1950) showed that the homotopy type of a simply connected 4-dimensional CW-complex is determined by its Γ-sequence. We now give a new proof of this result which only uses simple properties of the tower of categories in (2.11). In fact, our proof shows that Whitehead's result holds more generally for reduced complexes of homological dimension $\leqq 4$ where reduced complexes are defined as in (A), (B), (C), and (D) of § 1 respectively.

Let **C** be one of the categories **Top**, **CW-spaces**(R), $\mathbf{DA}_*$(flat), and **DL** as described in (A), (B), (C) and (D) of § 1 respectively. Recall that R is the ring of coefficients for **C** and that S denotes the 1-sphere in **C**. Moreover, for a reduced complex X in **C** we have the exact Γ-sequence (1.5). The first non-trivial Γ-group in this sequence, $\Gamma_2^S(s^{-1}X)$, has the following property.

(4.1) **Proposition.** *For reduced complexes X in* **C** *there is a natural isomorphism of R-modules*

$$\Gamma_C(\tilde{H}_2(X,R)) = \Gamma_2^S(s^{-1}X)$$

where Γ_C is the functor below which carries R-modules to R-modules.

We prove this result in (4.5) below.

$$\text{(4.2)} \qquad \Gamma_C(M) = \begin{cases} \Gamma(M) & \text{in case (A) and (B),} \\ M \otimes_R M & \text{in case(C),} \\ [M, M] & \text{in case(D).} \end{cases}$$

Here M is an R-module and $M \otimes_R M$ is the tensor product. Moreover, $[M, M]$ is the image of the homomorphism

$$[\ ,\ ]: M \underset{R}{\otimes} M \to M \underset{R}{\otimes} M$$

which carries $x \otimes y$ to $[x, y] = x \otimes y + y \otimes x$. Finally, Γ is Whitehead's quadratic functor defined by the following universal property.

(4.3) ***Definition.*** A function $f: A \to B$ between abelian groups is **quadratic** if $f(a) = f(-a)$ and if

$$[a, b]_f = f(a+b) - f(a) - f(b)$$

is bilinear in a and b. There is a quadratic function $\gamma: A \to \Gamma(A)$ such that for each quadratic function $f: A \to B$ there is a unique homomorphism $\bar{f}: \Gamma(A) \to B$ with $\bar{f}\gamma = f$. If $g: A' \to A$ is a homomorphism between abelian groups we obtain $\Gamma(g): \Gamma(A') \to \Gamma(A)$ by $\Gamma(g) = \overline{\gamma g}$. This shows that Γ is a functor which carries abelian groups to abelian groups. ||

If M is an R-module with $R \subset \mathbb{Q}$ then also $\Gamma(M)$ is an R-module. In fact, if multiplication by n, $n: M \to M$ is an isomorphism then $\Gamma(n) = n^2: \Gamma(M) \to \Gamma(M)$ is an isomorphism and hence multiplication by n is an isomorphism on $\Gamma(M)$. Now all functors in (4.2) are defined.

(4.4) ***Properties of*** $\boldsymbol{\Gamma}$. We define the 'Whitehead product'

$$[\ ,\ ]: A \otimes A \to \Gamma(A)$$

by $[a,b] = \gamma(a+b) - \gamma(a) - \gamma(b)$ and we define the homomorphisms

$$\tau: \Gamma(A) \to A \otimes A,$$
$$\sigma: \Gamma(A) \to A \otimes \mathbb{Z}/2$$

via the universal property by $\tau\gamma(a) = a \otimes a$ and $\sigma\gamma(a) = a \otimes 1$. We have $\sigma[a,b] = 0$ and $\tau[a,b] = a \otimes b + b \otimes a$. Hence $\tau\Gamma(A) \subset [A, A]$ where $\tau\Gamma(A)$ is the subgroup of $A \otimes A$ generated by $\{a \otimes a : a \in A\}$. Therefore the cokernel of τ is the exterior product $A \wedge A$. We have the exact sequences

$$A \otimes A \xrightarrow{[\ ,\ ]} \Gamma(A) \xrightarrow{\sigma} A \otimes \mathbb{Z}/2 \to 0,$$

$$\Gamma(A) \xrightarrow{\tau} A \otimes A \to A \wedge A \to 0$$

where τ and $[\ ,\ ]$ need not to be injective. As an abelian group we obtain $\Gamma(A)$ by the formulas

$$\Gamma(A \oplus B) = \Gamma(A) \oplus \Gamma(B) \oplus A \otimes B,$$
$$\Gamma(\mathbb{Z}) = \mathbb{Z},$$
$$\Gamma(\mathbb{Z}/n) = \mathbb{Z}/2n, \quad n \text{ even},$$
$$\Gamma(\mathbb{Z}/n) = \mathbb{Z}/n, \quad n \text{ odd},$$

where $\Gamma(\mathbb{Z})$ and $\Gamma(\mathbb{Z}/n)$ is a cyclic group generated by $\gamma(1)$. Moreover, the functor Γ commutes with direct limits of abelian groups. Since $\gamma(na) = n^2\gamma(a)$ we get $2\gamma(a) = [a,a]$ by definition of $[\ ,\ ]$ above. Therefore the composition

$$\cdot 2: \Gamma A \xrightarrow{\tau} A \otimes A \xrightarrow{[\ ,\ ]} \Gamma A$$

is multiplication by 2 (here we use the universal property). If multiplication by 2 is an isomorphism on A we see that τ is injective and thus $\tau: \Gamma(M) \cong [M, M]$ for an R-module M with $1/2 \in R \subset \mathbb{Q}$.

(4.5) *Proof of* (4.1). We only consider case(**A**) with $\mathbf{C} = \mathbf{Top}$. Let X be a CW-complex with $X^1 = *$. Then the Hopf-map $\eta: S^3 \to S^2$ gives us the map

$$\eta^*: \pi_2(X) \to \Gamma_3(X) = \Gamma_2^S(s^{-1}X)$$

with $\eta^*(\alpha) = \alpha \circ \eta$. This function is quadratic since we have the left distributivity law

$$\eta^*(\alpha+\beta)=\eta^*(\alpha)+\eta^*(\beta)+[\alpha,\beta],$$

where $[\alpha,\beta]$ is the Whitehead product. Hence η^* determines a homomorphism

$$\overline{\eta^*}:\Gamma(\pi_2(X))\to\Gamma_3(X),$$

where $\pi_2(X)=\tilde{H}_2(X,\mathbb{Z})$ by the Hurewicz theorem. Now $\bar{\eta}^*$ is an isomorphism. In fact, this is true if X is a one point union of 2-spheres (as follows from the Hilton–Milnor theorem). Now let $g:M=M(C_3,2)\to M(C_2,2)=X^2=Y$ be the attaching map of 3-cells in X which induces $d_3:C_3\to C_2$ in $C_*=\tilde{C}_*X$. Then we get (as in the proof (V.8.5)(5)) the commutative diagram

$$\begin{array}{ccccc}
\pi_3(M\vee Y)_2 & \xrightarrow{(g,1)_*} & \pi_3(Y) & \longrightarrow & \Gamma_3X\to 0\\
\uparrow\cong & & \uparrow\cong & & \uparrow\overline{\eta^*}\\
\Gamma(C_3)\oplus C_3\otimes C_2 & \xrightarrow{\bar{g}} & \Gamma(C_2) & \longrightarrow & \Gamma(\pi_2X)\to 0
\end{array}$$

with $\bar{g}=(\Gamma(d_3),[d_3,1])$. Since the rows of this diagram are exact we see that $\bar{\eta}^*$ is an isomorphism. In fact, the bottom row is exact for algebraic reasons (using properties of Γ) since $\pi_2X=H_2X=C_2/d_3C_3$. □

(4.6) ***Remark.*** Let $M(A,2)$ and $K(A,2)$ be a Moore space and an Eilenberg–Mac Lane space in degree 2 respectively. Then (4.1) and the exact Γ-sequence yield immediately the isomorphisms of abelian groups

$$\pi_3M(A,2)=\Gamma(A)=H_4(K(A,2),\mathbb{Z})$$

since $\pi_2M(A,2)=A=\pi_2K(A,2)$.

We now use the isomorphism (4.1) and the Γ-sequence (1.5) for the definition of the following

(4.7) ***Category of* Γ*-sequences*** denoted by **Γ*-sequences***[4]. Objects are the exact sequences ΓS,

$$H_4\to\Gamma_C(H_2)\to\pi_3\to H_3\to 0,$$

of R-modules where H_4 is a free R-module. Morphisms $f:\Gamma S\to\Gamma S'$ are triples $f=(f_4,f_3,f_2)$, $f_i:H_i\to H_i'$, of homomorphisms for which there exists φ such that the diagram

$$\begin{array}{ccccccc}
H_4 & \xrightarrow{b_4} & \Gamma_C(H_2) & \longrightarrow & \pi_2^S & \longrightarrow & H_3\longrightarrow 0\\
\downarrow f_4 & & \downarrow\Gamma_C(f_2) & & \downarrow\varphi & & \downarrow f_3\\
H_4' & \xrightarrow[b_4']{} & \Gamma_C(H_2') & \longrightarrow & (\pi_2^S)' & \longrightarrow & H_3'\longrightarrow 0
\end{array}$$

commutes in the category of R-modules.

Let $\mathbf{rC}^4$ be the full subcategory of $\mathbf{C}$ consisting of reduced complexes of homological dimension $\leqq 4$. By (1.5) we have an obvious functor

$$\Gamma S:\mathbf{rC}^4/\simeq \;\to \mathbf{\Gamma\text{-sequences}}^4 \tag{4.8}$$

which carries a reduced complex X to its Γ-sequence (1.5) which we denote by $\Gamma S(X)$.

(4.9) **Theorem.** *The functor ΓS is a detecting functor on* $\mathbf{rC}^4/\simeq$.

Compare (IV.1.5). Whence each Γ-sequence as in (4.7) is realizable and a homology homomorphism $f:\tilde{H}_*(X,R)\to\tilde{H}_*(X',R)$ with $X,\ X'\in\mathbf{rC}^4$ is realizable by a map $X\to X'$ in $\mathbf{C}$ if and only if f is compatible with the Γ-sequences. This implies.

(4.10) **Corollary.** *Homotopy types in* $\mathbf{rC}^4/\simeq$ *are canonically* 1–1 *corresponded to isomorphism classes of objects in the category* $\mathbf{\Gamma\text{-sequences}}^4$.

(4.11) ***Remark.*** In case (A), $\mathbf{C}=\mathbf{Top}$, theorem (4.9) describes exactly the result on simply connected 4-dimensional CW-complexes in J.H.C. Whitehead (1950). It is a remarkable fact that this result, which originally was considered as a highly topological one, is available in each of the cases (A), (B), (C), and (D) respectively described in § 1. Indeed, examples of this kind show that there is a striking similarity of solutions for the homotopy classification problems in various different categories $\mathbf{C}$ and they indicate that part of the homotopy classification problem is of an abstract nature which does not depend on the underlying category $\mathbf{C}$. This part is longing for the elaboration of 'algebraic homotopy'.

For the proof of (4.9) we restrict the tower of categories (2.11) to reduced complexes X in $\mathbf{C}$ with $X_4 = X$. This yields the tower of categories

$$\begin{array}{ccccc}
H^4\Gamma_4 & \xrightarrow{+} & \mathbf{rC}^4/\simeq & & \\
 & & \downarrow r & & \\
H^3\Gamma_3 & \xrightarrow{+} & (\mathbf{rH}^c_4)^4/\simeq & \longrightarrow & 0 \\
 & & \downarrow \lambda & & \\
 & & (\mathbf{rH}^c_3)^4/\simeq & \xrightarrow{\mathfrak{D}} & H^4\Gamma_3 \\
 & & \cap & & \\
 & & \mathbf{Chain}_R/\simeq & &
\end{array} \tag{4.12}$$

with $i\lambda r = \tilde{C}_*$. Here i is a full inclusion, the objects in $(\mathbf{rH}^c_3)^4$ are exactly the free chain complexes

$$C = (C_4 \to C_3 \to C_2) \tag{1}$$

in $\mathbf{Chain}_R$ with $C_i = 0$ for $i < 2$ and $i > 4$. Using 'Moore spaces' in $\mathbf{C}$ one readily shows that each such chain complex is realizable in $\mathbf{rC}^4$. Therefore we have by (VI.5.12) a transitive and effective action

$$H^4(C, \Gamma_{\mathbf{C}}(H_2 C)) \xrightarrow[\approx]{+} \mathrm{Real}_{\lambda r}(C) \tag{2}$$

on the set of all realizations of C in $\mathbf{rC}^4/\simeq$. Moreover, for $X \in \mathrm{Real}_{\lambda r}(C)$ and $X' \in \mathrm{Real}_{\lambda r}(C')$ and for a chain map $\xi: C \to C'$ in $\mathbf{Chain}_R/\simeq$ there exists a map $F: X \to X'$ in $\mathbf{rC}^4/\simeq$ with $\tilde{C}_*(F) = \xi$ if and only if the obstruction

$$\mathfrak{O}_{X,X'}(\xi) \in H^4(C, \Gamma_{\mathbf{C}}(H_2 C')) \tag{3}$$

vanishes. We use these facts, which are immediate properties of the tower (4.12), for the proof of (4.9). In addition, we use the following well-known short exact sequences (4) and (5) which are available since R is a principal ideal domain.

$$\mathrm{Ext}_R(H_3, \Gamma) \overset{\Delta}{\rightarrowtail} H^4(C, \Gamma) \overset{\mu}{\twoheadrightarrow} \mathrm{Hom}(H_4, \Gamma) \tag{4}$$

$$\bigoplus_{i=2,3} \mathrm{Ext}(H_i, H'_{i+1}) \overset{i}{\rightarrowtail} [C, C'] \overset{H}{\twoheadrightarrow} \mathrm{Hom}(H_*, H'_*) \tag{5}$$

Here Γ is an R-module and we set $H_i = H_i(C)$ and $H'_i = H_i(C')$, $i \in \mathbb{Z}$.

(4.13) *Proof of* (4.9). Let $\Gamma = \Gamma_{\mathbf{C}}(H_2 C)$. Using the Γ-sequence we have the function

$$b_4: \mathrm{Real}_{\lambda r}(C) \to \mathrm{Hom}(H_4, \Gamma) \tag{6}$$

which carries a realization X to the secondary boundary operator $b_4 X$ in the Γ-sequence. By definition of b_4 one checks that the function b_4 in (6) is μ-equivariant with respect to μ in (4) and with respect to the action in (2), that is

$$b_4(X + \alpha) = b_4(X) + \mu(\alpha). \tag{7}$$

Since μ is surjective this implies that b_4 in (6) is surjective. Again using the Γ-sequence we have the function

$$\pi: b_4^{-1}(b) \to \mathrm{Ext}_R(H_3, \mathrm{cok}(b)) \tag{8}$$

where $b \in \mathrm{Hom}(H_4, \Gamma)$. This function carries $X \in \mathrm{Real}_{\lambda r}(C)$ with $b_4 X = b$ to the extension element $\pi(X) = \{\pi_2^S(X)\}$ determined by the short exact sequence

$$\mathrm{cok}(b) \rightarrowtail \pi_2^S(X) \twoheadrightarrow H_3 \tag{9}$$

associated to X, see (1.5). Let $q: \Gamma \to \mathrm{cok}(b)$ be the quotient map. Then π is again equivariant with respect to the action of $\mathrm{Ext}(H_3, \Gamma)$ via (2), that is,

$$\pi(X + \Delta(\beta)) = \pi(X) + q_*\beta \tag{10}$$

for $\beta \in \mathrm{Ext}_R(H_3, \Gamma)$. Since q is surjective this also shows that the function π in (8) is surjective. We deduce from the surjectivity of (6) and (8) that the functor (4.8) satisfies the realizability condition with respect to objects. Next we consider the realizability of a morphism $f:\Gamma S(X) \to \Gamma S(X')$ where X and X' are objects in $\mathbf{rC}^4$ with cellular chain complexes $C = \tilde{C}_*(X)$ and $C' = \tilde{C}_*(X')$ respectively. By (5) we know that there is a chain map $\xi: C \to C'$ which induces f on homology. For the obstruction (3) and for μ in (4) one readily gets

$$\mu\mathfrak{O}_{X,X'}(\xi) = b_4' f_4 - \Gamma_C(f_2)b_4 = 0 \tag{11}$$

since f is a morphism in **Γ-sequences**[4]. By (11) and exactness in (4) the element $\Delta^{-1}\mathfrak{O}_{X,X'}(\xi)$ is defined. For $q':\Gamma_C(H_2') \twoheadrightarrow \mathrm{cok}(b_4')$ we get

$$q_*'\Delta^{-1}\mathfrak{O}_{X,X'}(\xi) = f_3^*\pi(X') - \Gamma_C(f_2)_*\pi(X) = 0 \tag{12}$$

where we use π in (8). Since for f there exists a φ as in (4.7) we see that the element (12) is trivial. We now use the inclusion i in (5). For an element $\alpha \in \mathrm{Ext}_R(H_3, H_4')$ we have the formula

$$\mathfrak{O}_{X,X'}(\xi + i\alpha) = \mathfrak{O}_{X,X'}(\xi) + \Delta(b_4')_*(\alpha). \tag{13}$$

Now the sequence

$$\mathrm{Ext}(H_3, H_4') \xrightarrow{(b_4')_*} \mathrm{Ext}(H_3, \Gamma_C H_2') \xrightarrow{q_*'} \mathrm{Ext}(H_3, \mathrm{cok}\, b_4') \tag{14}$$

is exact. Hence by (12) we can choose α with

$$(b_4')_*(\alpha) = \Delta^{-1}\mathfrak{O}_{X,X'}(\xi). \tag{15}$$

Therefore (13) shows

$$\mathfrak{O}_{X,X'}(\xi - i\alpha) = 0 \tag{16}$$

and thus there exists a realization $F: X \to X'$ with $\tilde{C}_*F = \xi - i\alpha$. Here $\xi - i\alpha$ induces f in homology. Whence F is a realization of f. This completes the proof of (4.9). □

We point out that in the proof above we proved simultaneously four different results which we now describe explicitly.

(A) Theorem. *Consider simply connected CW-spaces X with integral homology groups $H_i(X, \mathbb{Z}) = H_i$ where $H_i = 0$ for $i \geq 5$ and where H_4 is free abelian. The homotopy types of such CW-spaces in* **Top**/≃ *are classified by the isomorphism classes of exact sequences of abelian groups*

$$H_4 \to \Gamma(H_2) \to \pi_3 \to H_3 \to 0.$$

The boundary invariants in Baues (1985) show that theorem (A) is the start of an inductive classification of simply connected n-dimensional polyhedra, $n \geqq 4$. In a similar way one has boundary invariants for reduced complexes in **C**, we will describe details of this program elsewhere.

Example. There are exactly 5 different homotopy types of simply connected CW-complexes in **Top**/$\simeq$ with the homology groups

$$H_2 = \mathbb{Z}/2, \quad H_3 = \mathbb{Z}/2, \quad H_4 = \mathbb{Z},$$

and $H_i = 0$ for $i \geqq 5$. In fact, all such homotopy types are given by pairs (b, π) with

$$b \in \mathrm{Hom}(\mathbb{Z}, \Gamma(\mathbb{Z}/2)) = \mathbb{Z}/4,$$
$$\pi \in \mathrm{Ext}(\mathbb{Z}/2, \Gamma(\mathbb{Z}/2)/b\mathbb{Z}) = U_b.$$

Hence $U_b = \mathbb{Z}/2$, 0, $\mathbb{Z}/2$, 0 for $b = 0$, 1, 2, 3. The pairs (b, π) with $b = 1$ and $b = 3$ correspond to each other by the isomorphism in **Γ-sequence**[4] given by $-1: H_4 \to H_4$.

(B) **Theorem.** *Let R be a subring of $\mathbb{Q}$ and consider simply connected CW-spaces X with integral homology groups $H_i(X, \mathbb{Z}) = H_i$ where H_i is an R-module which is free for $i = 4$ and which is trivial for $i \geqq 5$. The homotopy types of such CW-spaces in* **Top**/$\simeq$ *are classified by the isomorphism classes of exact sequences of R-modules*

$$H_4 \to \Gamma(H_2) \to \pi_3 \to H_3 \to 0.$$

For $R = \mathbb{Z}$ this gives us exactly the result in (A).

(C) **Theorem.** *Let R be a principal ideal domain and consider chain algebras A over R in* $\mathbf{DA}_*$(flat) *with $H_0(A) = R$ and with*

$$\mathrm{Tor}_i^A(R, R) = \tilde{H}_i$$

where $\tilde{H}_i$ is a free R-module for $i = 4$ and where $\tilde{H}_i = 0$ for $i \geqq 5$. The homotopy types of such chain algebras in Ho $\mathbf{DA}_*$(flat) *are classified by the isomorphism classes of exact sequences of R-modules*

$$\tilde{H}_4 \xrightarrow[b_4]{} \tilde{H}_2 \bigotimes_R \tilde{H}_2 \to \pi_2^T \to \tilde{H}_3 \to 0.$$

Recall that $\pi_2^T(A) = H_2(A)$ is the homology of the underlying chain complex of A. The isomorphism class of the exact sequence in (C) is completely determined by b_4 if $\tilde{H}_3$ is a free R-module, in particular, if R is a field.

(D) **Theorem.** *Let R be a field of characteristic zero and consider chain Lie algebras L over R in* **DL** *with $H_0 L = 0$ and with*

$$\mathrm{Tor}_i^{UL}(R, R) = \tilde{H}_i,$$

where $\tilde{H}_i = 0$ for $i \geqq 5$. The homotopy types of such chain Lie algebras in $Ho(\mathbf{DL})$ are classified by the isomorphism classes of exact sequences

$$\tilde{H}_4 \xrightarrow[b_4]{} [\tilde{H}_2, \tilde{H}_2] \to \pi_2^S \to \tilde{H}_3 \to 0.$$

Since R is a field the isomorphism class of the exact sequence in (D) is completely determined by b_4. Recall that $\pi_2^S(L) = H_2(L)$ is the homology of the underlying chain complex of L and that for $R = \mathbb{Q}$ we have $\Gamma(\tilde{H}_2) = [\tilde{H}_2, \tilde{H}_2]$.

(4.14) ***Remark.*** Theorem (A) above is the result in J.H.C. Whitehead (1950). Moreover, Theorem (D) is readily obtained by use of minimal models in $\mathbf{DL}_1$, in fact, b_4 determines exactly the differential in the minimal model $(L(V), d) \sim L$ with $V = s^{-1}\tilde{H}_*$ and

$$d = b_4 : V_3 \to L(V)_2 = V_2 \oplus [V_1, V_1].$$

The theorems (B) and (C) seem to be new.

Using the functors in § 3 we obtain by (3.5) obvious connections between the exact sequences in (A), (B), (C), and (D) above. For example for X as in (A) and $A = SC_*\Omega X$ as in (C) we have $\tilde{H}_i(A) = H_i(X) = H_i$ and $\alpha = SC_*\Omega(.)$ induces

$$\alpha = \tau : \Gamma_3 X = \Gamma(H_2) \to \Gamma_3 A = H_2 \otimes H_2 \qquad (4.15)$$

where τ is the homomorphism in (4.4). Hence by (3.5) we have the commutative diagram of abelian groups ($R = \mathbb{Z}$)

$$\begin{array}{ccccccccc} H_4 & \longrightarrow & \Gamma(H_2) & \longrightarrow & \pi_3 & \longrightarrow & H_3 & \longrightarrow & 0 \\ \| & & \downarrow{\scriptstyle\tau} & \circledast & \downarrow & & \| & & \\ H_4 & \longrightarrow & H_2 \otimes H_2 & \longrightarrow & \pi_2^T & \longrightarrow & H_3 & \longrightarrow & 0 \end{array} \qquad (4.16)$$

Since the rows are exact the bottom row is completely determined (up to isomorphism) by the top row; in fact, the diagram $\circledast$ is a push out diagram in the category of abelian groups. By (A) the top row determines the homotopy type of a space X and by (C) the bottom row determines the homotopy type of the chain algebra $A = C_*\Omega X$. The map $\pi_3 \to \pi_2^T$ in (4.16) is the Hurewicz homomorphism

$$\pi_3 = \pi_3 X = \pi_2 \Omega X \to H_2 \Omega X = \pi_2^T(A) = \pi_2^T.$$

Similar results as in this section can be obtained for reduced complexes of homological dimension $\leqq 5$ though these are not classified by the Γ-sequence (1.5). We derive these results again from properties of the tower of categories in (2.8). Indeed, there are many further applications of the various towers of categories.

BIBLIOGRAPHY

Adams, J.F., 'Four applications of self-obstruction invariants'. *J. London Math. Soc.*, **31** (1956), 148–59.

Adams, J.F. & Hilton, P.J., 'On the chain algebra of a loop space'. *Comment. M. Helv.*, **30** (1956), 305–30.

Anderson, D.W., 'Fibrations and geometric realizations'. *Bull. AMS*, **54** (1978), 765–88.

Arkowitz, M., 'The generalized Whitehead product'. *Pacific J. Math.*, **12** (1962), 7–23.

– 'Commutators and cup products'. *Ill. J. Math.*, **8** (1964), 571–81.

Barcus, W.D. & Barratt, M.G., 'On the homotopy classification of the extensions of a fixed map'. *Trans. AMS*, **88** (1958), 57–74.

Barratt, M.G., 'Homotopy ringoids and homotopy groups'. *Q.J. Math. Oxford* (2), **5** (1954), 271–90.

Baues, H.J., 'Relationen für primäre, Homotopieoperationen und eine verallgemeinerte EHP-Sequenz'. *Ann. Scien. Ec. Norm. Sup.*, **8** (1975), 509–33.

– *Obstruction Theory (Springer Verlag) Lecture Notes in Math.*, **628** (1977), 387 pages.

– *'Geometry of Loop spaces and the cobar construction'. Memoirs of the AMS 230* (1980), 170 pages. American Mathematical Society.

– *Commutator Calculus and Groups of Homotopy Classes. London Math. Soc. Lecture Note Series, 50* (1981), 160 pages. Cambridge University Press.

– *On the Homotopy Classification Problem.* Preprint (1983), Max-Planck-Institut für Mathematik in Bonn, MPI/SFB 83–25.

– 'The Chains on the loops and 4-dimensional homotopy types'. *Astérisque*, **113–14** (*Homotopie Algebrique and Algèbre Local*) (1984), 44–59.

– On Homotopy Classification Problems of J.H.C. Whitehead. *Lecture Notes in Math, 1172* (Springer Verlag), (*Algebraic Topology, Göttingen 1984 Proceedings*) (1985), 17–55.

– Topologie. Jahrbuch der Max-Plank-Gesellschaft (1986)

Baues, H.J. & Lemaire, J.M., 'Minimal models in Homotopy theory'. *Math. Ann.*, **225** (1977), 219–42.

Baues, H.J. & Wirsching, G., 'The cohomology of small categories'. *J. Pure Appl. Algebra*, **38** (1985), 187–211.

Blakers, A.L. & Massey, W.S., 'The homology groups of a triad II'. *Ann. of Math.*·**55** (1952), 192–201.

Bott, R. & Samelson, H., 'On the Pontryagin product in spaces of paths'. *Comment. M. Helv.*, **27** (1953), 320–37.

Bousfield, A.K., 'The localization of spaces with respect to homology'. *Topology*, **14** (1975), 133–50.
Bousfield, A.K. & Gugenheim, V.K.A.M., *'On PL De Rham theory and rational homotopy type'*. *Memoirs of the AMS*, **179** (1976).
Bousfield, A.K. & Friedlander, E.M., Homotopy Theory of Γ-spaces, Spectra, and Bisimplicial Sets. *Lecture Notes in Math. 658* (Springer Verlag), *Geometric Applications of homotopy Theory* **11** (1978), 80–130.
Bousfield, A.K. & Kan, D.M., *Homotopy Limits, Completions and Localizations Lecture Notes in Math. 304* (Springer Verlag), Berlin–Heidelberg–New York (1972).
Brown, K.S., 'Abstract homotopy theory and generalized sheaf cohomology'. *Trans. AMS*, **186** (1973), 419–58.
Brown, R., 'Non-abelian cohomology and the homotopy classification of maps'. *Astérisque*, **113–14** (1984), 167–72.
Brown, R. & Heath, P.R., 'Coglueing homotopy equivalences'. *Math. Z.*, **113** (1970), 313–25.
Brown, R. & Higgings, P.J., Crossed Complexes and Chain Complexes with Operators. Preprint, University College of North Wales Bangor, Gwynedd. U.K. (1984).
– 'Crossed complexes and non-abelian extensions'. *Proc. Int. Conf. on Category Theory, Gummerbach, 1981. Lecture Notes* (Springer).
Brown, R. & Huebschmann, J.: Identities among Relations. "*Brown and Thickstun (1982), Low Dimensional Topology*". *Proceedings 1*, Cambridge University Press.
Cappel, S.E. & Shaneson, J.J.: 'On 4-dimensional surgery and applications. *Comment. Math. Helv.*, **46** (1971) 500–28.
Cartan, H. & Ellenberg, S., *Homological Algebra.* Princeton University Press 1956.
Chang, S.C., 'Successive homology operations and their applications'. *Cahiers de Top et Geom. Diff.*, Vol. VIII (1965), pp. 1–5.
Chow, S.-K., 'Cohomology operations and homotopy type II. *Scienta Sinica*, Vol XIII (1964), 1033–43.
Cenkl, B. & Porter, R., *Foundations of De Rham Theory.* Preprint Northeastern University Boston, Massachusetts 02115 (1985), 151 pages.
Didierjean, G., Groupes d'homotopie du monoide des équivalences d'homotopie fibres. C.R. Acad. Sc. Paris t. **292** (1981).
Dieck, tom T., Kamps, K.H. & Puppe, D., *Homotopietheorie, Lecture Notes in Math. 157* (Springer Verlag) (1970).
Dold, A., 'Partitions of unity in the theory of fibrations. *Ann. of Math.*, **78** (1963), 223–55.
– *Halbexakte Homotopiefunktoren. Lecture Notes in Math. 12* (Springer Verlag) (1966).
– *Lectures on Algebraic Topology* (Springer Verlag) (1972).
Dold, A. & Lashof, R., 'Principal quasifiberings and fiber homotopy equivalence of bundles'. *Ill. J. Math.*, **3** (1959), 285–305.
Dold, A. & Whitney, H., 'Classification of oriented sphere bundles over a 4-complex'. *Ann. of Math.*, **69** (1959), 667–77.
Dror, E. & Zabrodsky, A., 'Unipotency and nilpotency in homotopy equivalences'. *Topology* **18** (1979), 187–97.
Dwyer, W.G., 'Tame homotopy theory'. *Topology*, **18** (1979) 321–38.
Dwyer, W. & Friedlander, E., 'Etale K-theory and arithmetic'. *Bull. AMS*, **6** (1982), 453–5.
Dwyer, W.G. & Kan, D.M., 'Function complexes for diagrams of simplicial sets. *Proc. Konink. Neder. Akad.*, **86** (1983), 139–50.
– 'Function complexes in homotopical algebra'. *Topology*, **19** (1980), 427–40.
– 'Calculating simplicial localizations'. *J. Pure Appl. Algebra*, **18** (1980), 17–35.
Dwyer, W.G. & Kan, D.M., 'Simplicial localizations of categories *J. Pure Appl. Algebra*, **17** (1980), 267–84.
– 'Homotopy theory and simplicial groupoids'. *Proc. Konink. Neder. Akad.*, **87** (1984), 379–89.

Dyer, N.M., 'Homotopy classification of (π, m)-complexes'. *J. Pure Appl. Algebra*, **7** (1976), 249–82.
Eckmann, B., Homotopie et dualite. Colloque de topologie algebrique Louvain (1956)
– 'Groupes d'homotopie et dualité. *Bull. Soc. Math.*, France, **86** (1958), 271–81.
– 'Homotopy and cohomology theory'. *Proceedings of the International Congress of Math.* (1962).
Eckmann, B. & Hilton, P.J., 'Groupes d'homotopie et dualité'. *Comptes Rendus*, **246** (1958), 2444–7, 2555–8, and 2911–3.
– 'On the homology and homotopy, decomposition of continuous maps. *Proceedings Nat. Acad. Science*, **45** (1959), 372–5.
– 'Composition functors and spectral sequences'. *Comment. M. Helv.*, **41** (1966), 187–221.
Edwards, D.A. & Hastings, H.M., *Čech and Steenrod Homotopy Theories. Lecture Notes in Math. 542* (Springer Verlag) (1976).
Eggar, M.H., Ex-Homotopy theory, *Composito Math.*, **27** (2) (1973), 185–95.
Ellis, G.J., Crossed Modules and their Higher Dimensional Analogues. University of Wales Ph.D. thesis.
Federer, H., 'A study of function spaces by spectral sequences'. *Trans. AMS* (1956), 340–61.
Fritsch, R. & Latch, D.M., 'Homotopy inverses for Nerve'. *Math. Z.*, **177** (2) (1981), 147–80.
Gabriel, P. & Zisman, M., '*Calculus of fractions and homotopy theory*'. *Ergebnisse der Mathematik und ihrer Grenzgebiete 35*, Springer-Verlang, 1967.
Ganea, T., 'A generalization of the homology and homotopy suspension'. *Comment. Math. Helv.*, **39** (1965), 295–322.
Gray, B., *Homotopy Theory.* Academic Press (1975).
Grivel, P.P., 'Formes differentielles et suites spectrales'. *Ann. Inst. Fourier*, **29** (1979).
Grothendieck, A., *Pursuing Stakes* (preprint).
Gugenheim, V.K.A.M. & Munkholm, H.J., 'On the extended functoriality of Tor and Cotor'. *J. Pure and Appl. Algebra*, **4** (1974), 9–29.
Halperin, S., 'Lectures on minimal models'. *Memoire de la societe math. de France*, 9/10, SMF (1983).
Halperin, S. & Watkiss, C., *Relative Homotopical Algebra.* Lille Publ., IRMA
Halperin, S. & Stasheff, J., 'Obstructions to homotopy equivalences'. *Advances in Math.*, **32** (1979), 233–79.
Haefliger, A., *Rational Homotopy of the Space of Sections of a Nilpotent Bundle.* Preprint.
Hartl. M., *The Secondary Cohomology of the Category of Finitely Generated Abelian Groups.* Preprint, Bonn (1985), Displomarbeit Mathematisches Institut der Universität Bonn.
Hastings, M.H., 'Fibrations of compactly generated spaces'. *Mich J. Math.*, **21** (1974), 243–51.
Heller, A., Stable homotopy categories'. *Bull. AMS*, **74** (1968), 28–63.
Helling, B., Homotopieklassifikation in der Kategorie der differentiellen Algebren Diplomarbeit, Math. Inst. der Universität Bonn (1984).
Hilton, P., *Homotopy Theory and Duality.* Nelson (1965) Gordon Breach.
– *General Cohomology theory and K-theory. London Math. Soc. Lecture Note 1* (1971).
– *An Introduction to Homotopy Theory.* Cambridge University Press (1953).
Hilton, P.J., Milslin, G. & Roitberg, J., *Localizations of Nilpotent Groups and Spaces.* North Holland Math. Studies 15, Amsterdam (1975).
Hilton, P.J., Mislin, G., Roitberg, J. & Steiner, R., On Free Maps and Free Homotopies into Nilpotent Spaces. *Lecture Notes in Math. 673. Algebraic Topology, Proceedings*, Vancover 1977, 202–18, Springer-Verlag, 1978.
Hilton, P.J. & Stammbach, U., *A Course in Homological Algebra.* Springer GTM 4, New York 1971.
Hu, S.T., *Homotopy Theory*, Academic Press, New York and London, 1959.
Huber, P.J., 'Homotopy theory in general categories', *Math. Annalen*, **144** (1961), 361–85.

Huber, M. & Meyer, W., 'Cohomology theories and infinite CW-complexes'. *Comment. Math. Helv.*, **53** (1978), 239–57.
Husemoller, D., Moore, J.C. & Stasheff, J., 'Differential homological algebra and homogeneous spaces'. *J. Pure and Appl. Algebra*, **5** (1974), 113–85.
Illusie, L., *Complexe cotangent et deformations* II. *Lect Notes in Math.* 283 (Springer Verlag).
James, I.M., *General Topology and Homotopy Theory*. Springer Verlag, Berlin (1984).
– 'Ex Homotopy theory I'. *Illinois J. of Math.*, **15** (1971), 329–45.
James, I.M. & Thomas, E., 'On the enumeration of cross sections. *Topology*, **5** (1966), 95–114.
Jardine, J.F., Simplicial Presheaves, preprint (1985), Univ. of Western Ontario.
Johnstone, P.T., *Topos Theory*. Academic Press London 1977.
Kamps, K.H., 'Kan-Bedingungen und abstrakte Homotopietheorie'. *Math. Z.*, **124** (1972), 215–36.
– 'Fundamentalgruppoid and Homotopien'. *Arch. Math.*, **24** (1973), 456–60.
Kahn, P.J., 'Self-equivalences of $(n-1)$-connected $2n$-manifolds. *Bull. Amer. Math. Soc.*, **72** (1966), 562–6.
Kan, D.M.: 'Abstract homotopy I, II'. *Proc. Nat. Acad. Sci. USA*, **41** (1955), 1092–6: **42** (1956), 255–8.
– 'On homotopy theory and C.s.s. groups'. *Ann. of Math.*, **68** (1958), 38–53.
– 'A relation between CW-complexes and free C.s.s. groups'. *Amer. J. Math.*, **81** (1959), 521–28.
Kelly, G.M. & Street, R., R., Review of the Elements of 2-categories. *Lecture Notes in Math*, **420** (1974), 75–103 (Springer-Verlag).
Legrand, A., *Homotopie des espaces de sections. Lecture Notes in Math.* 941 (Springer Verlag), 1982.
Lehmann, D., 'Théorie homotopique des formes differentielles (d'apres Sullivan). *Asterisque*, **45** (1977).
Lemaire, J.M., *Algèbras connexes et homologic des espaces de lacets. Lecture Notes in Math.* 422. (Springer Verlag) (1979).
Loday, J.L., 'K-théorie algébrique et représentations de groupes'. *Ann. Sc. Ec. Norm. Sup.*, **9** (1976), 309–77.
Mac Lane, S., *Homology*, Grundlehren 114, Springer 1967.
Mac Lane, S. & Whitehead, J.H.C., 'On th 3-type of a complex'. *Proc. Nat. Acad. Sciences, USA*, **36** (1950), 41–8.
Magnus, W., Karrass, A. & Solitar, D., '*Combinatorial group theory*'. *Pure and Appl. Math.* **13**, Interscience, New York, 1966.
Marcum, J.H., 'Parameter constructions in homotopy theory', *Ann. Acad. Brasil. Cienc.* (1976), **48**, 387–402.
Massey, W.S., Exact couples in algebraic topology I–V. *Ann. of Math.*, **56** (1952), 363–96; **57** (1953), 248–86.
Mc Clendon, J.F., 'Higher order twisted cohomology operations'. *Inventiones Math.*, **7** (1969), 183–214.
– 'Reducing towers of principal fibrations'. *Nagoya Math. J.*, **54** (1974), 149–64.
Milgram, J.R., 'The bar construction and abelian H-spaces'. *Ill. J. Math.*, **11** (1967), 242–50.
Milnor, J., 'On spaces having the homotopy type of a CW-complex'. *Trans. Amer. Soc.*, **90** (1959) 272–80.
– 'On axiomatic homology theory'. *Pacific J. Math.*, **12** (1962), 377–41.
– *Morse Theory*, Princeton Univ. Press (1963).
Milnor, J. & Moore, J.C., 'On the structure of Hopf Algebras'. *Ann. of Math.* (2), **81** (1965), 211–64.
Mimura, M., Nishida, G. & Toda, 'Localization of CW-complexes and its applications'. *J. Math. Soc., Japan*, **23** (1971), 593–624.

Mitchell, B., 'Rings with several objects'. *Advances in Math.*, **8** (1972), 1–161.

Moore, J.C., Seminaire H. Carlan 1954 – 55 Exp. 3.

– 'Differential homological algebra'. *Proc. Int. Cong. Math.*, **1** (1970), 1–5.

Munkholm, H.J., 'DGA algebras as in Quillen model category, relations to shm maps'. *J. Pure Appl. Algebra*, **13** (1978) 221–32.

Neisendorfer, J., 'Lie algebras, coalgebras, and rational homotopy theory for nilpotent spaces'. *Pacific J. Math.*, **74** (2) (1978), 249–60.

Nomura, Y., 'A non-stable secondary operation and homotopy classification of maps'. *Osaka J. Math.*, **6** (1969), 117–34.

– 'Homotopy equivalences in a principal fiber space'. *Math. Z.*, **2** (1966), 380–8.

Oka, S., Sawashita, N., & Sugawara, M., 'On the group of self equivalences of a mapping cone'. *Hiroshima Math. J.*, **4** (1974), 9–28.

Pontrjagin, L., 'A classification of mappings of the 3-dimensional complex into the 2-dimensional sphere'. *Rec. Math. (Math. Sbornik) N.S.*, **9** (51) (1941), 331–63.

Puppe, D., 'Homotopiemengen und ihre induzierten Abbildungen Γ. *Math. Z.*, **69** (1958), 299–344.

Puppe, V., 'A remark on homotopy fibrations'. *Manuscripta Math.*, **12** (1974), 113–20.

Quillen, D.G., *Homotopical Algebra, Lecture Notes in Math.* 43 (Springer Verlag) 1967.

– Higher Algebraic K-theory I. *Springer Lecture Notes* 341 (1973), 85–147.

– 'An application of simplicial profinite groups'. *Comment. Math. Helv.* **44** (1969), 45–60.

– 'Rational homotopy theory'. *Ann. of Math.*, **90** (1969), 205–95.

Quinn, F., *Isotopy of 4-manifolds.* Preprint (1985) Virginia state university.

Ratcliffe, J.G., 'Free and projective crossed modules'. *J. London Math. Soc.* (2), **22** (1980) 66–74.

Roos, J.E., 'Sur les foncteurs dérivés de, $\varprojlim$, Applications'. *Compte rendue Acad. Sc. Paris*, **252** (1961), 3702–4.

Rutter, J.W., 'A homotopy classification of maps into an induced fibre space'. *Topology*, **6** (1967), 379–403.

– 'Self equivalences and principal morphisms'. *Proc. London Math. Soc.*, **20** (1970), 644–58.

– 'Groups of self homotopy equivalences of induced spaces. *Comment. Math. Helv.*, **45** (1970), 236–55.

Samelson, H., 'On the relation between the, Whitehead and Pontryagin product'. *Amer. J. Math.*, **75** (1953) 744–52.

Sawashita, N., 'On the group of self equivalences of the product of spheres'. *Hiroshima Math. J.*, **5** (1970), 69–86.

Scheerer, H., 'Lokalisierungen von Abbildungsräumen'. *Comp. Math*, **40** (1980), 269–81.

– 'Arithmeticity of groups of fibre homotopy equivalence classes'. *Manuscripta Math.*, **31** (1980), 413–24.

– 'Fibrewise P-universal nilpotent fibrations'. *Proc. Roy. Soc. Edinburgh*, **98A** (1984).

Schellenberg, B., 'The group of homotopy self-equivalences of some compact CW-complexes'. *Math. Ann.*, **200** (1973), 253–66.

Shih, W., 'On the group $E(X)$ of Homotopy Equivalence maps'. *Bull. Amer. Math. Soc.*, **70** (1964).

Shitanda, Y., 'Sur la théorie d'homotopie abstract'. *Memoirs of the Faculty of Science, Kyushu University Ser.* **A38** (1984), 183–98.

Shiraiwa, K., 'On the homotopy type of an A_n^3-polyhedron, $n \geqq 3$'. *Amer. J. of Math.*, **76** (1954), 235–45.

Silveira, F.E.A., Rational Homotopy Theory of Fibrations. Thèse genène (1979).

Spanier, E., *Algebraic Topology.* McGraw Hill (1966).

Steenrod, N.E., 'Product of cocycles and extensions of mappings'. *Ann. of Math.*, **48** (1947), 290–320.

Strøm, A., Note on cofibrations II. *Math. Scand.*, **22** (1968) 130–42.

– 'The homotopy category is a homotopy category', *Arch. Math.*, **23** (1972), 435–41.

Sullivan, D., *Geometric Topology – 1: Localization, Periodicity and Galois Symmerty*. MIT Press (1970).
– 'Infinitesimal computations in topology'. *Public. Math.*, **47** (1977), *Inst. Hautes Etudes Sc. Paris.*
Switzer, R.M., 'Counting elements in homotopy sets'. *Math. Z.*, **178** (1981), 527–54.
Tanré, D, *Homotopie Rationnelle; Modèles de Chen, Quillen, Sullivan. Lecture Notes in Math.* 1025 (Springer) (1983).
Thomason, R.W., Cat as a Closed Model Category, Preprint, MIT, Cambridge, Massachusetts.
Toda, H., 'Composition methods in homotopy groups of spheres'. *Ann. Math. Studies*, 49.
Tsukiyama, K., 'On the group of fibre homotopy-equivalences'. *Hiroshima Math. J.*, **12** (1982), 349–76.
– 'Self-homotopy-equivalences of a space with two nonvanishing homotopy groups'. *Proceedings AMS*, **79** (1980), 134–8.
Varadarajan, K., 'Numerical invariants in homotopical algebra I, II'. *Can J. Math.*, **27** (1975), 901–34, 935–60.
Vogt, R.M., 'Homotopy limits and colimits'. *Math. Z.*, **134** (1973), 11–52.
Waldhausen, F., *Algebraic K-theory of Space*, Preprint (1984), Bielefeld.
Wall, C.T.C., 'Finiteness conditions for CW-complexes II'. *Proc. Roy. Soc.*, **295** (1966), 129–39.
Watts, Ch. E., 'A Homology Theory for small categories'. *Proc. of the Conf. on Categorical Algebra.* La Jolla. CA 1965.
Whitehead, G.W., 'On mappings into group like spaces'. *Comment. Math. Helv.* **21** (1954), 320–38.
Whitehead, J.H.C., 'On simply connected 4-dimensional polyhedra'. *Comment Math. Helv.*, **22** (1949), 48–92.
– 'A certain exact sequence'. *Ann. Math.*, **52** (1950), 51–110.
– 'Combinatorial homotopy II'. *Bull. AMS*, **55** (1949); 213–45.
– 'Algebraic homotopy theory'. *Proc. Int. Congress of Mathematicians, Harvard*, **2** (1950), 354–57.
– 'Simple homotopy types'. *Amer. J. Math.*, **72** (1950), 1–57.
Wilkerson, C.W., 'Applications of minimal simplicial groups'. *Topology*, **15** (1976), 111–30.

INDEX